Biodegradative Bacteria

Hideaki Nojiri • Masataka Tsuda
Masao Fukuda • Yoichi Kamagata
Editors

Biodegradative Bacteria

How Bacteria Degrade, Survive, Adapt, and Evolve

Springer

Editors
Hideaki Nojiri
Biotechnology Research Center
The University of Tokyo
Tokyo, Japan

Masao Fukuda
Department of Bioengineering
Nagaoka University of Technology
Nagaoka, Niigata, Japan

Masataka Tsuda
Graduate School of Life Sciences
Tohoku University
Sendai, Miyagi, Japan

Yoichi Kamagata
Bioproduction Research Institute
National Institute of Advanced Industrial Science and Technology
Sapporo, Hokkaido, Japan

ISBN 978-4-431-54519-4 ISBN 978-4-431-54520-0 (eBook)
DOI 10.1007/978-4-431-54520-0
Springer Tokyo Heidelberg New York Dordrecht London

Library of Congress Control Number: 2013950987

Printed on acid-free paper

Springer is part of Springer Science+Business Media (www.springer.com)

Preface

From a macroscopic standpoint, the environment where we live maintains constancy to some extent. This means the environment has a buffering capacity for various types of impacts. Bacterial metabolic capacity, which plays an important role in the global material cycle, contributes largely to the buffering capacity for huge and unintended release of various chemicals. Recently, however, the prosperity and globalization of material civilization has led not only to severe local contamination by hazardous chemicals, but also to continuous increment of contaminant concentrations worldwide. To solve such urgent global issues, bacterial functions that are involved in biodegradation of hazardous chemicals have been analyzed. The term "biodegradative bacteria" refers to the bacteria that have the ability to degrade such xenobiotic (man-made) and/or hazardous chemicals.

Since the 1980s, following on the development of microbial genetics, molecular genetic research techniques have been applied to the investigation of the function of biodegradative bacteria. This research has resulted in the cloning of many genes and gene clusters involved in the biodegradation of various types of contaminated chemicals. Similarly, many enzymes involved in unique biodegradative reactions with recalcitrant hazardous chemicals have been analyzed to understand why biodegradative bacteria can decompose corresponding chemicals. These analyses have provided an understanding of the structural bases of such novel reactions for xenobiotic chemicals. As readers may know, the recent development of genome-wide investigative techniques has largely promoted the comprehensive interpretation of a whole-cell system of biodegradative bacteria. The degradative capacity of bacteria for xenobiotic compounds may be developed through the adaptive change of native enzymatic function and/or recruitment of genes or gene clusters accounting for the new metabolic capacities. Therefore, analyses of biodegradative bacteria can be said to have partially dealt with the rapid evolution process of bacteria in natural ecosystems. From the application point of view, to remediate actual contaminated sites by the biodegradative capacity of bacteria, it is quite important to know or to control the behavior of biodegradative bacteria in the environment. These are quite difficult issues, however, because there are numerous bacterial strains and diverse environmental conditions in actual contaminated sites. Thus, in situ analyses of biodegradative bacterial function should include investigation by recently developed molecular ecological techniques.

As mentioned earlier, analyses of biodegradative bacteria include various types of studies, such as genetics, enzymology, genomics, cell physiology, ecology, and evolutional biology. In another words, the targets investigated in research on biodegradative bacteria include single molecules, single cell systems, bacterial consortia (interaction with surrounding microorganisms), and interaction with surrounding biotic and abiotic materials. Such complexity makes the research on biodegradative bacteria difficult but quite interesting.

Biodegradative Bacteria consists of three parts, each consisting of five to seven chapters: (1) Genetic and genomic systems, (2) Degradative enzyme systems, and (3) Bacterial behavior in natural environmental systems. The reader will find that the scope of the various chapters is not uniform. In our opinion, they all are fascinating, however. We hope this book will contribute to readers' knowledge of how far biodegradative bacteria research has progressed and how various recent technical innovations have led to such progress. To all the authors who in spite of their numerous other duties found time to review the novel work in their field, we express our gratitude. We are also grateful to Springer Japan and especially to Kaoru Hashimoto and Yoshiko Shikano of the editorial staff for their support and counsel.

Tokyo, Japan — Hideaki Nojiri
Sendai, Japan — Masataka Tsuda
Sapporo, Japan — Yoichi Kamagata
Nagaoka, Japan — Masao Fukuda

Contents

Part II Enzyme Systems

Part III Bacterial Behavior in Natural Environmental Systems

Part I
Genetic and Genomic Systems

Chapter 1
Rhodococcus Multiple-Enzyme and Parallel-Degradation System for Aromatic Compounds

Masao Fukuda

Abstract Aromatic hydrocarbons are aerobically degraded by a variety of bacteria via enzyme systems including ring-hydroxylating dioxygenases and monooxygenases and extradiol and intradiol ring-cleavage dioxygenases. The degradation enzyme genes of gram-negative bacteria usually constitute a single gene cluster. Multiple gene clusters of degradation enzyme genes were found in the gram-positive *Rhodococcus* strain RHA1. Multiple isozyme genes with 2,3-dihydroxybiphenyl dioxygenase activity have been reported in several biphenyl/PCB-degrading *Rhodococcus* strains including RHA1, suggesting the multiple-enzyme pathways in *Rhodococcus* strains. Transcriptional induction of multiple isozymes and functional involvement of multiple isozymes in degradation of not only biphenyl and PCBs but benzene and alkylbenzenes have been reported in RHA1. A couple of two-component regulatory systems are responsible for transcriptional induction of multiple isozymes. This multiple-enzyme pathway is thought to support the strong PCB degradation activity and the wide range of degradation activity of RHA1. The multiple-enzyme system in combination with the regulatory system with a broad inducing-substrate spectrum appears to provide the parallel-degradation system conferring the tough degradation activity toward a broad range of aromatics, suggesting the advantage of the multiple-enzyme system.

Keywords Acidovorax • Biphenyl • Dioxygenase • Extradiol dioxygenase • Isozyme • Polychlorinated biphenyls (PCBs) • *Pseudomonas* • *Rhodococcus* • Ring-cleavage dioxygenase • Ring-hydroxylating dioxygenase

M. Fukuda (✉)
Department of Bioengineering, Nagaoka University of Technology,
Kamitomioka 1603-1, Nagaoka, Niigata 940-2188, Japan
e-mail: masao@vos.nagaokaut.ac.jp

H. Nojiri et al. (eds.), *Biodegradative Bacteria: How Bacteria Degrade, Survive, Adapt, and Evolve*, DOI 10.1007/978-4-431-54520-0_1, © Springer Japan 2014

1.1 Aromatic Degradation Pathway Enzymes and Genes

Aromatic hydrocarbons are usually hydroxylated to generate dihydrodiol intermediates by ring-hydroxylating dioxygenases at the initial step as illustrated in Fig. 1.1. In some cases, dihydrodiol intermediates are formed by duplicate hydroxylation reactions by ring-hydroxylating monooxygenases. Dihydrodiol intermediates are converted to diol intermediates by dehydrogenases. Then the aromatic rings of diols are cleaved by extradiol or intradiol ring-cleavage dioxygenases. Extradiol and intradiol ring-cleavage dioxygenases cleave the C–C bonds neighboring to and between two vicinal hydroxyl groups, respectively. The cleaved fragments formed via ring cleavage are trimmed further to form intermediates such as pyruvate and acetyl-CoA, which are incorporated into a TCA cycle. Bacterial aerobic degradation enzyme systems of aromatic hydrocarbons had been characterized well in gram-negative aerobic bacteria such as the genera *Pseudomonas*, *Burkholderia*, and *Sphingomonas*.

The degradation enzyme genes of these gram-negative degraders usually constitute a single gene cluster specialized for degradation of the target substrate. Biphenyl catabolic enzyme pathway has been characterized in detail, because it is available to degrade polychlorinated biphenyls (PCBs), which are well-known hazardous

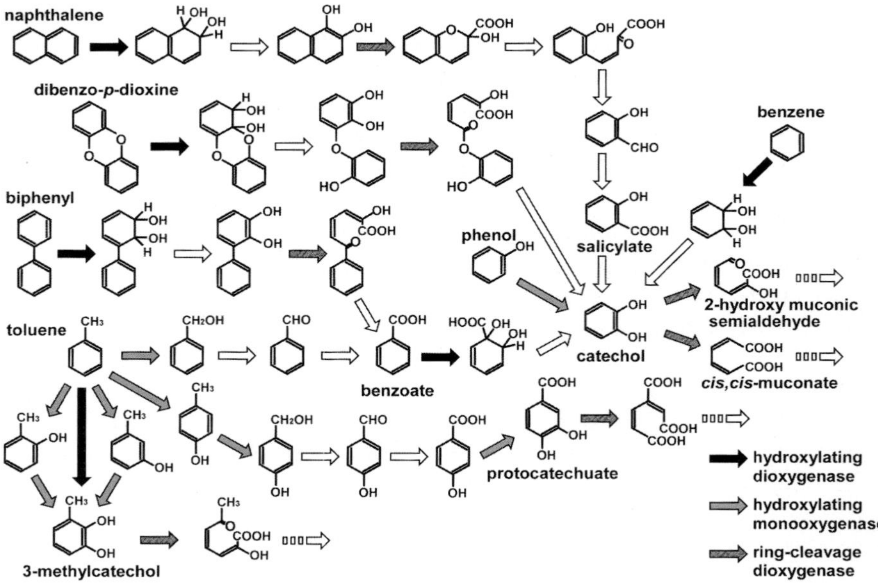

Fig. 1.1 Oxygenases involved in aromatic degradation pathways. The degradation enzyme steps for the representative aromatic hydrocarbons are presented. Oxygenases are involved in the key steps responsible for the initial hydroxylation and intermediate ring cleavage. *Black* and *gray arrows* represent hydroxylating dioxygenases and monooxygenases, respectively. *Shaded arrows* represent ring-cleavage dioxygenases. *Open arrows* indicate the other enzymes

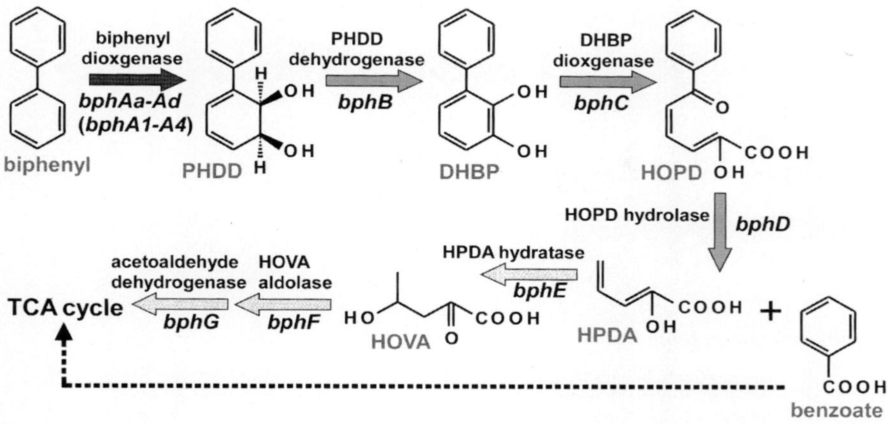

Fig. 1.2 Enzymes and genes of biphenyl degradation pathway. The enzyme and gene names of each step in the biphenyl degradation pathway are presented *above* and *below* an *arrow*, respectively. HOVA is converted to pyruvate and acetaldehyde, which is converted to acetyl-CoA by acetaldehyde dehydrogenase. Benzoate is degraded through the independent benzoate pathway via 1,2-dihydroxybenzoate, catechol and 3-oxoadipate, to succinate and acetyl-CoA. Pyruvate, succinate, and acetyl-CoA formed from biphenyl are utilized in TCA cycle. Compound names are indicated below the corresponding compounds and abbreviated as follows: *DHBP* 2,3-dihydroxybiphenyl, *HOVA* 4-hydroxy-2-oxovalerate, *HOPD* 2-hydroxy-6-oxo-6-phenylhexa-2,4-dienoate, *HPDA* 2-hydroxypenta-2,4-dienoate, *PHDD* 3-phenylcyclohexa-3,5-diene-1,2-diol (dihydrodiol)

contaminants in the environment. Although PCBs are not the inducing substance for the biphenyl pathway, they are cometabolized by the biphenyl pathway in the presence of biphenyl. In the case of biphenyl catabolic enzyme pathway (Fig. 1.2) the upper catabolic pathway enzymes are responsible for the transformation of biphenyl to pentadienoate and benzoate, and the lower catabolic pathway enzymes are responsible for that of pentadienoate to pyruvate and acetyl-CoA. Benzoate is transformed via catechol, *cis, cis*-muconate, and 3-oxoadipate through the specialized benzoate pathway to succinyl-CoA and acetyl-CoA. Pyruvate, succinyl-CoA, and acetyl-CoA generated by the lower catabolic and benzoate pathways are incorporated into the citric acid cycle and used as carbon and energy sources. In the gram-negative biphenyl-degrading strains, *Acidovorax* sp. KKS102 (Ohtsubo et al. 2001) and *Pseudomonas pseudoalcaligenes* KF707 (Watanabe et al. 2003), all the upper and lower catabolic pathway enzymes are encoded by a single gene cluster (Fig. 1.3a, b), although their gene organizations are different from each other. Both the gene clusters contain the regulatory genes responsible for the transcriptional induction of catabolic enzymes and encode a single respective enzyme gene for each step, indicating that a single enzyme is responsible for each catabolic step.

In the strain *Rhodococcus jostii* RHA1 the upper and lower catabolic pathway enzyme genes are distributed among multiple separate gene clusters (Fig. 1.3a) (McLeod et al. 2006; Takeda et al. 2010). These gene clusters code the multiple sets of isozymes that differ in amino acid sequence but catalyze the same chemical reaction.

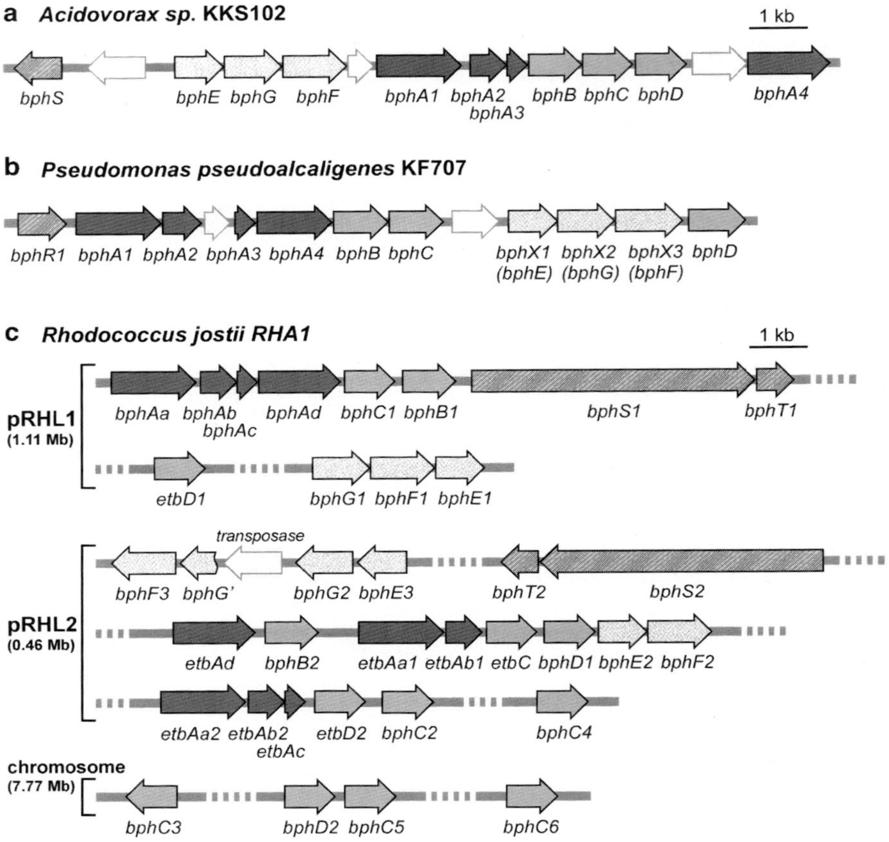

Fig. 1.3 Gene organizations of biphenyl degradation genes. Biphenyl degradation gene clusters of gram-negative *Acidovorax* sp. KKS102 (**a**), *Pseudomonas pseudoalcaligenes* KF707 (**b**), and gram-positive *Rhodococcus jostii* RHA1 (**c**) are presented. The sizes of linear plasmids and a chromosome in RHA1 are indicated in *parentheses*. Genes coding biphenyl dioxygenase subunits are presented by *black arrows*. *Gray arrows* indicate PHDD dehydrogenase, DHBP dioxygenase, or HOPD hydrolase genes. *Spotted arrows* specify HPDA hydratase, HOVA aldolase, or acetaldehyde dehydrogenase genes. Regulatory genes are implied by *shaded arrows*. Compound names are abbreviated as Fig. 1.2

In the upper pathway two sets of the first- to third-step enzyme genes are separately distributed in the gene clusters of linear plasmids, pRHL1 and pRHL2. The first-step enzyme, biphenyl dioxygenase, is composed of four subunits including large and small subunits of terminal dioxygenase responsible for catalytic reaction and ferredoxin and reductase subunits responsible for electron transfer from NADH. The sets of the large and small subunits are encoded by *bphAabphAb*, *etbAa1etbAb1*, and *etbAa2etbAb2*. The deduced amino acid sequences of *etbAa1etbAb1* and *etbAa2etbAb2* are the same. The ferredoxin subunits are encoded by *bphAc* and *etbAc*, and the reductase subunits are by *bphAd* and *etbAd*.

The second-step dehydrogenases are encoded by *bphB1* and *bphB2*. The third-step 2,3-dihydroxybiphenyl dioxygenases are encoded by *bphC1*, *etbC*, *bphC2*, and so on. The fourth-step hydrolase is encoded by *etbD1*, *bphD1*, *etbD2*, and *bphD2*. The lower pathway enzymes are encoded by four gene sets including *bphGbphF1bphE1*, *bphE2bphF2*, *bphE3bphG2*, and *bphG'bphF3*. The *bphG'* is truncated homolog of *bphG* and is estimated to be inactive. In addition to these there are two sets of regulatory genes including *bphS1bphT1* and *bphS2bphT2*. Therefore at least a couple of isozyme genes are found for each catabolic step of the upper and lower pathway, suggesting that *R. jostii* RHA1 has the multiple-enzyme pathway.

1.2 Multiplicity of Extradiol Dioxygenase Genes Found in Biphenyl-Degrading *Rhodococcus* Strains

The multiplicity of extradiol dioxygenase genes was first reported in *Rhodococcus globerulus* P6 (Asturias and Timmis 1993). Three 2,3-dihydroxybiphenyl 1,2-dioxygenase (DHBP dioxygenase) genes were found in the P6 strain. One was the *bphC1* gene encoding 291 amino acid residues that has good similarity with the previously reported DHBP dioxygenase genes of type I.3 extradiol dioxygenases (Fig. 1.4). The other two genes, *bphC2* and *bphC3,* encoded enzymes of 190 amino acid residues and were classified in type I.1 (Eltis and Bolin 1996). Because DHBP dioxygenases are usually composed of around 300 amino acid residues, the sizes of these enzymes are unique. Their amino acid sequences of the enzymes encoded by *bphC2* and *bphC3* and the carboxyl-terminal half of type I.3 enzyme have some similarities. The amino acid sequences of P6 BphC2 and BphC3 enzymes are aligned with that of *Acidovorax* sp. KKS102 BphC type I.3 enzyme in Fig. 1.5. The secondary structure of KKS102 BphC estimated from its three-dimensional (3D) structure solved by X-ray crystallography is presented above the sequence alignment. The 3D structure of KKS102 BphC suggested that BphC enzyme is composed of two similar domains and each domain contains a couple of motifs primarily consisting of β–α–β–β–β structures (Senda et al. 1996). The ferrous ion required for catalytic activity resides only in the carboxyl-terminal domain indicating that the amino-terminal domain is not involved in the catalytic activity. The secondary structure of the carboxyl-terminal domain of KKS102 BphC staring at the residue 136 shows good similarity with the predicted secondary structure of P6 BphC2 (Fig. 1.5). In addition the key amino acid residues of KKS102 BphC responsible for the ligands of a ferrous ion (H146, H210 and E259), the substrate binding (Y250 and H241), and the catalytic base (H195) are all conserved in P6 BphC2 and P6 BphC3. Thus the type I.1 enzyme appears to constitute a single-domain subfamily of type I enzyme and corresponds to the carboxyl-terminal domain of enzymes of the type I.3 subfamily (Bergdoll et al. 1998).

The significant multiplicity of DHBP dioxygenases was then reported in *R. erythropolis* TA421, which was isolated from a termite ecosystem (Maeda et al. 1995). Seven DHBP dioxygenase genes including *bphC1* to *bphC7* were found in

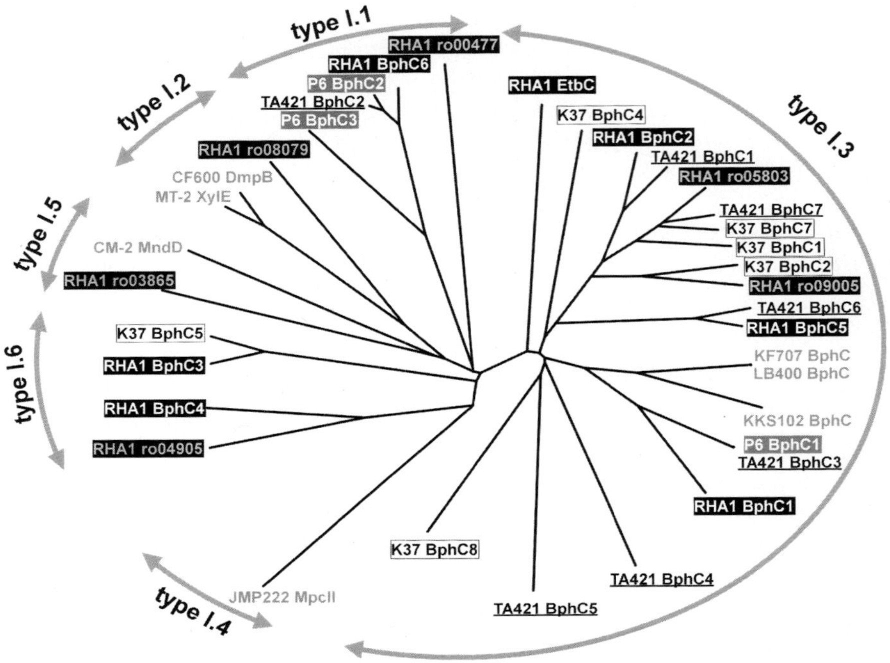

Fig. 1.4 Phylogenetic tree of type I DHBP dioxygenases from *Rhodococcus* strains. The phylogenetic tree of type I DHBP dioxygenases from *Rhodococcus* strains and the representative ones was generated by the neighbor-joining method with Clustal W and drawn using the TreeView program. Subtypes are indicated by *peripheral arcs*. DHBP dioxygenases from *R. erythropolis* TA421 are *underlined*. Those from *R. rhodochrous* K37 are *boxed*. Those from *R. globerulus* P6 are highlighted with *gray background*. Those from *R. jostii* RHA1 are highlighted with *black background*. The remaining type I enzymes of RHA1 found in the genome sequence are partly highlighted with *black background* using *gray characters*. The representative type I extradiol dioxygenases are presented using *gray characters* and their abbreviations are as follows: CF600 DmpB, *Pseudomonas* sp. CF600 catechol 2,3-dioxygenase; CM-2 MndD, *Arthrobacter globiformis* CM-2 3,4-dihydroxyphenylacetate 2,3-dioxygenase; JMP222 MpcII, *Cupriavidus necator* (*Alcaligenes eutrophus*) JMP 222 metapyrocatechase II; KF707 BphC, *Pseudomonas pseudoalcaligenes* KF707 2,3-dihydroxybiphenyl-1,2-dioxygenase; KKS102 BphC, *Acidovorax* (*Pseudomonas*) sp. KKS102 2,3-dihydroxybiphenyl-1,2-dioxygenase; LB400 BphC, *Burkholderia xenovorans* LB400 2,3-dihydroxybiphenyl-1,2-dioxygenase; MT-2 XylE, *Pseudomonas putida* mt-2 catechol 2,3-dioxygenase

TA421 by screening of the *Escherichia coli* gene library against DHBP dioxygenase activity. Recombinant clones expressing DHBP dioxygenase convert DHBP to HOPD and develop yellow color on the plate containing DHBP. Among the seven genes, six genes are included in type I.3 and *bphC2* belongs to type I.1 (Fig. 1.4). The *bphC2* to *bphC4* genes were encoded in a 500-kb large linear plasmid, pTA421(Kosono et al. 1997). Loss of pTA421 resulted in the deficiency in growth on biphenyl and degradation of PCBs, suggesting that at least one of these three DHBP dioxygenase genes is involved in biphenyl metabolism. The TA421 *bphC1*

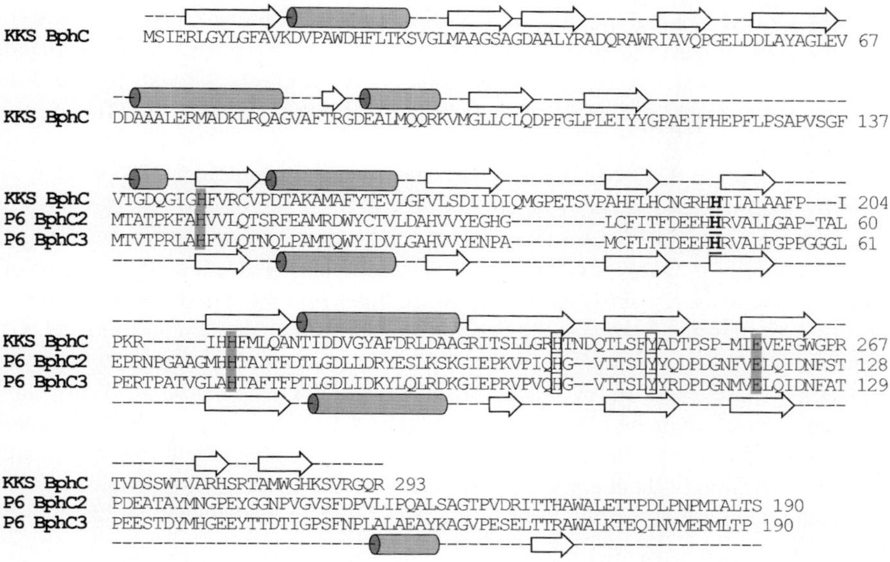

Fig. 1.5 Amino acid sequence and secondary structure alignment of type I.1 and type I.3 DHBP dioxygenases. Amino acid sequences of *R. globerulus* P6 BphC2 and BphC3 dioxygenases belonging to type I.1 are aligned with *Acidovorax* sp. KKS102 BphC based on the results of sequence alignment by clustal W program and their secondary structures. The secondary structure of KKS102 BphC was estimated from the results of X-ray crystallographic analysis and is presented above the amino acid sequence alignment. That of P6 BphC2 was predicted by the JPRED3 program (http://www.compbio.dundee.ac.uk/www-jpred/) and is presented below the sequence alignment. *Horizontal cylinders* and *arrows* represent the regions of helices and β strands, respectively. An *underlined bold character* indicates the amino acid residue required for the enzyme activity as a catalytic base. *Shaded characters* represent the residues that provide ligands to a ferrous ion in the active center. *Boxed residues* specify the key residues responsible for substrate binding

had mostly the same gene sequence as the P6 *bphC1*. The TA421 *bphC2* showed high amino acid sequence similarity to the P6 *bphC2* gene. Southern hybridization experiments using TA421 *bphC* gene probes indicated the presence of the orthologs of each TA421 *bphC* gene in the strain P6. These results suggest that TA421 and P6 share the similar set of DHBP dioxygenase genes.

Using the same screening procedure, seven DHBP dioxygenase genes including *bphC1* to *bphC6* and *etbC* were found in *R. jostii* RHA1 (Shimizu et al. 2001; Sakai et al. 2002). Among the seven genes, four genes are included in type I.3, one is in type I.1, and two belongs to type I.6, implying the differences with the strains TA421 and P6 (Fig. 1.4). The entire genome sequencing of RHA1 revealed additional six extradiol dioxygenase genes as illustrated in Fig. 1.4. The thirteen extradiol dioxygenase genes in RHA1 are distributed among all the subtypes. Northern blot and DNA array analysis indicated that *bphC1* and *etbC* were induced transcriptionally during the growth on biphenyl (Sakai et al. 2002; Goncalves et al. 2006) and *bphC5* was induced in the presence of cholesterol (Van der Geize et al. 2007). These results suggest that *bphC1* and *etbC* are involved in biphenyl catabolism and *bphC5* is responsible for cholesterol catabolism. Accordingly the *bphC5* gene was reannotated as *hsaC* coding the 3,4-dihydroxy-9,10-seconandrost-1,3,5(10)-triene-9,17-dione dioxygenase.

R. rhodochrous K37 has eight DHBP dioxygenase genes including *bphC1* to *bphC8* with distinct diversity (Taguchi et al. 2004). In addition to type I.3 and I.6 enzymes indicated in Fig. 1.4, K37 has type II extradiol dioxygenases including *bphC3* and *bphC6*. Only the *bphC8* was transcriptionally induced in the presence of biphenyl and *bphC1*, *bphC2*, and *bphC7* were transcriptionally induced in the presence of testosterone. The type II extradiol dioxygenases have neither domains nor duplicate motifs based on the three-dimensional structures of protocatechuate 4,5-dioxygenase encoded by *ligAB* of *Sphingomonas paucimobilis* SYK-6 and appear to have originated from the ancestor independent from type II enzymes (Sugimoto et al. 1999). However, the active centers of the type I and II enzymes are very similar. Both of them contain a ferrous ion coordinated by two histidines and a glutamate and have a histidine near the substrate, which is estimated to be a Schiff base to draw a proton out of the substrate. The type I and II extradiol dioxygenases appear to have evolved convergently to catalyze the same kind of reaction. In conclusion *Rhodococcus* strains appear to have seven or more DHBP dioxygenase genes and at least some of them are thought to be involved in metabolism of biphenyl.

1.3 Functional Involvement of Multiple Isozymes in Biphenyl Degradation

Functional involvement of DHBP dioxygenase isozymes was examined in the strains P6 and RHA1. Among the three DHBP dioxygenase isozymes in P6, two enzyme proteins encoded by *bphC1* and *bphC2* were induced and detected by Western blot analysis during the growth on biphenyl (McKay et al. 2003). Purified P6 DHBP dioxygenases prepared from *E. coli* recombinant strains showed different substrate specificities toward chlorinated DHBPs as illustrated in Fig. 1.6a (McKay et al. 2003; Vaillancourt et al. 2003). In the strain RHA1 the two DHBP dioxygenase isozymes encoded by *bphC1* and *etbC* were induced during the growth on biphenyl (Hauschild et al. 1996). Enzyme activities toward both DHBP and catechol, which are diol metabolites of biphenyl and benzene, respectively, were indicated by activity stain after electrophoretic separation. The results of activity assay using the purified enzymes of *bphC1* and *etbC* suggest that EtbC has broader substrate spectrum than BphC1 (Fig. 1.6b). Higher relative activity of BphC1 against catechol may suggest that BphC1 is more responsible for metabolism of single-ring aromatics.

Two types of ring-cleavage-product hydrolases were purified from RHA1 grown on biphenyl (Hatta et al. 1998). One enzyme was specific to HOPD, which was the

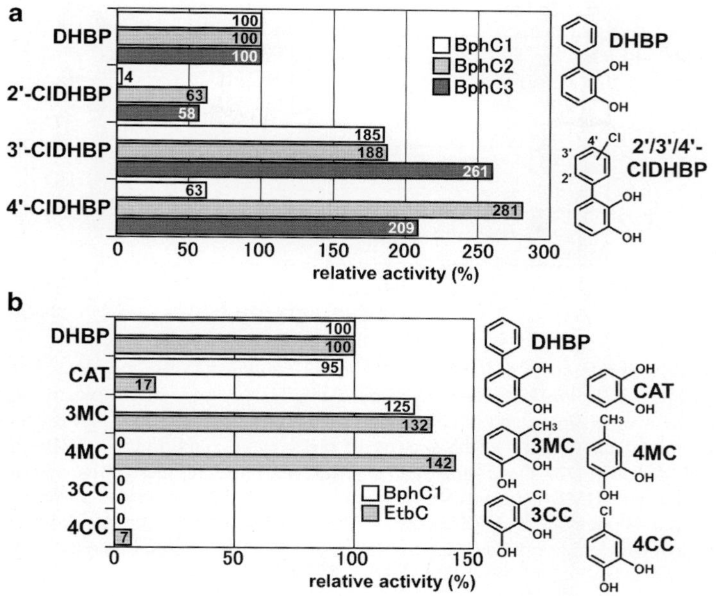

Fig. 1.6 Substrate preferences of P6 (**a**) and RHA1 (**b**) DHBP dioxygenases involved in biphenyl degradation. The relative activity toward each substrate is displayed by a *vertical bar* and the exact value is presented on the *bar*. It is expressed as the percentage ratio of the activity to the activity toward DHBP. The results of DHBP dioxygenases encoded by *bphC1* (BphC1, *open bars*), *bphC2* (BphC2, *gray bars*), and *bphC3* (BphC3, *black bars*) of *R. globerulus* P6 and those encoded by *bphC1* (BphC1, *open bars*) and *etbC* (EtbC, *gray bars*) of *R. jostii* RHA1 are presented in the panels A and B, respectively. Substrate names are abbreviated as follows: *DHBP* 2,3-dihydroxybiphenyl, *2′-ClDHBP* 2′-chloro-2,3-dihydroxybiphenyl, *3′-ClDHBP* 3′-chloro-2,3-dihydroxybiphenyl, *4′-ClDHBP* 4′-chloro-2,3-dihydroxybiphenyl, *CAT* catechol, *3MC* 3-methylcatechol, *4MC* 4-methylcatechol, *3CC* 3-chlorocatechol, *4CC* 4-chlorocatechol. Substrate structures are presented on the *right*. Substrate structures are displayed on the *right*

ring-cleavage product of DHBP generated from biphenyl. The other one was specific to 2-hydroxy-6-oxohepta-2,4-dienoate (HOHD), which was the ring-cleavage product of 3-methylcatechol generated from toluene. Then one HOPD hydrolase gene, *bphD1*, and two HOHD hydrolase genes, *etbD1* and *etbD2*, were cloned (Yamada et al. 1998). RNA slot blot hybridization analysis indicated that these three hydrolase genes are induced during the growth on biphenyl, benzene, toluene, and ethylbenzene. The *bphD1*, *etbD1*, and *etbD2* genes expressed in *E. coli* exhibited activities specific to HOPD, HOHD, and HOHD, respectively, and the latter two also showed activity to 2-hydroxy muconic semialdehyde (HMSA) generated from benzene (Fig. 1.7). The amino-terminal amino acid sequences of the purified enzymes, however, agreed with those of *bphD1* and *etbD1*. The *etbD2* gene appears to have failure in expression in the posttranscriptional process. These results suggest that *bphD1* is responsible for metabolism of biphenyl and *etbD1* is accountable for that of toluene and benzene.

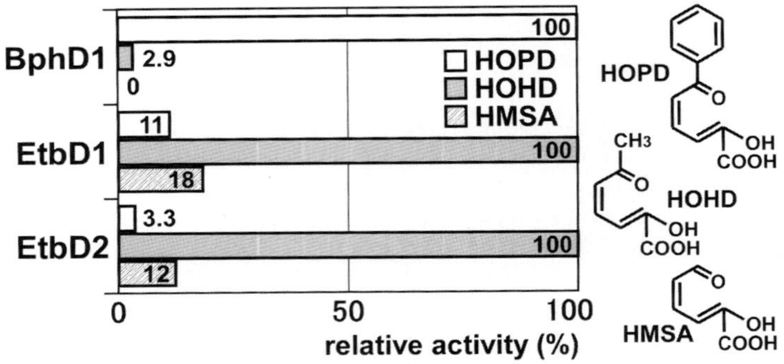

Fig. 1.7 Substrate preferences of RHA1 BphD1, EtbD1, and EtbD2 ring-cleavage-product hydrolases encoded by *bphD1, etbD1*, and *etbD1*, respectively. The relative activity toward each substrate is displayed by a *vertical bar* and the exact value is presented on the *bar*. It is expressed as the percentage ratio of the activity to the activity of the same enzyme toward DHBP (BphD1) or HOHD (EtbD1 and EtbD2). *Open, gray*, and *shaded bars* represent the relative activities toward HOPD, HOHD, and HMSA, respectively. Substrate names are abbreviated as follows: *HOPD* 2-hydroxy-6-oxo-6-phenylhexa-2,4-dienoate, *HOHD* 2-hydroxy-6-oxohepta-2,4-dienoic acid, *HMSA* 2-hydroxymuconic semialdehyde. Substrate structures are presented on the *right*

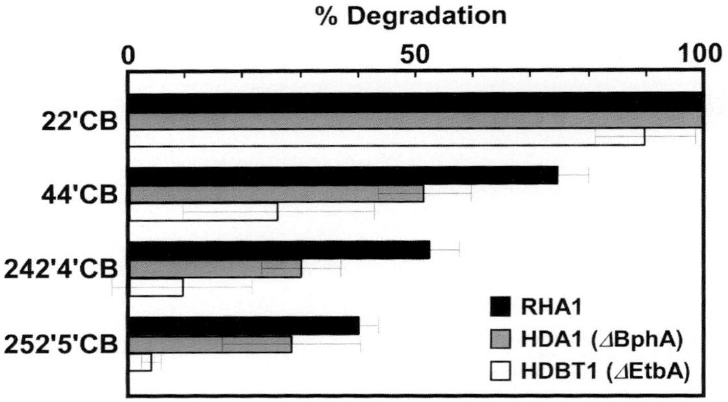

Fig. 1.8 Involvement of BphA and EtbA biphenyl dioxygenases in PCB degradation by *R. jostii* RHA1. The degradation activities of the wild-type strain (*black bars*) and the deletion mutant strains of *bphA* (*gray bars*) and *etbA* (*white bars*) are presented. Cells were incubated with a mixture of chlorinated biphenyl congeners (1 mM each) for 10 h and the ethyl acetate extracts were subjected to gas chromatography–mass spectrometry. Percentages of degradation were estimated from the remaining amount of each congener after incubation and that after incubation with autoclaved cells

The functional involvement of two biphenyl dioxygenase isozymes was examined by determining their activities of the mutant strains against PCB congeners (Iwasaki et al. 2006). Figure 1.8 presents the results on 2,2′-, 4,4′-, 2,4,2′,4′- and 2,5,2′,5′-chlorobiphenyls. The terminal dioxygenase subunit gene, *bphAa*, was inactivated in the HDA1 mutant strain to eliminate the BphA enzyme activity from

RHA1. Because there are two sets of *etbAa* and *etbAb* genes, *etbAa1Ab1* and *etbAa2Ab2*, coding the large and small subunits of a terminal dioxygenase with the same amino acid sequences, both the *etbAa1* and *etbAa2* subunit genes were inactivated in the HDBT1 mutant strain to remove the EtbA enzyme activity. Elimination of either of the dioxygenases resulted in apparent reduction of degradation activity except for 2,2′-chlorobiphenyl. Reduction of degradation activity of the EtbA-deficient strain HDBT1 (ΔEtbA) was always more significant than that of the BphA-lacking strain HDA1 (ΔBphA). These results suggest that both the BphA and EtbA dioxygenases are involved in PCB degradation and EtbA is more responsible for degradation than BphA. Broad substrate range of BphA and EtbA including biphenyl, naphthalene, phenanthrene, dibenzofuran, dibenzo-*p*-dioxin, benzene, toluene, and ethylbenzene was indicated using the recombinant *Rhodococcus* strains expressing either BphA or EtbA (Iwasaki et al. 2007).

1.4 The Regulatory System for Induction of Multiple Isozymes

Based on DNA microarray and quantitative reverse-transcription PCR analyses (Goncalves et al. 2006), transcriptional expression of multiple isozymes in each step of upper biphenyl metabolic pathway was indicated in RHA1 as presented in the panel A of Fig. 1.9 (Goncalves et al. 2006). A couple of isozymes encoded by *bphAa* to *bphAd* genes and *etbAa* to *etbAd* genes are involved in the first step. There are two sets of *etbAa* and *etbAb* genes, *etbAa1Ab1* and *etbAa2Ab2*, as mentioned above. The second and third steps also have a couple of isozymes encoded by *bphB1* and *bphB2* genes and *bphC1* and *etbC* genes, respectively. Three isozymes encoded by *bphD1*, *etbD1*, and *etbD2* genes are involved in the fourth step. Functional involvement of these genes except for *bphB1* and *bphB2* genes has been already described above.

These enzyme genes are distributed among five operons, which are under the control of duplicate two-component regulatory systems encoded by *bphS1T1* and *bphS2T2* genes (Takeda et al. 2010) as illustrated in the panel B of Fig. 1.9. Reporter assay experiments using the *luxAB* luciferase gene connected to the *bphAa* promoter as a reporter indicated that both the *bphS1T1*- and *bphS2T2*-coding regulatory systems activate the *bphAa* promoter in the presence of benzene or alkylbenzenes such as ethylbenzene, toluene, isopropylbenzene, or xylenes (Fig. 1.10). In the presence of biphenyl only the *bphS1T1*-coding system activates the reporter, suggesting some difference between *bphS1T1*- and *bphS2T2*-coding systems in the inducing-substrate preference. Higher degradation activity of PCBs was observed in the presence of ethylbenzene than in the presence of biphenyl, suggesting transcription activation by the *bphS1T1*- plus *bphS2T2*-coding systems is superior to that by the sole *bphS1T1*-coding system (Seto et al. 1996). Transcriptional activation by the *bphS1T1*-coding system was also observed with the reporter containing either of the *etbD1*, *etbAd*, *etbAa1*, or *etbAa2* promoters (Takeda et al. 2004). The knowledge obtained from the previously characterized two-component regulatory systems

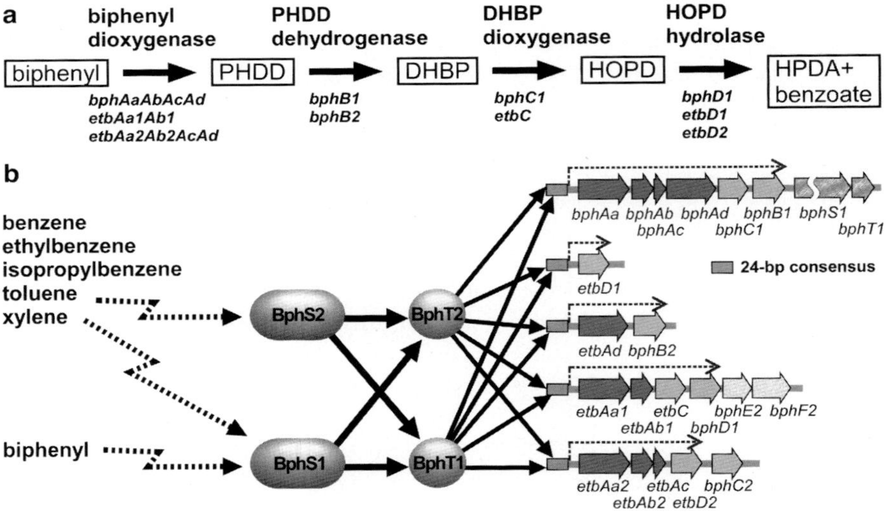

Fig. 1.9 The isozymes of biphenyl degradation steps (**a**) and the duplicated two-component regulatory systems (**b**) in *R. jostii* RHA1. (**a**) The isozyme genes involved in each step of the upper biphenyl degradation pathway are presented. The enzyme and gene names of each step are presented *above* and *below* an *arrow*, respectively. Compound names are abbreviated as Fig. 1.2. (**b**) The signal transduction pathways of the duplicated two-component regulatory systems, BphS1T1 and BphS2T2, which are responsible for transcriptional induction of biphenyl degradation enzyme genes in *R. jostii* RHA1, are presented. The sensor kinases, BphS1 and BphS2, activate the response regulators BphT1 and BphT2 by phosphorylation in the presence of an inducing substrate such as benzene and alkylbenzenes as illustrated. Biphenyl is the inducing substrate not for BphS2 but for BphS1. Phosphorylated BphT1 and BphT2 activate the five promoters of biphenyl degradation gene operons that share the 24-bp consensus sequence indicated by a *shaded box*. A *thin dashed line* originating from the 24-bp consensus sequence represents the direction of transcription

suggests that the *bphS1*- and *bphS2-coding* sensor kinases phosphorylate *bphT1*- and *bphT2*-coding response regulators, which simultaneously activate the transcription of the five degradation enzyme operons (Table 1.1). Cross talk between the sensor kinases and the heterologous response regulators was revealed using *bphS1T2*- and *bphS2T1*-coding hybrid systems (Takeda et al. 2010). These five operons share the 24-bp consensus sequence located around the nucleotide 28–30-bp upstream from their transcription start (Figs. 1.9b and 1.11) (Shimodaira et al. 2012). The 24-bp consensus is also found in the upstream from the degradation genes of other *Rhodococcus* strains including *R. opacus* B4, *R. erythropolis* BD2, and *Rhodococcus* sp. TFB (Fig. 1.11), and is thought to be the binding site of the specific transcription complex containing the *bphT*-coding response regulators or the response regulator itself. Thus the strain RHA1 has the multiple-enzyme system for biphenyl metabolism, which is induced by the multiple regulatory systems.

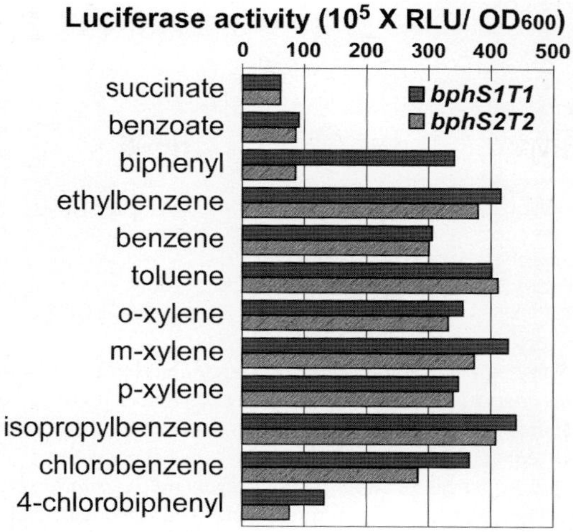

Fig. 1.10 Inducing substrates of the duplicated two-component regulatory systems, BphS1T1 and BphS2T2. Substrate-dependent relative luciferase activities conferred by the *luxAB* reporter genes under the control of BphS1T1 (*black bars*) or BphS2T2 (*shaded bars*) are presented. The *luxAB* reporter gene was connected to the *bphAa* promoter in a reporter plasmid, which was introduced in a host strain *R. erythropolis* IAM1399 containing a plasmid expressing either BphS1T1 or BphS2T2. The recombinant strain was grown in the presence of one of the substrates indicated and subjected to the luciferase activity measurement using a luminometer. Succinate and benzoate were employed as non-inducing control substrates. The activity (RLU/OD_{600}) was expressed as relative light unit (RLU) per milliliter per unit of the final OD_{600}

Table 1.1 Involvement of the *bphST* regulatory system in the growth of RHA1 on a variety of aromatics

	Growth[a]						
Substrate	Wild (RHA1)	*bphS1* mutant	*bphS2* mutant	*bphS1S2* mutant	*bphT1* mutant	*bphT2* mutant	*bphT1T2* mutant
Biphenyl	+	−	+	−	+	+	−
Ethylbenzene	+	+	+	−	+	+	−
Toluene	+	+	+	−	+	+	−
Benzene	+	+	+	−	+	+	−
Isopropylbenzene	+	+	+	−	+	+	−
o-xylene	+	+	+	−	+	+	−
Terephthalate	+	+	+	+	+	+	+

[a]− no growth, + growth

1.5 The Advantages and Disadvantages of the Multiple-Enzyme System

Consequently the multiple-enzyme system in combination with the regulatory system with a broad inducing-substrate spectrum, which was uncovered in the strain RHA1, appears to provide the parallel-degradation system conferring the tough

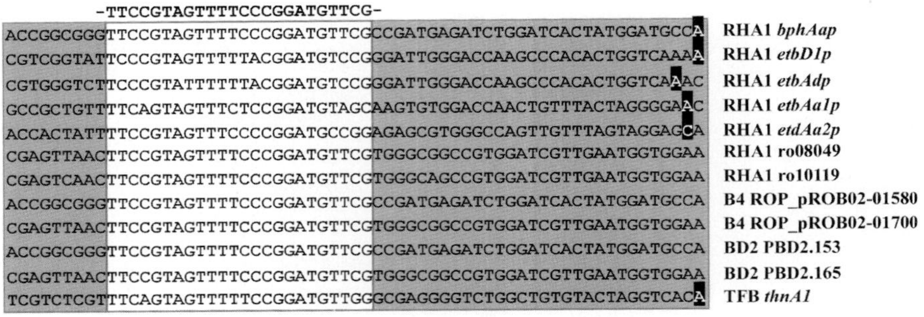

Fig. 1.11 The consensus sequence among the BphST-regulated promoters and their homologs in aromatic-degrading *Rhodococcus* strains. The promoter sequences regulated by BphST in *R. jostii* RHA1 are aligned with the homologous sequences found in other aromatic-degrading *Rhodococcus* strains. The conserved sequences corresponding to the consensus, which is presented on the *top*, are in an *open box*. The sequences outside the consensus are in *gray boxes*. The transcriptional starts are *highlighted*. The sequences are extracted from the upstream from *bphAa* of *R. jostii* RHA1 (CP000432), *etbD1* of RHA1 (CP000432), *etbAa1* of RHA1 (CP000433), *etbAa2* of RHA1 (CP000433), *etbAd* of RHA1 (CP000433), ro08049 of RHA1 (CP000432), ro10119 of RHA1 (CP000433), ROP_pROB02-01580 of *R. opacus* B4 (AP011117), ROP_pROB02-01700 of *R. opacus* B4 (AP011117), PBD2.153 of *R. erythropolis* BD2 (AY223810), PBD2.165 of *R. erythropolis* BD2 (AY223810), and *thnA1* of *Rhodococcus* sp. TFB (32)

degradation activity toward a broad range of aromatics, suggesting the advantage of the multiple-enzyme system. Table 1.1 clearly illustrates the broad substrate range of the parallel-degradation system. Loss of degradation activity by the multiple-enzyme system in the double knockouts of *bphS1* and *bphS2* or *bphT1* and *bphT2* resulted in the growth deficiency on biphenyl, benzene, and alkylbenzenes. In addition the multiple-enzyme system with the broad induction system is expected to give superior chances to evolve the degradation systems for new substrates with accelerated evolution velocity. Recently we found that the competition between BphA and EtbA dioxygenases makes RHA1 unable to grow on stylene, suggesting the disadvantage of the multiple-enzyme system (unpublished results). However, the multiple-enzyme system in *Rhodococcus* strains definitely seem to provide the superior degradation activity and growth ability on a wide range of aromatics.

References

Asturias JA, Timmis KN (1993) Three different 2,3-dihydroxybiphenyl-1,2-dioxygenase genes in the gram-positive polychlorobiphenyl-degrading bacterium *Rhodococcus globerulus* P6. J Bacteriol 175:4631–4640

Bergdoll M, Eltis LD, Cameron AD, Dumas P, Bolin JT (1998) All in the family: structural and evolutionary relationships among three modular proteins with diverse functions and variable assembly. Protein Sci 7:1661–1670

Eltis LD, Bolin JT (1996) Evolutionary relationships among extradiol dioxygenases. J Bacteriol 178:5930–5937

Goncalves ER, Hara H, Miyazawa D, Davies JE, Eltis LD, Mohn WW (2006) Transcriptomic assessment of isozymes in the biphenyl pathway of *Rhodococcus* sp. strain RHA1. Appl Environ Microbiol 72:6183–6193

Hatta T, Shimada T, Yoshihara T, Yamada A, Masai E, Fukuda M, Kiyohara H (1998) Meta-fission product hydrolases from a strong PCB degrader *Rhodococcus* sp. RHA1. J Ferment Bioeng 85:174–179

Hauschild JE, Masai E, Sugiyama K, Hatta T, Kimbara K, Fukuda M, Yano K (1996) Identification of an alternative 2,3-dihydroxybiphenyl 1,2-dioxygenase in *Rhodococcus* sp. strain RHA1 and cloning of the gene. Appl Environ Microbiol 62:2940–2946

Iwasaki T, Miyauchi K, Masai E, Fukuda M (2006) Multiple-subunit genes of the aromatic-ring-hydroxylating dioxygenase play an active role in biphenyl and polychlorinated biphenyl degradation in *Rhodococcus* sp. strain RHA1. Appl Environ Microbiol 72:5396–5402

Iwasaki T, Takeda H, Miyauchi K, Yamada T, Masai E, Fukuda M (2007) Characterization of two biphenyl dioxygenases for biphenyl/PCB degradation in A PCB degrader, *Rhodococcus* sp. strain RHA1. Biosci Biotechnol Biochem 71:993–1002

Kosono S, Maeda M, Fuji F, Arai H, Kudo T (1997) Three of the seven *bphC* genes of *Rhodococcus erythropolis* TA421, isolated from a termite ecosystem, are located on an indigenous plasmid associated with biphenyl degradation. Appl Environ Microbiol 63:3282–3285

Maeda M, Chung SY, Song E, Kudo T (1995) Multiple genes encoding 2,3-dihydroxybiphenyl 1,2-dioxygenase in the gram-positive polychlorinated biphenyl-degrading bacterium *Rhodococcus erythropolis* TA421, isolated from a termite ecosystem. Appl Environ Microbiol 61:549–555

McKay DB, Prucha M, Reineke W, Timmis KN, Pieper DH (2003) Substrate specificity and expression of three 2,3-dihydroxybiphenyl 1,2-dioxygenases from *Rhodococcus globerulus* strain P6. J Bacteriol 185:2944–2951

McLeod MP, Warren RL, Hsiao WW, Araki N, Myhre M, Fernandes C, Miyazawa D, Wong W, Lillquist AL, Wang D, Dosanjh M, Hara H, Petrescu A, Morin RD, Yang G, Stott JM, Schein JE, Shin H, Smailus D, Siddiqui AS, Marra MA, Jones SJ, Holt R, Brinkman FS, Miyauchi K, Fukuda M, Davies JE, Mohn WW, Eltis LD (2006) The complete genome of *Rhodococcus* sp. RHA1 provides insights into a catabolic powerhouse. Proc Natl Acad Sci USA 103:15582–15587

Ohtsubo Y, Delawary M, Kimbara K, Takagi M, Ohta A, Nagata Y (2001) BphS, a key transcriptional regulator of bph genes involved in polychlorinated biphenyl/biphenyl degradation in *Pseudomonas* sp. KKS102. J Biol Chem 276:36146–36154

Sakai M, Masai E, Asami H, Sugiyama K, Kimbara K, Fukuda M (2002) Diversity of 2,3-dihydroxybiphenyl dioxygenase genes in a strong PCB degrader, *Rhodococcus* sp. strain RHA1. J Biosci Bioeng 93:421–427

Senda T, Sugiyama K, Narita H, Yamamoto T, Kimbara K, Fukuda M, Sato M, Yano K, Mitsui Y (1996) Three-dimensional structures of free form and two substrate complexes of an extradiol ring-cleavage type dioxygenase, the BphC enzyme from *Pseudomonas* sp. strain KKS102. J Mol Biol 255:735–752

Seto M, Ida M, Okita N, Hatta T, Masai E, Fukuda M (1996) Catabolic potential of multiple PCB transformation systems in *Rhodococcus* sp. strain RHA1. Biotechnol Lett 18:1305–1308

Shimizu S, Kobayashi H, Masai E, Fukuda M (2001) Characterization of the 450-kb linear plasmid in a polychlorinated biphenyl degrader, *Rhodococcus* sp. strain RHA1. Appl Environ Microbiol 67:2021–2028

Shimodaira J, Furusawa Y, Miyazawa Y, Kasai D, Miyauchi K, Masai E, Fukuda M (2012) The 24-bp consensus sequence responsible for regulation of the BphS1T1 two-component system in a hybrid promoter. J Biosci Bioeng 113:279–285

Sugimoto K, Senda T, Aoshima H, Masai E, Fukuda M, Mitsui Y (1999) Crystal structure of an aromatic ring opening dioxygenase LigAB, a protocatechuate 4,5-dioxygenase, under aerobic conditions. Structure 7:953–965

Taguchi K, Motoyama M, Kudo T (2004) Multiplicity of 2,3-dihydroxybiphenyl dioxygenase genes in the Gram-positive polychlorinated biphenyl degrading bacterium *Rhodococcus rhodochrous* K37. Biosci Biotechnol Biochem 68:787–795

Takeda H, Hara N, Sakai M, Yamada A, Miyauchi K, Masai E, Fukuda M (2004) Biphenyl-inducible promoters in a polychlorinated biphenyl-degrading bacterium, *Rhodococcus* sp. RHA1. Biosci Biotechnol Biochem 68:1249–1258

Takeda H, Shimodaira J, Yukawa K, Hara N, Kasai D, Miyauchi K, Masai E, Fukuda M (2010) Dual two-component regulatory systems are involved in aromatic compound degradation in a polychlorinated-biphenyl degrader, *Rhodococcus jostii* RHA1. J Bacteriol 192:4741–4751

Vaillancourt FH, Haro MA, Drouin NM, Karim Z, Maaroufi H, Eltis LD (2003) Characterization of extradiol dioxygenases from a polychlorinated biphenyl-degrading strain that possess higher specificities for chlorinated metabolites. J Bacteriol 185:1253–1260

Van der Geize R, Yam K, Heuser T, Wilbrink MH, Hara H, Anderton MC, Sim E, Dijkhuizen L, Davies JE, Mohn WW, Eltis LD (2007) A gene cluster encoding cholesterol catabolism in a soil actinomycete provides insight into *Mycobacterium tuberculosis* survival in macrophages. Proc Natl Acad Sci USA 104:1947–1952

Watanabe T, Fujihara H, Furukawa K (2003) Characterization of the second LysR-type regulator in the biphenyl-catabolic gene cluster of *Pseudomonas pseudoalcaligenes* KF707. J Bacteriol 185:3575–3582

Yamada A, Kishi H, Sugiyama K, Hatta T, Nakamura K, Masai E, Fukuda M (1998) Two nearly identical aromatic compound hydrolase genes in a strong polychlorinated biphenyl degrader, *Rhodococcus* sp. strain RHA1. Appl Environ Microbiol 64:2006–2012

Chapter 2
Appearance and Evolution of γ-Hexachlorocyclohexane-Degrading Bacteria

Yuji Nagata, Michiro Tabata, Satoshi Ohhata, and Masataka Tsuda

Abstract α-Proteobacterial strains belonging to the so-called sphingomonads group degrade various highly recalcitrant compounds, including xenobiotics, but generally each strain degrades only a limited set of compounds, suggesting that sphingomonads tend to be specialists for the degradation of extremely recalcitrant compounds. In this chapter, the appearance and evolution of γ-hexachlorocyclohexane (γ-HCH/γ-BHC/lindane)-degrading bacteria belonging to sphingomonads will be discussed on the basis of the structure and function of their genomes and mobile genetic elements. γ-HCH is a typical, completely man-made chlorinated pesticide that has caused serious environmental problems due to its toxicity and long persistence in upland soils. The genome sequence of an archetypal γ-HCH-degrading strain, *Sphingobium japonicum* UT26, and its comparison with those of other closely related γ-HCH-degrading and non-γ-HCH-degrading sphingomonad strains revealed that (1) these γ-HCH-degraders appeared independently and in parallel at geographically different areas by recruiting the "specific" *lin* genes into strains having the core functions of sphingomonads; (2) various sphingomonad-specific plasmids and the insertion sequence IS*6100* play important roles in the recruitment and dissemination of the "specific" *lin* genes; and (3) transposition of IS*6100* causes genome rearrangements including deletion and inversion of DNA sequences and fusion and resolution of replicons.

Keywords Biodegradation • Environmental pollutant • Genome • Mobile genetic element • Plasmid • Insertion sequence • Sphingomonads • Xenobiotics • γ-Hexachlorocyclohexane

---

Y. Nagata (✉) • M. Tabata • S. Ohhata • M. Tsuda
Graduate School of Life Sciences, Tohoku University,
2-1-1 Katahira, Sendai 980-8577, Japan
e-mail: aynaga@ige.tohoku.ac.jp

H. Nojiri et al. (eds.), *Biodegradative Bacteria: How Bacteria Degrade, Survive, Adapt, and Evolve*, DOI 10.1007/978-4-431-54520-0_2, © Springer Japan 2014

Abbreviations

Chr	Chromosome
IS	Insertion sequence
PEG	Polyethylene glycol
POPs	Persistent organic pollutants
TBDR	TonB-dependent receptor
TCA	Tricarboxylic acid
t-HCH	Technical-HCH
γ-HCH	γ-Hexachlorocyclohexane

2.1 Introduction

Microorganisms have developed new metabolic pathways for various chemical compounds, including xenobiotics, in order to exploit new carbon sources and to detoxify toxic compounds (Janssen et al. 2005). The mechanism by which they adapt to xenobiotics is of great interest. Furthermore, an improved understanding of the molecular mechanisms underlying the degradation of xenobiotics could be applied for successful bioremediation (de Lorenzo 2008). For these purposes, many bacterial strains that utilize xenobiotics have been isolated and characterized (Megharaj et al. 2011). Among them, the former genus *Sphingomonas*, which belongs to α-proteobacteria, forms one of the most important groups, because they degrade various compounds, including highly recalcitrant ones (Stolz 2009). This group of bacteria is also becoming increasingly interesting in the field of environmental microbiology, because strains belonging to this former genus are widely distributed in nature. Furthermore, they have important properties that can be exploited for use in industrial biotechnology (White et al. 1996).

Taxonomically, the former genus *Sphingomonas* is complicated. This genus was previously divided into four genera, *Sphingomonas sensu stricto*, *Sphingobium*, *Novosphingobium*, and *Sphingopyxis* (Takeuchi et al. 2001), and then Yabuuchi et al. regrouped the species of the four genera again as the single genus *Sphingomonas* (Yabuuchi et al. 2002). However, the four genera can still be recognized, and we here refer to these four *Sphingomonas*-related groups collectively as sphingomonads. The isolation of various sphingomonads which harbor different metabolic pathways for the degradation of a wide variety of xenobiotic compounds suggests that the members of sphingomonads have the ability to adapt more quickly or more efficiently to the degradation of new compounds in the environment than members of other bacterial genera (Stolz 2009). In this chapter, we will focus on the sphingomonad strains that assimilate γ-hexachlorocyclohexane (γ-HCH/ γ-BHC/lindane) aerobically and discuss the appearance and evolution of such bacterial strains on the basis of the structure and function of their genomes and mobile genetic elements.

2.2 Aerobic γ-HCH-Degrading Bacteria

γ-HCH is a completely man-made chlorinated pesticide that has caused serious environmental problems due to its toxicity and long persistence in upland soils (Lal et al. 2010; Vijgen et al. 2011). HCH is chemically synthesized by the process of photochlorination of benzene. The synthesized product is called technical-HCH (t-HCH) and consists of five isomers, namely, α-(60–70 %), γ-(12–16 %), β-(10–12 %), δ-(6–10 %), and ε-(3–4 %) (Vijgen et al. 2011). Among these HCH isomers in t-HCH, only γ-HCH has insecticidal activity, and thus γ-HCH is generally purified from the other isomers in t-HCH and used as the insecticide lindane (>99 % purity). However, we must take all the HCH isomers into consideration in the cleanup of HCH-contaminated sites, because the remaining isomers are also improperly discarded as "HCH muck" and have caused environmental problems. Although the use of γ-HCH has been banned, many HCH (γ- and other isomers)-contaminated sites are still present around the world (Lal et al. 2010; Vijgen et al. 2011). Furthermore, the α-, β-, and γ-HCH isomers were recently included as new persistent organic pollutants (POPs) at the Stockholm Convention, and thus HCH contamination continues to be a major global concern (Vijgen et al. 2011). Only 60 years after the first release of γ-HCH into the environment, a number of bacterial strains that aerobically assimilate γ-HCH have been isolated from geographically distinct locations (Lal et al. 2010). This fact suggests that these bacteria acquired their γ-HCH-degrading abilities within a relatively short period.

Sphingobium japonicum UT26 was isolated from an upland experimental field to which γ-HCH had been applied once a year for 12 years (Senoo and Wada 1989; Imai et al. 1989). UT26 utilizes γ-HCH as a sole source of carbon and energy under aerobic conditions and is the first and archetypal strain from which the *lin* genes for the utilization of γ-HCH have been cloned and characterized (Nagata et al. 2007). The γ-HCH degradation pathway in UT26 is shown in Fig. 2.1, and the *lin* genes are listed in Table 2.1. In UT26, γ-HCH is converted to β-ketoadipate via reactions catalyzed by dehydrochlorinase (LinA) (Imai et al. 1991; Nagata et al. 1993a), haloalkane dehalogenase (LinB) (Nagata et al. 1993b; Nagata et al. 1997), dehydrogenase (LinC) (Nagata et al. 1994), reductive dechlorinase (LinD) (Miyauchi et al. 1998), ring-cleavage dioxygenase (LinE) (Miyauchi et al. 1999), and maleylacetate reductase (LinF) (Endo et al. 2005), and β-ketoadipate is further converted to succinyl-CoA and acetyl-CoA by succinyl-CoA:3-oxoadipate CoA transferase (LinGH) and β-ketoadipyl CoA thiolase (LinJ), respectively (Nagata et al. 2007; Nagata et al. 2011). The last two compounds are metabolized in the citrate/tricarboxylic acid (TCA) cycle, which is a part of the central metabolic pathways. The *linD* and *linE* genes constitute an operon and their expression is regulated by a LysR-type transcriptional regulator (LinR) (Miyauchi et al. 2002).

The *lin* genes for the upstream and downstream pathways (Fig. 2.1: *linA* to *linF*) are "specific" for the γ-HCH-degrading pathway. In particular, the *linA* gene is unique, because it does not show significant similarity to any sequences in the

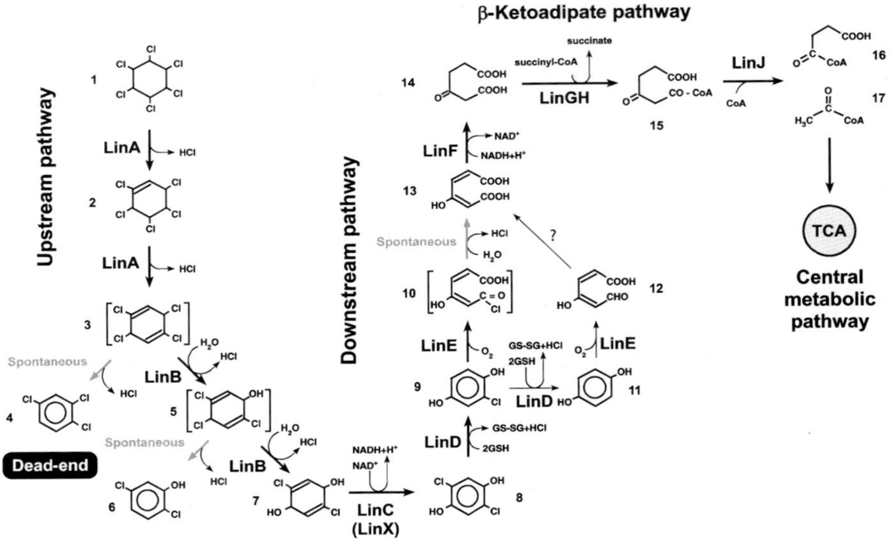

Fig. 2.1 Degradation pathway of γ-HCH in UT26. Compounds: *1*, γ-hexachlorocyclohexane (γ-HCH); *2*, γ-pentachlorocyclohexene; *3*, 1,3,4,6-tetrachloro-1,4-cyclohexadiene; *4*, 1,2,4-trichlorobenzene; *5*, 2,4,5-trichloro-2,5-cyclohexadiene-1-ol; *6*, 2,5-dichlorophenol; *7*, 2,5-dichloro-2,5-cyclohexadiene-1,4-diol; *8*, 2,5-dichlorohydroquinone; *9*, chlorohydroquinone; *10*, acylchloride; *11*, hydroquinone; *12*, γ-hydroxymuconic semialdehyde; *13*, maleylacetate; *14*, β-ketoadipate; *15*, 3-oxoadipyl-CoA; *16*, succinyl-CoA; *17*, acetyl-CoA. TCA, citrate/tricarboxylic acid cycle

databases except for the almost identical *linA* genes (>90 % identical) from other bacterial strains (Nagata et al. 2007; Lal et al. 2010). On the other hand, the β-ketoadipate pathway is one of the major pathways for the degradation of aromatic compounds, and the genes for this pathway are often carried by many environmental bacterial strains (Harwood and Parales 1996). In addition to the *lin* genes encoding these catabolic enzymes, genes for an ABC transporter system are also necessary for the utilization of γ-HCH in UT26 (Endo et al. 2007). This ABC transporter system consists of four components, LinK, LinL, LinM, and LinN, and is involved in the tolerance toward a toxic metabolite, 2,5-dichlorophenol (compound 6 in Fig. 2.1) (Endo et al. 2007). It was suggested that LinKLMN is not a simple efflux pump of toxic compounds, but is involved in the integrity of the outer membrane (Endo et al., unpublished data). However, the substrate(s) of this transporter system remain(s) unknown.

Aerobic γ-HCH-degrading bacterial strains other than UT26 have been isolated around the world, and most of them, at least those that have been adequately characterized, are sphingomonads (Lal et al. 2010). Furthermore, all of these analyzed strains have almost identical *lin* genes and identical IS*6100* insertion sequences (Lal et al. 2006). Although *Sphingobium indicum* B90A from India (Kumari et al. 2002) and *Sphingobium francense* Sp+from France (Ceremonie et al. 2006) have been

Table 2.1 *lin* genes involved in the γ-HCH utilization in *S. japonicum* UT26S

Gene	A.A. residues	G+C content (%)	Function	Expression	Location[a]	Reference
linA	156	53.9	Dehydrochlorinase	Constitutive	Chr1_1860686–1861156_c	Imai et al. (1991), Nagata et al. (1993a)
linB	296	62.5	Halidohydrolase (haloalkane dehalogenase)	Constitutive	Chr1_1966541–1967431	Nagata et al. (1993b, 1997)
linC	250	64.3	Dehydrogenase	Constitutive	Chr1_566609–567361_c	Nagata et al. (1994)
linD	346	60.8	Reductive dechlorinase (glutathione *S*-transferase)	Inducible	pCHQ1_110947–111987_c	Miyauchi et al. (1998)
linE	321	60.1	Ring-cleavage dioxygenase	Inducible	pCHQ1_114235–115200_c	Miyauchi et al. (1999)
linR	303	61.3	LysR-family transcriptional regulator	NA	pCHQ1_115332–116243	Miyauchi et al. (2002)
linF	352	68.1	Maleylactate reductase	NA	Chr2_562332–563390_c	Endo et al. (2005)
linG	156	62.9	Acyl-CoA transferase, alpha subunit	NA	Chr2_603108–603755	Nagata et al. (2007, 2011)
linH	212	63	Acyl-CoA transferase, beta subunit	NA	Chr2_603755–604393	Nagata et al. (2007, 2011)
linI	267	70.2	IclR-family transcriptional regulator	NA	Chr2_602168–602971_c	Nagata et al. (2007, 2011)
linJ	403	68.8	Thiolase	NA	Chr2_600921–602132_c	Nagata et al. (2007, 2011)
linK	316	65.9	Putative ABC transporter system, inner membrane protein	NA	Chr1_19347–20477	Endo et al. (2007)
linL	282	65.7	Putative ABC transporter system, ATPase	NA	Chr1_20477–21325	Endo et al. (2007)
linM	320	65.6	Putative ABC transporter system, periplasmic protein	NA	Chr1_21329–22291	Endo et al. (2007)
linN	202	67.8	Putative ABC transporter system, lipoprotein	NA	Chr1_22299–22907	Endo et al. (2007)

c encoding on complementary strand, *NA* not analyzed

[a]Base coordinates on Chr 1 (AP010803), Chr 2 (AP010804), and pCHQ1 (AP010805)

analyzed in detail, their genome sequences are unavailable. In this chapter, we will describe mainly the structure of the UT26 genome and its comparison with the genomes of other closely related non-γ-HCH-degrading sphingomonad strains and three γ-HCH-degraders, *Sphingobium* sp. MI1205 from Japan (Ito et al. 2007), *Sphingomonas* sp. MM-1 from India (Tabata et al. 2011), and *Sphingobium* sp. TKS from Japan (Ohhata et al., unpublished data).

2.3 Genomes of γ-HCH-Degrading Bacteria

The complete genome sequence of *S. japonicum* UT26 revealed five replicons composed of two circular chromosomes (Chr), Chr 1 (3,514,822 bp, 64.8 % G+C, 3,529 ORFs) and Chr 2 (681,892 bp, 65.9 % G+C, 589 ORFs) and three circular plasmids, pCHQ1 (190,974 bp, 63.0 % G+C, 224 ORFs), pUT1 (31,776 bp, 63.7 % G+C, 44 ORFs), and pUT2 (5,398 bp, 61.0 % G+C, 8 ORFs) (Nagata et al. 2010; Nagata et al. 2011). Chr 1 and Chr 2 have one and two copies of rRNA gene (*rrn*) operons, respectively. Because the *rrn*-containing large replicons in bacteria are usually defined as chromosomes, the two large replicons were designated Chr 1 and Chr 2. However, the border between "chromosome" and "plasmid" is vague. Although pNL2 of *Novosphingobium aromaticivorans* DSM 12444 and Lpl of *Novosphingobium* sp. PP1Y (D'Argenio et al. 2011) have no *rrn* operon and thus were categorized as plasmids, their putative replication origins are similar to those of Chr 2 in *Sphingobium japonicum* UT26 and *Sphingobium chlorophenolicum* L-1 (Copley et al. 2011), suggesting these replicons are derived from a common ancestral *repABC* family plasmid (Fig. 2.2a, see below). Chr 1 and Chr 2 carry 51 and 4 tRNA genes, respectively, while the other replicons carry no tRNA genes. Out of the 206 bacterial essential genes proposed by Gil et al. (Gil et al. 2004), ten genes for fructose-1, 6-bisphosphate aldolase, 6-phosphofructokinase, ribose 5-phosphate isomerase, *sn*-glycerol-3-phosphate acyltransferase, hypoxanthine phosphoribosyltransferase, pantothenate kinase, adenylyl transferase, nicotinamide phosphoribosyltransferase, pyridoxal kinase, and thiamine pyrophosphokinase were not found in the UT26 genome, and all the other 196 genes were located on Chr 1. Furthermore, the putative replication origin of Chr 1 was found to be of α-proteobacterial-chromosome type (Brassinga and Marczynski 2001; Sibley et al. 2006); this is an upstream region of the uroporphyrinogen decarboxylase (*hemE*) gene with multiple DnaA boxes [TT(A/T)TNCACA] (Schaper and Messer 1995; Nagata et al. 2011). These results clearly indicated that Chr 1 is a main chromosome of UT26. However, some important genes, e.g., the *cysC* and *cysD* genes for cysteine biosynthesis, were found only on Chr 2 (Fig. 2.3), suggesting that Chr 2 is also essential for some fundamental cell functions. The putative *repA* (for replication initiation protein) genes of Chr 2 and pCHQ1 constitute clusters with putative partitioning genes, *parA* and *parB* (Fig. 2.2a, b). The existence of this type of cluster is a feature of the

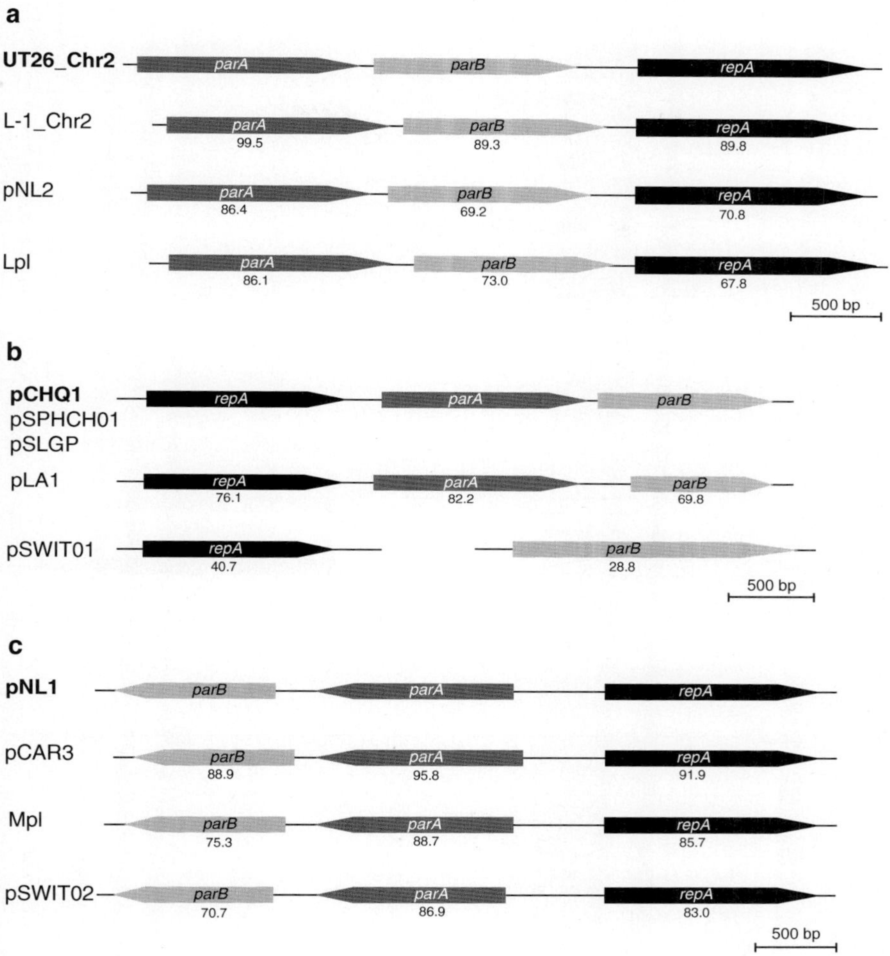

Fig. 2.2 Proposed *repA*, *parA*, and *parB* regions of the sphingomonad plasmids/subreplicons whose complete sequences were available (Table 2.3). Their types were classified on the basis of RepA similarity: (**a**), UT26_Chr 2 type; (**b**), pCHQ1 type; (**c**), pNL1 type; (**d**), pUT1 type; (**e**), pUT2 type; and (**f**), pLB1 type. Numbers under the genes show amino acid identity (%) with RepA, ParA, and ParB of representative replicons of each type shown in bold at the top of each panel

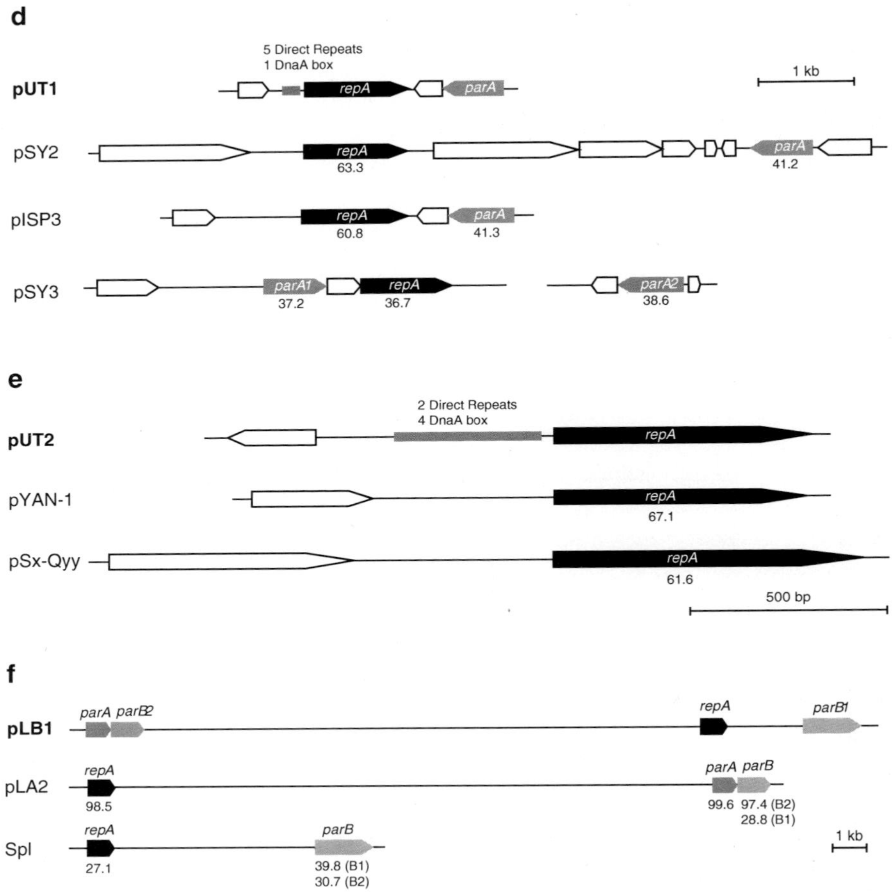

Fig. 2.2 (continued)

so-called *repABC* family plasmids, which are widely distributed among α-proteobacteria (Cevallos et al. 2008), suggesting that Chr 2 and pCHQ1 are originated from *repABC* family plasmids. The replication origin of pUT1 was proposed to be located at the region upstream of the putative *repA* gene, because direct repeats and a DnaA box were found in this region (Fig. 2.2d), which are typical features for the replication origins of iteron-type plasmids (Chattoraj 2000). The replication origin of pUT2 was also proposed to be located at the region upstream of the putative *repA* gene (Fig. 2.2e). However, pUT2 seems to be a member of the IncP-9/pM3 family (Krasowiak et al. 2006).

Sixty-three genes encoding TonB-dependent receptor (TBDR)-like proteins were found in the UT26 genome: 43 on Chr 1, 18 on Chr 2, and 2 on pCHQ1.

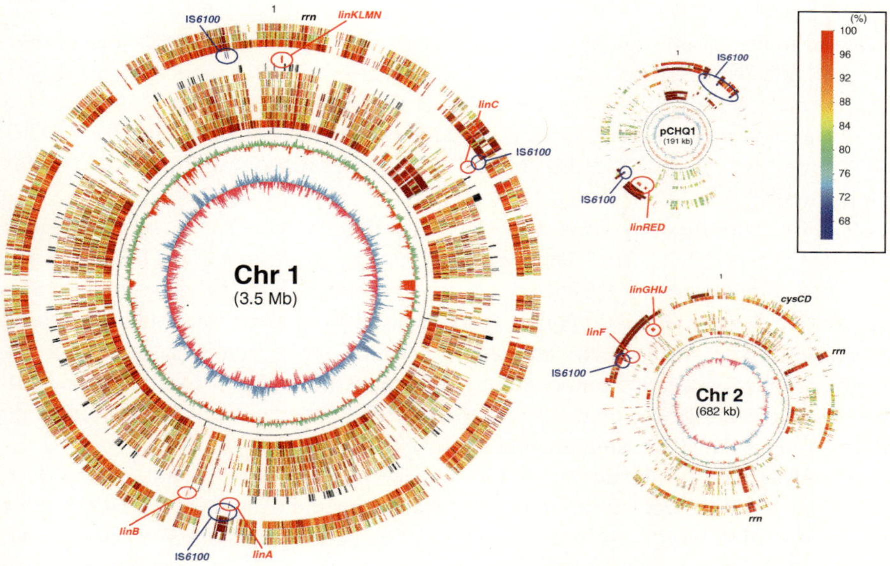

Fig. 2.3 Three main replicons of UT26. Each replicon is drawn in *circular*. Note that the size scale depends on the replicons. The top positions of the three replicons have been defined as putative replication origins (position 1). Positions of the *lin* genes, IS*6100*, and bacterial essential genes proposed by Gil et al. are marked with *red*, *blue*, and *back bars*, respectively. Results of GC skew and G+C content are shown *inside* of the *black scale circles*: GC skew (*inside*), the parts higher and lower than zero were colored with *cyan* and *magenta*, respectively; G+C content (*outside*), the parts higher and lower than average of each replicon were colored with *green* and *red*, respectively. Result of a blastN search of each region of the UT26 genome toward genome sequences of seven other non-γ-HCH-degrading sphingomonad strains, *Sphingobium chlorophenolicum* L-1, *Sphingomonas* sp. SKA58, *Sphingobium* sp. SYK-6, *Sphingomonas wittichii* RW1, *Sphingopyxis alaskensis* RB2256, *Novosphingobium aromaticivorans* DSM 12444, and *Novosphingobium* sp. PP1Y, is shown *outside* of the *black scale circles* in this order. The region whose homologous sequence was found in the other strains (L-1, SKA58, SYK-6, RW1, RB2256, DSM 12444, and PP1Y from *inside* to *outside*) was colored in the gradient depending on the level of similarity as shown in explanatory note. Result of the same blastN search of each region of the UT26 genome toward draft genome sequences of three γ-HCH-degrading sphingomonad strains, *Sphingobium* sp. TKS, *Sphingobium* sp. MI1205, and *Sphingomonas* sp. MM-1 were shown *outside* of the *circle* for the positions of IS*6100* in this order. See text for detail. This figure was drawn by ArcWithColor (http://www.ige.tohoku.ac.jp/joho/gmProject/gmdownload.html)

Although TBDRs in other Gram(−) bacteria have originally been identified as outer membrane proteins that mediate transport of siderophores into the periplasm (Ferguson and Deisenhofer 2002), outer membrane polysaccharide-binding proteins have sequence similarity with TBDRs (Cheng et al. 1995; Reeves et al. 1996). Furthermore, the involvement of TBDR-like proteins in the uptake of polyethylene glycol (PEG) (Tani et al. 2007), alkyl sulfate esters (Kahnert et al. 2002), and alginate (Hashimoto et al. 2005) has also been suggested. The dioxin-mineralizing bacterium

Sphingomonas wittichii RW1 was also found to have many genes for TBDR-like proteins in its genome, and it was suggested that these proteins serve as transporters of aromatic hydrocarbons and/or their breakdown products (Miller et al. 2010). These facts strongly suggest that TBDR-like proteins are necessary for the utilization of various compounds and that the possession of many genes for TBDR-like proteins ensures the UT26 and RW1 degradation abilities of various compounds.

The *lin* genes for the γ-HCH degradation are dispersed on Chr 1 (*linA*, *linB*, *linC*, and *linKLMN*), Chr 2 (*linF* and *linGHIJ*), and pCHQ1 (*linRED*) (Table 2.1 and Fig. 2.3), suggesting that UT26 did not simply acquire all the *lin* genes at once as a sequential cluster. This is in contrast to the case of aromatic compound-degrading *Pseudomonas* strains, which can acquire a whole set of genes necessary for the degradation of aromatic compounds by conjugative transfer of plasmids and integrative and conjugative elements (Davison 1999; Maeda et al. 2003; Sota et al. 2006; Yano et al. 2010; Shintani and Nojiri 2013).

Putative genes for the degradation of toluene/phenol, chlorophenol, anthranilate, and homogentisate were also found in the UT26 genome (Nagata et al. 2011). These genes constitute four clusters for the degradation of the respective compounds, and each cluster contains all the genes necessary for conversion of each compound to metabolites in the central metabolic pathway, strongly suggesting that UT26 can utilize these compounds. However, the gene repertoire of UT26 for the degradation of recalcitrant compounds is relatively small, compared with other environmental bacterial strains, such as the versatile pollutant degrader *Cupriavidus necator* JMP134 (Lykidis et al. 2010) and *Burkholderia multivorans* ATCC 17616 (Ohtsubo et al. unpublished data), which is known to possess versatile catabolic activities (Stanier et al. 1966; Yuhara et al. 2008). Although more analysis is necessary to estimate the critical metabolic versatility of UT26, it can be concluded at this stage that UT26 is a "specialist" for the degradation of γ-HCH, but not a "generalist" for the degradation of various compounds. It is known that sphingomonad strains degrade various extremely recalcitrant compounds, such as dioxin-related compounds, chlorinated herbicides, lignin-related compounds, polyaromatic hydrocarbons, and PEG, but each strain degrades only a limited set of compounds (Stolz 2009). In other words, sphingomonads tend to be specialists for the degradation of highly recalcitrant compounds, not generalists like *Cupriavidus* and *Burkholderia* strains.

Comparative genomics often provide some insight into the evolution of bacterial genomes. Thus, the genome of UT26 was compared with those of seven other non-γ-HCH-degrading sphingomonad strains whose genome sequences were available in September, 2012 (Table 2.2): *Sphingobium chlorophenolicum* L-1 (pentachlorophenol degrader; Copley et al. 2011), *Sphingomonas* sp. SKA58 (representative marine oligotroph), *Sphingobium* sp. SYK-6 (lignin degrader; Masai et al. 2012), *Sphingomonas wittichii* RW1 (dioxin degrader; Miller et al. 2010), *Sphingopyxis alaskensis* RB2256 (representative marine oligotroph; Lauro et al. 2009), *Novosphingobium aromaticivorans* DSM 12444 (degrader of a wide variety of aromatic hydrocarbons), and *Novosphingobium* sp. PP1Y (degrader of a wide variety of aromatic hydrocarbons; D'Argenio et al. 2011). The 16S rRNA gene analysis

Table 2.2 Sphingomonad strains whose genome sequences were available[a]

Strain	Total size (bp)	Chr 1 (bp)	UT26_Chr 2-type replicon (bp)	pCHQ1-type plasmid (bp)	pNL1-type plasmid (bp)	pUT1-type plasmid (bp)	pUT2-type plasmid (bp)	Other-type plasmid (bp)	Number of *rrn* operon	Reference
Sphingobium japonicum UT26S (γ-HCH-degrader)	4,424,862	3,514,822	681,892	pCHQ1 190,974		pUT1 31,776	pUT2 5,398		3	Nagata et al. (2011)
Sphingobium chlorophenolicum L-1 (pentachlorophenol degrader)	4,573,221	3,080,818	1,368,670	pSPHCH01 123,733					3	Copley et al. (2011)
Sphingomonas sp. SKA58 (representative marine oligotroph)	ca. 3,900,000 (41 contigs)								(3)	
Sphingobium sp. SYK-6 (lignin degrader)	4,348,133	4,199,332		pSLGP 148,801					2	Masai et al. (2012)
Sphingomonas wittichii RW1 (dioxin degrader)	5,915,246	5,382,261		pSWIT01 310,228	pSWIT02 222,757				2	Miller et al. (2010)
Sphingopyxis alaskensis RB2256 (representative marine oligotroph)	3,373,713	3,345,170						F plasmid 28,543	1	Lauro et al. (2009)
Novosphingobium aromaticivorans DSM 12444 (degrader of wide variety of aromatic hydrocarbons)	4,233,314	3,561,584	pNL2 487,268		pNL1 184,462				3	Unpublished
Novosphingobium sp. PP1Y (degrader of wide variety of aromatic hydrocarbons)	5,313,905	3,911,486	Lpl 192,103		Mpl 1,161,602			Spl 48,714	3	D'Argenio et al. (2011)

[a]In September, 2012

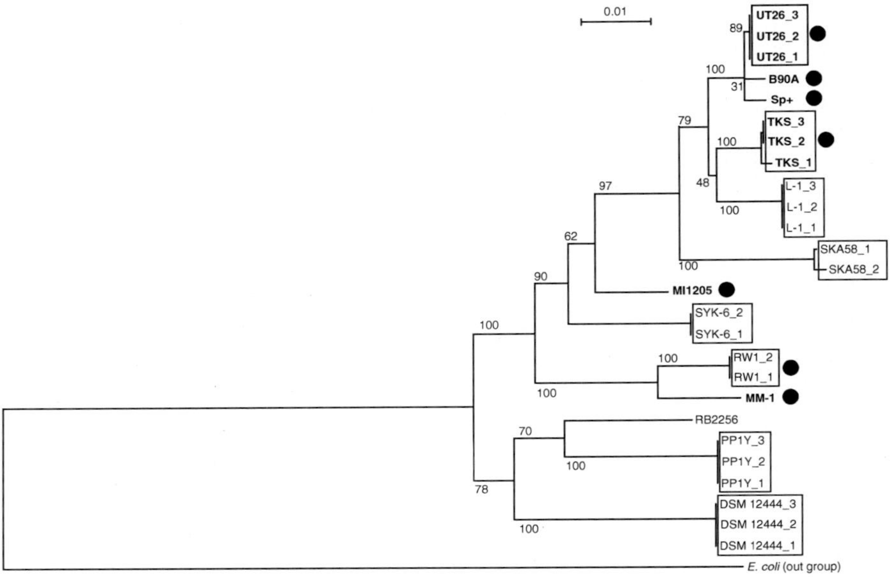

Fig. 2.4 Neighbor-joining phylogenetic tree of 16S rRNA genes of sphingomonad strains. Neighbor-joining phylogenetic tree of the conserved sites (1,385 nucleotides) in 16S rRNA genes of 13 sphingomonad strains, *S. japonicum* UT26S (UT26_1, SJA_C1-r0010; UT26_2, SJA_C2-r0010; UT26_3, SJA_C2-r0040), *Sphingobium indicum* B90A (B90A, NR_042943), *Sphingobium francense* Sp+(Sp+, NR_042944), *Sphingobium* sp. TKS (TKS_1, TKS_2, and TKS_3: unpublished data), *Sphingobium chlorophenolicum* L-1 (L-1_1, Sphch_R0043; L-2_2, Sphch_R0058; L-1_3, Sphch_R0067), *Sphingomonas* sp. SKA58 (SKA58_1, SKA58_r00366; SKA58_2, SKA58_r18278), *Sphingobium* sp. MI1205 (MI1205: unpublished data), *Sphingobium* sp. SYK-6 (SYK6_1, SLG_r0030; SYK6_2, SLG_r0060), *Sphingomonas wittichii* RW1 (RW1_1, Swit_R0031; RW1_2, Swit_R0040), *Sphingomonas* sp. MM-1 (MM-1: unpublished data), *Sphingopyxis alaskensis* RB2256 (RB2256, Sala_R0048), *N. aromaticivorans* DSM 12444 (DSM_1, Saro_R0065; DSM_2, Saro_R0059; DSM_3, Saro_R0053), and *Novosphingobium* sp. PP1Y (PPY_1, PP1Y_AR03; PPY_2, PP1Y_AR23 ;PPY_3, PP1Y_AR65), was constructed using MAFFT program (http://mafft.cbrc.jp/alignment/software/) and visualized by Njplot software. 16S rRNA gene (*rrsE*: gene ID 7437018) of *Escherichia coli* str. K-12 substr. W3110 (*E. coli*) was used as an out-of-group sequence. Bootstrap values calculated from 1,000 resampling using neighbor-joining are shown at the respective nodes. *Length of lines reflects* relative evolutionary distances among the sequences. *Sphingomonas* sp. SKA58 should be *Sphingobium* sp. SKA58 on the basis of comprehensive 16S rDNA analysis. However, we used *Sphingomonas* for the strain according to the database in order to avoid confusion. γ-HCH-degraders are bolded. Strains having IS*6100* are marked with *black circles*

suggested that these seven strains are phylogenetically similar to UT26, with L-1 showing the highest similarity, followed in order by SKA58, SYK-6, RW1, RB2256, DSM 12444, and PP1Y (Fig. 2.4). A blastN search of each region of the UT26 genome was conducted toward genome sequences of the seven other strains. The region whose homologous sequence was found in the other strains was colored in the gradient depending on the level of similarity (Fig. 2.3). Because all the strains have *rrn* genes homologous to those of UT26, the three positions of the *rrn* operons were colored red for all seven strains (Fig. 2.3). It was also clear that, among these seven strains, only RW1 had the IS*6100* insertion sequence (Fig. 2.3). Although a more detailed examination of each region would certainly provide critical information, our more general analysis revealed that the UT26 genome consists of regions conserved among these sphingomonad strains as well as specific regions unique to UT26 (Fig. 2.3). All of the abovementioned 196 bacterial essential genes proposed by Gil et al. (Gil et al. 2004) are dispersed on the conserved regions of Chr 1 (Fig. 2.3). The G+C contents of the specific regions of Chr 1 are relatively low (Fig. 2.3). One more important point is that the *linKLMN* and *linGHIJ* genes are located on relatively conserved regions among sphingomonads, while the "specific" *lin* genes (*linA*, *linB*, *linC*, *linRED*, and *linF*) are located within unique regions of UT26 (Fig. 2.3). Most of the specific *lin* genes (*linA*, *linC*, *linRED*, and *linF*) are highly associated with IS*6100* (Fig. 2.3), suggesting an important role of IS*6100* in recruitment of the specific *lin* genes.

Recently, the draft genome sequences of three γ-HCH-degraders were determined: *Sphingobium* sp. MI1205 from Japan (Ito et al. 2007), *Sphingomonas* sp. MM-1 from India (Tabata et al. 2011), and *Sphingobium* sp. TKS from Japan (Ohhata et al. unpublished data). These three strains are phylogenetically dispersed among closely related strains (Fig. 2.4). A blastN search of each region of the UT26 genome was also conducted toward the draft genome sequences of the three γ-HCH-degrading strains (Fig. 2.3). Among these γ-HCH-degraders, regions for the specific *lin* genes (*linA*, *linB*, *linC*, and *linRED*) are conserved. As for *linF*, some strains have only part of *linF*, and such strains may have another *linF* functional homologue (Ohhata et al., unpublished data). On the other hand, most of the other regions unique to UT26 are not conserved among γ-HCH-degraders (Fig. 2.3). In other words, UT26 and the three γ-HCH-degraders are similar with respect to γ-HCH degradation ability, but their overall gene repertoires are relatively different. In addition to suggesting their phylogenetic divergence (Fig. 2.4), this fact strongly suggested that more than one γ-HCH-degrader was emerged independently. At least, it is unlikely that there was a single ancestral γ-HCH-degrader that spread around the world.

2.4 Mobile Genetic Elements of γ-HCH-Degrading Bacteria

Although only the *linRED* cluster is located on a plasmid in UT26, other "specific" *lin* genes are also located on plasmids in other γ-HCH-degrading strains: *linA*, *linB*, and *linE* in Sp+ (Ceremonie et al. 2006), *linA* in B90 (Kumari et al. 2002), and *linA* to *linF* in MM-1 (Tabata et al. 2011). Furthermore, we successfully isolated by the exogenous plasmid isolation technique from HCH-contaminated soil a conjugative plasmid, pLB1, that carries an IS*6100*-composite transposon containing two copies of *linB* (Miyazaki et al. 2006). These facts strongly suggested that plasmids have important roles in the distribution of the specific *lin* genes under environmental conditions. Catabolic gene clusters for the degradation of recalcitrant compounds are often located on plasmids and disseminated by conjugation (Davison 1999; Maeda et al. 2003; Sota et al. 2006; Yano et al. 2010; Shintani and Nojiri 2013). Although the structural dynamism of several catabolic plasmids for the degradation of aromatic compounds has been analyzed in detail, the information on plasmids carrying *lin* genes is fragmented (Ceremonie et al. 2006; Tabata et al. 2011; Shintani and Nojiri 2013). In fact, there are still only a limited number of sphingomonad plasmids/subreplicons for which the complete sequences are available (Table 2.3). As mentioned above, UT26 has three plasmids, pCHQ1, pUT1, and pUT2, and it was strongly suggested that Chr 2 also has a plasmid-type replication machinery. Based on the similarity of the putative RepA proteins and the structure of putative replication origins, the sphingomonads plasmids/subreplicons could be classified (Fig. 2.2 and Table 2.3). However, further analysis will be needed for the strict classification, because most of their putative RepA proteins showed a very low level of similarity to those of well-studied plasmids (e.g., IncP-1, F, IncP-7, and IncP-9 plasmids). It is interesting that the size and gene contents of UT26_Chr 2-type replicons are highly divergent (Table 2.3), suggesting that this type of replicon underwent relatively dynamic rearrangements. The putative regions for the replication and partition of pCHQ1, pSPHCH01, and pSLGP are identical (Fig. 2.2b). A detailed comparison of these pCHQ1-type plasmids should provide clues about the structural dynamism of sphingomonad plasmids. Among the four plasmid-type replicons in UT26, only pCHQ1 carries a putative gene cluster for the type IV conjugative transfer system. Conjugal transfer of pCHQ1 from UT26 to *Sphingomonas paucimobilis* IAM 12578 was indeed detected in our previous study (Nagata et al. 2006). Although only a few cases were experimentally proved (Nagata et al. 2006; Miyazaki et al. 2006), many sphingomonad plasmids may be self-transmissible (Shintani and Nojiri 2013). A narrow host range of sphingomonad plasmids is also suggested (Nagata et al. 2006; Miyazaki et al. 2006). Active DNA rearrangements between plasmids and host chromosomes may cause the "mosaic" genetic structure of sphingomonads.

IS*6100* is often found in close proximity to *lin* genes in HCH-degrading strains (Boltner et al. 2005; Fuchu et al. 2008) (Fig. 2.4), and copies of IS*6100* are also located near the *linA*, *linC*, *linRED*, and *linF* genes in the UT26 genome (Fig. 2.3). IS*6100* is 880 bp in size and carries the *tnp* gene encoding transposase and 14-bp terminal inverted repeats (IR) at both ends (Mahillon and Chandler 1998). A comparison of the flanking regions of the 13 copies of IS*6100* in UT26, which are dispersed on Chr 1, Chr 2, pCHQ1, and pUT1, revealed five pairs of identical 8-bp sequences (Nagata et al., unpublished data). IS*6100* is a "replicative" IS, and on the basis of its transposition mechanism, the various near past events caused by IS*6100* can be proposed (Fig. 2.5). Not only simple transposition with inversion (Fig. 2.5a) but also transposition accompanying the replicon fusion (Fig. 2.5b, c) seem to have occurred. The important point is that DNA exchange between replicons is possible by these events. In addition to these near past events, IS*6100* seems to be involved in the current gene rearrangements in UT26. The *linA*, *linC*, and *linRED* genes are genetically unstable, and spontaneous deletion mutants of the regions containing these genes could easily be obtained. The deleted regions of YO5 (*linA*-deletion

Table 2.3 Sphingomonad plasmids/subreplicons whose complete sequences were available[a]

Name	Size (bp)	Host	Host features	Type	Accession number	Reference
Chr 2	681,892	*Sphingobium japonicum* UT26	γ-HCH[b] degradation	UT26_Chr 2/*repABC*	AP010804, NC_014013	Nagata et al. (2011)
Chr 2	1,368,670	*Sphingobium chlorophenolicum* L-1	PCP[c] degradation	UT26_Chr 2/*repABC*	CP002799, NC_015594	Copley et al. (2011)
pNL2	487,268	*Novosphingobium aromaticivorans* DSM 12444	Aromatic compounds degradation	UT26_Chr 2/*repABC*	CP000677, NC_009427	Unpublished
Lpl	192,103	*Novosphingobium* sp. PP1Y	Aromatic compounds degradation	UT26_Chr 2/*repABC*	FR856860, NC_015579	D'Argenio et al. (2011)
pCHQ1	190,974	*Sphingobium japonicum* UT26	γ-HCH degradation	pCHQ1/*repABC*	AP010805, NC_014007	Nagata et al. (2011)
pSPHCH01	123,733	*Sphingobium chlorophenolicum* L-1	PCP degradation	pCHQ1/*repABC*	CP002800, NC_015595	Copley et al. (2011)
pSLGP	148,801	*Sphingobium* sp. SYK-6	Lignin degradation	pCHQ1/*repABC*	AP012223, NC_015976	Masai et al. (2012)
pLA1	188,476	*Novosphingobium pentaromativorans* US6-1	Benzo(a)pyrene degradation	pCHQ1/*repABC*	AGFM01000122, NZ_AGFM01000122	Luo et al. (2012)
pSWIT01	310,228	*Sphingomonas wittichii* RW1	Dioxin degradation	pCHQ1?	CP000700, NC_009507	Miller et al. (2010)
pNL1	184,462	*Novosphingobium aromaticivorans* DSM 12444	Aromatic compounds degradation	pNL1/iteron	CP000676, NC_009426	Romine et al. (1999)
pCAR3	254,797	*Novosphingobium* sp. KA1	Carbazole degradation	pNL1/iteron	AB270530, NC_008308	Shintani et al. (2007)
Mpl	1,161,602	*Novosphingobium* sp. PP1Y	Aromatic compounds degradation	pNL1/iteron	FR856861, NC_015583	D'Argenio et al. (2011)
pSWIT02	222,757	*Sphingomonas wittichii* RW1	Dioxin degradation	pNL1/iteron	CP000701, NC_009508	Miller et al. (2010)
pUT1	31,776	*Sphingobium japonicum* UT26	γ-HCH degradation	pUT1/iteron	AP010806, NC_014005	Nagata et al. (2011)

(continued)

Table 2.3 (continued)

Name	Size (bp)	Host	Host features	Type	Accession number	Reference
pSY2	18,779	*Sphingobium chungbukense* DJ77	Aromatic compounds degradation	pUT1/iteron	JN180627, NC_016000	Yeon and Kim (2011)
pISP3	43,398	*Sphingomonas* sp. MM-1	γ-HCH degradation	pUT1/iteron	AB549721, NC_013970	Tabata et al. (2011)
pSY3	35,735	*Sphingobium chungbukense* DJ77	Aromatic compounds degradation	pUT1?	GU811235	Yeon and Kim (2009)
pUT2	5,398	*Sphingobium japonicum* UT26	γ-HCH degradation	pUT2/IncP-9/pM3	AP010807, NC_014009	Nagata et al. (2011)
pYAN-1	5,182	*Sphingobium yanoikuyae* JCM 7371		pUT2	AB265740, NC_008246	Ochou et al. (2008)
pSx-Qyy	5,683	*Sphingomonas xenophaga* QYY	Decolorize BAA[d]	pUT2	AY787146, NC_006826	Unpublished
pLB1	65,998	Unidentified soil bacterium (*Sphingobium japonicum* UT26)	γ-HCH degradation	pLB1	AB244976, NC_008330	Miyazaki et al. (2006)
pLA2	60,085	*Novosphingobium pentaromativorans* US6-1	Benzo(a)pyrene degradation	pLB1	AGFM01000123, NZ_AGFM01000123	Luo et al. (2012)
Spl	48,714	*Novosphingobium* sp. PP1Y	Aromatic compounds degradation	pLB1?	FR856859, NC_015582	D'Argenio et al. (2011)
pA1	46,557	*Sphingomonas* sp. strain A1	Super-channel	IncP-1β	AB231906, NC_007353	Harada et al. (2006)
F plasmid	28,543	*Sphingopyxis alaskensis* RB2256	Marine oligotroph		CP000357, NC_008036	Lauro et al. (2009)
pYAN-2	4,924	*Sphingobium yanoikuyae* JCM 7371			AB265741, NC_008247	Ochou et al. (2008)
pSM103mini	4,328	*Sphingopyxis macrogoltabida*	PEG[e] degradation		AB534719, NC_014466	Tani et al. (2011)

[a]In September, 2012
[b]γ-hexachlorocyclohexane
[c]Pentachlorophenol
[d]1-Amino-4-bromoanthraquinone-2-sulfonic acid
[e]Polyethylene glycol

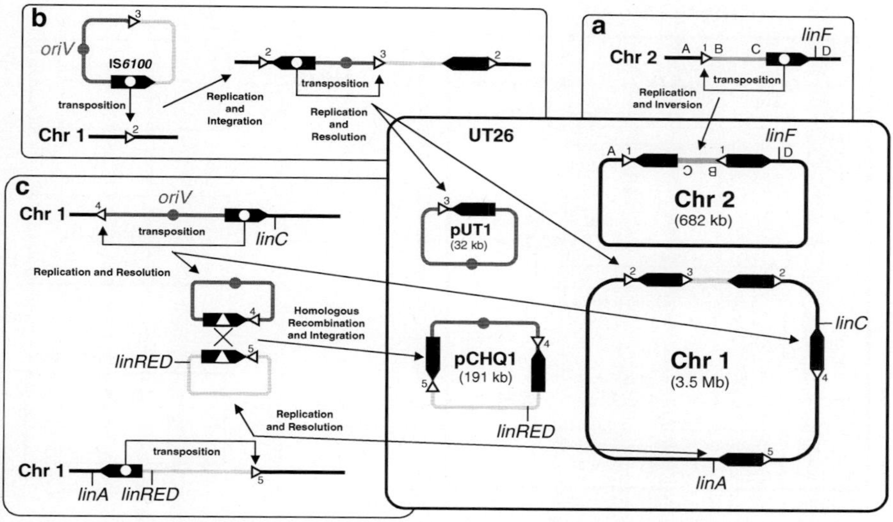

Fig. 2.5 Proposed events caused by transposition and homologous recombination of IS*6100*. *Black pentagons* and *white triangles* with numbers indicate IS*6100* and 8-bp target sites, respectively. *Triangles* with the same number mean identical sequence. *Black pentagons* marked with *white circle* and *triangle* indicate IS*6100* which transposed and caused homologous recombination, respectively. Intramolecular transposition can cause inversion (**a**). Intermolecular transposition and the following intramolecular one resulted in DNA exchange between replicons (**b**). Intramolecular transposition also caused deletion, and the deleted region was circlized. The deleted region which has *oriV* could be maintained in the host as an independent replicon (**c**). The deleted region without *oriV* also could be maintained in the host by homologous recombination with other replicons (**c**)

mutant) (Nagata et al. 1999) and UT72 (*linC*-deletion mutant) (Nagata et al. 1994) were estimated using tiling microarrays, and their exact deleted regions were determined using PCR followed by sequencing analysis of the PCR products (Nagata et al. 2011). The deletion in YO5 can be explained by an intrachromosomal single homologous recombination event between two copies of IS*6100* (Fig. 2.6a). On the other hand, the deletion in UT72 can be explained by replicative transposition of IS*6100* from one side of *linC* to the other side (Fig. 2.6b). Although an unknown rearrangement seems to have occurred at one of the ends of the deleted region, IS*6100* also seems to have been involved in the deletion of *linRED* (Fig. 2.6c). Furthermore, our recent use of the *sacB*-based IS entrapment technique indeed revealed the transposability of endogenous IS*6100* in UT26 (Nagata et al., unpublished data). These results strongly suggested that transposition of IS*6100* as well as a homologous recombination event between two copies of IS*6100* are involved in the genome rearrangements in UT26.

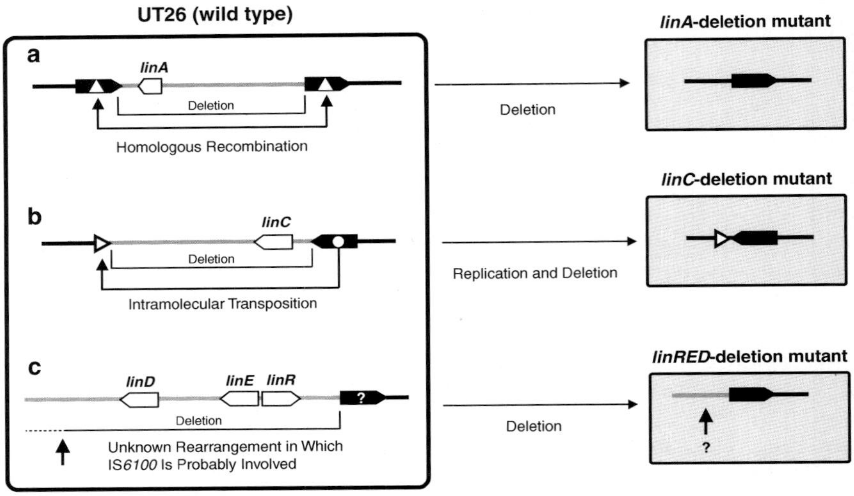

Fig. 2.6 Proposed deletion mechanisms of *lin*-deletion mutants, YO5 (*linA*-deletion mutant), UT72 (*linC*-deletion mutant), and UT116 (*linRED*-deletion mutant). *Black pentagons* and *white triangles* indicate IS*6100* and target sites, respectively. *Black pentagons* marked with *white circle* and *triangle* indicate IS*6100* which transposed and caused homologous recombination, respectively. See text for detail

2.5 Conclusions

On the basis of the genome structure of UT26, it is strongly suggested that UT26 is a "specialist" for the degradation of γ-HCH and potentially may have a physiological background that supports the degradation of various recalcitrant compounds, although experimental confirmation is necessary. Comparison of the UT26 genome sequence with those of other non-γ-HCH-degrading and γ-HCH-degrading sphingomonad strains suggested that the γ-HCH-degrading bacteria were emerged through the recruitment of "specific" *lin* genes into an ancestral strain that had core functions of sphingomonads, such as the LinKLMN-type ABC transporter system and β-ketoadipate pathway (Fig. 2.7). TBDR-like proteins and other unknown factors may also be involved in the core functions. The important conclusion at this stage is that more than one γ-HCH-degrader was born independently and in parallel around the world (Fig. 2.7). Plasmids, most of which are unique to sphingomonads, and IS*6100* seem to be involved in the recruitment of the specific *lin* genes. Furthermore, IS*6100* is involved in the genome rearrangements, including deletion, recombination, and fusion of replicons. Further critical genome analysis of strains even more closely related to UT26 will provide us with insights into more detailed mechanisms for creating various γ-HCH-degrading strains. The genomic structural information of

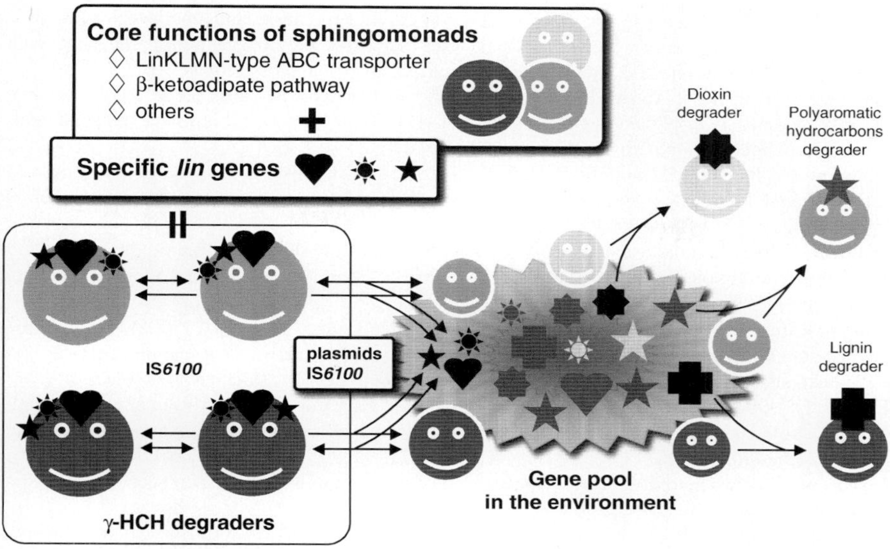

Fig. 2.7 Proposed model for the appearance and evolution of γ-HCH-degraders. See text for detail

UT26 will also be useful for practical biotechnological applications in the future. In theory, for example, we could construct an engineered strain for bioremediation that maintains *lin* genes more stably and expresses higher HCH degradation activity.

Acknowledgments This work was supported in part by Grant-in-Aids for Scientific Research from Ministry of Education, Culture, Sports, Science, and Technology, and Ministry of Agriculture, Forestry, and Fisheries of Japan.

References

Boltner D, Moreno-Morillas S, Ramos JL (2005) 16S rDNA phylogeny and distribution of *lin* genes in novel hexachlorocyclohexane-degrading *Sphingomonas* strains. Environ Microbiol 7:1329–1338

Brassinga AK, Marczynski GT (2001) Replication intermediate analysis confirms that chromosomal replication origin initiates from an unusual intergenic region in *Caulobacter crescentus*. Nucleic Acids Res 29:4441–4451

Ceremonie H, Boubakri H, Mavingui P, Simonet P, Vogel TM (2006) Plasmid-encoded γ-hexachlorocyclohexane degradation genes and insertion sequences in *Sphingobium francense* (ex-*Sphingomonas paucimobilis* Sp+). FEMS Microbiol Lett 257:243–252

Cevallos MA, Cervantes-Rivera R, Gutierrez-Rios RM (2008) The *repABC* plasmid family. Plasmid 60:19–37

Chattoraj DK (2000) Control of plasmid DNA replication by iterons: no longer paradoxical. Mol Microbiol 37:467–476

Cheng Q, Yu MC, Reeves AR, Salyers AA (1995) Identification and characterization of a Bacteroides gene, *csuF*, which encodes an outer membrane protein that is essential for growth on chondroitin sulfate. J Bacteriol 177:3721–3727

Copley SD, Rokicki J, Turner P, Daligault H, Nolan M, Land M (2011) The whole genome sequence of *Sphingobium chlorophenolicum* L-1: insights into the evolution of the pentachlorophenol degradation pathway. Genome Biol Evol 4:184–198

D'Argenio V, Petrillo M, Cantiello P, Naso B, Cozzuto L, Notomista E, Paolella G, Di Donato A, Salvatore F (2011) De novo sequencing and assembly of the whole genome of *Novosphingobium* sp. strain PP1Y. J Bacteriol 193:4296

Davison J (1999) Genetic exchange between bacteria in the environment. Plasmid 42:73–91

de Lorenzo V (2008) Systems biology approaches to bioremediation. Curr Opin Biotechnol 19:579–589

Endo R, Kamakura M, Miyauchi K, Fukuda M, Ohtsubo Y, Tsuda M, Nagata Y (2005) Identification and characterization of genes involved in the downstream degradation pathway of γ-hexachlorocyclohexane in *Sphingomonas paucimobilis* UT26. J Bacteriol 187:847–853

Endo R, Ohtsubo Y, Tsuda M, Nagata Y (2007) Identification and characterization of genes encoding a putative ABC-type transporter essential for utilization of γ-hexachlorocyclohexane in *Sphingobium japonicum* UT26. J Bacteriol 189:3712–3720

Ferguson AD, Deisenhofer J (2002) TonB-dependent receptors-structural perspectives. Biochim Biophys Acta 1565:318–332

Fuchu G, Ohtsubo Y, Ito M, Miyazaki R, Ono A, Nagata Y, Tsuda M (2008) Insertion sequence-based cassette PCR: cultivation-independent isolation of γ-hexachlorocyclohexane-degrading genes from soil DNA. Appl Microbiol Biotechnol 79:627–632

Gil R, Silva FJ, Pereto J, Moya A (2004) Determination of the core of a minimal bacterial gene set. Microbiol Mol Biol Rev 68:518–537

Harada KM, Aso Y, Hashimoto W, Mikami B, Murata K (2006) Sequence and analysis of the 46.6-kb plasmid pA1 from *Sphingomonas* sp. A1 that corresponds to the typical IncP-1β plasmid backbone without any accessory gene. Plasmid 56:11–23

Harwood CS, Parales RE (1996) The β-ketoadipate pathway and the biology of self-identity. Annu Rev Microbiol 50:553–590

Hashimoto W, He J, Wada Y, Nankai H, Mikami B, Murata K (2005) Proteomics-based identification of outer-membrane proteins responsible for import of macromolecules in *Sphingomonas* sp. A1: alginate-binding flagellin on the cell surface. Biochemistry 44:13783–13794

Imai R, Nagata Y, Senoo K, Wada H, Fukuda M, Takagi M, Yano K (1989) Dehydrochlorination of γ-hexachlorocyclohexane (γ-BHC) by γ-BHC-assimilating *Pseudomonas paucimobilis*. Agric Biol Chem 53:2015–2017

Imai R, Nagata Y, Fukuda M, Takagi M, Yano K (1991) Molecular cloning of a *Pseudomonas paucimobilis* gene encoding a 17-kilodalton polypeptide that eliminates HCl molecules from γ-hexachlorocyclohexane. J Bacteriol 173:6811–6819

Ito M, Prokop Z, Klvana M, Ohtsubo Y, Tsuda M, Damborsky J, Nagata Y (2007) Degradation of β-hexachlorocyclohexane by haloalkane dehalogenase LinB from γ-hexachlorocyclohexane-utilizing bacterium *Sphingobium* sp. MI1205. Arch Microbiol 188:313–325

Janssen DB, Dinkla IJ, Poelarends GJ, Terpstra P (2005) Bacterial degradation of xenobiotic compounds: evolution and distribution of novel enzyme activities. Environ Microbiol 7:1868–1882

Kahnert A, Mirleau P, Wait R, Kertesz MA (2002) The LysR-type regulator SftR is involved in soil survival and sulphate ester metabolism in *Pseudomonas putida*. Environ Microbiol 4:225–237

Krasowiak R, Sevastsyanovich Y, Konieczny I, Bingle LE, Thomas CM (2006) IncP-9 replication initiator protein binds to multiple DNA sequences in *oriV* and recruits host DnaA protein. Plasmid 56:187–201

Kumari R, Subudhi S, Suar M, Dhingra G, Raina V, Dogra C, Lal S, van der Meer JR, Holliger C, Lal R (2002) Cloning and characterization of *lin* genes responsible for the degradation of hexachlorocyclohexane isomers by *Sphingomonas paucimobilis* strain B90. Appl Environ Microbiol 68:6021–6028

Lal R, Dogra C, Malhotra S, Sharma P, Pal R (2006) Diversity, distribution and divergence of *lin* genes in hexachlorocyclohexane-degrading sphingomonads. Trends Biotechnol 24:121–130

Lal R, Pandey G, Sharma P, Kumari K, Malhotra S, Pandey R, Raina V, Kohler HP, Holliger C, Jackson C, Oakeshott JG (2010) Biochemistry of microbial degradation of hexachlorocyclohexane and prospects for bioremediation. Microbiol Mol Biol Rev 74:58–80

Lauro FM, McDougald D, Thomas T, Williams TJ, Egan S, Rice S, DeMaere MZ, Ting L, Ertan H, Johnson J, Ferriera S, Lapidus A, Anderson I, Kyrpides N, Munk AC, Detter C, Han CS, Brown MV, Robb FT, Kjelleberg S, Cavicchioli R (2009) The genomic basis of trophic strategy in marine bacteria. Proc Natl Acad Sci USA 106:15527–15533

Luo YR, Kang SG, Kim SJ, Kim MR, Li N, Lee JH, Kwon KK (2012) Genome sequence of benzo(a)pyrene-degrading bacterium *Novosphingobium pentaromativorans* US6-1. J Bacteriol 194:907

Lykidis A, Perez-Pantoja D, Ledger T, Mavromatis K, Anderson IJ, Ivanova NN, Hooper SD, Lapidus A, Lucas S, Gonzalez B, Kyrpides NC (2010) The complete multipartite genome sequence of Cupriavidus necator JMP134, a versatile pollutant degrader. PLoS One 5:e9729

Maeda K, Nojiri H, Shintani M, Yoshida T, Habe H, Omori T (2003) Complete nucleotide sequence of carbazole/dioxin-degrading plasmid pCAR1 in *Pseudomonas resinovorans* strain CA10 indicates its mosaicity and the presence of large catabolic transposon Tn*4676*. J Mol Biol 326:21–33

Mahillon J, Chandler M (1998) Insertion sequences. Microbiol Mol Biol Rev 62:725–774

Masai E, Kamimura N, Kasai D, Oguchi A, Ankai A, Fukui S, Takahashi M, Yashiro I, Sasaki H, Harada T, Nakamura S, Katano Y, Narita-Yamada S, Nakazawa H, Hara H, Katayama Y, Fukuda M, Yamazaki S, Fujita N (2012) Complete genome sequence of *Sphingobium* sp. strain SYK-6, a degrader of lignin-derived biaryls and monoaryls. J Bacteriol 194:534–535

Megharaj M, Ramakrishnan B, Venkateswarlu K, Sethunathan N, Naidu R (2011) Bioremediation approaches for organic pollutants: a critical perspective. Environ Int 37:1362–1375

Miller TR, Delcher AL, Salzberg SL, Saunders E, Detter JC, Halden RU (2010) Genome sequence of the dioxin-mineralizing bacterium *Sphingomonas wittichii* RW1. J Bacteriol 192:6101–6102

Miyauchi K, Suh SK, Nagata Y, Takagi M (1998) Cloning and sequencing of a 2,5-dichlorohydroquinone reductive dehalogenase gene whose product is involved in degradation of γ-hexachlorocyclohexane by *Sphingomonas paucimobilis*. J Bacteriol 180:1354–1359

Miyauchi K, Adachi Y, Nagata Y, Takagi M (1999) Cloning and sequencing of a novel *meta-cleavage* dioxygenase gene whose product is involved in degradation of γ-hexachlorocyclohexane in *Sphingomonas paucimobilis*. J Bacteriol 181:6712–6719

Miyauchi K, Lee HS, Fukuda M, Takagi M, Nagata Y (2002) Cloning and characterization of *linR*, involved in regulation of the downstream pathway for γ-hexachlorocyclohexane degradation in *Sphingomonas paucimobilis* UT26. Appl Environ Microbiol 68:1803–1807

Miyazaki R, Sato Y, Ito M, Ohtsubo Y, Nagata Y, Tsuda M (2006) Complete nucleotide sequence of an exogenously isolated plasmid, pLB1, involved in γ-hexachlorocyclohexane degradation. Appl Environ Microbiol 72:6923–6933

Nagata Y, Hatta T, Imai R, Kimbara K, Fukuda M, Yano K, Takagi M (1993a) Purification and characterization of γ-hexachlorocyclohexane (γ-HCH) dehydrochlorinase (LinA) from *Pseudomonas paucimobilis*. Biosci Biotech Biochem 57:1582–1583

Nagata Y, Nariya T, Ohtomo R, Fukuda M, Yano K, Takagi M (1993b) Cloning and sequencing of a dehalogenase gene encoding an enzyme with hydrolase activity involved in the degradation of γ-hexachlorocyclohexane in *Pseudomonas paucimobilis*. J Bacteriol 175:6403–6410

Nagata Y, Ohtomo R, Miyauchi K, Fukuda M, Yano K, Takagi M (1994) Cloning and sequencing of a 2,5-dichloro-2,5-cyclohexadiene-1,4-diol dehydrogenase gene involved in the degradation of γ-hexachlorocyclohexane in *Pseudomonas paucimobilis*. J Bacteriol 176:3117–3125

Nagata Y, Miyauchi K, Damborsky J, Manova K, Ansorgova A, Takagi M (1997) Purification and characterization of a haloalkane dehalogenase of a new substrate class from a γ-hexachlorocyclohexane-degrading bacterium, *Sphingomonas paucimobilis* UT26. Appl Environ Microbiol 63:3707–3710

Nagata Y, Futamura A, Miyauchi K, Takagi M (1999) Two different types of dehalogenases, LinA and LinB, involved in γ-hexachlorocyclohexane degradation in *Sphingomonas paucimobilis* UT26 are localized in the periplasmic space without molecular processing. J Bacteriol 181:5409–5413

Nagata Y, Kamakura M, Endo R, Miyazaki R, Ohtsubo Y, Tsuda M (2006) Distribution of γ-hexachlorocyclohexane-degrading genes on three replicons in *Sphingobium japonicum* UT26. FEMS Microbiol Lett 256:112–118

Nagata Y, Endo R, Ito M, Ohtsubo Y, Tsuda M (2007) Aerobic degradation of lindane (γ-hexachlorocyclohexane) in bacteria and its biochemical and molecular basis. Appl Microbiol Biotechnol 76:741–752

Nagata Y, Ohtsubo Y, Endo R, Ichikawa N, Ankai A, Oguchi A, Fukui S, Fujita N, Tsuda M (2010) Complete genome sequence of the representative γ-hexachlorocyclohexane-degrading bacterium *Sphingobium japonicum* UT26. J Bacteriol 192:5852–5853

Nagata Y, Natsui S, Endo R, Ohtsubo Y, Ichikawa N, Ankai A, Oguchi A, Fukui S, Fujita N, Tsuda M (2011) Genomic organization and genomic structural rearrangements of *Sphingobium japonicum* UT26, an archetypal γ-hexachlorocyclohexane-degrading bacterium. Enzyme Microb Technol 49:499–508

Ochou M, Saito M, Kurusu Y (2008) Characterization of two compatible small plasmids from *Sphingobium yanoikuyae*. Biosci Biotechnol Biochem 72:1130–1133

Reeves AR, D'Elia JN, Frias J, Salyers AA (1996) A *Bacteroides thetaiotaomicron* outer membrane protein that is essential for utilization of maltooligosaccharides and starch. J Bacteriol 178:823–830

Romine MF, Stillwell LC, Wong KK, Thurston SJ, Sisk EC, Sensen C, Gaasterland T, Fredrickson JK, Saffer JD (1999) Complete sequence of a 184-kilobase catabolic plasmid from *Sphingomonas aromaticivorans* F199. J Bacteriol 181:1585–1602

Schaper S, Messer W (1995) Interaction of the initiator protein DnaA of *Escherichia coli* with its DNA target. J Biol Chem 270:17622–17626

Senoo K, Wada H (1989) Isolation and identification of an aerobic γ-HCH-decomposing bacterium from soil. Soil Sci Plant Nutr 35:79–87

Shintani M, Nojiri H (2013) Mobile genetic elements (MGEs) carrying catabolic genes. In: Malik A, Elisabeth Grohmann E, Madalena Alves M (eds) Management of microbial resources in the environment. Springer, Netherlands, pp 167–214

Shintani M, Urata M, Inoue K, Eto K, Habe H, Omori T, Yamane H, Nojiri H (2007) The *Sphingomonas* plasmid pCAR3 is involved in complete mineralization of carbazole. J Bacteriol 189:2007–2020

Sibley CD, MacLellan SR, Finan T (2006) The *Sinorhizobium meliloti* chromosomal origin of replication. Microbiology 152:443–455

Sota M, Yano H, Ono A, Miyazaki R, Ishii H, Genka H, Top EM, Tsuda M (2006) Genomic and functional analysis of the IncP-9 naphthalene-catabolic plasmid NAH7 and its transposon Tn*4655* suggests catabolic gene spread by a tyrosine recombinase. J Bacteriol 188:4057–4067

Stanier RY, Palleroni NJ, Doudoroff M (1966) The aerobic pseudomonads: a taxonomic study. J Gen Microbiol 43:159–271

Stolz A (2009) Molecular characteristics of xenobiotic-degrading sphingomonads. Appl Microbiol Biotechnol 81:793–811

Tabata M, Endo R, Ito M, Ohtsubo Y, Kumar A, Tsuda M, Nagata Y (2011) The *lin* genes for γ-hexachlorocyclohexane degradation in *Sphingomonas* sp. MM-1 proved to be dispersed across multiple plasmids. Biosci Biotechnol Biochem 75:466–472

Takeuchi M, Hamana K, Hiraishi A (2001) Proposal of the genus *Sphingomonas sensu stricto* and three new genera, *Sphingobium*, *Novosphingobium* and *Sphingopyxis*, on the basis of phylogenetic and chemotaxonomic analyses. Int J Syst Evol Microbiol 51:1405–1417

Tani A, Charoenpanich J, Mori T, Takeichi M, Kimbara K, Kawai F (2007) Structure and conservation of a polyethylene glycol-degradative operon in sphingomonads. Microbiology 153: 338–346

Tani A, Tanaka A, Minami T, Kimbara K, Kawai F (2011) Characterization of a cryptic plasmid, pSM103mini, from polyethylene-glycol degrading *Sphingopyxis macrogoltabida* strain 103. Biosci Biotechnol Biochem 75:295–298

Vijgen J, Abhilash PC, Li YF, Lal R, Forter M, Torres J, Singh N, Yunus M, Tian C, Schaffer A, Weber R (2011) Hexachlorocyclohexane (HCH) as new Stockholm Convention POPs-a global perspective on the management of Lindane and its waste isomers. Environ Sci Pollut Res Int 18:152–162

White DC, Sutton SD, Ringelberg DB (1996) The genus *Sphingomonas*: physiology and ecology. Curr Opin Biotechnol 7:301–306

Yabuuchi E, Kosako Y, Fujiwara N, Naka T, Matsunaga I, Ogura H, Kobayashi K (2002) Emendation of the genus *Sphingomonas* Yabuuchi et al. 1990 and junior objective synonymy of the species of three genera, *Sphingobium*, *Novosphingobium* and *Sphingopyxis*, in conjunction with *Blastomonas ursincola*. Int J Syst Evol Microbiol 52:1485–1496

Yano H, Miyakoshi M, Ohshima K, Tabata M, Nagata Y, Hattori M, Tsuda M (2010) Complete nucleotide sequence of TOL plasmid pDK1 provides evidence for evolutionary history of IncP-7 catabolic plasmids. J Bacteriol 192:4337–4347

Yeon SM, Kim YC (2009) Characterization of plasmid pSY3 in *Sphingobium chungbukense* DJ77. J Microbiol 47:796–800

Yeon SM, Kim YC (2011) Complete sequence and organization of the *Sphingobium chungbukense* DJ77 pSY2 plasmid. J Microbiol 49:684–688

Yuhara S, Komatsu H, Goto H, Ohtsubo Y, Nagata Y, Tsuda M (2008) Pleiotropic roles of iron-responsive transcriptional regulator fur in *Burkholderia multivorans*. Microbiology 154:1763–1774

Chapter 3
Diversity of 2,4-Dichlorophenoxyacetic Acid (2,4-D)-Degradative Genes and Degrading Bacteria

Wataru Kitagawa and Yoichi Kamagata

Abstract Biodegradation of an anthropogenic herbicide, 2,4-dichlorophenoxyacetic acid (2,4-D), was documented in the 1950s, the early age of its usage. The 2,4-D-degradative genes in bacteria have long been studied since the first report was published in the 1980s. The well-known 2,4-D catabolic gene cluster *tfdABCDFE* on the plasmid pJP4 was originally isolated from *Cupriavidus pinatubonensis* JMP134 and is found to be widely distributed among the phylum *Proteobacteria*. 2,4-D catabolic gene cluster in pJP4 possesses some distinguished features for the effective degradation and dissemination, e.g., complete gene set located on one site, duplication of the toxic chlorophenol degradation pathway genes, and its self-transmissibility. The dioxygenase gene, *tfdA*, that catalyzes the initial step of 2,4-D degradation is a distinctive gene in this system on the point that neither similar nucleotide sequence gene nor similar function enzyme was discovered in the environment. *tfdA* has been used as a good marker or reporter gene for identifying or detecting the 2,4-D degradation system. However, two decades later, a novel 2,4-D degradation gene system, *cad* gene cluster, was identified from *Bradyrhizobium* sp. strain HW13 in the 2000s. This finding and subsequent studies changed our whole view of 2,4-D metabolism as *tfd* gene cluster had long been believed to be an exclusive system responsible for 2,4-D degradation. The initial dioxygenase gene *cadA* is completely different from *tfdA* and is a member of aromatic ring hydroxylation dioxygenase genes. In contrast to *tfd* gene carriers, it is notable that the first organism carrying *cadA* was discovered from non-2,4-D-contaminated pristine environment. To date, *cad* system is known to be widespread in the genera *Bradyrhizobium* and *Sphingomonas* and perhaps more diverse organisms. In this chapter, we overview the fundamental features, distribution, and competitiveness of those two representative 2,4-D gene systems based upon genetic and biochemical studies together with culture-independent molecular community analyses.

W. Kitagawa (✉) • Y. Kamagata
Bioproduction Research Institute, National Institute of Advanced Science and Technology (AIST), 2-17-2-1, Tsukisamu-Higashi, Toyohira-ku, Sapporo 062-8517, Japan

Graduate School of Agriculture, Hokkaido University, Sapporo, Japan
e-mail: w-kitagawa@aist.go.jp

H. Nojiri et al. (eds.), *Biodegradative Bacteria: How Bacteria Degrade, Survive, Adapt, and Evolve*, DOI 10.1007/978-4-431-54520-0_3, © Springer Japan 2014

Keywords 2,4-D • Biodegradation • Diversity

3.1 Introduction

2,4-Dichlorophenoxyacetic acid (2,4-D) has been widely used as an agricultural chemical since its introduction in the 1940s. It has phytohormone auxin activity, and it exerts selective herbicidal action on broad leaf weeds by disrupting their endogenous hormone system, particularly in the meristem (Inoue et al. 1988). The advantages of 2,4-D and its related compounds have made it a globally used treatment for rice and other gramineous crops. Rapid environmental degradation of 2,4-D by indigenous microorganisms was documented in the early days of its usage. As early as the 1950s, degradation of 2,4-D by certain bacteria, via 2,4-dichlorophenol (2,4-DCP) as an intermediate, was proposed (Steenson and Walker 1957). Since then, numerous bacteria that degrade and mineralize 2,4-D have been isolated, mainly from 2,4-D-treated agricultural, urban, and industrial soils and sediments (Bhat et al. 1994; Don and Pemberton 1981; Ka et al. 1994; Mae et al. 1993; Suwa et al. 1996; Hoffmann et al. 2003). Due to the nature of these bacteria as readily cultivable organisms, the biodegradation pathway of 2,4-D and its corresponding genes have been comprehensively studied. In a historical context, 2,4-D is one of the oldest, massively used artificial chemicals and is also the first chemical of which the biodegradation system was elucidated (Duxbury et al. 1970). It goes without saying that the well-known 2,4-D-degrading strain, *Cupriavidus pinatubonensis* JMP134 (formerly known as *Ralstonia eutropha*, and more formerly, *Alcaligenes eutrophus*), and its 2,4-D-degradative gene (*tfd*), which carries plasmid pJP4, have been extremely valuable in the study of 2,4-D and have also been an important model system for microbial degradation of xenobiotics (Don and Pemberton 1985; Don et al. 1985; Neilson et al. 1994; You and Ghosal 1995; Filer and Harker 1997; Trefault et al. 2002; DiGiovanni et al. 1996; Leveau et al. 1999). Complete genome sequences of the pJP4 plasmid and chromosome were determined and published in 2004 and 2010, respectively (Trefault et al. 2004; Lykidis et al. 2010).

In this chapter, we summarize 2,4-D-degradative genes and 2,4-D-degrading bacteria for the *tfd* gene system in pJP4 and an alternative "*cad*" gene system, and we speculate on their diversity and evolutionary relationships. Finally, we discuss the possibility of the presence of unknown 2,4-D-utilizing bacteria that have been proposed and identified by advanced molecular biological techniques such as stable isotope probing (SIP).

3.2 2,4-D Degradation by *tfd* Genes

The 2,4-D-mineralizing strain, JMP134, that harbors pJP4 was originally isolated in Australia, and since then, the strain has been extensively studied for decades. The 2,4-D-degradative gene, *tfd* (two, four-dichlorophenoxyacetic acid), is encoded on the self-transmissible pJP4 plasmid, and a degradation pathway by *tfd* gene products has

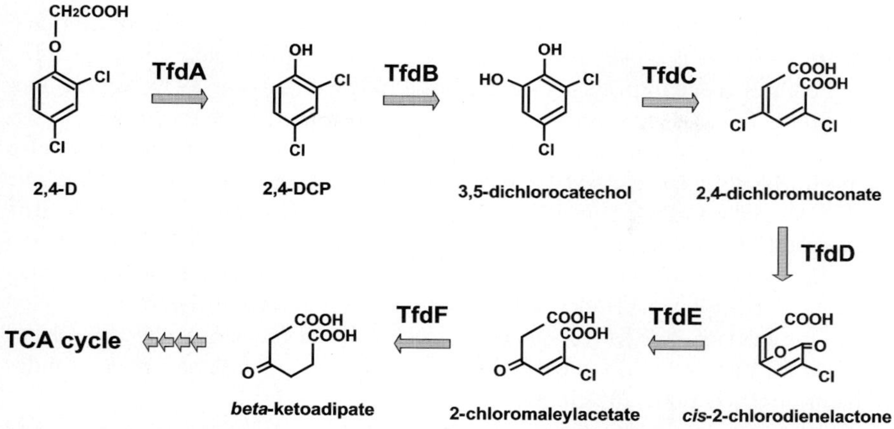

Fig. 3.1 Proposed degradation pathway of 2,4-D in *Cupriavidus pinatubonensis* JMP134 (pJP4) *TfdA* 2,4-D dioxygenase, *TfdB* 2,4-DCP monooxygenase, *TfdC* chlorocatechol 1,2-dioxygenase, *TfdD* chloromuconate cycloisomerase, *TfdE* dienelactone hydrolase, *TfdF* maleylacetate reductase

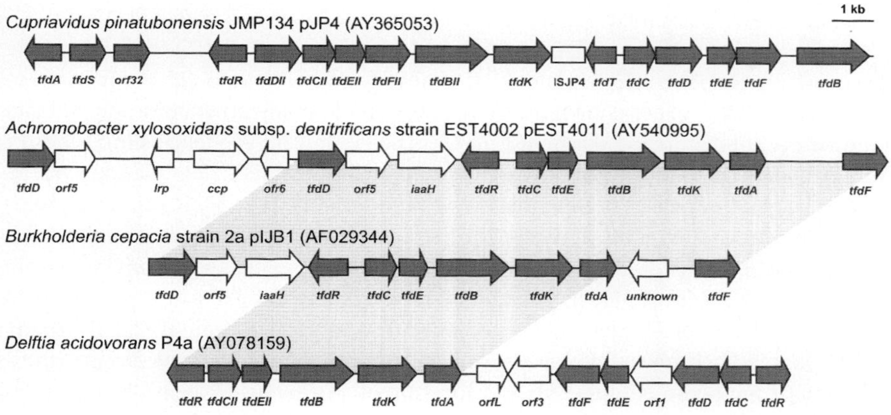

Fig. 3.2 The organization of the 2,4-D-degradative gene cluster. Genes involved in 2,4-D degradation are filled in *gray*. Highly conserved regions found between pIJB1 and pEST4011 and between pIJB1 and P4a are shaded in *light gray*. Accession numbers are shown in *parentheses*

been proposed (Fig. 3.1). The pathway can be roughly divided into the following three steps: (1) 2,4-D is transformed to 2,4-dichlorophenol (2,4-DCP) by α-ketoglutarate-dependent 2,4-D dioxygenase encoded by *tfdA*; (2) 2,4-DCP is subsequently hydroxylated by 2,4-DCP hydroxylase encoded by *tfdB* to form 3,5-dichlorocatechol (3,5-DCC); and (3) 3,5-DCC is further metabolized through an intradiol ring cleavage pathway encoded by *tfdCDEF*. Both the *tfdB* and *tfdCDEF* units exist in duplicate, whereas the initial oxygenase gene, *tfdA*, exists as a single copy in pJP4 (Laemmli et al. 2000) (Fig. 3.2). Because 2,4-DCP and 3,5-DCC are more cytotoxic than 2,4-D, the duplication and expression of the latter two gene units is important in enabling rapid consumption of these toxic intermediates (Ledger et al. 2006).

Transcription of *tfd* genes is intricately regulated by at least three transcriptional regulatory proteins, two identical LysR-type TfdR and TfdS proteins, and one IclR-type protein encoded by *orf32* (You and Ghosal 1995; Harker et al. 1989; Kaphammer et al. 1990; Trefault et al. 2009). In addition to these genes, the incomplete LysR-type regulator gene *tfdT*, containing an insertion of the ISJP4 sequence, is also found in the *tfd* gene cluster (Leveau and van der Meer 1996, 1997). Thus, the overall regulation of *tfd* genes is still not completely understood. However, it is known that TfdR/S positively regulate the transcription of *tfdBCDEF* and the second *tfdBCDEF* genes with 2,4-dichloromuconate as the inducing molecule, produced by TfdC (Filer and Harker 1997). In contrast, transcription of *tfdA* is controlled by the product of *orf32* (Trefault et al. 2009). These transcriptional differences between *tfdA* and other *tfd* genes might also prevent toxic intermediate accumulation during 2,4-D mineralization.

Another *tfd* gene in the cluster is *tfdK*, whose product is involved in the 2,4-D uptake. TfdK is a membrane protein, with significant similarity to a major facilitator superfamily (MFS) transporter, and has the typical 12 transmembrane regions. At high concentrations of 2,4-D, the host organism can utilize the substrate with or without TfdK because it enters cells by diffusion. However, at low concentrations, TfdK benefits the host's ability of uptake and utilization of 2,4-D (Leveau et al. 1998).

The most distinguishing feature of the *tfd* system is the initial dioxygenase *tfdA* gene and its function. Despite nucleotide and estimated amino acid sequences of *tfdB* and *tfdCDEF* genes showing certain similarity to many other chlorophenol and chlorocatechol degradation systems, *tfdA* has no significant sequence similarities or similar functions to those that have been found in any aromatic degradation system. Therefore, *tfdA* serves as a good marker or reporter gene (i.e., PCR primer, hybridization probe, and enzyme activity) for identifying the 2,4-D degradation system. In fact, the 2,4-D-degrading bacteria that have been isolated for many years are those that have recruited this system. The IncP-1β plasmid pJP4 in strain JMP134 contains the complete gene cluster needed for mineralization of 2,4-D. Many 2,4-D-degrading bacteria have been isolated to date, but only a few strains (plasmids) have the complete gene set at a location other than pJP4 (and its close relatives). For example, an IncP-1β plasmid, pIJB1, from *Burkholderia cepacia* strain 2a (Poh et al. 2002) and an IncP-1δ plasmid, pEST4011, from *Achromobacter xylosoxidans* subsp. *denitrificans* strain EST4002 (Vedler et al. 2004) harbor complete gene sets. Since these two IncP-1 plasmids lack some parts of their transfer genes, both are not self-transmissible plasmid (Sen et al. 2010). A complete gene set was also identified in the chromosomal DNA of an alkali-tolerant strain, *Delftia acidovorans* P4a (Fig. 3.2). Each of the corresponding *tfd* genes share sequence similarity, but the organization of the cluster (gene order and multiplicity) is diverse, as commonly observed in other incomplete 2,4-D gene clusters (Fulthorpe et al. 1995; Vallaeys et al. 1999). In many other 2,4-D degraders, the complete gene set might exist, separated on multiple loci of the genome, or the microbial community may mineralize the substrate. The pJP4-type *tfd* gene has achieved great success on the predominant environmental distribution for several reasons. First, as mentioned above, pJP4 has a complete gene set for 2,4-D mineralization at a single locus.

Second, *tfdB* and *tfdCDEF* units, which play important roles in decomposing toxic intermediates, are duplicative. The multiplicity and expression of those genes are shown to accelerate the host organism's growth on 2,4-D (Leveau et al. 1999; Laemmli et al. 2000; Pérez-Pantoja et al. 2000). Third, pJP4 is a multicopy, broad-host range, and a self-transmissible plasmid. There are approximately five copies of pJP4 in JMP134 (Trefault et al. 2002). Because it has a complete gene set, pJP4 confers recipient microorganisms with the ability to utilize 2,4-D as the sole carbon and energy source. In 2,4-D-disturbed fields, the pJP4 system could rapidly spread over to indigenous microflora (Neilson et al. 1994; DiGiovanni et al. 1996; Newby et al. 2000; Bathe et al. 2004). As summarized above, genetic studies on 2,4-D degradation have long been performed exclusively on *tfd*, but this was changed by the discovery of a new 2,4-D-degradative "*cad*" gene.

3.3 Discovery of a Novel 2,4-D-Degradative "*cad*" Gene in Pristine Soil Bacteria

Since their initial isolation, 2,4-D degraders have been easily obtained from 2,4-D-treated environments. However, Tiedje and his group put much effort into isolating degrading bacteria from non-contaminated pristine soils over the world (Fulthorpe et al. 1996; Kamagata et al. 1997). As a result, four strains were isolated from Hawaii, one strain from Saskatchewan, and one strain from Chile. One of these strains, HW1 from Hawaii, contained *tfdA*, but the remaining five strains did not. The five strains showed negative results for a *tfdA* internal sequence by PCR amplification as well as an α-ketoglutarate-dependent 2,4-D dioxygenase activity test. In addition, no hybridization to the genomic DNA of the five strains occurred using *tfdA*, *tfdB*, and *tfdC* nucleotide sequences as the probes. The 16S rRNA gene analysis indicated that HW1 was a β-*Proteobacteria*; however, the other five strains were *α-Proteobacteria*, in contrast to most of the other previously isolated 2,4-D degraders, which are β- or γ-*Proteobacteria*. These findings provided encouragement for the isolation of a novel, non-*tfd* type of 2,4-D-degrading system.

In order to identify a new 2,4-D degradation system, we selected *Bradyrhizobium* sp. strain HW13 for our further study (Kitagawa et al. 2002) because it showed the fastest growth among the five strains (doubling time: 16 h). This strain was isolated from buried, pristine soil from Volcanoes National Park, Hawaii. This soil was covered by lava flow 4,800 years ago, which separated the soil from human impact following Hawaiian colonization (Kamagata et al. 1997). A gene library for strain HW13 was constructed in *Escherichia coli* using a broad-host-range cosmid vector. The library was screened for 2,4-D-degrading activity, but neither *E. coli* nor *Pseudomonas putida* host strains showed 2,4-D conversion activity. We then transferred the gene library into an alternative host strain, *Sinorhizobium meliloti*, a member of α-*Proteobacteria*. *S. meliloti*, harboring the HW13 gene library, was screened for 2,4-D-degrading activity. Of the approximately 2,500 clones that were screened, only one clone showed 2,4-D degradation activity. Consumption of 2,4-D in the culture

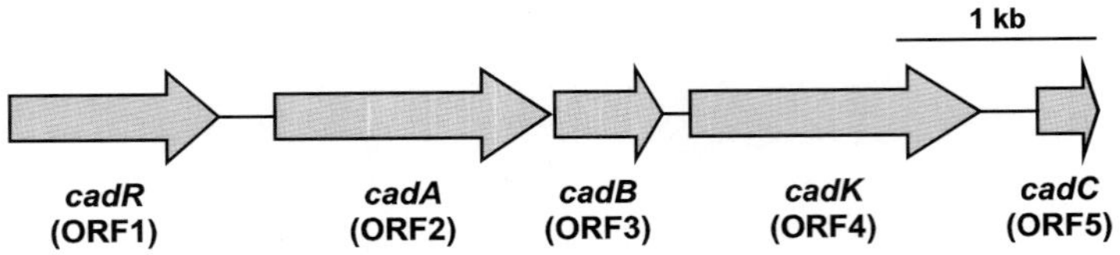

Fig. 3.3 Organization of the *cad* gene cluster in *Bradyrhizobium* sp. HW13

broth was accompanied by the accumulation of 2,4-DCP. These results indicated that the clone metabolized 2,4-D to 2,4-DCP, as had been found for strain JMP134. Five open reading frames (ORFs) were identified from the minimal nucleotide region required for the activity (Fig. 3.3). The deduced amino acid sequence of ORF1 showed similarity to the AraC/XylS type of transcriptional regulators. ORF2 and ORF3 showed similarity to large (iron-sulfur protein-α, ISPα) and small (ISPβ) subunits of aromatic ring hydroxylation dioxygenase (RHDO), with 46 % and 44 % sequence identities to TftA and TftB, respectively, of *B. cepacia* AC1100, corresponding to the large and small subunits of 2,4,5-trichlorophenoxyacetic acid (2,4,5-T) oxygenase, respectively (Danganan et al. 1995). ORF4 showed similarity to the 2,4-D transport protein, TfdK, of pJP4 (Leveau et al. 1998). ORF5 showed similarity to the [2Fe-2S]-type ferredoxin of the P-450 monooxygenase system. On the basis of these results, we inferred these ORFs to be chloroaryl ether degradation genes and designated ORFs 1–5 as *cadR*, *cadA*, *cadB*, *cadK*, and *cadC*, respectively. With the exception of *cadK*, no similarities were observed with the *tfd* genes. A putative ferredoxin reductase gene, which should work with *cadC*, was not found in the region; however, an unidentified ferredoxin reductase supplied by the host *S. meliloti* strain might compensate for the electron transport function. When these *cad* genes were expressed in *P. putida*, 2,4-D conversion activity was not observed. This might be because of the lack of compatible ferredoxin reductase in *P. putida*. From additional investigations, including frame shift mutation and northern hybridization, we concluded that *cadABC* is a multicomponent 2,4-D oxygenase gene and *cadR* is a transcriptional regulator gene, whose product induces *cadABKC* transcription in the presence of 2,4-D or 4-chlorophenoxyacetic acid as the effector molecule. The discovery that an RHDO type of enzyme catalyzes 2,4-D transformation to 2,4-DCP, which is a completely different enzyme from *tfdA*, was surprising. The phylogenetic tree of the ISPα protein of RHDO is shown in Fig. 3.4, and it indicates that CadA is a member of the benzoate dioxygenase subfamily. TftAB has also been reported to transform 2,4-D to 2,4-DCP; however, the host strain *B. cepacia* AC1100 does not convert 2,4-DCP any further, even more, it does not grow on 2,4-D. In contrast, *cadABC* degrades 2,4,5-T, but strain HW13 does not mineralize the substrate. This was the first identification of a 2,4-D-mineralizing bacterium that possesses degradative genes other than *tfdA* for growth. So far, no endogenous plasmid has been found in strain HW13; however, localization of the *cad* gene in the genome still remains unknown.

Since that study, the distribution of the *cad* gene among other 2,4-D degraders has been extensively investigated. 2,4-D-Degrading bacteria have been classified into three as follows (Kamagata et al. 1997): Class I, *tfdA*-positive, fast-growing β- and γ-*Proteobacteria*, mostly isolated from contaminated sites, e.g., *C. pinatubonensis*

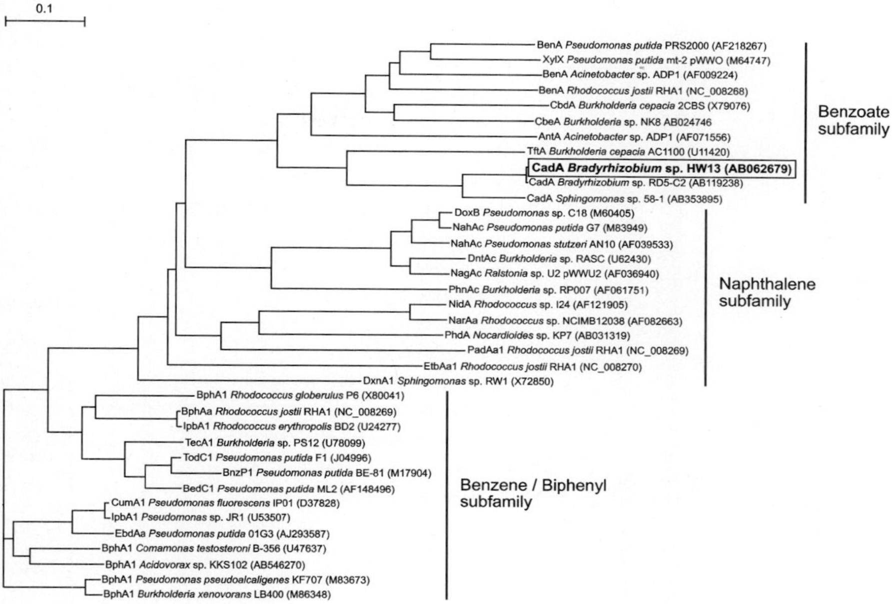

Fig. 3.4 The phylogenetic tree of aromatic ring hydroxylation dioxygenase (RHDO) of ISPα proteins. Accession numbers are shown in *parentheses*

JMP134; Class II, *tfdA*-negative, fast-growing α-*Proteobacteria*, isolated from contaminated sites, e.g., *Sphingomonas* sp. B6-10 (Fulthorpe et al. 1995); and Class III, *tfdA*-negative, slow-growing α-*Proteobacteria*, isolated from non-contaminated sites, e.g., *Bradyrhizobium* sp. strain HW13. Surprisingly, the presence and activity of the *cadA* gene were confirmed in all tested Class II and III degraders (*Sphingomonas* sp. B6-10, TFD26, TFD44, *Bradyrhizobium* sp. HWK12, and BTH) but not in the Class I strain, JMP134. An important observation is that the *cad* gene was widely distributed among Class II and III degraders, regardless of whether or not they were isolated from 2,4-D-contaminated environments. The discovery of *cad* genes has had great impact on the field of biodegradation research, in terms of genetic diversity.

3.4 Diversity, Distribution, and Evolution of the 2,4-D Genes

As mentioned above, the characteristic feature of 2,4-D degradation system is the initial oxygenation step. Two types of genes, *tfdA* and *cadA*, were identified; therefore, we can now classify the various 2,4-D-degrading genes into two major families on the basis of *tfdA* and *cadA* gene sequences. In this section, we discuss the diversity and evolution of the 2,4-D genes and also summarize the diversity and distribution of host microorganisms.

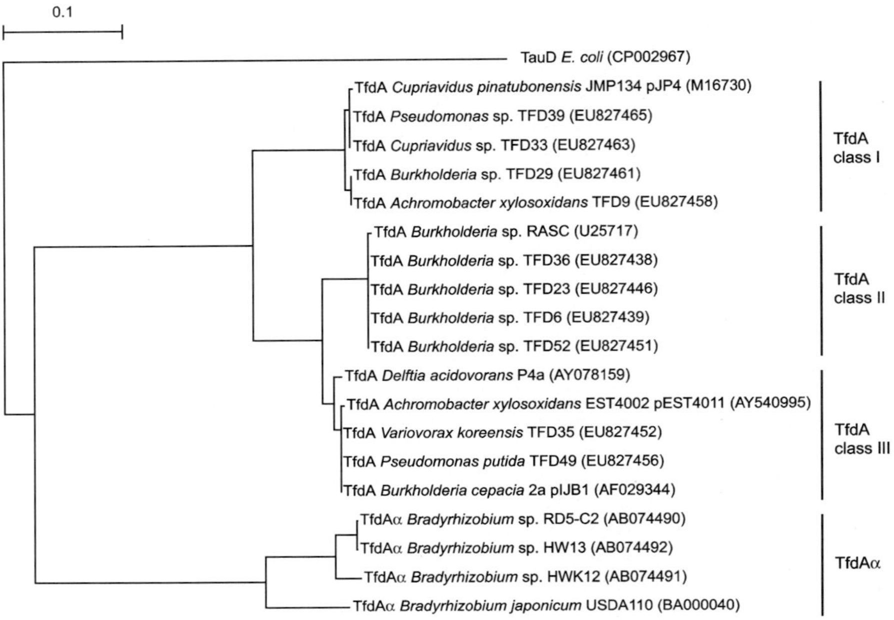

Fig. 3.5 The phylogenetic tree of TfdA proteins. All strains indicated are 2,4-D degraders, except for *Escherichia coli* and *Bradyrhizobium japonicum* USDA110. *E. coli* TauD (taurine dioxygenase) was used as an out-group. Accession numbers are shown in *parentheses*

3.4.1 *tfdA* Family

Phylogenetic studies of *tfdA* have mostly been performed using the partial nucleotide sequence of the gene, but recent accumulation of genetic information has enabled the analysis to be performed with the entire *tfdA* coding sequence (Bælum et al. 2010). A phylogenetic tree, based on the entire TfdA protein sequences, is shown in Fig. 3.5. As previously proposed, TfdAs fall into three subgroups at the genetic level, i.e., Classes I, II, and III (McGowan et al. 1998). Generally, Class I and III genes are distributed in a wide range of host microorganisms (mostly β- and γ-*Proteobacteria*) compared to Class II genes (genus *Burkholderia*). Most of the Class I and III *tfdA* genes are coded on self-transmissible plasmids such as pJP4 (Bælum et al. 2010), whereas Class II *tfdA* genes are coded on chromosomes, such as in the *Burkholderia tropica* strain RASC. Although only partial sequences were determined, large numbers of *tfdA* genes have been reported so far, including 2,4-D-degrading α-*Proteobacteria* (Lee et al. 2005).

Deduced amino acid sequence similarities between Classes I and II, Classes I and III, and Classes II and III were 82 %, 84 %, and 96 %, respectively. However, the functional differences between them remain unclear. A gene designated as *tfdAα* was placed in the *tfdA* phylogenetic tree. The gene was originally identified in 2,4-D-degrading *Bradyrhizobium* sp. by PCR amplification by using degenerated nucleotide sequence primers designed on the basis of sequences of *tfdA* of strains JMP134 and RASC. Nucleotide and deduced amino acid sequences of *tfdAα* of *Bradyrhizobium* sp. RD5-C2 showed 66 % and 47 % sequence similarity to that of the pJP4 type of *tfdA* (Classes I, II, and III). The recombinant TfdAα (MalE fusion protein) showed significantly weak α-ketoglutarate-dependent 2,4-D dioxygenase activity compared to TfdA of JMP134 (1/1,000 of specific activity) (Itoh et al. 2002). *Bradyrhizobium* sp. RD5-C2 was also found to harbor *cad* genes, suggesting that *tfdAα* does not contribute to the 2,4-D mineralization in the host strain; instead, the *cad* gene plays a role in 2,4-D degradation.

This is an important discovery when considering the evolution of *tfdA*. Recent genomic studies revealed that *tfdAα*-type genes are ubiquitously found in both 2,4-D-degrading and 2,4-D-non-degrading strains of *Bradyrhizobium*. In contrast, the pJP4 type of *tfdA* could not be ubiquitously found in any specific group of microorganisms. Interestingly, *tfdAα* is identified in *cad*-dependent 2,4-D degraders. Ancestral *tfdA* might have evolved in *cad*-dependent 2,4-D degraders, and horizontal gene transfer of *tfdA* to chlorophenol degraders might have resulted in a highly organized and efficient 2,4-D degradation system such as pJP4. For instance, *tfdA* and *tfdAα* share an ancestral origin. Distribution of *tfdA* or a *tfdA*-like gene among 2,4-D-non-degrading bacteria has also been reported (Hogan et al. 1997). The partial nucleotide sequences of *tfdA* obtained from these strains showed 78–100 % sequence identity with that of JMP134. This score is much higher than that obtained between *tfdAα* and *tfdA* of JMP134. They are not only from the phylum *Proteobacteria* but also from *Bacteroidetes*, *Firmicutes*, and *Actinobacteria*. The original substrate of ancestral TfdA or the substrate of a TfdA-like protein in 2,4-D non-degraders remains unknown. However, cinnamic acid has been proposed as a substrate. TfdA showed, to some extent, degradation activity for cinnamic acid, which is produced by plants and is widely and massively distributed in nature (Hotopp and Hausinger 2001).

3.4.2 cadA Family

Only three entire coding sequences of *cadA* family genes have been determined so far, namely, two *cadA* genes from *Bradyrhizobium* sp. ($cadA_{Br}$) (Kitagawa et al. 2002; Itoh et al. 2004) and the *cadA* gene from *Sphingomonas* sp. ($cadA_{Sp}$) (Shimojo et al. 2009). The nucleotide and deduced amino acid sequence identities between $cadA_{Br}$ and $cadA_{Sp}$ are 75 % and 86 %, respectively. In addition, *tftA* of *B. cepacia* AC1100 was found to be a member of this family (Fig. 3.4). The nucleotide and deduced amino acid sequence identities between these *cadA* and *tftA* genes are 66 % and 54 %, respectively. The *cad* gene products of HW13 and the *tft* gene products of AC1100 showed functional differences in their substrate preferences. The Cad system converts 2,4-D, 2,4,5-T, and phenoxyacetic acid (PAA) at almost equal efficiencies; on the other hand, the Tft system preferentially converts substrates in the following order: 2,4,5-T, 2,4-D, and PAA (Danganan et al. 1995). PCR screening and Southern hybridization analysis demonstrated the presence of the *cadA*-type gene in *Bradyrhizobium* spp. and *Sphingomonas* spp. (Kitagawa et al. 2002; Itoh et al. 2004). Although the host organisms are unknown, *cadA* and *cadA*-like genes are also detected in the environment, such as in activated sludge (see below). The *cad* gene may be more widespread than previously believed.

As described above, *cad* and *tft* are members of the aromatic RHDO, related to the benzoate dioxygenase subfamily (Fig. 3.4). RHDO introduces two oxygen atoms to an aromatic ring, resulting in hydroxylation of the aromatic ring. These results strongly indicate an ancestral origin of *cad* and *tft* belong to RHDO. The GC content of the *cad* gene in *Bradyrhizobium* sp. was different from that of host housekeeping genes, indicating that the *cad* gene may also have been acquired by horizontal gene transfer (Itoh et al. 2004). RHDOs are widely distributed in a diverse array of microorganisms; hence, it is very difficult to speculate on the direct ancestor and donor. How the ancestral RHDO became a functional 2,4-D etherase is an intriguing subject that is yet to be investigated.

3.5 Culture-Independent Studies on 2,4-D-Degrading Bacteria and 2,4-D-Degradative Genes

As discussed above, most 2,4-D degraders have been isolated from 2,4-D-amended environments, by a common practice that uses 2,4-D as the sole carbon and energy source in the isolation media along with an enrichment procedure. This method is simple and easy; however, only a limited group of bacteria, in terms of physiological, functional, and phylogenetic aspects, can be isolated. Some bacteria do not grow on synthetic media, even if they can degrade 2,4-D much faster than others. Some others may mineralize 2,4-D only in the presence of specific partner organisms. Even if there are many varieties of complete 2,4-D mineralizers in the environment, enrichment culture distorts the evaluation of the microbial and catabolic diversity (Dunbar et al. 1997). From that viewpoint, some unknown microorganisms in the environment might degrade 2,4-D much faster and might have unknown degrading enzymes. Culture-independent analyses would be a powerful and convincing strategy to better understand which microorganisms truly degrade 2,4-D and utilize it as the carbon source in situ as well as to determine which microorganisms or genes increase or emerge after 2,4-D is introduced to their environments. In this section, we summarize these studies by using advanced molecular methods.

3.5.1 Quantitative Detection of Multiple 2,4-D Oxygenase Genes

Before and after treatment of 2,4-D in activated sludge samples, quantitative detection of five types of 2,4-D initial oxygenase genes and microbial community analysis was performed to investigate the differential emergence and diversity (Lee et al. 2005). Prior to the investigation, the authors had identified a *cadA*-like gene from the sample by PCR screening, using degenerate primers based on *cadA* of HW13 and *tftA* of AC1100 nucleotide sequences. The nucleotide and deduced amino acid sequence of the *cadA*-like gene showed 64 % and 60 % sequence identities with *cadA* and 60 % and 54 % sequence identities with *tftA*. In addition to the four known 2,4-D oxidative genes (*tfdA* of JMP134 [type I], *tfdA* of RASC [type II], *cadA* of HW13 [type III], and *tftA* of AC1100 [type V]), the *cadA*-like gene was used as a target gene (type IV). From the original sample (before 2,4-D treatment), only type V was detected at low levels (8.8×10^3 copies/μg DNA), and the others were below the detection limit (10^2 copies). After 2,4-D amendment, the copy number of type V did not seem to be strongly affected; in contrast, the other four types emerged and increased in copy number. In the continuous culture of both low (10 mg/L) and high (300 mg/L) concentrations of 2,4-D, copy numbers of types I–IV were estimated to be approximately 10^7–10^8, 10^5–10^6, 10^3–10^4, and 10^5 copies/μg of DNA, respectively. In the repeated batch culture (subcultured 11 times, 300 mg/L of 2,4-D), only types I and III were detected at approximately 10^9 and 10^4 copies, respectively.

In this study, the emergence of gene types varied in response to the concentrations of 2,4-D and the 2,4-D amendment. Interestingly, even under highly selective conditions (enrichment by consecutive transfer to the fresh medium amended with high concentration of 2,4-D), the type III gene remained and coexisted with type I, although the type III gene copy number was five orders of magnitude lower than that of type I. This finding implies that a low level of *cadA* gene could coexist with *tfdA*, somehow functioning as a 2,4-D-degrading gene in 2,4-D-contaminated sites. It may also explain why *tfdA* dominates over *cadA* and, thus, why organisms harboring *tfdA* are always isolated from 2,4-D-contaminated sites. The quick response and overwhelming adaptation of the *tfdA* system that occurred under the experimental conditions were also verified by PCR-DGGE and 16S rRNA gene cloning analyses, i.e., apparent microbial community change was observed during the 2,4-D treatment. Several DGGE bands emerged, corresponding to the 16S rRNA genes of 2,4-D degraders isolated from the samples. Interestingly, only the type I degrader was isolated from the sample, as supported by the fact that the quantitative PCR results demonstrated that type I is always numerically highly dominant. Nucleotide sequence of another intense band in DGGE, detected only after treatment, did not match any 2,4-D-degrading isolates, indicating that unknown 2,4-D degraders, or at least bacteria whose growth were strongly and positively affected by 2,4-D, are present in the environment, and they would not grow under laboratory conditions. These culture-independent methods revealed that the diversity and dynamics of 2,4-D genes and communities, in turn, show biases caused by classical culture-dependent methods.

3.5.2 Application of Stable Isotope Probing

The SIP method has advanced the study of microbial ecology because it enables functions to be linked to the identities of organisms without their cultivation. SIP facilitates the identification of bacteria that assimilate a stable isotope-labeled compound from the complex surrounding microbial community. In the case of 2,4-D, a ^{13}C-labeled substrate (ring labeled) was introduced to soil samples (Cupples and Sims 2007). After treatment of labeled 2,4-D, total DNA was extracted and subjected to density gradient ultracentrifugation. Ultracentrifugation separated the light fraction from the heavy fraction, which contained ^{13}C-incorporated DNAs. Only 2,4-D-assimilating bacterial DNAs were labeled and recovered in the heavy fraction during the early stages of the cultivation. Analysis of the 16S rRNA gene in the heavy fraction revealed the presence of clones similar to the class *Comamonadaceae* (β-*Proteobacteria*), which have not been reported as 2,4-D degraders so far. This strongly indicates that these bacteria belong to a novel group of 2,4-D utilizers. The results also suggest that SIP can be a powerful tool for identifying unknown 2,4-D-degrading organisms.

3.6 Remarks

2,4-D is an anthropogenic xenobiotic compound that was first demonstrated to be biodegradable into TCA cycle compounds. A large number of studies have been conducted that aid our understanding of the entire mechanism of 2,4-D degradation. These studies cover a broad range of topics, including the degradation pathway, functions of each degradative gene, regulatory system, horizontal gene transfer, evolution of genes, distribution and diversity of the genes and host bacteria, and the effect of 2,4-D perturbation on the overall microbial community structure. The 2,4-D-degrading bacteria, *Cupriavidus pinatubonensis* JMP134, its plasmid pJP4, and *tfd* genes have long been a part of 2,4-D studies. However, a new door was opened about two decades ago, after the pJP4 type of *tfd* genes were found, i.e., *cadA* and *tfdAα* were found from *Bradyrhizobium* spp. and *Sphingomonas* spp. These findings had great impact on later 2,4-D studies, particularly in the context of their genetic diversity and evolution. Two intriguing objectives are (1) to understand the ancestral gene(s) of the anthropogenic chemical-degrading genes and their original substrate before the chemical emerged in the twentieth century and (2) to clarify the evolutionary (and/or recruiting) process of assembling the complete gene cluster. Genetic information currently being accumulated at an accelerated rate will shed new light on these important questions.

References

Bælum J, Jacobsen CS, Holben WE (2010) Comparison of 16S rRNA gene phylogeny and functional *tfdA* gene distribution in thirty-one different 2,4-dichlorophenoxyacetic acid and 4-chloro-2-methylphenoxyacetic acid degraders. Syst Appl Microbiol 33(2):67–70. doi:10.1016/j.syapm.2010.01.001

Bathe S, Lebuhn M, Ellwart JW, Wuertz S, Hausner M (2004) High phylogenetic diversity of transconjugants carrying plasmid pJP4 in an activated sludge-derived microbial community. FEMS Microbiol Lett 235(2):215–219. doi:10.1016/j.femsle.2004.04.038

Bhat MA, Tsuda M, Horiike K, Nozaki M, Vaidyanathan CS, Nakazawa T (1994) Identification and characterization of a new plasmid carrying genes for degradation of 2,4-dichlorophenoxyacetate from *Pseudomonas cepacia* CSV90. Appl Environ Microbiol 60(1):307–312

Cupples AM, Sims GK (2007) Identification of in situ 2,4-dichlorophenoxyacetic acid-degrading soil microorganisms using DNA-stable isotope probing. Soil Biol Biochem 39(1):232–238. doi:10.1016/j.soilbio.2006.07.011

Danganan CE, Shankar S, Ye RW, Chakrabarty AM (1995) Substrate diversity and expression of the 2,4,5-trichlorophenoxyacetic acid oxygenase from *Burkholderia cepacia* AC1100. Appl Environ Microbiol 61(12):4500–4504

DiGiovanni GD, Neilson JW, Pepper IL, Sinclair NA (1996) Gene transfer of *Alcaligenes eutrophus* JMP134 plasmid pJP4 to indigenous soil recipients. Appl Environ Microbiol 62(7):2521–2526

Don RH, Pemberton JM (1981) Properties of six pesticide degradation plasmids isolated from *Alcaligenes paradoxus* and *Alcaligenes eutrophus*. J Bacteriol 145(2):681–686

Don RH, Pemberton JM (1985) Genetic and physical map of the 2,4-dichlorophenoxyacetic acid-degradative plasmid pJP4. J Bacteriol 161(1):466–468

Don RH, Weightman AJ, Knackmuss HJ, Timmis KN (1985) Transposon mutagenesis and cloning analysis of the pathways for degradation of 2,4-dichlorophenoxyacetic acid and 3-chlorobenzoate in *Alcaligenes eutrophus* JMP134(pJP4). J Bacteriol 161(1):85–90

Dunbar J, White S, Forney L (1997) Genetic diversity through the looking glass: effect of enrichment bias. Appl Environ Microbiol 63(4):1326–1331

Duxbury JM, Tiedje JM, Alexande M, Dawson JE (1970) 2,4-D metabolism: enzymatic conversion of chloromaleylacetic acid to succinic acid. J Agric Food Chem 18(2):199. doi:10.1021/jf60168a029

Filer K, Harker AR (1997) Identification of the inducing agent of the 2,4-dichlorophenoxyacetic acid pathway encoded by plasmid pJP4. Appl Environ Microbiol 63(1):317–320

Fulthorpe RR, McGowan C, Maltseva OV, Holben WE, Tiedje JM (1995) 2,4-Dichlorophenoxyacetic acid-degrading bacteria contain mosaics of catabolic genes. Appl Environ Microbiol 61(9):3274–3281

Fulthorpe RR, Rhodes AN, Tiedje JM (1996) Pristine soils mineralize 3-chlorobenzoate and 2,4-dichlorophenoxyacetate via different microbial populations. Appl Environ Microbiol 62(4):1159–1166

Harker AR, Olsen RH, Seidler RJ (1989) Phenoxyacetic acid degradation by the 2,4-dichlorophenoxyacetic acid (TFD) pathway of plasmid pJP4: mapping and characterization of the TFD regulatory gene, *tfdR*. J Bacteriol 171(1):314–320

Hoffmann D, Kleinsteuber S, Muller RH, Babel W (2003) A transposon encoding the complete 2,4-dichlorophenoxyacetic acid degradation pathway in the alkalitolerant strain *Delftia acidovorans* P4a. Microbiology 149:2545–2556. doi:10.1099/mic.0.026260-0

Hogan DA, Buckley DH, Nakatsu CH, Schmidt TM, Hausinger RP (1997) Distribution of the *tfdA* gene in soil bacteria that do not degrade 2,4-dichlorophenoxyacetic acid (2,4-D). Microb Ecol 34(2):90–96. doi:10.1007/s002489900038

Hotopp JCD, Hausinger RP (2001) Alternative substrates of 2.4-dichlorophenoxyacetate/alpha-ketoglumate dioxygenase. J Mol Catal B: Enzym 15(4–6):155–162

Inoue T, Sakakibara F, Handa M, Suzuki T, Takeda M, Hayashi A, Murakoshi Z (1988) Herbicides. In: Inoue T (ed) Nouyaku-gaku. Hirokawa Publishing Co., Tokyo

Itoh K, Kanda R, Sumita Y, Kim H, Kamagata Y, Suyama K, Yamamoto H, Hausinger RP, Tiedje JM (2002) *tfdA*-like genes in 2,4-dichlorophenoxyacetic acid-degrading bacteria belonging to the *Bradyrhizobium-Agromonas-Nitrobacter-Afipia* cluster in alpha-proteobacteria. Appl Environ Microbiol 68(7):3449–3454. doi:10.1128/aem.68.7.3449-3454.2002

Itoh K, Tashiro Y, Uobe K, Kamagata Y, Suyama K, Yamamoto H (2004) Root nodule *Bradyrhizobium* spp. Harbor *tfdA* alpha and *cadA*, homologous with genes encoding 2,4-dichlorophenoxyacetic acid-degrading proteins. Appl Environ Microbiol 70(4):2110–2118. doi:10.1128/a-em.70.4.2110-2118.2004

Ka JO, Holben WE, Tiedje JM (1994) Genetic and phenotypic diversity of 2,4-dichlorophenoxyacetic acid (2,4-D)-degrading bacteria isolated from 2,4-D-treated field soils. Appl Environ Microbiol 60(4):1106–1115

Kamagata Y, Fulthorpe RR, Tamura K, Takami H, Forney LJ, Tiedje JM (1997) Pristine environments harbor a new group of oligotrophic 2,4-dichlorophenoxyacetic acid-degrading bacteria. Appl Environ Microbiol 63(6):2266–2272

Kaphammer B, Kukor JJ, Olsen RH (1990) Regulation of *tfdCDEF* by *tfdR* of the 2,4-dichlorophenoxyacetic acid degradation plasmid pJP4. J Bacteriol 172(5):2280–2286

Kitagawa W, Takami S, Miyauchi K, Masai E, Kamagata Y, Tiedje JM, Fukuda M (2002) Novel 2,4-dichlorophenoxyacetic acid degradation genes from oligotrophic *Bradyrhizobium* sp. strain HW13 isolated from a pristine environment. J Bacteriol 184(2):509–518. doi:10.1128/jb.184.2.509-518.2002

Laemmli CM, Leveau JHJ, Zehnder AJB, van der Meer JR (2000) Characterization of a second *tfd* gene cluster for chlorophenol and chlorocatechol metabolism on plasmid pJP4 in *Ralstonia eutropha* JMP134(pJP4). J Bacteriol 182(15):4165–4172. doi:10.1128/jb.182.15.4165-4172.2000

Ledger T, Pieper DH, Gonzalez B (2006) Chlorophenol hydroxylases encoded by plasmid pJP4 differentially contribute to chlorophenoxy acetic acid degradation. Appl Environ Microbiol 72(4):2783–2792. doi:10.1128/aem.72.4.2783-2792.2006

Lee TH, Kurata S, Nakatsu CH, Kamagata Y (2005) Molecular analysis of bacterial community based on 16S rDNA and functional genes in activated sludge enriched with 2,4-dichlorophenoxyacetic acid (2,4-d) under different cultural conditions. Microb Ecol 49(1):151–162. doi:10.1007/s00248-003-1035-6

Leveau JHJ, van der Meer JR (1996) The *tfdR* gene product can successfully take over the role of the insertion element-inactivated TfdT protein as a transcriptional activator of the *tfdCDEF* gene cluster, which encodes chlorocatechol degradation in *Ralstonia eutropha* JMP134(pJP4). J Bacteriol 178(23):6824–6832

Leveau JHJ, van der Meer JR (1997) Genetic characterization of insertion sequence ISJP4 on plasmid pJP4 from *Ralstonia eutropha* JMP134. Gene 202(1–2):103–114. doi:10.1016/s0378-1119(97)00460-5

Leveau JHJ, Zehnder AJB, van der Meer JR (1998) The tfdK gene product facilitates uptake of 2,4-dichlorophenoxyacetate by Ralstonia eutropha JMP134(pJP4). J Bacteriol 180(8):2237–2243

Leveau JHJ, Konig F, Fuchslin H, Werlen C, van der Meer JR (1999) Dynamics of multigene expression during catabolic adaptation of Ralstonia eutropha JMP134 (pJP4) to the herbicide 2,4-dichlorophenoxyacetate. Mol Microbiol 33(2):396–406. doi:10.1046/j.1365-2958.1999.01483.x

Lykidis A, Pérez-Pantoja D, Ledger T, Mavromatis K, Anderson IJ, Ivanova NN, Hooper SD, Lapidus A, Lucas S, González B, Kyrpides NC (2010) The complete multipartite genome sequence of *Cupriavidus necator* JMP134, a versatile pollutant degrader. PLoS One 5(3):e9729

Mae AA, Marits RO, Ausmees NR, Koiv VM, Heinaru AL (1993) Characterization of a new 2,4-dichlorophenoxyacetic acid degrading plasmid pEST4011: physical map and localization of catabolic genes. J Gen Microbiol 139:3165–3170

McGowan C, Fulthorpe R, Wright A, Tiedje JM (1998) Evidence for interspecies gene transfer in the evolution of 2,4-dichlorophenoxyacetic acid degraders. Appl Environ Microbiol 64(10):4089–4092

Neilson JW, Josephson KL, Pepper IL, Arnold RB, Digiovanni GD, Sinclair NA (1994) Frequency of horizontal gene transfer of a large catabolic plasmid (pJP4) in soil. Appl Environ Microbiol 60(11):4053–4058

Newby DT, Josephson KL, Pepper IL (2000) Detection and characterization of plasmid pJP4 transfer to indigenous soil bacteria. Appl Environ Microbiol 66(1):290–296

Pérez-Pantoja D, Guzmán L, Manzano M, Pieper DH, González B (2000) Role of $tfdC_ID_IE_IF_I$ and $tfdD_{II}C_{II}E_{II}F_{II}$ gene modules in catabolism of 3-chlorobenzoate by *Ralstonia eutropha* JMP134(pJP4). Appl Environ Microbiol 66(4):1602–1608. doi:10.1128/aem.66.4.1602-1608.2000

Poh RPC, Smith ARW, Bruce IJ (2002) Complete characterisation of Tn5530 from *Burkholderia cepacia* strain 2a (pIJB1) and studies of 2,4-dichlorophenoxyacetate uptake by the organism. Plasmid 48(1):1–12. doi:10.1016/s0147-619x(02)00018-5

Sen D, Yano H, Suzuki H, Król JE, Rogers L, Brown CJ, Top EM (2010) Comparative genomics of pAKD4, the prototype IncP-1δ plasmid with a complete backbone. Plasmid 63(2):98–107

Shimojo M, Kawakami M, Amada K (2009) Analysis of genes encoding the 2,4-dichlorophenoxyacetic acid-degrading enzyme from *Sphingomonas agrestis* 58–1. J Biosci Bioeng 108(1):56–59. doi:10.1016/j.jbiosc.2009.02.018

Steenson TI, Walker N (1957) The pathway of breakdown of 2:4-dichloro- and 4-chloro-2-methyl-phenoxyacetic acid by bacteria. J Gen Microbiol 16(1):146–155. doi:10.1099/00221287-16-1-146

Suwa Y, Wright AD, Fukimori F, Nummy KA, Hausinger RP, Holben WE, Forney LJ (1996) Characterization of a chromosomally encoded 2,4-dichlorophenoxyacetic acid alpha-ketoglutafate dioxygenase from Burkholderia sp strain RASC. Appl Environ Microbiol 62(7):2464–2469

Trefault N, Clement P, Manzano M, Pieper DH, González B (2002) The copy number of the catabolic plasmid pJP4 affects growth of *Ralstonia eutropha* JMP134 (pJP4) on 3-chlorobenzoate. FEMS Microbiol Lett 212(1):95–100. doi:Pii s0378-1097(02)00734-6 10.1016/s0378-1097(02)00734-6

Trefault N, De la Iglesia R, Molina AM, Manzano M, Ledger T, Pérez-Pantoja D, Sánchez MA, Stuardo M, González B (2004) Genetic organization of the catabolic plasmid pJP4 from *Ralstonia eutropha* JMP134 (pJP4) reveals mechanisms of adaptation to chloroaromatic pollutants and evolution of specialized chloroaromatic degradation pathways. Environ Microbiol 6(7):655–668. doi:10.1111/j.1462-2920.2004.00596.x

Trefault N, Guzman L, Perez H, Godoy M, Gonzalez B (2009) Involvement of several transcriptional regulators in the differential expression of *tfd* genes in *Cupriavidus necator* JMP134. Int Microbiol 12(2):97–106. doi:10.2436/20.1501.01.86

Vallaeys T, Courde L, Mc Gowan C, Wright AD, Fulthorpe RR (1999) Phylogenetic analyses indicate independent recruitment of diverse gene cassettes during assemblage of the 2,4-D catabolic pathway. FEMS Microbiol Ecol 28(4):373–382. doi:10.1111/j.1574-6941.1999.tb00591.x

Vedler E, Vahter M, Heinaru A (2004) The completely sequenced plasmid pEST4011 contains a novel IncP1 backbone and a catabolic transposon harboring *tfd* genes for 2,4-dichlorophenoxyacetic acid degradation. J Bacteriol 186(21):7161–7174. doi:10.1128/jb.186.21.7161-7174.2004

You IS, Ghosal D (1995) Genetic and molecular analysis of a regulatory region of the herbicide 2,4-dichlorophenoxyacetate catabolic plasmid pJP4. Mol Microbiol 16(2):321–331. doi:10.1111/j.1365-2958.1995.tb02304.x

Chapter 4
Genetic System of Organohalide-Respiring Bacteria

Taiki Futagami, Masatoshi Goto, and Kensuke Furukawa

Abstract Organohalide-respiring bacteria (OHRB) utilize halogenated organic compounds as terminal electron acceptors and are considered to be significantly important from both viewpoints of bioremediation and natural halogen cycle. Growth-linked bioremediation using OHRB has been successfully applied to removal of chlorinated solvents, e.g., tetrachloroethene is successively converted to trichloroethene, dichloroethenes, vinyl chloride, and nontoxic ethene. From OHRB, versatile reductive dehalogenases (RDases), which catalyze the reductive dehalogenation reaction, were purified and their corresponding genes have been identified. In this chapter, we present an overview of current understanding of organohalide respiration, showing the RDase genes and their associated genes are highly conserved in phylogenetically diverse OHRB.

Keywords Chlorinated solvent • Chloroethene • Halogen cycle • Halogenated organic compound • Organohalide • Organohalide respiration • Organohalide-respiring bacteria • Reductive dehalogenase • Reductive dehalogenation

Abbreviations

3Cl4OHPA	3-Chloro-4-hydroxyphenylacetate
4-OHPA	4-Hydroxyphenylacetate
CARD-FISH	Catalyzed reporter deposition-fluorescent in situ hybridization

T. Futagami (✉) • M. Goto
Department of Bioscience and Biotechnology, Kyushu University,
6-10-1, Hakozaki, Fukuoka 812-8581, Japan
e-mail: futagami@brs.kyushu-u.ac.jp

K. Furukawa
Department of Food and Bioscience, Faculty of Food Science and Nutrition,
Beppu University, Kitaishigaki 82, Beppu, Oita 874–8501, Japan

H. Nojiri et al. (eds.), *Biodegradative Bacteria: How Bacteria Degrade, Survive, Adapt, and Evolve*, DOI 10.1007/978-4-431-54520-0_4, © Springer Japan 2014

CD	Carbon dichloride
CF	Chloroform
cis-DCE	*cis*-1,2-dichloroethene
CP	Chlorophenol
DMSO	Dimethyl sulfoxide
EPR	Electron paramagnetic resonance
ETH	Ethene
IS	Insertion sequence
OHRB	Organohalide-respiring bacteria
PCBs	Polychlorinated biphenyls
PCE	Tetrachloroethene
RDase	Reductive dehalogenase
Tat	Twin-arginine translocation
TCB	Trichlorobenzene
TCE	Trichloroethene
TeCB	Tetrachlorobenzene
trans-DCE	*trans*-1,2-dichloroethene
VC	Vinyl chloride

4.1 Introduction

Organohalide-respiring bacteria (OHRB) utilize halogenated organic compounds as terminal electron acceptors and have been successfully applied to the detoxification of soil and groundwater contaminated with chlorinated ethenes (Fig. 4.1a). Growth-linked bioremediation using OHRB is a powerful technology for the removal of chlorinated solvents such as tetrachloroethene [perchloroethene (PCE)] and trichloroethene (TCE) in anaerobic environments. In contrast to aerobic microbial degradation processes, the reductive processes associated with organohalide respiration favor highly halogenated organic substrates. This is advantageous for degradation of compounds such as PCE, as removal of PCE through reductive dehalogenation is generally much more effective than oxygenative degradation.

In this chapter, we present an overview of current understanding of organohalide respiration by anaerobic bacteria. The ability to perform organohalide respiration is widespread among bacteria and to date has been described in the phyla *Chloroflexi* and *Firmicutes*, as well as the epsilon and gamma subdivisions of the phylum *Proteobacteria*. However, the genetic system controlling organohalide respiration is well conserved among the versatile OHRB. The key enzyme, reductive dehalogenase (RDase), contains an N-terminal Tat (twin-arginine translocation) signal sequence and two Fe-S cluster-binding motifs. In addition, biochemical studies have revealed that a corrinoid cofactor plays a significant role in the RDase-mediated redox reaction. The gene encoding RDase is found in a gene cluster that also contains a gene encoding a membrane-spanning protein that is believed to act as membrane anchor for RDase. Genes encoding a transcriptional regulator, chaperone,

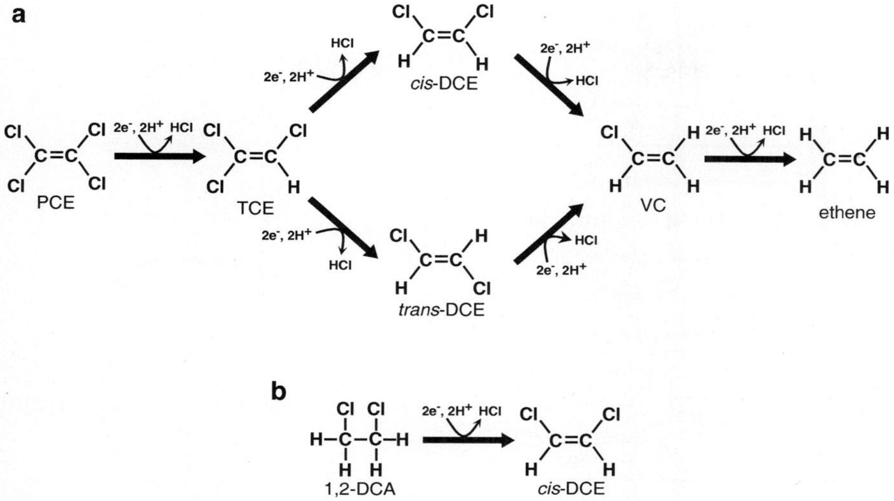

Fig. 4.1 Reductive dechlorination during organohalide respiration. Sequential dechlorination of chlorinated ethenes by hydrogenolytic dechlorination (**a**). Dichloroelimination of 1,2-trichloroethane (**b**). *PCE* tetrachloroethene, *TCE* trichloroethene, *DCE* dichloroethene, *VC* vinyl chloride, *1,2-DCA* 1,2-dichloroethane

transposase, and phage integrase are also frequently found near the RDase gene. Consequently, the transcriptional regulation and genetic rearrangement of these gene clusters and chaperone-associated maturation of RDase have been studied. Genetic events such as mutation and horizontal gene transfer were likely involved in the evolution of the RDase gene, as well as in the evolution of other degradation genes. An increasing number of studies have revealed that OHRB and RDase genes are present even in pristine environments, indicating that they also play a significant role in the dehalogenation of naturally produced organohalides, and not just artificial compounds.

4.2 Diversity of Organohalide-Respiring Bacteria

After the 3-chlorobenzoate-respiring bacterium *Desulfomonile tiedjei* DCB-1 was first isolated in 1984 (Shelton and Tiedje 1984; Deweerd and Suflita 1990), other versatile OHRB belonging to a wide range of bacterial phyla have been identified, including members of the *Firmicutes* (low G+C Gram-positive bacteria), the epsilon and gamma subdivisions of the phylum *Proteobacteria*, and the *Chloroflexi* (Fig. 4.2). Most OHRB are obligate anaerobic bacteria that require reducing conditions, the exception being the organohalide-respiring *Anaeromyxobacter dehalogenans* strains, which are facultative anaerobes (Cole et al. 1994; Sanford et al. 2002).

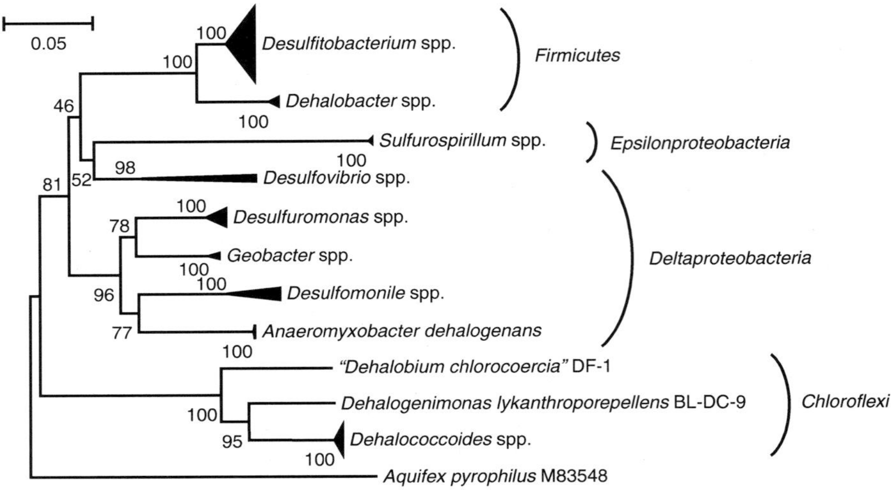

Fig. 4.2 Phylogenetic tree of organohalide-respiring bacteria. The neighbor joining (NJ) method was used to construct a phylogenetic tree of the 16S rRNA gene sequences obtained from the Hierarchy Browser of the Ribosomal Database Project (http://rdp.cme.msu.edu/). The 16S rRNA gene sequence of *Aquifex pyrophilus* was used as the out group. The *scale bar* represents 0.02 substitutions per nucleotide sequence

With respect to their auxotrophic character, the OHRB can be classified roughly into two types: (1) obligate OHRB and (2) OHRB capable of utilizing a range of electron acceptors. The former are able to grow only by organohalide respiration, whereas the latter can grow using electron acceptors other than organohalides. All of the isolates from the phylum *Chloroflexi*, including the genera *Dehalococcoides*, "*Dehalobium*," and *Dehalogenimonas*, are obligate OHRB (Fig. 4.2). In addition, isolates from the genus *Dehalobacter* are also obligate OHRB, but they are classified into the phylum *Firmicutes* and the genus *Desulfitobacterium*, members of which are able to utilize a range of electron acceptors, including fumarate, thiosulfate, sulfite, nitrate, nitrite, dimethyl sulfoxide (DMSO), sulfonate, trimethylamine *N*-oxide, As(V), Mn(IV), Fe(III), U(VI), Se(VI), and anthraquinone-2,6-disulfonate (a humic acid analog) (reviewed in Villemur et al. 2006). Non-dechlorinating *Desulfitobacterium* spp. strains have also been identified (van de Pas et al. 2001).

4.3 The Genus *Dehalococcoides*: A Key Bacterial Group for DCE- and VC-Reductive Dechlorination

The genus *Dehalococcoides* is an important member of the OHRB because these bacteria are able to completely detoxify chlorinated ethenes by converting them to ethene. In this process, PCE is successively converted to TCE, dichloroethenes [*cis-1*,2-dichloroethene (*cis*-DCE) or *trans*-1,2-dichloroethene (*trans*-DCE)], vinyl chloride (VC), and nontoxic ethene through a process of hydrogenolytic dehalogenation (Fig. 4.1a).

A number of OHRB genera are capable of respiring with PCE and TCE, including *Dehalobacter*, "*Dehalobium*," *Dehalococcoides*, *Desulfitobacterium*, *Desulfuromonas*, *Geobacter*, and *Sulfurospirillum*. In contrast, DCE and VC respirers have thus far been found solely in the genus *Dehalococcoides* (reviewed in Smidt and de Vos 2004; Löffler and Edwards 2006; Hiraishi 2008; Tiehm and Schmidt 2011). Because VC is a well-known carcinogen and causes liver cancer (Kielhorn et al. 2000), the accumulation of VC during the degradation process is of serious concern. The dechlorination of PCE by OHRB is also important because oxygenative degradation of PCE is generally difficult and thus rarely reported. Some examples of aerobic degradation of PCE are available, such as that mediated by toluene-*o*-xylene monooxygenase of *Pseudomonas stutzeri* OX1 and the cytochrome P450 system of the white-rod fungus *Trametes versicolor* (Ryoo et al. 2000; Marco-Urrea et al. 2006, 2009).

The first member of the genus *Dehalococcoides* to be isolated was *Dehalococcoides mccartyi* 195 (formerly "*Dehalococcoides ethenogenes*" 195), which was described in 1997 (Maymó-Gatell et al. 1997; Löffler et al. 2012). *D. mccartyi* strains CBDB1, FL2, BAV1, VS, GT, and MB have been isolated since that time. Strains 195 and FL2 dechlorinate PCE and TCE, respectively, to ethene (Maymó-Gatell et al. 1997; Löffler et al. 2000; He et al. 2005). However, these strains are unable to use VC as a growth-supporting electron acceptor, and the dechlorination of VC to ethene is thus a cometabolic process (Maymó-Gatell et al. 1999). In contrast, three other *D. mccartyi* strains, BAV1, VS, and GT, can use VC as an electron acceptor and thereby efficiently dechlorinate VC to ethene (He et al. 2003; Cupples et al. 2003; Müller et al. 2004; Sung et al. 2006). The strains CBDB1 and MB predominantly dechlorinate PCE to *trans*-DCE (Cheng and He 2009; Marco-Urrea et al. 2011). Strain CBDB1 was isolated based upon its ability to respire with chlorobenzenes and dioxins, such as 1,2,3-trichlorobenzene, 1,2,4-trichlorobenzene, 2,3-dichloro-*p*-dibenzodioxin, and 2,3,7,8-tetrachloro-p-dibenzodioxin (Adrian et al. 2000; Bunge et al. 2003).

Many molecular ecological studies have shown that the *Dehalococcoides*-like *Chloroflexi* and their close relatives inhabit a wide range of anaerobic terrestrial and marine ecosystems. The "*Dehalobium*" and *Dehalogenimonas* belong to the class "*Dehalococcoidetes*" (subphylum II) of the phylum *Chloroflexi*, as does *Dehalococcoides* (Hugenholtz and Stackebrandt 2004; Yamada et al. 2006) (Fig. 4.2). "*Dehalobium chlorocoercia*" DF-1 predominantly dechlorinates PCE and TCE to *trans*-DCE rather than to *cis*-DCE (Miller et al. 2005). Strain DF-1 was isolated based upon its ability to dechlorinate chlorobenzenes as well as polychlorinated biphenyls (PCBs). For example, cultures containing DF-1 are able to dechlorinate hexachlorobenzene, pentachlorobenzene, 1,2,3,5-tetrachlorobenzene, and 1,3,5-trichloro benzene (Wu et al. 2002; May et al. 2008). On the other hand, *Dehalogenimonas lykanthroporepellens* strains BL-DC-8 and BL-DC-9 dechlorinate polychlorinated aliphatic alkanes, including 1,2,3-trichloropropane, 1,2-dichloropropane, 1,1,2,2-tetrachloroethane, 1,1,2-trichloroethane, and 1,2-dichloroethane (Moe et al. 2009; Yan et al. 2009).

4.4 The Substrate Specificity of Reductive Dehalogenase

Reductive dehalogenase is a key enzyme in the organohalide respiratory chain, acting as a terminal reductase to catalyze the dehalogenation reaction. RDases from a number of genera, including *Desulfitobacterium*, *Dehalobacter*, *Dehalococcoides*, and *Sulfurospirillum*, have been purified and functionally characterized (Fig. 4.3a). Importantly, the substrate spectrum of each OHRB is not dependent on the bacterial strain but rather is more likely dependent upon the type of RDase. Studies of RDases

a

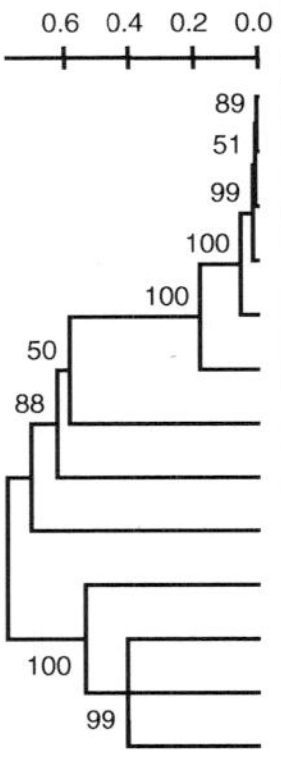

Protein name	Host organism	Enzymatic reaction	Accession number	Reference
PceA	*Desulfitobacterium hafniense* Y51	PCE→TCE→*cis*-DCE	AAW80323	Suyama et al. 2002
PceA	*Desulfitobacterium hafniense* TCE1	PCE→TCE→*cis*-DCE	CAD28792	Maillard et al. 2003
PceA	*Dehalobacter restrictus*PER-K23	PCE→TCE→*cis*-DCE	CAD28790	Maillard et al. 2003
PceA	*Desulfitobacterium hafniense* PCE-S	PCE→TCE→*cis*-DCE	AAO60101	Miller et al. 1997, Maillard et al. 2003
DcaA	*Desulfitobacterium dichloroeliminans* DCA1	PCE→*cis*-DCE	CAJ75430	Marzorati et al. 2007
CprA5	*Desulfitobacterium hafniense* PCP-1	3,5-dichlorophenol→3-CP	AAQ54585	Thibodeau et al. 2004
PrdA	*Desulfitobacterium* sp. KBC1	PCE→TCE	BAE45338	Tsukagoshi et al. 2006
PceA	*Sulfurospirillum multivorans*	PCE→TCE→*cis*-DCE	AAC60788	Neumann et al. 1998
CprA	*Desulfitobacterium dehalogenans*JW/IU-DC1	3Cl4OHPA→4-OHPA	AAD44542	Van de Pas et al. 1999
CbrA	*Dehalococcoides mccartyi* CBDB1	1,2,3,4-TeCB→1,2,4-TCB	CAI82345	Adrian et al. 2007
TceA	*Dehalococcoides mccartyi* 195	TCE→*cis*-DCE→VC→ETH	AAW39060	Magnuson et al. 2000
BvcA	*Dehalococcoides mccartyi* BAV1	*cis*-DCE→VC→ETH	AAT64888	Krajmalnik-Brown et al. 2004
VcrA	*Dehalococcoides mccartyi* VS	*cis*-DCE→VC→ETH	AAQ94119	Müller et al. 2004

b

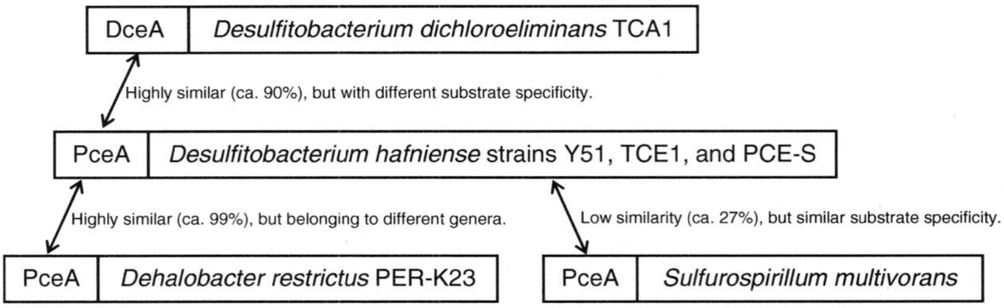

Fig. 4.3 Phylogenetic comparison of amino acid sequences of the functionally characterized reductive dehalogenases (**a**). The NJ method was used to build the phylogenetic tree. The substrates shown in the "Enzymatic reaction" column that utilize chloroethenes were confirmed not to dechlorinate the other chloroethenes. However, it should be noted that the catalytic reactions shown here are representative examples, e.g., the PceA RDases from *Sulfurospirillum multivorans* and *Desulfitobacterium hafniense* PCE-S also debrominate brominated ethenes (Ye et al. 2010). In addition, CprA and CbrA are known as *ortho*-chlorophenol dehalogenase and chlorobenzene dehalogenase, respectively. *PCE* tetrachloroethene, *TCE* trichloroethene, *DCE* dichloroethene, *VC* vinyl chloride, *ETH* ethene, *CP* chlorophenol, *3Cl4OHPA* 3-chloro-4-hydroxyphenylacetate, *4-OHPA* 4-hydroxyphenylacetate, *TeCB* tetrachlorobenzene, *TCB* trichlorobenzene. Schematic relationships among the RDases from *Desulfitobacterium*, *Dehalobacter*, and *Sulfurospirillum* (**b**). Three relationships were found: highly similar (ca. 90 %), but with different substrate specificity; highly similar (ca. 99 %), but belonging to different genera; low similarity (ca. 27 %), but similar substrate specificity

isolated from *Desulfitobacterium*, *Dehalobacter*, and *Sulfurospirillum* illustrate the relationship between the RDase and OHRB (Fig. 4.3b).

The genus *Desulfitobacterium* is ubiquitous at contaminated sites (reviewed in Villemur et al. 2006). Most of the isolates are capable of reductively dechlorinating chloroethenes and/or chlorophenols, and the corresponding RDase enzymes have been identified and characterized. *Desulfitobacterium hafniense* strains Y51, TCE1, PCE-S, and JH1 dechlorinate PCE to *cis*-DCE via TCE, but do not dechlorinate chlorophenols (Suyama et al. 2001; Miller et al. 1997; Gerritse et al. 1999; Fletcher et al. 2008). These strains produce PCE/TCE RDases (PceA) that are nearly 99 % similar based on amino acid sequences (Fig. 4.3a). In contrast, *Desulfitobacterium dehalogenans* IW/IU-DC1 and *D. hafniense* DCB-2 dechlorinate chlorophenols but not chloroethenes (Madsen and Licht 1992; Utkin et al. 1994; Christiansen and Ahring 1996). Strains IW/IU-DC1 and DCB-2 produce an *ortho*-chlorophenol RDase (CprA) (Fig. 4.3a).

Dehalobacter restrictus strain PER-K23 dechlorinates PCE to *cis*-DCE via TCE in a reaction mediated by a PceA enzyme that is highly similar (99 %) to the enzyme produced by *Desulfitobacterium* (Maillard et al. 2003) (Fig. 4.3a, b). As is the case with *Desulfitobacterium*, *Dehalobacter* is classified in the phylum *Firmicutes*, but *Dehalobacter* is phylogenetically distant from the genus *Desulfitobacterium* (Fig. 4.1), indicating that horizontal gene transfer might have occurred between species of these two genera (see Sect. 4.8).

Sulfurospirillum multivorans also dechlorinates PCE to *cis*-DCE via TCE using PceA (Fig. 4.3a, b) (Scholz-Muramatsu et al. 1995). However, the PceA produced by *Sulfurospirillum multivorans* shows only 27 % sequence identity to the PceA enzymes produced by *Desulfitobacterium* and *Dehalobacter* (Neumann et al. 1996, 1998). In contrast, DcaA, which is produced by *Desulfitobacterium dichloroeliminans* DCA1 and was identified from a 1,2-dichloroethane (1,2-DCA)-contaminated enrichment culture, shows a higher similarity (90 %) to the PceA enzymes from *Desulfitobacterium* and *Dehalobacter* than to the PceA from *S. multivorans* (Marzorati et al. 2007). The DcaA enzyme catalyzes a different reaction, however, dechlorinating 1,2-DCA to *cis*-DCE through a reductive dehalogenation reaction called dichloroelimination that simultaneously removes the adjacent chlorine atoms (De Wildeman et al. 2003; Marzorati et al. 2007) (Fig. 4.1b). An analysis of evolution rates using the method of Nei and Gojobori (Nei and Gojobori 1986) indicated that DcaA evolved due to positive selection. Both the enrichment culture and *D. dichloroeliminans* DCA1 were isolated from a site that had been contaminated with 1,2-DCA for more than 30 years (De Wildeman et al. 2003; Marzorati et al. 2006, 2007).

The functions of four *Dehalococcoides* spp. RDases (TceA, VcrA, BvcA, and CbrA) have been characterized to date (Fig. 4.3a). TceA catalyzes the dechlorination of TCE to ethene (Magnuson et al. 1998). The VcrA and BvcA dehalogenases isolated from strains VS and BAV1, respectively, catalyze the dechlorination of *cis-DCE* to ethene via VC (Müller et al. 2004; Krajmalnik-Brown et al. 2004). Alternatively, CbrA, identified from strain CBDB1, catalyzes the reductive dechlorination of chlorobenzenes such as 1,2,3-trichlorobenzene (1,2,3-TCB) (Adrian et al. 2007).

Most of the OHRB RDases thus far characterized reductively dechlorinate chlorinated aliphatic or aromatic compounds, such as chloroethenes, chlorophenols, and chlorobenzenes. The fact that the known substrates for these enzymes consist primarily of chlorinated organic compounds may be due to study bias. It should be noted that OHRB are also able to utilize brominated and iodinated compounds. The PceA enzymes were first identified as PCE- and TCE-reductive dehalogenases. However, further study revealed that the PceA enzymes from *S. multivorans* and *D. hafniense* PCE-S also catalyze the reductive debromination of brominated ethenes (Ye et al. 2010).

Because the type of RDase has a significant influence on the dehalogenation capability of OHRB, detecting the specific RDase genes rather than the 16S rRNA gene is necessary for assessing and monitoring bioremediation potential in a given environment. Detection of both RDase and 16S rRNA genes of OHRB using techniques such as real-time PCR is now well established (reviewed in Cupples 2008). Active *Dehalococcoides* populations can be detected also through fluorescent in situ hybridization (FISH) and catalyzed reporter deposition (CARD)-FISH of 16S rRNA (Aulenta et al. 2004; Dijk et al. 2008; Fazi et al. 2008). Recent advances now enable researchers to detect and monitor OHRB using various omics technologies as well (reviewed in Maphosa et al. 2010), for monitoring not only the expression of the RDase gene but also the expression of other key genes. For example, DNA-chip technology is no longer limited to the whole-genome microarray analysis of OHRB, but can be used to detect the hydrogenase genes expressed by hydrogen-producing and hydrogen-consuming microbes (Marshall et al. 2012). This is a significant advance since hydrogen is a key electron donor for obligate hydrogenotrophic OHRB such as *Dehalococcoides* and *Dehalobacter*.

4.5 Biochemical Properties of Reductive Dehalogenase

Corrinoid, which is a derivative of vitamin B_{12} (cobalamin), is a cofactor located at the catalytic center of the RDase enzyme. Involvement of the corrinoid cofactor in the dehalogenation reaction was predicted through studies in which corrinoid was reversibly inactivated with propyl iodides (Neumann et al. 1994). The importance of corrinoid to the activity of *S. multivorans* PceA was also confirmed by the discovery that mutants incapable of dechlorinating PCE cannot synthesize corrinoid (Siebert et al. 2002). The corrinoid cofactor of *S. multivorans* PceA was purified and structurally identified as norpseudovitamin B_{12} (Kräutler et al. 2003). Furthermore, the activity of the PceA from *S. multivorans* was found to be enhanced by an estimated 4,800-fold over the nonenzymatic cofactor-dependent reaction (Glod et al. 1997; Neumann et al. 2002; McCauley et al. 2005). The corrinoid cofactor of *Dehalobacter restrictus* PER-K23 PceA has also been isolated, and its properties were found to be the same as those of commercially available cobalamin (Maillard et al. 2003). A cobalamin-binding domain "DXHXXG…SXL…GG," which is found in subsets of cobalamin-dependent methyltransferases and isomerases, has also been found in seven RDase homologues from *Dehalococcoides* strains and the CbrA enzyme produced by strain CBDB1 has a truncated cobalamin-binding domain; however, in many cases no cobalamin-binding domain has been identified (Ludwig and Matthews 1997; Hölscher et al. 2004; Adrian et al. 2007).

Schemes involving one corrinoid and two Fe-S clusters have been proposed as feasible reaction mechanisms for reductive dechlorination by OHRB. For instance, PCE is reduced through one-electron transfer from the Co(I) corrinoid, producing the trichlorovinyl radical (Neumann et al. 1996; Holliger et al. 1998, 2003; Banerjee and Ragsdale 2003; McCauley et al. 2005; Diekert et al. 2005). All of the functionally characterized RDase sequences contain two highly conserved Fe-S cluster-binding motifs. These two Fe-S clusters are hypothesized to be involved in the redox activation of the corrinoid cofactor. The presence of Fe-S clusters in the PceA produced by *D. restrictus* PER-K23 and in the CprA produced by *D. dehalogenans* IW/IU-DC1 was experimentally confirmed through electron paramagnetic resonance (EPR) analyses, which demonstrated that the former enzyme contains two [4Fe-4S] clusters and that the latter contains one [4Fe-4S] and one [3Fe-4S] cluster (Schumacher et al. 1997; van de Pas et al. 1999; Maillard et al. 2003).

4.6 The Reductive Dehalogenase Gene Cluster

RDase-encoding genes are always organized in an operon with at least one gene encoding a protein containing a 2- or 3-transmembrane domain (Fig. 4.4). Suyama et al. (2002) reported that the mature PceA RDase of *D. hafniense* Y51, from which the N-terminal Tat sequence has been cleaved, is localized in the periplasmic space, while the unprocessed protein with the Tat sequence present is found in the cytoplasmic fraction. Moreover, the authors reported that the RDase did not possess any membrane-associated domain structures. Given these facts, B proteins (e.g., PceB; Fig. 4.5) are believed to act as membrane anchors for RDases. The localization of the PceA of *S. multivorans* was characterized microscopically using the freeze-fracture replica immunogold labeling technique (John et al. 2006). The results of that study indicated that localization of PceA depends upon the electron acceptor, though the mechanism remains unknown. When *S. multivorans* is grown with fumarate as the electron acceptor, PceA localizes in the cytoplasm or associates with the membrane side facing the cytoplasm. In contrast, when the same strain is grown with either PCE or TCE as an electron acceptor, most of the PceA localizes along the periplasmic side of the cytoplasmic membrane.

The constitution and order of the *pceA*, *pceB*, *pceC*, and *pceT* genes is highly conserved among the *Desulfitobacterium* and *Dehalobacter* strains that dechlorinate PCE to *cis*-DCE (Fig. 4.4). The PceC protein has not been functionally characterized, but is similar to the NirI/NosR family of membrane-binding transcriptional regulators that are known to be involved in nitrous oxide respiration (Cuypers et al. 1992) (Fig. 4.5). NosR has also been characterized as an electron-donating protein (Wunsch and Zumft 2005). The PceC protein is composed of a 6-transmembrane

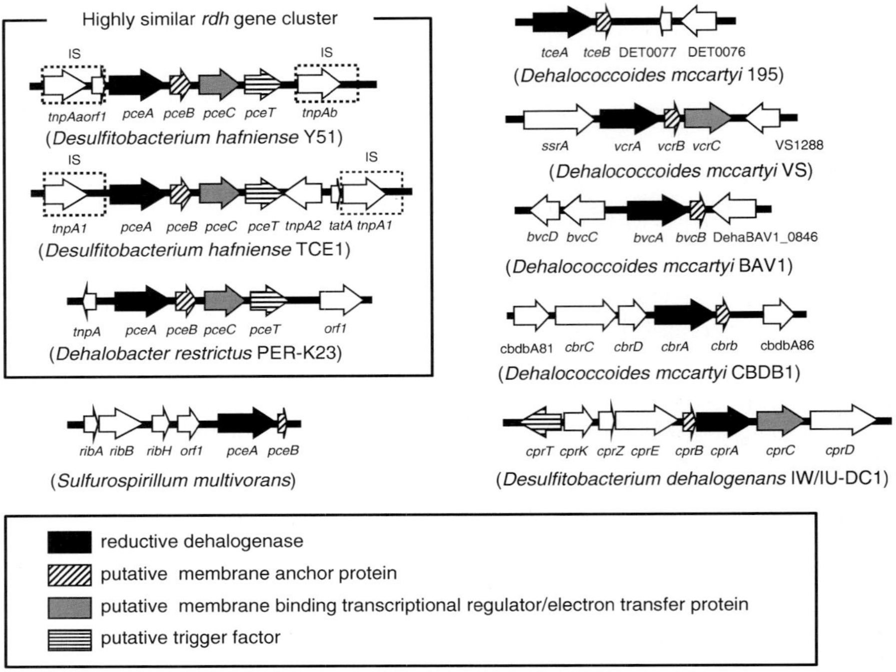

Fig. 4.4 Comparison of reductive dehalogenase gene clusters. An identical pattern indicates similar genes, as described in the *inset*. GenBank accession numbers are as follows. *Desulfitobacterium hafniense* Y51 *pce* genes: AY706985; *Desulfitobacterium hafniense* TCE1 *pce* genes: AJ439608; *Desulfitobacterium hafniense* PCE-S *pce* genes: AY216592; *Dehalobacter restrictus* PER-K23 *pce* genes: AJ439607; *Sulfurospirillum multivorans pce* genes: AF022812; *Dehalococcoides mccartyi* 195 *tce* genes: CP000027; *Dehalococcoides mccartyi* VS *vcr* genes: AY322364; *Dehalococcoides mccartyi* CBDB1 *cbr* genes: AJ965256; *Desulfitobacterium dehalogenans strain* IW/IU-DC1 *cpr* genes: AF115542. The genome information for *Dehalococcoides mccartyi* strains VS, BAV1, and CBDB1 was also obtained from the Joint Genome Institute (http://jgi.doe.gov/)

domain, a flavin mononucleotide (FMN)-binding domain, and a poly ferredoxin-like domain (Fig. 4.5), implying that PceC localizes in the cell membrane in order to donate electrons to PceA. The importance of PceC for organohalide respiration is supported by the fact that the protein is returned as a reciprocal "best hit" in comparisons of the genome sequences of *D. hafniense* Y51 and *D. mccartyi* 195 (Nonaka et al. 2006). In addition, the NirI/NosR-like protein-encoding genes are conserved, as is the *cprC* gene in *D. dehalogenans* IW/IU-DC1 and the *vcrC* gene in *D. mccartyi* VS (Smidt et al. 2000; Müller et al. 2004). Moreover, the *pceABC* genes of strain Y51 and the *vcrABC* genes of strain VS are cotranscribed (Müller et al. 2004; Furukawa et al. 2005; Futagami et al. 2006a, b), indicating that the C proteins might play a role in organohalide respiration.

On the other hand, the *pceT* gene is not cotranscribed in strain Y51 (Furukawa et al. 2005; Futagami et al. 2006b), but might be involved in the maturation of PceA as a trigger factor (Fig. 4.5). The RDase contains a conserved Tat signal sequence in the N-terminal region, indicating that the enzyme is localized across

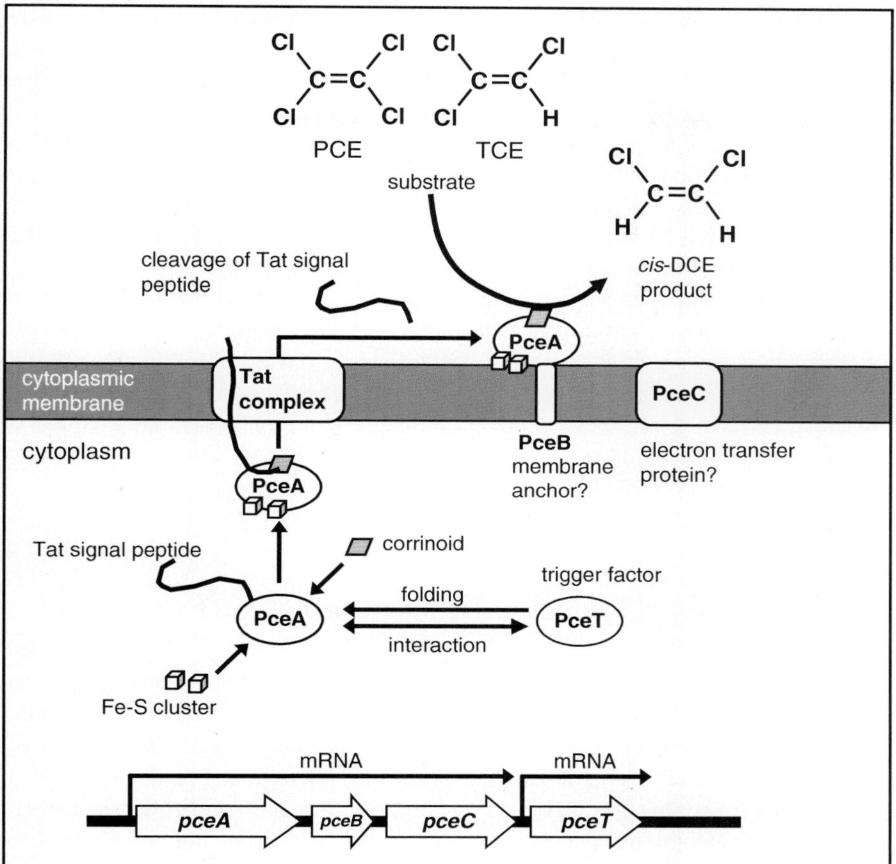

	PceA	PceB	PceC	PceT
Domain structure	Tat signal sequence (1)	transmembrane domain (3)	transmembrane domain (6)	Bacterial trigger factor protein C-terminus (1)
	Fe-S cluster binding motif (2)		FMN-binding domain (1)	FKBP domain (1)
			poly ferredoxin-like domain (1)	Bacterial trigger factor protein N-terminus (1)

Fig. 4.5 Schematic representation of the maturation and localization of PCE-/TCE-reductive dehalogenase (PceA) in *Desulfitobacterium hafniense* Y51. The domain constitution of each protein is summarized in the *inset*. The number of each domain is shown in *parentheses*. *Tat* twin-arginine translocation system, *PCE* tetrachloroethene, *TCE* trichloroethene, *cis-DCE cis*-1,2-dichloroethene

the plasma membrane. Due to this Tat system, PceA should localize correctly only when the PceA precursor has been previously folded properly with its cofactors (reviewed in Palmer and Berks 2012). Thus, PceT may contribute to the correct folding of the PceA precursor protein during Tat-mediated secretion.

Recombinant PceT has peptidyl-prolyl *cis–trans* isomerase and chaperone activity, and co-immunoprecipitation assay results showed that PceT interacts with the Tat signal sequence of PceA in strain Y51, indicating that PceT helps mediate correct folding of the precursor PceA (Morita et al. 2009). The PceT enzyme from *D. hafniense* TCE1 has also been characterized and was shown to efficiently aid in the solubilization of PceA during heterologous expression using an *Escherichia coli* strain lacking both the trigger factor and DnaK chaperones (Maillard et al. 2011). These results confirm that PceT is involved in mediating the correct folding of the precursor of PceA.

The larger components of the RDase gene cluster have been identified in chlorophenol-respiring *Desulfitobacterium* strains. For example, *D. dehalogenans* IW/IU-DC1 possesses an *ortho*-chlorophenol RDase gene cluster containing eight genes: *cprT*, *cprK*, *cprZ*, *cprE*, *cprB*, *cprA*, *cprC*, and *cprD* (Fig. 4.4) (Smidt et al. 2000). CprK is a CRP-FNR (cAMP-binding protein/fumarate nitrate reduction regulatory protein) family transcriptional regulator, whereas CprD and CprE are putative GroEL-type molecular chaperones.

4.7 Regulation of Reductive Dehalogenase Gene Expression

Transcription of the *ortho*-chlorophenol RDase-encoding *cprA* gene in *Desulfitobacterium* is regulated by CprK, which is similar to CRP-FNR family proteins. The *cpr* gene cluster responds at the transcription level to the presence of chlorophenols (Smidt et al. 2000; Gábor et al. 2008; Bisaillon et al. 2011). Both in vivo and in vitro studies have revealed that high-affinity interaction between chlorinated aromatic compounds and a CprK effector domain triggers binding of CprK to an upstream target DNA sequence called a dehalobox "TTAAT-N4-ATTAA," which closely resembles the FNR box (Pop et al. 2004, 2006; Gábor et al. 2006, 2008; Joyce et al. 2006; Mazon et al. 2007). Joyce et al. (2006) determined X-ray crystal structures of the oxidized form of *D. hafniense* DCB-2 CprK bound to a 3-chloro-4-hydroxyphenylacetate ligand and of the reduced form of *D. dehalogenans* IW/IU-DC1 CprK (both proteins are 89 % identical) without the ligand, thus enabling identification of the allosteric changes induced by ligand binding.

Long-term regulation of the *pceA* gene in *S. multivorans* has also been demonstrated (John et al. 2009). The authors of that study reported that transcription of *pceA* decreases during subcultivation. After 35 subcultivations (approximately 105 generations), no *pceA* transcripts, PceA protein, or PceA activity could be detected. Biosynthesis of catalytically active PceA could be restored to a level before the subcultivation within about 50 h (approximately three generations) by the addition of PCE or TCE to the culture medium. These results indicated that a novel type of long-term regulation of *pceA* gene expression exists in *S. multivorans*.

Expression of the *Dehalococcoides* RDase genes *bvcA* and *cbrA* has been detected in cells cultivated in the presence of VC and 1,2,3-TCB, respectively (Krajmalnik-Brown et al. 2004; Adrian et al. 2007). These genes are located near by the putative transcriptional regulator genes. Within the *bvc* and *cbr* gene clusters of *D. mccartyi* strains BAV1 and CBDB1, the *bvcC-bvcD* and *cbrC-cbrD* genes encode a putative two-component system consisting of a sensor signal transduction histidine kinase and a response regulator (Fig. 4.4) (Adrian et al. 2007; McMurdie et al. 2009).

Expression of RDase gene is not always regulated, however (e.g., the *pceA* gene of *D. hafniense* Y51 is constitutively transcribed in cultures grown in media containing various electron acceptors such as fumarate, TCE, nitrate, or DMSO) (Peng et al. 2012). The level of PceA expression in populations of *D. hafniense* Y51 is affected by the emergence of non-dechlorinating variants that have lost the *pceABCT* genes or *pceABC* promoter region through genetic rearrangements (Futagami et al. 2006a, b) (see Sect. 4.8).

4.8 Genetic Rearrangement of the Reductive Dehalogenase Gene Cluster

Degradative genes are frequently found on mobile DNA elements (reviewed in van der Meer et al. 1992; Tsuda et al. 1999; Liang et al. 2012). The RDase gene cluster is no exception, and rearrangements involving mobile genetic elements, such as gene duplication, the formation of chimeric genes, and gene transfer, are believed to have played a role in the evolution of organohalide respiration.

Comparative sequence analyses have revealed vestiges of chromosomal rearrangement. The *pceABCT* gene cluster of *D. hafniense* strains Y51 and TCE1 is surrounded by two nearly identical copies of an insertion sequence (IS) element that include a gene encoding the IS*256* type transposase (Fig. 4.4). The *pceABCT* genes of strain TCE1 share a 99.7 % identity with those of strain Y51. The direct repeat sequences "CTGAACCA" and "TTTTTATA" are found just upstream of the first IS and downstream of the second IS in strains Y51 and TCE1, respectively (Maillard et al. 2005; Futagami et al. 2006a). Thus, the *pceABCT* gene cluster seems to be inserted into the chromosome as a composite transposon in these two strains. In addition, a circular molecule carrying an entire *pceABCT* gene cluster and two terminal IS copies has been described, indicating that the catabolic transposon can still function and be excised from the chromosome. The *pceABCT* gene cluster, including the intergenic regions, is highly conserved among *Desulfitobacterium* and *Dehalobacter* sp., suggesting horizontal transfer between these genera. The recent acquisition of the *pceABCT* gene cluster was supported by proteomic analyses of *D. hafniense* TCE1, the results of which revealed that the expression of proteins involved in stress responses and associated regulation pathways increases in the presence of PCE, suggesting that strain TCE1 is still incompletely adapted to PCE respiration and that this strain is thus not fully suited to PCE respiration (Prat et al. 2011).

During subculturing, strain Y51 was also found to spontaneously give rise to two types of non-PCE dechlorinating variants (Futagami et al. 2006a). One variant was generated from deletion of the left IS, which contains a promoter region of the *pceABC* gene

Table 4.1 Comparison of the genomes of organohalide dehalorespiring bacteria

Name	Size (Mbp)	G+C (%)	rRNA operon	Predicted CDS	RDase gene	Plasmid
Desulfitobacterium hafniense Y51	5.7	47	6	5,060	2	0
Desulfitobacterium hafniense DCB-2	5.3	48	5	5,042	7	0
Anaeromyxobacter dehalogenans 2CP-C	5.0	75	6	4,346	2	0
Geobacter lovleyi SZ	3.9	55	2	3,476	2	1
Dehalococcoides mccartyi 195	1.5	49	1	1,590	17	0
Dehalococcoides mccartyi CBDB1	1.4	47	1	1,385	32	0
Dehalococcoides mccartyi BAV1	1.3	47	1	1,371	11	0
Dehalococcoides mccartyi VS	1.4	47	1	2,096	36	0
Dehalogenimonas lykanthroporepellens BL-DC-9	1.7	55	1	1,720	17	0

The data were summarized from Kube et al. (2005); Seshadri et al. (2005); Nonaka et al. (2006); Thomas et al. (2008); McMurdie et al. (2009); Kim et al. (2012), and Wagner et al. (2012). *Dehalogenimonas lykanthroporepellens* BL-DC-9 sequence information was obtained from the Joint Genome Institute (http://www.jgi.doe.gov/)

cluster (Fig. 4.4). Transcription of the *pceABC* genes was thus abolished in this variant, and accordingly the PCE-dechlorination capability. The other variant arose from homologous recombination between the left IS and right IS, resulting in excision of the entire *pceABCT* gene cluster. Thus, in the absence of chloroethenes, several modes of genetic rearrangement occur around the *pceABCT* gene cluster in strain Y51.

The PCE non-dechlorinating variants of strain Y51 predominate in the presence of chloroform (CF) because CF significantly inhibits the growth of wild type strain Y51 but not the non-dechlorinating variants (Futagami et al. 2006b). Moreover, CF-mediated inhibition of dechlorination by *Sulfurospirillum* and *Dehalococcoides* has also been reported (Neumann et al. 1996; Maymó-Gatell et al. 2001; Duhamel et al. 2002). On the other hand, studies have shown that CF can act as an electron acceptor. Growth-linked dechlorination of CF to carbon dichloride (CD) was observed in the enrichment culture containing *Dehalobacter* (Grostern et al. 2010). Dechlorination of CF to CD and fermentation of CD to acetate, hydrogen, and carbon dioxide has also been observed in *Dehalobacter*-dominated cultures (Lee et al. 2012).

Large numbers of putative RDase genes have been identified in the genome sequences of OHRB (Table 4.1). From genomic information of *Dehalococcoides* strains, recombination appears to have taken place between two RDase genes (DhcVS1399 and DhcVS1427) of *D. mccartyi* strain VS, resulting in formation of the apparently chimeric gene, DET1535 of strain 195 (McMurdie et al. 2009). Analyses of codon usage in the *vcrA* and *bvcA* genes of *D. mccartyi* strains VS and BAV1, respectively, showed that these genes are highly unusual and are characterized by a low G+C content at the third position (McMurdie et al. 2007). The comparatively high degree of abnormal codon usage in the *vcrA* and *bvcA* genes suggests that the evolutionary history of these genes is quite different than that of most other *Dehalococcoides* genes. These data also suggest that mobile elements played an important role in the arrangement and consequently the evolution of the RDase genes in *Dehalococcoides*.

The *tceAB* gene cluster of *D. mccartyi* 195 is located in a putative integrated element (Seshadri et al. 2005). The gene DET0076, which is located downstream of the *tceAB* genes, encodes a protein that is highly similar to resolvase (Fig. 4.4). The *vcrABC* genes of *D. mccartyi* VS are also embedded in a horizontally acquired genomic island (McMurdie et al. 2011). The *ssrA* gene encodes a site-specific recombinase (Müller et al. 2004) (Fig. 4.4). The *vcrABC*-containing genomic islands obtained from *Dehalococcoides* enrichment cultures have been sequenced. Using available *Dehalococcoides* phylogenomic data, it can be estimated that these *ssrA*-specific genomic islands are at least as old as the *Dehalococcoides* group itself, which in turn far predates human civilization, indicating that it took place before emergence of anthropogenic chemicals. The *vcrABC*-containing genomic islands represent a recently acquired subset of a diverse collection of *ssrA*-specific mobile elements that are a major contributor to strain-level diversity in *Dehalococcoides*, and may have been throughout its evolution. The high degree of similarity between *vcrABC* sequences is quantitatively consistent with recent horizontal acquisition driven by ~100 years of industrial pollution with chloroethenes (McMurdie et al. 2011). Moreover, transcriptional analysis of a *Dehalococcoides-containing* microbial consortium uncovered evidence of prophage activation (Waller et al. 2012).

D. mccartyi FL2 was isolated from a pristine environment, suggesting that OHRB might be able to survive on naturally occurring organohalides (Löffler et al. 2000). In fact, more than 4,500 different natural organohalides have been identified in samples from biotic and abiotic sources (Gribble 2003, 2010). RDase homologous genes were detected in marine subsurface environment using a degenerate primer set designed based on *Dehalococcoides* RDase genes (Krajmalnik-Brown et al. 2004; Futagami et al. 2009). The deepest sediment in which RDase homologous genes were detected was formed ca. 460,000 years ago, suggesting that OHRB existed before industrial activity (Aoike 2007; Futagami et al. 2009). Molecular ecological studies have shown that the density of *Dehalococcoides*-like *Chloroflexi* in terrestrial pristine environments is proportional to the quantity of natural organochlorines, suggesting that these bacteria play a significant role in natural halogen cycles (Krzmarzick et al. 2012).

4.9 Conclusions and Future Perspectives

Organohalide respiration has received considerable attention because of its important role in the remediation of environments polluted with chlorinated organic chemicals. Investigations of the physiology of OHRB and the key RDase enzymes have provided crucial background information for the establishment of OHRB bioremediation technologies. An increasing number of studies have also begun to focus on the evolutionary history of organohalide respiration. Genetic events such as mutation and the gene transfer, as well as the selective force imparted by anthropogenic organohalides released into the environment, likely have played a significant role in the evolution of microbial organohalide respiration.

Recent progress in genomic technologies has enabled researchers to undertake experiments using global approaches in order to move toward a more comprehensive understanding of organohalide respiration. Currently, the complete genome sequences of the genera *Anaeromyxobacter*, *Desulfitobacterium*, *Dehalococcoides*, *Dehalogenimonas*, and *Geobacter* are available (Table 4.1). However, more in-depth studies of OHRB (e.g., determination of the substrates of the large number of uncharacterized RDase homologues) require the establishment and application of genetic engineering tools tailored to the study of organohalide respiration. Several such tools, such as gene recombination using thermosensitive plasmids, gene disruption using transposon, and in vitro expression of active RDases, have been described (Smidt et al. 1999, 2001; Kimoto et al. 2010).

For those interested in obtaining a deeper understanding of organohalide respiration, we strongly recommend several additional resources, including the excellent volume edited by Häggblom and Bossert (2003), and a number of recent review articles (Holliger et al. 1998; Smidt and de Vos 2004; Villemur et al. 2006; Löffler and Edwards 2008; Hiraishi 2008; Futagami et al. 2008; Maphosa et al. 2010; Tiehm and Schmidt 2011).

Acknowledgment The research group of K.F. is supported by grants from the Japan Society for the Promotion of Science (JSPS).

References

Adrian L, Szewzyk U, Wecke J, Gorisch H (2000) Bacterial dehalorespiration with chlorinated benzenes. Nature 408:580–583. doi:10.1038/35046063

Adrian L, Rahnenfuhrer J, Gobom J, Hölscher T (2007) Identification of a chlorobenzene reductive dehalogenase in *Dehalococcoides* sp. strain CBDB1. Appl Environ Microbiol 73:7717–7724. doi:10.1128/AEM.01649-07

Aoike K (2007) CDEX laboratory operation report: CK06-06 D/V *Chikyu* shakedown cruise offshore Shimokita. http://sio7.jamstec.go.jp/JAMSTEC-exp-report/902/CK06-06_CR.pdf

Aulenta F, Rossetti S, Majone M, Tandoi V (2004) Detection and quantitative estimation of *Dehalococcoides* spp. in a dechlorinating bioreactor by a combination of fluorescent in situ hybridisation (FISH) and kinetic analysis. Appl Microbiol Biotechnol 64:206–212. doi:10.1007/s00253-003-1503-4

Banerjee R, Ragsdale SW (2003) The many faces of vitamin B_{12}: catalysis by cobalamin-dependent enzymes. Annu Rev Biochem 72:209–247. doi:10.1146/annurev.biochem.72.121801.161828

Bisaillon A, Beaudet R, Lepine F, Villemur R (2011) Quantitative analysis of the relative transcript levels of four chlorophenol reductive dehalogenase genes in *Desulfitobacterium hafniense* PCP-1 exposed to chlorophenols. Appl Environ Microbiol 77:6261–6264. doi:10.1128/AEM.00390-11

Bunge M, Adrian L, Kraus A, Opel M, Lorenz WG, Andreesen JR, Gorisch H, Lechner U (2003) Reductive dehalogenation of chlorinated dioxins by an anaerobic bacterium. Nature 421:357–360. doi:10.1038/nature01237

Cheng D, He J (2009) Isolation and characterization of "*Dehalococcoides*" sp. strain MB, which dechlorinates tetrachloroethene to *trans*-1,2-dichloroethene. Appl Environ Microbiol 75:5910–5918. doi:10.1128/AEM.00767-09

Christiansen N, Ahring BK (1996) Introduction of a de novo bioremediation activity into anaerobic granular sludge using the dechlorinating bacterium DCB-2. Antonie Van Leeuwenhoek 69:61–66

Cole JR, Cascarelli AL, Mohn WW, Tiedje JM (1994) Isolation and characterization of a novel bacterium growing via reductive dehalogenation of 2-chlorophenol. Appl Environ Microbiol 60:3536–3542

Cupples AM (2008) Real-time PCR quantification of *Dehalococcoides* populations: methods and applications. J Microbiol Methods 72:1–11. doi:10.1016/j.mimet.2007.11.005

Cupples AM, Spormann AM, McCarty PL (2003) Growth of a *Dehalococcoides*-like microorganism on vinyl chloride and *cis*-dichloroethene as electron acceptors as determined by competitive PCR. Appl Environ Microbiol 69:953–959. doi:10.1128/AEM.69.2.953-959.2003

Cuypers H, Viebrock-Sambale A, Zumft WG (1992) NosR, a membrane-bound regulatory component necessary for expression of nitrous oxide reductase in denitrifying *Pseudomonas stutzeri*. J Bacteriol 174:5332–5339

De Wildeman S, Diekert G, Van Langenhove H, Verstraete W (2003) Stereoselective microbial dehalorespiration with vicinal dichlorinated alkanes. Appl Environ Microbiol 69:5643–5647. doi:10.1128/AEM.69.9.5643-5647.2003

Deweerd KA, Suflita JM (1990) Anaerobic aryl reductive dehalogenation of halobenzoates by cell extracts of "*Desulfomonile tiedjei*". Appl Environ Microbiol 56:2999–3005

Diekert G, Gugova D, Limoges B, Robert M, Savéant J-M (2005) Electroenzymatic reactions. Investigation of a reductive dehalogenase by means of electrogenerated redox cosubstrates. J Am Chem Soc 127:13583–13588. doi:10.1021/ja053403d

Dijk JA, Breugelmans P, Philips J, Haest PJ, Smolders E, Springael D (2008) Catalyzed reporter deposition-fluorescent in situ hybridization (CARD-FISH) detection of *Dehalococcoides*. J Microbiol Methods 73:142–147. doi:10.1016/j.mimet.2008.01.012

Duhamel M, Wehr SD, Yu L, Rizvi H, Seepersad D, Dworatzek S, Cox EE, Edwards EA (2002) Comparison of anaerobic dechlorinating enrichment cultures maintained on tetrachloroethene, trichloroethene, *cis*-dichloroethene and vinyl chloride. Water Res 36:4193–4202. doi:10.1016/S0043-1354(02)00151-3

Fazi S, Aulenta F, Majone M, Rossetti S (2008) Improved quantification of *Dehalococcoides* species by fluorescence in situ hybridization and catalyzed reporter deposition. Syst Appl Microbiol 31:62–67. doi:10.1016/j.syapm.2007.11.001

Fletcher KE, Ritalahti KM, Pennell KD, Takamizawa K, Löffler FE (2008) Resolution of culture *Clostridium bifermentans* DPH-1 into two populations, a *Clostridium* sp. and tetrachloroethene-dechlorinating *Desulfitobacterium hafniense* strain JH1. Appl Environ Microbiol 74:6141–6143. doi:10.1128/AEM.00994-08

Furukawa K, Suyama A, Tsuboi Y, Futagami T, Goto M (2005) Biochemical and molecular characterization of a tetrachloroethene dechlorinating *Desulfitobacterium* sp. strain Y51: a review. J Ind Microbiol Biotechnol 32:534–541. doi:10.1007/s10295-005-0252-z

Futagami T, Tsuboi Y, Suyama A, Goto M, Furukawa K (2006a) Emergence of two types of non-dechlorinating variants in the tetrachloroethene-halorespiring *Desulfitobacterium* sp. strain Y51. Appl Microbiol Biotechnol 70:720–728. doi:10.1007/s00253-005-0112-9

Futagami T, Yamaguchi T, Nakayama S, Goto M, Furukawa K (2006b) Effects of chloromethanes on growth of and deletion of the *pce* gene cluster in dehalorespiring *Desulfitobacterium hafniense* strain Y51. Appl Environ Microbiol 72:5998–6003. doi:10.1128/AEM.00979-06

Futagami T, Goto M, Furukawa K (2008) Biochemical and genetic bases of dehalorespiration. Chem Rec 8:1–12. doi:10.1002/tcr.20134

Futagami T, Morono Y, Terada T, Kaksonen AH, Inagaki F (2009) Dehalogenation activities and distribution of reductive dehalogenase homologous genes in marine subsurface sediments. Appl Environ Microbiol 75:6905–6909. doi:10.1128/AEM.01124-09

Gábor K, Veríssimo CS, Cyran BC, Ter Horst P, Meijer NP, Smidt H, de Vos WM, van der Oost J (2006) Characterization of CprK1, a CRP/FNR-type transcriptional regulator of halorespiration from *Desulfitobacterium hafniense*. J Bacteriol 188:2604–2613. doi:10.1128/JB.188.7.2604-2613.2006

Gábor K, Hailesellasse Sene K, Smidt H, de Vos WM, van der Oost J (2008) Divergent roles of CprK paralogues from *Desulfitobacterium hafniense* in activating gene expression. Microbiology 154:3686–3696. doi:10.1099/mic.0.2008/021584-0

Gerritse J, Drzyzga O, Kloetstra G, Keijmel M, Wiersum LP, Hutson R, Collins MD, Gottschal JC (1999) Influence of different electron donors and acceptors on dehalorespiration of tetrachloroethene by *Desulfitobacterium frappieri* TCE1. Appl Environ Microbiol 65:5212–5221

Glod G, Angst W, Holliger C, Schwarzenbach RP (1997) Corrinoid-mediated reduction of tetrachloroethene, trichloroethene, and trichlorofluoroethene in homogeneous aqueous solution: reaction kinetics and reaction mechanisms. Environ Sci Technol 31:253–260. doi:10.1021/es9603867

Gribble GW (2003) Natural production of organohalogen compounds. Springer, Vienna

Gribble GW (2010) Naturally occurring organohalogen compounds – a comprehensive update. Springer, Vienna

Grostern A, Duhamel M, Dworatzek S, Edwards EA (2010) Chloroform respiration to dichloromethane by a *Dehalobacter* population. Environ Microbiol 12:1053–1060. doi:10.1111/j.1462-2920.2009.02150.x

Häggblom MM, Bossert ID (2003) Dehalogenation: microbial processes and environmental applications. Kluwer, Boston

He J, Ritalahti KM, Yang KL, Koenigsberg SS, Löffler FE (2003) Detoxification of vinyl chloride to ethene coupled to growth of an anaerobic bacterium. Nature 424:62–65. doi:10.1038/nature01717

He J, Sung Y, Krajmalnik-Brown R, Ritalahti KM, Löffler FE (2005) Isolation and characterization of *Dehalococcoides* sp. strain FL2, a trichloroethene (TCE)- and 1,2-dichloroethene-respiring anaerobe. Environ Microbiol 7:1442–1450. doi:10.1111/j.1462-2920.2005.00830.x

Hiraishi A (2008) Biodiversity of dehalorespiring bacteria with special emphasis on polychlorinated biphenyl/dioxin dechlorinators. Microbes Environ 23:1–12. doi:10.1264/jsme2.23.1

Holliger C, Wohlfarth G, Diekert G (1998) Reductive dechlorination in the energy metabolism of anaerobic bacteria. FEMS Microbiol Rev 22:383–398. doi:10.1111/j.1574-6976.1998.tb00377.x

Holliger C, Regeard C, Diekert G (2003) Dehalogenation by anaerobic bacteria. In: Häggblom MM, Bossert ID (eds) Dehalogenation: microbial processes and environmental applications. Kluwer, Boston

Hölscher T, Krajmalnik-Brown R, Ritalahti KM, Von Wintzingerode F, Gorisch H, Löffler FE, Adrian L (2004) Multiple nonidentical reductive-dehalogenase-homologous genes are common in *Dehalococcoides*. Appl Environ Microbiol 70:5290–5297. doi:10.1128/AEM.70.9.5290-5297.2004

Hugenholtz P, Stackebrandt E (2004) Reclassification of *Sphaerobacter thermophilus* from the subclass *Sphaerobacteridae* in the phylum *Actinobacteria* to the class *Thermomicrobia* (emended description) in the phylum *Chloroflexi* (emended description). Int J Syst Evol Microbiol 54:2049–2051. doi:10.1099/ijs.0.03028-0

John M, Schmitz RP, Westermann M, Richter W, Diekert G (2006) Growth substrate dependent localization of tetrachloroethene reductive dehalogenase in *Sulfurospirillum multivorans*. Arch Microbiol 186:99–106. doi:10.1007/s00203-006-0125-5

John M, Rubick R, Schmitz RP, Rakoczy J, Schubert T, Diekert G (2009) Retentive memory of bacteria: long-term regulation of dehalorespiration in *Sulfurospirillum multivorans*. J Bacteriol 191:1650–1655. doi:10.1128/JB.00597-08

Joyce MG, Levy C, Gábor K, Pop SM, Biehl BD, Doukov TI, Ryter JM, Mazon H, Smidt H, van den Heuvel RH, Ragsdale SW, van der Oost J, Leys D (2006) CprK crystal structures reveal mechanism for transcriptional control of halorespiration. J Biol Chem 281:28318–28325. doi:10.1074/jbc.M602654200

Kielhorn J, Melber C, Wahnschaffe U, Aitio A, Mangelsdorf I (2000) Vinyl chloride: still a cause for concern. Environ Health Perspect 108:579–588. doi:10.1289/ehp.00108579

Kim SH, Harzman C, Davis JK, Hutcheson R, Broderick JB, Marsh TL, Tiedje JM (2012) Genome sequence of *Desulfitobacterium hafniense* DCB-2, a Gram-positive anaerobe capable of dehalogenation and metal reduction. BMC Microbiol 12:21. doi:10.1186/1471-2180-12-21

Kimoto H, Suye S, Makishima H, Arai J, Yamaguchi S, Fujii Y, Yoshioka T, Taketo A (2010) Cloning of a novel dehalogenase from environmental DNA. Biosci Biotechnol Biochem 74:1290–1292. doi:10.1271/bbb.100027

Krajmalnik-Brown R, Hölscher T, Thomson IN, Saunders FM, Ritalahti KM, Löffler FE (2004) Genetic identification of a putative vinyl chloride reductase in *Dehalococcoides* sp. strain BAV1. Appl Environ Microbiol 70:6347–6351. doi:10.1128/AEM.70.10.6347-6351.2004

Kräutler B, Fieber W, Ostermann S, Fasching M, Ongania K-H, Gruber K, Kratky C, Mikl C, Siebert A, Diekert G (2003) The cofactor of tetrachloroethene reductive dehalogenase of *Dehalospirillum multivorans* is norpseudo-B_{12}, a new type of a natural corrinoid. Helv Chim Acta 86:3698–3716. doi:10.1002/hlca.200390313

Krzmarzick MJ, Crary BB, Harding JJ, Oyerinde OO, Leri AC, Myneni SC, Novak PJ (2012) Natural niche for organohalide-respiring *Chloroflexi*. Appl Environ Microbiol 78:393–401. doi:10.1128/AEM.06510-11

Kube M, Beck A, Zinder SH, Kuhl H, Reinhardt R, Adrian L (2005) Genome sequence of the chlorinated compound-respiring bacterium *Dehalococcoides* species strain CBDB1. Nat Biotechnol 23:1269–1273. doi:10.1038/nbt1131

Lee M, Low A, Zemb O, Koenig J, Michaelsen A, Manefield M (2012) Complete chloroform dechlorination by organochlorine respiration and fermentation. Environ Microbiol 14:883–894. doi:10.1111/j.1462-2920.2011.02656.x

Liang B, Jiang J, Zhang J, Zhao Y, Li S (2012) Horizontal transfer of dehalogenase genes involved in the catalysis of chlorinated compounds: evidence and ecological role. Crit Rev Microbiol 38:95–110. doi:10.3109/1040841X.2011.618114

Löffler FE, Edwards EA (2006) Harnessing microbial activities for environmental cleanup. Curr Opin Biotechnol 17:274–284. doi:10.1016/j.copbio.2006.05.001

Löffler FE, Sun Q, Li J, Tiedje JM (2000) 16S rRNA gene-based detection of tetrachloroethene-dechlorinating *Desulfuromonas* and *Dehalococcoides* species. Appl Environ Microbiol 66:1369–1374. doi:10.1128/AEM.66.4.1369-1374.2000

Löffler FE, Yan J, Ritalahti KM, Adrian L, Edwards EA, Konstantinidis KT, Muller JA, Fullerton H, Zinder SH, Spormann AM (2012) *Dehalococcoides mccartyi* gen. nov., sp. nov., obligate organohalide-respiring anaerobic bacteria, relevant to halogen cycling and bioremediation, belong to a novel bacterial class, *Dehalococcoidetes* classis nov., within the phylum *Chloroflexi*. Int J Syst Evol Microbiol. doi:10.1099/ijs.0.034926-0

Ludwig ML, Matthews RG (1997) Structure-based perspectives on B_{12}-dependent enzymes. Annu Rev Biochem 66:269–313. doi:10.1146/annurev.biochem.66.1.269

Madsen T, Licht D (1992) Isolation and characterization of an anaerobic chlorophenol-transforming bacterium. Appl Environ Microbiol 58:2874–2878

Magnuson JK, Stern RV, Gossett JM, Zinder SH, Burris DR (1998) Reductive dechlorination of tetrachloroethene to ethene by a two-component enzyme pathway. Appl Environ Microbiol 64:1270–1275

Magnuson JK, Romine MF, Burris DR, Kingsley MT (2000) Trichloroethene reductive dehalogenase from *Dehalococcoides ethenogenes*: sequence of *tceA* and substrate range characterization. Appl Environ Microbiol 66:5141–5147. doi:10.1128/AEM.66.12.5141-5147.2000

Maillard J, Schumacher W, Vazquez F, Regeard C, Hagen WR, Holliger C (2003) Characterization of the corrinoid iron-sulfur protein tetrachloroethene reductive dehalogenase of *Dehalobacter restrictus*. Appl Environ Microbiol 69:4628–4638. doi:10.1128/AEM.69.8.4628-4638.2003

Maillard J, Regeard C, Holliger C (2005) Isolation and characterization of Tn-*Dha*1, a transposon containing the tetrachloroethene reductive dehalogenase of *Desulfitobacterium hafniense* strain TCE1. Environ Microbiol 7:107–117. doi:10.1111/j.1462-2920.2004.00671.x

Maillard J, Genevaux P, Holliger C (2011) Redundancy and specificity of multiple trigger factor chaperones in *Desulfitobacteria*. Microbiology 157:2410–2421. doi:10.1099/mic.0.050880-0

Maphosa F, de Vos WM, Smidt H (2010) Exploiting the ecogenomics toolbox for environmental diagnostics of organohalide-respiring bacteria. Trends Biotechnol 28:308–316. doi:10.1016/j.tibtech.2010.03.005

Marco-Urrea E, Gabarrell X, Sarra M, Caminal G, Vicent T, Reddy CA (2006) Novel aerobic perchloroethylene degradation by the white-rot fungus *Trametes versicolor*. Environ Sci Technol 40:7796–7802. doi:10.1021/es0622958

Marco-Urrea E, Aranda E, Caminal G, Guillén F (2009) Induction of hydroxyl radical production in *Trametes versicolor* to degrade recalcitrant chlorinated hydrocarbons. Bioresour Technol 100:5757–5762. doi:10.1016/j.biortech.2009.06.078

Marco-Urrea E, Nijenhuis I, Adrian L (2011) Transformation and carbon isotope fractionation of tetra- and trichloroethene to *trans*-dichloroethene by *Dehalococcoides* sp. strain CBDB1. Environ Sci Technol 45:1555–1562. doi:10.1021/es1023459

Marshall IP, Berggren DR, Azizian MF, Burow LC, Semprini L, Spormann AM (2012) The hydrogenase chip: a tiling oligonucleotide DNA microarray technique for characterizing hydrogen-producing and -consuming microbes in microbial communities. ISME J 6:814–826. doi:10.1038/ismej.2011.136

Marzorati M, Borin S, Brusetti L, Daffonchio D, Marsilli C, Carpani G, de Ferra F (2006) Response of 1,2-dichloroethane-adapted microbial communities to ex-situ biostimulation of polluted groundwater. Biodegradation 17:143–158. doi:10.1007/s10532-005-9004-z

Marzorati M, de Ferra F, Van Raemdonck H, Borin S, Allifranchini E, Carpani G, Serbolisca L, Verstraete W, Boon N, Daffonchio D (2007) A novel reductive dehalogenase, identified in a contaminated groundwater enrichment culture and in *Desulfitobacterium dichloroeliminans* strain DCA1, is linked to dehalogenation of 1,2-dichloroethane. Appl Environ Microbiol 73:2990–2999. doi:10.1128/AEM.02748-06

May HD, Miller GS, Kjellerup BV, Sowers KR (2008) Dehalorespiration with polychlorinated biphenyls by an anaerobic ultramicrobacterium. Appl Environ Microbiol 74:2089–2094. doi:10.1128/AEM.01450-07

Maymó-Gatell X, Chien Y, Gossett JM, Zinder SH (1997) Isolation of a bacterium that reductively dechlorinates tetrachloroethene to ethene. Science 276:1568–1571. doi:10.1126/science.276.5318.1568

Maymó-Gatell X, Anguish T, Zinder SH (1999) Reductive dechlorination of chlorinated ethenes and 1,2-dichloroethane by "*Dehalococcoides ethenogenes*" 195. Appl Environ Microbiol 65:3108–3113

Maymó-Gatell X, Nijenhuis I, Zinder SH (2001) Reductive dechlorination of *cis*-1,2-dichloroethene and vinyl chloride by "*Dehalococcoides ethenogenes*". Environ Sci Technol 35:516–521. doi:10.1021/es001285i

Mazon H, Gabor K, Leys D, Heck AJ, van der Oost J, van den Heuvel RH (2007) Transcriptional activation by CprK1 is regulated by protein structural changes induced by effector binding and redox state. J Biol Chem 282:11281–11290. doi:10.1074/jbc.M611177200

McCauley KM, Pratt DA, Wilson SR, Shey J, Burkey TJ, van der Donk WA (2005) Properties and reactivity of chlorovinylcobalamin and vinylcobalamin and their implications for vitamin B_{12}-catalyzed reductive dechlorination of chlorinated alkenes. J Am Chem Soc 127:1126–1136. doi:10.1021/ja048573p

McMurdie PJ, Behrens SF, Holmes S, Spormann AM (2007) Unusual codon bias in vinyl chloride reductase genes of *Dehalococcoides* species. Appl Environ Microbiol 73:2744–2747. doi:10.1128/AEM.02768-06

McMurdie PJ, Behrens SF, Müller JA, Göke J, Ritalahti KM, Wagner R, Goltsman E, Lapidus A, Holmes S, Löffler FE, Spormann AM (2009) Localized plasticity in the streamlined genomes of vinyl chloride respiring *Dehalococcoides*. PLoS Genet 5:e1000714. doi:10.1371/journal.pgen.1000714

McMurdie PJ, Hug LA, Edwards EA, Holmes S, Spormann AM (2011) Site-specific mobilization of vinyl chloride respiration islands by a mechanism common in *Dehalococcoides*. BMC Genomics 12:287. doi:10.1186/1471-2164-12-287

Miller E, Wohlfarth G, Diekert G (1997) Comparative studies on tetrachloroethene reductive dechlorination mediated by *Desulfitobacterium* sp. strain PCE-S. Arch Microbiol 168:513–519. doi:10.1007/s002030050529

Miller GS, Milliken CE, Sowers KR, May HD (2005) Reductive dechlorination of tetrachloroethene to *trans*-dichloroethene and *cis*-dichloroethene by PCB-dechlorinating bacterium DF-1. Environ Sci Technol 39:2631–2635. doi:10.1021/es048849t

Moe WM, Yan J, Nobre MF, da Costa MS, Rainey FA (2009) *Dehalogenimonas lykanthroporepellens* gen. nov., sp. nov., a reductively dehalogenating bacterium isolated from chlorinated solvent-contaminated groundwater. Int J Syst Evol Microbiol 59:2692–2697. doi:10.1099/ijs.0.011502-0

Morita Y, Futagami T, Goto M, Furukawa K (2009) Functional characterization of the trigger factor protein PceT of tetrachloroethene-dechlorinating *Desulfitobacterium hafniense* Y51. Appl Microbiol Biotechnol 83:775–781. doi:10.1007/s00253-009-1958-z

Müller JA, Rosner BM, Von Abendroth G, Meshulam-Simon G, McCarty PL, Spormann AM (2004) Molecular identification of the catabolic vinyl chloride reductase from *Dehalococcoides* sp. strain VS and its environmental distribution. Appl Environ Microbiol 70:4880–4888. doi:10.1128/AEM.70.8.4880-4888.2004

Nei M, Gojobori T (1986) Simple methods for estimating the numbers of synonymous and non-synonymous nucleotide substitutions. Mol Biol Evol 3:418–426

Neumann A, Scholz-Muramatsu H, Diekert G (1994) Tetrachloroethene metabolism of *Dehalospirillum multivorans*. Arch Microbiol 162:295–301. doi:10.1007/BF00301854

Neumann A, Wohlfarth G, Diekert G (1996) Purification and characterization of tetrachloroethene reductive dehalogenase from *Dehalospirillum multivorans*. J Biol Chem 271:16515–16519. doi:10.1074/jbc.271.28.16515

Neumann A, Wohlfarth G, Diekert G (1998) Tetrachloroethene dehalogenase from *Dehalospirillum multivorans*: cloning, sequencing of the encoding genes, and expression of the *pceA* gene in *Escherichia coli*. J Bacteriol 180:4140–4145

Neumann A, Siebert A, Trescher T, Reinhardt S, Wohlfarth G, Diekert G (2002) Tetrachloroethene reductive dehalogenase of *Dehalospirillum multivorans*: substrate specificity of the native enzyme and its corrinoid cofactor. Arch Microbiol 177:420–426. doi:10.1007/s00203-002-0409-3

Nonaka H, Keresztes G, Shinoda Y, Ikenaga Y, Abe M, Naito K, Inatomi K, Furukawa K, Inui M, Yukawa H (2006) Complete genome sequence of the dehalorespiring bacterium *Desulfitobacterium hafniense* Y51 and comparison with *Dehalococcoides ethenogenes* 195. J Bacteriol 188:2262–2274. doi:10.1128/JB.188.6.2262-2274.2006

Palmer T, Berks BC (2012) The twin-arginine translocation (Tat) protein export pathway. Nat Rev Microbiol 10:483–496. doi:10.1038/nrmicro2814

Peng X, Yamamoto S, Vertès AA, Keresztes G, Inatomi K, Inui M, Yukawa H (2012) Global transcriptome analysis of the tetrachloroethene-dechlorinating bacterium *Desulfitobacterium hafniense* Y51 in the presence of various electron donors and terminal electron acceptors. J Ind Microbiol Biotechnol 39:255–268. doi:10.1007/s10295-011-1023-7

Pop SM, Kolarik RJ, Ragsdale SW (2004) Regulation of anaerobic dehalorespiration by the transcriptional activator CprK. J Biol Chem 279:49910–49918. doi:10.1074/jbc.M409435200

Pop SM, Gupta N, Raza AS, Ragsdale SW (2006) Transcriptional activation of dehalorespiration. Identification of redox-active cysteines regulating dimerization and DNA binding. J Biol Chem 281:26382–26390. doi:10.1074/jbc.M602158200

Prat L, Maillard J, Grimaud R, Holliger C (2011) Physiological adaptation of *Desulfitobacterium hafniense* strain TCE1 to tetrachloroethene respiration. Appl Environ Microbiol 77:3853–3859. doi:10.1128/AEM.02471-10

Ryoo D, Shim H, Canada K, Barbieri P, Wood TK (2000) Aerobic degradation of tetrachloroethylene by toluene-*o*-xylene monooxygenase of *Pseudomonas stutzeri* OX1. Nat Biotechnol 18:775–778. doi:10.1038/77344

Sanford RA, Cole JR, Tiedje JM (2002) Characterization and description of *Anaeromyxobacter dehalogenans* gen. nov., sp. nov., an aryl-halorespiring facultative anaerobic myxobacterium. Appl Environ Microbiol 68:893–900. doi:10.1128/AEM.68.2.893-900.2002

Scholz-Muramatsu H, Neumann A, Meßmer M, Moore E, Diekert G (1995) Isolation and characterization of *Dehalospirillum multivorans* gen. nov., sp. nov., a tetrachloroethene-utilizing, strictly anaerobic bacterium. Arch Microbiol 163:48–56. doi:10.1007/BF00262203

Schumacher W, Holliger C, Zehnder AJ, Hagen WR (1997) Redox chemistry of cobalamin and iron-sulfur cofactors in the tetrachloroethene reductase of Dehalobacter restrictus. FEBS Lett 409:421–425. doi:10.1016/S0014-5793(97)00520-6

Seshadri R, Adrian L, Fouts DE, Eisen JA, Phillippy AM, Methe BA, Ward NL, Nelson WC, Deboy RT, Khouri HM, Kolonay JF, Dodson RJ, Daugherty SC, Brinkac LM, Sullivan SA, Madupu R, Nelson KE, Kang KH, Impraim M, Tran K, Robinson JM, Forberger HA, Fraser CM, Zinder SH, Heidelberg JF (2005) Genome sequence of the PCE-dechlorinating bacterium *Dehalococcoides ethenogenes*. Science 307:105–108. doi:10.1126/science.1102226

Shelton DR, Tiedje JM (1984) Isolation and partial characterization of bacteria in an anaerobic consortium that mineralizes 3-chlorobenzoic acid. Appl Environ Microbiol 48:840–848

Siebert A, Neumann A, Schubert T, Diekert G (2002) A non-dechlorinating strain of *Dehalospirillum multivorans*: evidence for a key role of the corrinoid cofactor in the synthesis of an active tetrachloroethene dehalogenase. Arch Microbiol 178:443–449. doi:10.1007/s00203-002-0473-8

Smidt H, de Vos WM (2004) Anaerobic microbial dehalogenation. Annu Rev Microbiol 58:43–73. doi:10.1146/annurev.micro.58.030603.123600

Smidt H, Song D, van Der Oost J, de Vos WM (1999) Random transposition by Tn*916* in *Desulfitobacterium dehalogenans* allows for isolation and characterization of halorespiration-deficient mutants. J Bacteriol 181:6882–6888

Smidt H, van Leest M, van der Oost J, de Vos WM (2000) Transcriptional regulation of the *cpr* gene cluster in *ortho*-chlorophenol-respiring *Desulfitobacterium dehalogenans*. J Bacteriol 182:5683–5691. doi:10.1128/JB.182.20.5683-5691.2000

Smidt H, van der Oost J, de Vos WM (2001) Development of a gene cloning and inactivation system for halorespiring *Desulfitobacterium dehalogenans*. Appl Environ Microbiol 67:591–597. doi:10.1128/AEM.67.2.591-597.2001

Sung Y, Ritalahti KM, Apkarian RP, Löffler FE (2006) Quantitative PCR confirms purity of strain GT, a novel trichloroethene-to-ethene-respiring *Dehalococcoides* isolate. Appl Environ Microbiol 72:1980–1987. doi:10.1128/AEM.72.3.1980-1987.2006

Suyama A, Iwakiri R, Kai K, Tokunaga T, Sera N, Furukawa K (2001) Isolation and characterization of *Desulfitobacterium* sp. strain Y51 capable of efficient dehalogenation of tetrachloroethene and polychloroethanes. Biosci Biotechnol Biochem 65:1474–1481. doi:10.1271/bbb.65.1474

Suyama A, Yamashita M, Yoshino S, Furukawa K (2002) Molecular characterization of the PceA reductive dehalogenase of *Desulfitobacterium* sp. strain Y51. J Bacteriol 184:3419–3425. doi:10.1128/JB.184.13.3419-3425.2002

Thibodeau J, Gauthier A, Duguay M, Villemur R, Lépine F, Juteau P, Beaudet R (2004) Purification, cloning, and sequencing of a 3,5-dichlorophenol reductive dehalogenase from *Desulfitobacterium frappieri* PCP-1. Appl Environ Microbiol 70:4532–4537. doi:10.1128/AEM.70.8.4532-4537.2004

Thomas SH, Wagner RD, Arakaki AK, Skolnick J, Kirby JR, Shimkets LJ, Sanford RA, Löffler FE (2008) The mosaic genome of *Anaeromyxobacter dehalogenans* strain 2CP-C suggests an aerobic common ancestor to the delta-proteobacteria. PLoS One 3:e2103. doi:10.1371/journal.pone.0002103

Tiehm A, Schmidt KR (2011) Sequential anaerobic/aerobic biodegradation of chloroethenes–aspects of field application. Curr Opin Biotechnol 22:415–421. doi:10.1016/j.copbio.2011.02.003

Tsuda M, Tan HM, Nishi A, Furukawa K (1999) Mobile catabolic genes in bacteria. J Biosci Bioeng 87:401–410. doi:10.1016/S1389-1723(99)80086-3

Tsukagoshi N, Ezaki S, Uenaka T, Suzuki N, Kurane R (2006) Isolation and transcriptional analysis of novel tetrachloroethene reductive dehalogenase gene from *Desulfitobacterium* sp. strain KBC1. Appl Microbiol Biotechnol 69:543–553. doi:10.1007/s00253-005-0022-x

Utkin I, Woese C, Wiegel J (1994) Isolation and characterization of *Desulfitobacterium dehalogenans* gen. nov., sp. nov., an anaerobic bacterium which reductively dechlorinates chlorophenolic compounds. Int J Syst Bacteriol 44:612–619. doi:10.1099/00207713-44-4-612

van de Pas BA, Smidt H, Hagen WR, van der Oost J, Schraa G, Stams AJ, de Vos WM (1999) Purification and molecular characterization of *ortho*-chlorophenol reductive dehalogenase, a key enzyme of halorespiration in *Desulfitobacterium dehalogenans*. J Biol Chem 274:20287–20292. doi:10.1074/jbc.274.29.20287 van de Pas BA, Harmsen HJ, Raangs GC, de Vos WM, Schraa G, Stams AJ (2001) A *Desulfitobacterium* strain isolated from human feces that does not dechlorinate chloroethenes or chlorophenols. Arch Microbiol 175:389–394. doi:10.1007/s002030100276

van der Meer JR, de Vos WM, Harayama S, Zehnder AJ (1992) Molecular mechanisms of genetic adaptation to xenobiotic compounds. Microbiol Rev 56:677–694

Villemur R, Lanthier M, Beaudet R, Lépine F (2006) The *Desulfitobacterium* genus. FEMS Microbiol Rev 30:706–733. doi:10.1111/j.1574-6976.2006.00029.x

Wagner DD, Hug LA, Hatt JK, Spitzmiller MA, Padilla-Crespo E, Ritalahti KM, Edwards EA, Konstantinidis KT, Löffler FE (2012) Genomic determinants of organohalide-respiration in *Geobacter lovleyi*, an unusual member of the *Geobacteraceae*. BMC Genomics 13:200. doi:10.1186/1471-2164-13-200

Waller AS, Hug LA, Mo K, Radford DR, Maxwell KL, Edwards EA (2012) Transcriptional analysis of a *Dehalococcoides*-containing microbial consortium reveals prophage activation. Appl Environ Microbiol 78:1178–1186. doi:10.1128/AEM.06416-11

Wu Q, Milliken CE, Meier GP, Watts JE, Sowers KR, May HD (2002) Dechlorination of chlorobenzenes by a culture containing bacterium DF-1, a PCB dechlorinating microorganism. Environ Sci Technol 36:3290–3294. doi:10.1021/es0158612

Wunsch P, Zumft WG (2005) Functional domains of NosR, a novel transmembrane iron-sulfur flavoprotein necessary for nitrous oxide respiration. J Bacteriol 187:1992–2001. doi:10.1128/JB.187.6.1992-2001.2005

Yamada T, Sekiguchi Y, Hanada S, Imachi H, Ohashi A, Harada H, Kamagata Y (2006) *Anaerolinea thermolimosa* sp. nov., *Levilinea saccharolytica* gen. nov., sp. nov. and *Leptolinea tardivitalis* gen. nov., sp. nov., novel filamentous anaerobes, and description of the new classes *Anaerolineae* classis nov. and *Caldilineae* classis nov. in the bacterial phylum *Chloroflexi*. Int J Syst Evol Microbiol 56:1331–1340. doi:10.1099/ijs.0.64169-0

Yan J, Rash BA, Rainey FA, Moe WM (2009) Isolation of novel bacteria within the *Chloroflexi* capable of reductive dechlorination of 1,2,3-trichloropropane. Environ Microbiol 11:833–843. doi:10.1111/j.1462-2920.2008.01804.x

Ye L, Schilhabel A, Bartram S, Boland W, Diekert G (2010) Reductive dehalogenation of brominated ethenes by *Sulfurospirillum multivorans* and *Desulfitobacterium hafniense* PCE-S. Environ Microbiol 12:501–509. doi:10.1111/j.1462-2920.2009.02093.x

Chapter 5
Mobile Catabolic Genetic Elements in Pseudomonads

Masataka Tsuda, Yoshiyuki Ohtsubo, and Hirokazu Yano

Abstract Bacterial capabilities to degrade various recalcitrant compounds are often encoded on mobile genetic elements (MGEs) such as transposons, plasmids, and integrative and conjugative elements (ICEs). The movement of the transposons and consequently induced rearrangements of genome in a cell and the intercellular transfer of the latter two MGEs greatly facilitate rapid adaptation and evolution of the host cells and wide dissemination of catabolic genes and gene clusters in a variety of phylogenetically distinct environmental bacteria. This chapter summarizes how MGEs participate in the emergence of degradative bacteria based on the earlier and recent findings and analysis of catabolic MGEs in pseudomonads.

Keywords DNA rearrangements • Horizontal gene transfer • Integrative and conjugative elements • Microbial degradation of recalcitrant compounds • Mobile genetic elements • Plasmids • Pseudomonads • Transposons

5.1 Introduction

A number of bacterial strains able to completely degrade various environmental pollutants as the sole sources of carbon and energy have been isolated around the world. The genetic basis underlying the aerobic degradation of recalcitrant aromatic hydrocarbons (e.g., toluene, xylenes, and naphthalene) and their halogenated compounds has been studied in details. Comparative sequence analyses for catabolic genes and their flanking regions have revealed the worldwide distribution of

---

M. Tsuda (✉) • Y. Ohtsubo • H. Yano
Graduate School of Life Sciences, Tohoku University,
2-1-1 Katahira, Sendai 980-8577, Japan
e-mail: mtsuda@ige.tohoku.ac.jp

H. Nojiri et al. (eds.), *Biodegradative Bacteria: How Bacteria Degrade, Survive, Adapt, and Evolve*, DOI 10.1007/978-4-431-54520-0_5,

catabolic gene clusters of common origins in phylogenetically distinct bacterial taxa. DNA segments containing catabolic genes often exhibit structural flexibility such as duplication, deletion, and translocation. It is now known that these features of catabolic gene clusters result from their presence near or within mobile genetic elements (MGEs) such as insertion sequences (ISs) and transposons. Furthermore, catabolic gene clusters are often located within transferable MGEs such as plasmids and genomic islands. This allows the horizontal transfer of the catabolic gene clusters to other bacteria, thus facilitating their rapid adaptation to the polluted environments.

Bacterial MGEs encoding other phenotypic traits (e.g., resistance to various antibiotics and heavy metals, pathogenicity, and symbiosis) are also recognized as major agents driving rapid adaptation and the evolution of host cells (Frost et al. 2005; Sørensen et al. 2005; Thomas and Nielsen 2005; Norman et al. 2009). This chapter summarizes how catabolic gene clusters are assembled in representative MGEs and also addresses the molecular mechanisms underlying their intracellular and intercellular movements. Since many of previously published reviews have focused on bacterial catabolic MGEs (Tsuda et al. 1999; Top and Springael 2003; van der Meer and Sentchilo 2003; Nojiri et al. 2004; Kivisaar 2004; Williams et al. 2004; Dennis 2005; Sota et al. 2008; Heuer and Smalla 2012; Shintani and Nojiri 2013), the present chapter emphasizes the progress of our recent studies.

5.2 Catabolic Plasmids of the Genus *Pseudomonas*

At the beginning of 1970s, several plasmids were discovered that encode the complete pathway enzymes for recalcitrant aromatic compounds such as toluene/xylenes, naphthalene, and 2,4-dichlorophenoxyacetic acid (2,4-D). Since then, a number of plasmids involved in the catabolism of various recalcitrant compounds from natural and man-made origins, collectively designated "catabolic plasmids," have been reported (Shintani and Nojiri 2013). Many of these plasmids were found in pseudomonad (the γ-proteobacterial genus *Pseudomonas* and its related genera, some of which now belong to taxonomically different classes such as *Alpha*- and *Betaproteobacteria*). One of traditionally convenient methods for classifying plasmids is based on incompatibility, a phenomenon in which two different plasmids with very similar replication/maintenance systems cannot coexist in the same cell, such that either one of plasmids will randomly and mutually excludes the other from the cell. The plasmids capable of residing in *Pseudomonas* cells have been classified into at least 14 incompatibility (Inc) groups (Thomas and Haines 2004), and the associated catabolic plasmids thus far analyzed belong to the IncP-1, P-2, P-7, and P-9 groups (Sota et al. 2008; Shintani and Nojiri 2013). In general, there is a correlation between the replication-host range of plasmids and the corresponding Inc groups. The IncP-1 plasmids are maintained in the phylum *Proteobacteria*, while the IncP-2 and IncP-7 plasmids can be maintained only in *Pseudomonas* (Durland and Helinski 1978; Hansen and Olsen 1978; Yano et al. 2010; Yano et al. 2012).

The IncP-9 catabolic plasmids can be additionally maintained in at least *Escherichia coli*, but they exhibit temperature sensitivity for their maintenance in this host (Tsuda et al. 1989; Tsuda and Iino 1990).

5.3 Catabolic Transposons

Microbial transposons can be defined as discrete DNA fragments capable of moving from one site to another without the requirement of genetic homology between donor and target sites (Chandler and Mahillon 2002; Grindley 2002; Roberts et al. 2008). The transposase (Tnp or TnpA) encoded by a transposon catalyzes specific and single- or double-stranded cleavage of the transposon ends at the donor site and nonspecific and double-stranded cleavage at the target site. Cleavage at the latter site generally produces a short (2–20 bases, depending on the transposons) single-stranded protrusion at the 5′ end. Then, the transfer of the cleaved transposon to such a target site, followed by the filling-in of the cohesive ends and subsequent ligation, gives rise to the duplication of target sequence so as to flank the inserted transposon. Most transposons carry terminal inverted repeat (IR) sequences with sizes of more than 25 bp and less than 50 bp, depending on the transposons, and such sequences are crucial for recognition and DNA cleavage by the transposases. Single- and double-stranded cleavage of transposon ends at the donor site prior to the transfer of transposon DNA to the target site defines the replicative and non-replicative modes, respectively, of transposition, and most transposons prefer only one of the two modes. The transposases mediate both inter- and intramolecular transpositions. Intramolecular transposition of replicative transposons results in either deletion or inversion of a replicon where the transposon is located. This accounts for the duplication or deletion of catabolic genes in naturally occurring plasmids (Król et al. 2012; Sen et al. 2011).

Well-characterized bacterial transposons have been classified into several classes, as based on the organization of transposition-related genes and sites and on the transposition mechanisms. Three classes of transposons carrying the genes for the catabolism of recalcitrant compounds (catabolic transposons) have been reported thus far: classes I, class II, and conjugative transposons. The class I transposons carrying the catabolic genes (Sota et al. 2008; Shintani and Nojiri 2013) are flanked by two (nearly) identical copies of an insertion sequence (IS) with direct or inverted orientation. Most of such class I catabolic transposons are less than 15 kb in sizes and only rarely does each of these transposons carry all the genes necessary for complete catabolism. This insufficiency suggests the accidental formation of the class I catabolic transposon by the insertion of two copies of an IS element in the vicinity of the catabolic gene cluster. Nonetheless, IS elements play important roles in the catabolic gene activation and the amplification of copy number of catabolic genes (e.g., Devers et al. 2008).

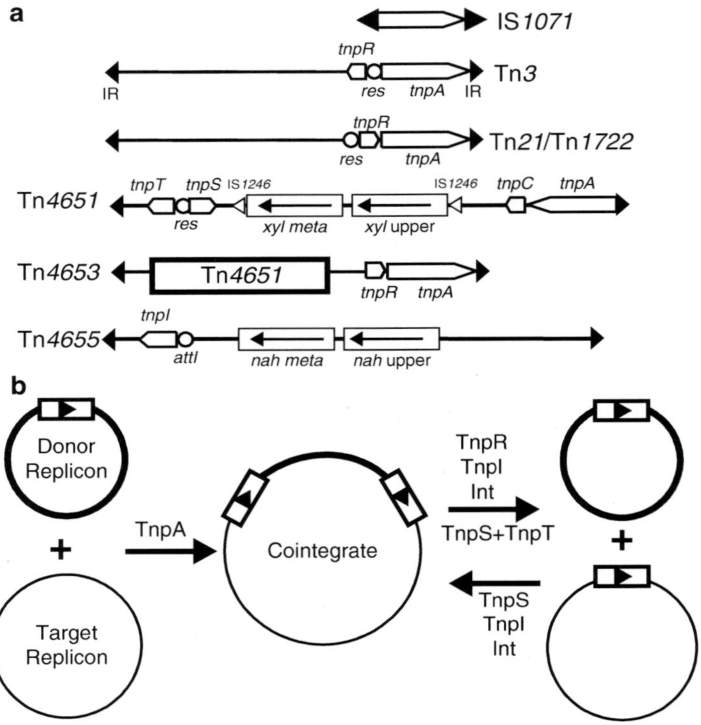

Fig. 5.1 Representative class II transposons and their transposition. (**a**). Structures of representative groups of class II transposons and a few class II catabolic transposons. *Closed triangles* and *circles* indicate the terminal IRs and the specific recombination sites for resolution, respectively. (**b**). Two-step transposition mechanism of class II transposons. *Box* and *triangle* indicate the transposon and the specific recombination site for resolution, respectively. See the text for details

Class II catabolic transposons are usually large (>37 kb in sizes), and each of these transposons governs the complete catabolic pathway (e.g., Fig. 5.1a) (Tsuda et al. 1999; Sota et al. 2008). The class II transposons, which also carry a number of genes for resistance to heavy metals (e.g., mercury) and various antibiotics, transpose by the replicative mode; in the intermolecular transposition, the transposase (TnpA) generates the cointegrate of the donor and target molecules connected by directly repeated copies of transposon, one at each junction (Fig. 5.1b) (Grindley 2002). Most class II transposons additionally encode, in the vicinity of *tnpA* genes, site-specific recombination systems consisting of site-specific recombinases (resolvases) and their target resolution (*res*) sites. Many such recombinases have a conserved serine residue at the N-terminal region, and the class II transposon-encoded serine recombinases (TnpR proteins) can generally carry out only resolution, but not inversion or intermolecular recombination. Several class II transposons encode tyrosine recombinases (TnpI or TnpS protein) (Rajeev et al. 2009), which have a conserved tyrosine residue at the C-terminal region, for the resolution of TnpA-mediated cointegrates at particular *att* sequences. Tyrosine recombinases can mediate both intramolecular and intermolecular recombination, but the so-called accessory sequences in their respective resolution sites are present to coordinate the direction of recombination such that resolution occurs most efficiently (Vanhooff et al. 2006).

The TnpA and TnpR functions encoded by different class II transposons often exhibit mutual interchangeability, indicating the phylogenetic relationship between transposons (Grindley 2002). These transposons can therefore be divided into several groups on the basis of the functional interchangeability of TnpR proteins (e.g., Tn*3* and Tn*21* groups); moreover, the functional interchangeability of TnpA proteins allows further subdivision of the latter group into Tn*21* and Tn*1722* subgroups (Fig. 5.1a) (Liebert et al. 1999).

5.4 Catabolic Integrative and Conjugative Elements (ICEs)

Some catabolic transposons are the members of conjugative transposons, a group of ICEs (van der Meer and Sentchilo 2003; Juhas et al. 2009; Toleman and Walsh 2011). The intact and functional ICE encodes a set of genes for tyrosine recombinase (Int), its auxiliary protein [recombination directionality factor (RDF) represented by excisionase (Xis)], and intercellular conjugal transfer. Such an ICE can be excised at its ends (*attL* and *attR*) from the donor site by the Int and RDF functions (Rajeev et al. 2009). The excised molecule is circularized, nicked at an *oriT* site, and the single-stranded DNA synthesized by the rolling circle-type replication is transferred to another cell by the ICE-encoded conjugal machinery. The transferred DNA in the recipient is then converted to double-stranded DNA and thereafter is inserted into the recipient genome by the Int function. Conjugative transposons can be distinguished from the other groups of ICEs by their capacity to be inserted into random target sites.

5.5 TOL and NAH Plasmids

A group of plasmids, collectively designated TOL plasmids, are primarily from *Pseudomonas* strains, and they encode a series of enzymes for the complete and aerobic degradation of toluene and its methyl derivatives, xylenes (Fig. 5.2a) (Sota et al. 2008). These compounds are dioxygenated and thereafter converted to benzoate or its methyl derivatives by several enzymes produced from catabolic *xyl* genes constituting the *xyl* upper pathway operon. The intermediates are then converted to catechol or its methyl derivatives and then to the central metabolites by enzymes encoded by the *xyl meta* pathway operon. Expression of the two operons is transcriptionally regulated positively by the *xylR* and *xylS* products, respectively.

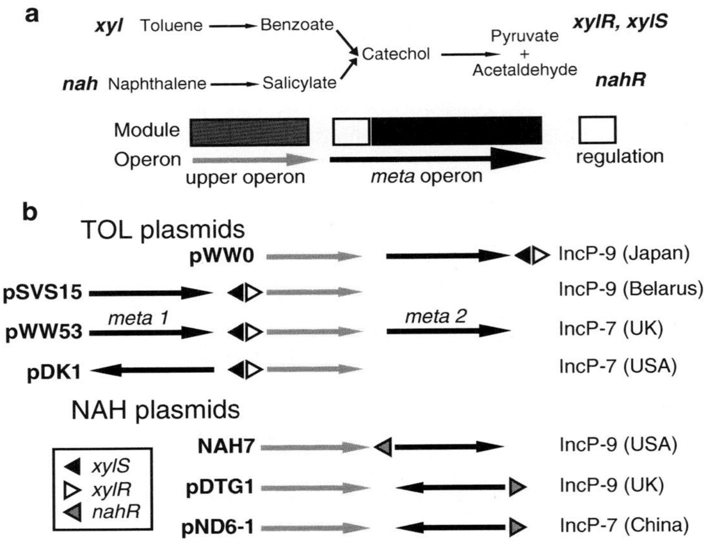

Fig. 5.2 Bacterial TOL and NAH plasmids and their specified aerobic degradation of toluene and naphthalene. (**a**). Simplified degradation pathways of toluene and naphthalene and their responsible gene clusters. Transcriptional regulatory genes are indicated at the *right side*. (**b**). Various TOL and NAH plasmids described in this chapter. Only the catabolic gene clusters are depicted for simplicity. At the *right side* of each plasmid are indicated its Inc group and the country, where the host strain was isolated. See the text for details. Plasmid pSVS15 is described in Sentchilo et al. (2000)

The gene organization of *xyl* genes in each operon is conserved among various TOL plasmids although they exhibit the structural diversity in terms of the following: sizes (from 78 to 270 kb), plasmid backbones, and the copy numbers and relative orientations of the four *xyl* gene transcriptional units (Fig. 5.2b) (Williams and Worsey 1976; Chatfield and Williams 1986; Sentchilo et al. 2000).

A group of plasmids governing the complete and aerobic degradation of naphthalene are designated NAH plasmids (Fig. 5.2) (Sota et al. 2008). The naphthalene-catabolic (*nah*) genes are organized into three transcriptional units: the *nah* upper operon for the conversion of naphthalene to salicylate, the *nah meta* operon for the conversion of salicylate to catechol and then to the central metabolites, and *nahR* for the transcriptional activation of the two operons. It is noteworthy that the *xyl* and *nah meta* operons share the evolutionarily common set of genes involved in the conversion of catechol to the central metabolites (Fig. 5.2a). NAH plasmids also exhibit the structural diversity with sizes ranging from 75 to 200 kb (Fig. 5.2b) (Dunn and Gunsalus 1973; Izmalkova et al. 2006; Ono et al. 2007; Sevastsyanovich et al. 2008).

5.6 An Archetypal IncP-9 TOL Plasmid pWW0 and Its Catabolic Transposons

A self-transmissible and 117-kb IncP-9 plasmid, pWW0, from *P. putida* mt-2 (Williams and Murray 1974; Nakazawa 2002) is the best-characterized TOL plasmid whose *xyl* genes have been analyzed extensively together with their transcriptional regulation (Ramos et al. 1997; Williams et al. 2004; Riuz et al. 2004). Such pioneer studies were greatly advanced by use of in vivo-constructed IncP-1 plasmid derivatives with the insertion of all of the *xyl*-containing fragments from pWW0 (Lehrbach et al. 1982). At least two such fragments were later identified as transposons, a 56-kb transposon, Tn*4651*, and its encompassing 70-kb transposon, Tn*4653* (Fig. 5.1a) (Tsuda and Iino 1987, 1988). The *xyl* gene clusters in Tn*4651* are located within a 39-kb segment that carries, at its extremities, two directly repeated copies of IS*1246*, and the reciprocal recombination between the two copies gives rise to an internal deletant of Tn*4651* to generate Tn*4652* (Jeenes and Williams 1982; Tsuda and Iino 1987). The experimental results from which the IS*1246*-flanked segment behaves as a composite transposon have not yet been obtained.

The replicative transposition of Tn*4651* and the phylogenetic relationship of its TnpA protein with those of other transposases led to the classification of Tn*4651* as a class II transposon (Tsuda and Iino 1987; Tsuda et al. 1989). However, the organization of genes and sites for the Tn*4651* transposition significantly differs from those of typical class II transpositions (Fig. 5.1a). The *tnpA* product shows moderate similarity (20–24 % identity) with other transposases of the class II transposons, and this gene is transcribed in an inward direction from a 46-bp IR (Hõrak and Kivisaar 1999). A series of studies by Kivisaar's group clarified the regulation of the Tn*4652* transposase activity in *P. putida* cells (Kivisaar 2004). A 120-amino acid product, TnpC, that is encoded just downstream of *tnpA*, positively controls the amount of TnpA at the posttranscriptional level, although the detailed mechanism for such control remains unclear (Hõrak and Kivisaar 1999). The Tn*4651* transposition is apparently a stationary phase-specific event, and the transcription from the Tn*4651 tnpA* promoter is indeed controlled positively by the stationary-specific sigma factor, RpoS (Ilves et al. 2001). A *P. putida* nucleoid-associated protein (NAP), IHF (integration host factor), is required for the transposition (Hõrak and Kivisaar 1998). This positive regulation by IHF is due to the transcriptional activation of the *tnpA* gene and the facilitation of TnpA binding to the IRs. The specific binding of IHF to the *tnpA* promoter region and IRs might result in the conformational change of the DNA regions to promote transcription and the cointegration reaction, respectively (Teras et al. 2000). Higher amounts of RpoS as well as IHF in *P. putida* stationary-phase cells provide favorable conditions for Tn*4651* transposition (Ilves et al. 2004). In contrast, another *P. putida* NAP, Fis (factor for inversion stimulation), exerts a negative effect on Tn*4651* transposition (Teras et al. 2009). Fis out-competes IHF from the *tnpA*-proximal end of Tn*4651*, leading to the abolishment of TnpA binding to the IR. Therefore, the Tn*4651* transposition in *P. putida* is regulated by relative amounts of the two sequence-specific NAPs that are encoded by the chromosome.

The Tn*4651*-specified resolution system is unique in that (1) its efficient resolution requires the two gene products, TnpS and TnpT, and (2) the 2.4-kb *tnpT-res-tnpS* region is 48 kb apart from the *tnpA* gene (Fig. 5.1a) (Tsuda and Iino 1987; Tsuda et al. 1989). [Although the original term of "*res*" (Tsuda and Iino 1987) was changed to "*rst*" in our recent article (Yano et al. 2013), the original term is used in this review.] TnpS belongs to the site-specific tyrosine recombinase family and shows structural similarity to Cre resolvase rather than phage integrases (Genka et al. 2002). In contrast, TnpT does not show any significant amino acid sequence similarity with other proteins directly or indirectly involved in recombination. The resolution of the Tn*4651*-mediated cointegrate requires both TnpS and TnpT (Yano et al. 2013), while the intermolecular recombination at *res* does not seem to be significantly dependent on TnpT (Fig. 5.1b) (Genka et al. 2002). In this sense, TnpT is an RDF. However, the predicted secondary structure of TnpT is unique when compared with those of other RDFs. The fully functional *res* region in Tn*4651* is located within a 136-bp segment, which contains a 33-bp inverted repeat (core site) and 20-bp inverted repeat. Strand exchange takes place within the core site. TnpT specifically binds to the 20-bp inverted repeat. The efficient resolution of the Tn*4651*-mediated cointegrate also requires two sequences that flank the core site. In other known resolution systems, the "accessory" sequence is located at only one side of the core site. Therefore, the Tn*4651* resolution system is unique, and our preliminary study (unpublished data) suggests the requirement of (an) unknown host protein(s) for resolution.

Three other Tn*4651*-like and experimentally transposable elements, collectively designated the Tn*4652* group of class II transposons (Grindley 2002), have been reported: a mercury-resistance transposon, Tn*5041*, from *Pseudomonas* sp. (Kholodii et al. 1997); a carbazole-catabolic transposon, Tn*4676*, from *P. resinovorans* IncP-7 plasmid pCAR1 (Maeda et al. 2003); and a cryptic transposon, Tn*4661*, from a *P. aeruginosa* IncP-7 drug-resistance plasmid Rms148 and from a *P. aeruginosa* C pathogenicity island PAGI-4 (Yano et al. 2013). The presence in the *P. aeruginosa* PA14 chromosome of a Tn*4651*-like sequence showing very high similarity with the Tn*4661* part from its one IR to the *res* site suggests that the *res* sites in the Tn*4652* group work as hot spots for TnpS-mediated site-specific recombination, leading to DNA shuffling among related MGEs. The *tnpST* modules lacking the *tnpAC* modules are located on the plasmids and chromosomes of *Pseudomonas* and *Xanthomonas*, and the former type of module might play an additional role in the stable maintenance of plasmids and chromosomes in the *Pseudomonas*-related genera (Yano et al. 2013).

Tn*4653* is a class II transposon and encodes its own cointegration system (Fig. 5.1a). Its 38-bp IRs and *tnpA* gene show very extensive sequence similarity with those of the Tn*1722*-subgroup of transposons (Tsuda et al. 1989). The Tn*4653* and Tn*1722* TnpA proteins indeed exhibit functional interchangeability. Just upstream of the Tn*4653 tnpA* gene is *tnpR*, a gene encoding a site-specific serine recombinase, which can complement the *tnpR* mutation of Tn*1722*. However, the cointegrate formed by the Tn*4653* derivative devoid of the Tn*4651* resolution system was unable to be resolved by supply in *trans* of the Tn*1722 tnpR* gene.

This nonreciprocal mode of complementation is due to the disruption of the ancestrally "intact" *res* site of Tn*4653* (Fig. 5.1a) (Tsuda et al. 1989). The transposons in the Tn*402*/Tn*5053* group of class II transposons are preferentially inserted into the *res* sites of the Tn*1722* subgroup of transposons (Minakhina et al. 1999), and complete sequence determination of pWW0 (Greated et al. 2002) revealed that the insertion of a Tn*5053*-type transposon into the "intact" *res* site in the Tn*4653* ancestor led to the loss of *res* function (Sota et al. 2008).

5.7 An Archetypal IncP-9 NAH Plasmid NAH7 and Its Catabolic Transposon

The best-characterized plasmid among the NAH plasmids is NAH7, which is a self-transmissible and 82-kb IncP-9 plasmid from *P. putida* G7 (Dunn and Gunsalus 1973). All the *nah* genes for the degradation of naphthalene are loaded on a 39-kb defective transposon, Tn*4655*, which is activated only when the Tn*4653* or Tn*1722* TnpA protein is expressed in the same cell (Fig. 5.1a) (Tsuda and Iino 1990). Although Tn*4655* has 38-bp IRs that are very similar to those of Tn*4653* and Tn*1722*, the *res-tnpR-tnpA* region is missing in the element. Tn*4655* instead encodes a unique site-specific resolution system, TnpI-*attI*. TnpI is a phage integrase-like tyrosine recombinase, and it can mediate the resolution of the Tn*4655*-mediated cointegrate as well as the intermolecular recombination between two molecules, each carrying one *attI* copy (Sota et al. 2006b). No *tnpI-attI* cluster has been found in any of the other groups of transposons. However, the homologous *tnpI-attI* cluster is located near the *nah* operons on three other NAH plasmids, pDTG1 (IncP-9), pNAH20 (IncP-9), and pND-1 (IncP-7) (Fig. 5.2b) (Dennis and Zylstra 2004; Li et al. 2004; Heinaru et al. 2009). These three NAH plasmids lack the class II catabolic transposons. Therefore, the TnpI-*attI* system might have originally played a role in the site-specific insertion of *nah* operon(s) into appropriate replicons rather than the resolution of the class II transposon-mediated cointegrate. It is at present unpredictable how the two *nah* operons and the *tnpI-attI* cluster are merged into the ancestral form of Tn*1722* to generate Tn*4655*.

Sequence comparison of three self-transmissible IncP-9 catabolic plasmids, pWW0, NAH7, and pDTG1, revealed a 39-kb backbone structure of IncP-9 plasmids (Sota et al. 2006b). pNAH20 with a size of 83 kb is considered to be a close relative of pDTG1 (Heinaru et al. 2009). The backbone mainly encodes the basic plasmid functions for replication, maintenance, and conjugation. The structural diversification of the three plasmids must have occurred by the independent insertion of class II catabolic transposons or catabolic gene clusters at different and dispensable positions in the ancestral IncP-9 plasmid backbone (Sota et al. 2006b). A comprehensive analysis of various IncP-9 NAH plasmids also suggested independent insertion events of catabolic gene clusters or transposons into an ancestral IncP-9 plasmid rather than divergent evolution from a single catabolic and ancestral plasmid (Sevastsyanovich et al. 2008).

pWW0 and NAH7 can be maintained in *P. aeruginosa* and *E. coli* at 30 °C but not at 42 °C (Tsuda and Iino 1988, 1990). Such plasmid phenotypes can facilitate the escape of catabolic transposons from the "fragile" replicons to other co-residing and "robust" replicons. These two plasmids, as well as other self-transmissible plasmids, carry the genes for DNA transfer and replication (Dtr) and mating pair formation (Mpf) systems. The former system governs the formation of the relaxase-single-stranded DNA (ssDNA) complex at the *oriT* site for initiating the rolling circle-type and conjugation-specific DNA replication, while the latter system is involved in the formation of the type IV secretion apparatus for the transfer of the relaxase-ssDNA complex into the recipient cell. Each *dtr* operon in broad-host-range plasmids is transcribed from its *oriT*-containing region, from which another operon with three genes is also transcribed divergently. The definite function of the latter operon remains unknown; however, our analysis of such an operon (*traD* operon) of NAH7 revealed that the three gene products are not essential to the conjugation but yet are required for efficient conjugation (Miyazaki et al. 2008). The most interesting finding is that the NAH7 derivative with a deletion mutation of the last gene (*traF*) in the operon is proficient in conjugal transfer from *Pseudomonas* to *Pseudomonas* and *E. coli* and from *E. coli* to *E. coli*, but is deficient in conjugal transfer from *E. coli* to *Pseudomonas*. This indicates that the *traF* product apparently works as a host-range modifier in the conjugal transfer of NAH7. It is able to obtain the suppressor mutants of *P. putida* that are able to receive the NAH7Δ*traF* derivative by conjugation from *E. coli*, and one of these identified genes globally controls the physiological states of the cell (Inoue et al. 2013). A detailed analysis of such a gene and elucidation of the function of the *traF* gene product will contribute to clarifying the host range of plasmid-specified conjugation systems.

5.8 IncP-7 TOL Plasmids

Plasmid pWW53 is a non-transmissible IncP-7 TOL plasmid from *P. putida* MT53 that was isolated in Great Britain (Keil et al. 1985a). Although pWW53 carries *xyl* gene clusters that are highly similar to those of pWW0, its unique property is the presence of two structurally homologous but functionally distinguishable *xyl meta* operons that flank an *xyl* upper operon (Fig. 5.2b) (Gallegos et al. 1997). The *xyl* containing regions on pWW53 could translocate to other plasmids such as RP4 (Keil et al. 1987), and one such region covering one set of upper and *meta* operons inserted in RP4 has been proven to be a 37-kb functional transposon, Tn*4656*, as a member of the Tn*1722* subgroup (Fig. 5.3a) (Tsuda and Genka 2001). The complete determination of the 108-kb genome of pWW53 revealed a complicated structure with two sets of *res-tnpR-tnpA* clusters and five IRs (IR1 to IR5) of class II transposons (Fig. 5.3a) (Yano et al. 2007). All of the *xyl* genes as well as the fragment governing the replication and partition systems are located within an 86-kb transposon, Tn*4657*. This transposon encompasses three class II transposons. The first transposon is Tn*4658*, a 5.7-kb and Tn*1722*-subgroup transposon that

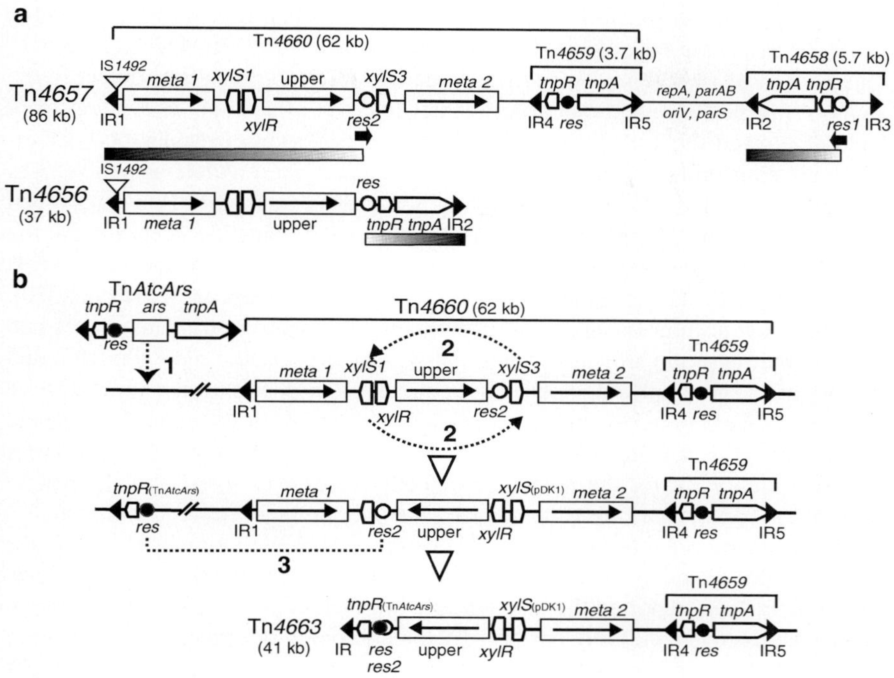

Fig. 5.3 Schematic structures of pWW53 and pDK1 with emphasis on class II transposons and rearrangements of two plasmids. (**a**). Structure of Tn*4657* on pWW53 (*top figure*). Site-specific inversion between *res1* and *res2* by the Tn*4658*-specified TnpR protein gives rise to Tn*4656* (*bottom figure*). (**b**). Structure of Tn*4663* on pDK1 (*bottom figure*) and a three-process model for the generation of Tn*4663* from the Tn*4660*-containing pWW53-like ancestral plasmid. The first two processes with unknown order are (1) the insertion of Tn*AtcArs*-like transposon in the ancestral plasmid and (2) the intramolecular inversion of the upper operon-containing region between *xylS1* and *xylS3*. The last process is the site-specific deletion of the *meta* 1-containing region between the *res2* site and the Tn*AtcArs*-specified *res* site by the TnpR protein encoded by the latter transposon

shares one IR (IR3) with Tn*4657*. The second one is an internally located 3.7-kb transposon, Tn*4659*, whose transposition-related genes and site are organized in the order of *tnpR-res-tnpA*. The last one is a 62-kb and *xyl*-containing transposon, Tn*4660*, which shares IRs with the *tnpA*-proximal IR of Tn*4659* (IR5) and the Tn*4658*-distal IR of Tn*4657* (IR1). Tn*4656* per se is not found on the determined genome of pWW53, and it consists of the two separate and inversely oriented segments of pWW53: the 3.7-kb and IR2-*tnpA-tnpR-res* (*res1*) segment from Tn*4658* and the 33-kb and IR1-*meta* 1 operon-upper operon-*res* (*res2*) segment from Tn*4657*. The Tn*4658*-encoded TnpR can mediate the inversion of the 50-kb segment located between the two inversely oriented *res* sites (*res1* and *res2*) to give rise to Tn*4656* in the MT53 population (Fig. 5.3a) (Yano et al. 2007). Such an inversion event mediated by the class II-encoded serine recombinase has not been detected in other naturally occurring plasmids. The Tn*4659* transposase can only mediate its own transposition among the pWW53-loaded transposons, while the Tn*4658* transposase can catalyze the transposition of Tn*4656*, Tn*4657*, and Tn*4658*. An insertion of IS*1492* in the Tn*4659*-distal IR of Tn*4660* results in defect in the Tn*4660* transposition, and our experimental removal of such an IS element from Tn*4660* allowed the Tn*4658* TnpA-mediated transposition of the Tn*4660* derivative with concomitant generation of target duplication. Tn*4660* on pWW53 is flanked by target duplication, suggesting a previous transposition event. When these observations are taken into consideration, one plausible scenario for the establishment of the present structure of pWW53 is sequential transposition events of Tn*4660* and Tn*4658* into the 76-kb IncP-7 backbone sequence (Yano et al. 2010). Therefore, the accumulation of multiple copies of transposition-related genes and sites on one plasmid can lead to their subsequent site-specific recombination and the formation of complex embedded structures of transposons. Such structures are also exemplified by Tn*510* and Tn*511* on IncP-1 plasmids, R906 and R772, respectively (Petrovski and Stanisich 2011). The DNA shuffling among transposons is obviously important for the revival of defective transposons and aids the persistence of the transposon core genes "*res-tnpR-tnpA*" in the bacterial population.

The structural organization of the *xyl* operons on pWW53 may be favorable for the host under selective conditions, but unfavorable under nonselective conditions. Keil et al. (1985a) reported the deletion of a segment containing the upper operon under nonselective condition, possibly by the homology-dependent recombination between the two *meta* operons. The evolutionary advantage of carrying two similar *meta* operons is currently unclear. But we can speculate that one of the two *xyl meta* operons can divergently evolve (or has already evolved) to be a novel operon that encodes the enzymes capable of degrading new substrates. It should be noted that the pWW53-type *xyl*-operon structure is not exceptional in naturally occurring TOL plasmids. This is exemplified by the TOL plasmids in *Pseudomonas* strains from various oil-contaminated sites (Keil et al. 1985b; Sentchilo et al. 2000). The Inc groups of such plasmids remain unknown.

Another IncP-7 TOL plasmid, pDK1, resides in *P. putida* HS1 from the United States (Kunz and Chapman 1981; Shaw and Williams 1988). This 129-kb plasmid differs from pWW53 in its self-transmissibility and possession of only one set of *xyl* upper and *meta* operons (Fig. 5.3b) (Yano et al. 2010). The gene products of the respective operons show remarkable similarity with those of the upper and *meta* 2 operons in pWW53, but the relative orientations of the two operons differ between the two plasmids (Fig. 5.2b). pDK1 carries three apparently intact class II transposons, Tn*4659*, Tn*4662*, and Tn*4663*. Tn*4662*, with a size of 7.2 kb, is located far from the *xyl* gene clusters and is considered as a Tn*5501*-related transposon in the class II transposons. Tn*4663*, with a size of 41 kb, carries the two *xyl* operons and is transposable to RP4 (Shaw and Williams 1988). This transposon has three copies of Tn*21*-like 38-bp IRs, one of which is shared by the nested Tn*4659* element. The Tn*4659*-encoded and Tn*21*-like transposase must have been involved in the transposition of Tn*4663*. The Tn*4659*-distal end of Tn*4663* is occupied by a 0.7-kb sequence very similar to that of the *res-tnpR*-IR cluster from a Tn*3*-like arsenate-resistance transposon, Tn*AtcArs*. On the basis of further detailed analysis of pDK1 and comparison with pWW53, the following three-step recombination event has been proposed for the formation of *xyl* gene clusters on pDK1 from the *xyl* containing region on an ancestor of pWW53 (Fig. 5.3b) (Yano et al. 2010). The first two steps, in unknown order, are the transposition of Tn*AtcArs* into the pDK1 ancestor with the pWW53-type *xyl* cluster and the homology-dependent intramolecular inversion between very similar *xyl*S genes (*xylS1* and *xylS3*); the last step is the site-specific resolution between the *res2* and *res* sites in the Tn*AtcArs*-like transposon by the resolvase encoded by this transposon. This plausible model is considered as another example by which the insertion events of more than two different class II transposons into one replicon play an important role in the structural diversification of catabolic plasmids and transposons.

The presence of *xyl* and *nah* operons of the same evolutionary origins on IncP-7 and IncP-9 plasmids strongly suggests the frequent exchange of catabolic operons between the two Inc groups of plasmids. This is reinforced by the finding that even the IncP-7 replicon itself is transposable as a part of a transposon (Fig. 5.3a) (Yano et al. 2007). In fact, a NAH plasmid containing fused replicons of the two Inc groups has been reported (Izmalkova et al. 2006).

Although a mini-replicon of pDK1 encoding its replication system was able to replicate in four *Pseudomonas* strains tested, as was also the case with two other IncP-7 plasmids (pWW53 and pCAR1), the successful conjugal transfer of pDK1 was only detected when a *P. putida* HS1 derivative cured of pDK1 and a *P. fluorescens* strain were used as recipient strains, but this was not observed with the use of two other *P. putida* strains as the recipient strains (Yano et al. 2010). These findings indicated that the host range of pDK1, contrary to that of many other plasmids, is restricted by the conjugation system rather than by the replication system. The molecular mechanisms defining this unique property of pDK1 are still unclear.

5.9 IncP-1 Catabolic Plasmids and IS*1071*

Bacterial plasmids responsible for the complete degradation of various "man-made" recalcitrant compounds belong to the broad-host-range and self-transmissible IncP-1 plasmids (especially IncP-1β plasmids, one subgroup of IncP-1 plasmids classified on the basis of the phylogenetic relationship of replication machinery systems) (Top and Springael 2003). The IncP-1β catabolic plasmids have been well characterized, and such examples are pJP4 (for the degradation of 2,4-D) (Trefault et al. 2004), pUO1 (haloacetates) (Sota et al. 2003), and pADP-1 (atrazine) (Martinez et al. 2001). The detailed comparison of these plasmids revealed that the phenotypic diversity of IncP-1β plasmids primarily results from the insertion of various gene clusters at two distinct regions on the 41-kb IncP-1β backbone: one cluster with catabolic genes in the Tn*21*-type transposon or its remnant between *oriV* and *trfA* and the other with various phenotypic genes in Tn*402*-type transposon or its remnant between *parA* and the *traI* operon (Dennis 2005; Sota et al. 2007, 2013).

Various recalcitrant compound-degrading genes on the above described IncP-1β plasmids and on the genomes of several environmental β-proteobacterial strains are located within IS*1071*-composite transposons or in the vicinity of IS*1071* (Sota et al. 2002, 2008; Król et al. 2012). This suggests that IS*1071* plays important roles in the recruitment of catabolic genes into IncP-1β plasmids. Although there was initially little evidence for the actual involvement of IS*1071* transposition in the insertion of catabolic genes in the plasmids, clear evidence was recently provided in the case of an IS*1071*-composite transposon carrying a 3-chloroaniline oxidation gene cluster (Tn*6063*) on an IncP-1β plasmid, pWLD7 (Król et al. 2012).

Regardless of its nomenclature, IS*1071* encodes a transposase classified as that of a member of class II transposons (Sota et al. 2006a). However, the unique properties of IS*1071* [the much longer (110 bp) size of IRs, the low (20 %) similarity of TnpA with those of the other class II transposons, and the complete absence of resolution system] indicate that this MGE is phylogenetically distinct from the other canonical members in the class II transposons. The final product of IS*1071* transposition is indeed a cointegrate of donor and target molecules connected by two copies of IS*1071*. Such a cointegrate can be resolved to the donor molecule and the IS*1071*-inserted target molecule by the homology-dependent intramolecular recombination system specified by the host chromosome. It is of interest why the IS*1071* elements thus far identified are (1) located close to genes involved in the degradation of various recalcitrant compounds and (2) transposable in *Betaproteobacteria* but not in *Alpha-* or *Gammaproteobacteria* (Sota et al. 2006a).

5.10 ICEs Carrying Catabolic Genes

The proteobacterial *bph* gene clusters responsible for the complete degradation of PCB/biphenyl are in most cases located on the chromosomes. Among such various *bph* gene clusters, those from three different proteobacterial strains were reported to be located on ICEs. The first example is from *Cupriavidus oxalaticus* A5, and the *bph* cluster has been reported to reside in a 55-kb element, Tn*4371* (Springael et al. 1993), which encodes proteins commonly associated with conjugative transposons: Int for excision from and integration into the genome and Dtr and Mpf systems for conjugative transfer (Toussaint et al. 2003; van Houdt et al. 2012). Unfortunately, the intercellular, but not intracellular, transfer of Tn*4371* has not yet been demonstrated. The second example is the *P. putida* KF715-derived *bph-sal* element (Nishi et al. 2000), whose 90-kb segment containing the *bph* gene cluster, as well as the salicylate-degrading gene, can move to other *Pseudomonas* chromosomes; however, the detailed gene organization of this element remains unknown. The last example is ICE$_{KKS102}$*4677*, which is located on the chromosome of a β-proteobacterial strain, *Acidovorax* sp. KKS102 (Ohtsubo et al. 2012a; Ohtsubo et al. 2012b). This 62-kb element is conjugally transferable among the chromosomes of three tested proteobacterial strains, *Burkholderia multivorans* (*Betaproteobacteria*), *Sphingobium japonicum* (*Alphaproteobacteria*), and *P. putida* (*Gammaproteobacteria*) with the preferred target site of 5′-GATTTTAAG-3′. ICE$_{KKS102}$*4677* carries the genes for integration and excision, conjugal transfer, and replication-partition proteins (*rep-parAB*). The *int* gene in the integrated form is located at one extreme end and is transcribed in the inward direction. The *int* transcriptional level is much higher in the circular form than in the integrated form, probably due to the read-through transcription from the outwardly directed promoter located at the other end of the element in the circular form. This higher level of transcription enables efficient integration into the genome of the circular form into the recipient cell. The role of Rep and ParAB proteins in the mobility and/or maintenance is still unclear although the functional ParA homolog encoded by a genomic island of *P. aeruginosa* PA14 (PAPI-1) is known to play an important role in the maintenance of this island in the genome (Qiu et al. 2006).

ICE$_{KKS102}$*4677* and Tn*4371* show very high resemblance to each other in terms of gene composition and organization. More than 35 Tn*4371*-related ICE-like elements are located on the genomes of various environmental bacterial strains belonging to the β- and γ-proteobacterial classes (Ryan et al. 2009; van Houdt et al. 2012), although the experimental evidence for their intercellular transfer is at present limited to that of ICE$_{KKS102}$*4677*. The 103-kb ICE$_{clc}$ element (in the pKL102/PAGI family of ICEs) (Klockgether et al. 2007) from *P. knackmussii* B13 encodes the enzymes for the degradation of 3-chlorocatechol and aminophenol (Ravatn et al. 1998; Gaillard et al. 2006), and its excision from the chromosome and transfer to the recipient genome, as well as the regulatory mechanisms for mobility, have been investigated in detail (Miyazaki et al. 2012). However, the *parA* and *int* genes of this element show weak (less than 26 % at the amino acid sequence level) similarity with the corresponding genes from ICE$_{KKS102}$*4677* in spite of the higher similarity (>55 %) of the common catabolic genes shared by the two ICEs. The Int protein of ICE$_{KKS102}$*4677* shows very low (<20 %) similarity to that of a well-characterized drug-resistance ICE, R391, in the STX/R391 family (Toleman and Walsh 2011).

A 232-kb chromosomal region of *Delftia* sp. Cs1-4 containing all of the genes for the complete degradation of phenanthrene is considered as an ICE-like genomic island (*phn* island), since this region additionally encodes the conjugal transfer system and DNA mobilization function (e.g., Int) (Hickey et al. 2012). However, its excision from the chromosome and intercellular transfer has not yet been confirmed.

5.11 Perspectives

At the present genomic era of microbiological researches, we will be able to add a very large number of microbial catabolic MGEs to the previous and experimentally analyzed list of MGEs. The *xyl* and *nah meta* operons must have been formed by combining the common gene cluster for the degradation of catechol to the TCA cycle intermediates with the gene cluster for toluate dioxygenase and the gene for salicylate hydroxylase, respectively. We do not currently know how the common gene cluster has been fused with each unique gene or gene cluster. It is also unclear how various catabolic gene clusters have been incorporated into the evolutionarily common class II transposon. Some clues to answering these unresolved questions may be obtained by comprehensive and comparative analyses of the large number of newly identified catabolic MGEs. The detailed investigation of Tn*4651* has revealed the importance of host-specified genetic determinants for transposition. Our analysis of IS*1071* also indicated the dependence of its transposability on the host factor(s). Such unknown factors and their respective roles will be elucidated by the establishment of appropriate in vitro systems for the transposition reaction. The microbial catabolic MGEs thus far analyzed have been limited to those from strains that are both culturable and capable of completely degrading compounds. However, recent genomic sequencing of a number of culturable strains reveals various strains containing only portions of catabolic pathway genes that are insufficient for complete degradation. Additionally, more than 99 % of environmental microbes still remain unculturable at present by conventional microbiological techniques. Therefore, it is expected that a much higher number of catabolic MGEs with structurally novel features must be present in the natural environments. Our functional screening of activated sludge-derived metagenomic DNA for the expression of catechol 2,3-dioxygenase (C23O) genes indeed led to the computational detection of a 37-kb plasmid that carries only the genes for phenol hydroxylase and C23O (Suenaga et al. 2009). One method of capturing transferable catabolic MGEs without the isolation of their host strains is referred to as "endogenous plasmid isolation technique"; this technique involves the use of an appropriate culturable host strain, as the recipient, and the environmental microbial community, as the donor, in a traditional conjugation experiment (Hill et al. 1992; Top et al. 1994). Our application of this technique resulted in the successful isolation from polluted soils of (1) two naphthalene-degrading plasmids and (2) a novel-type plasmid carrying only one gene for the degradation of γ-hexachlorocyclohexane (Ono et al. 2007; Miyazaki et al. 2006). Much more comprehensive analysis using these two culture-independent methods also promises to shed more light on catabolic MGEs in various microbial communities in the natural environments.

Acknowledgments We are grateful to all of our collaborators. Our works were supported by Grants-in-Aid from the Ministry of Education, Culture, Sports, Science and Technology, Japan.

References

Chandler M, Mahillon J (2002) Insertion sequences revisited. In: Craig NL, Craigie R, Gellert M, Lambowitz AM (eds) Mobile DNA II. ASM, Washington, DC, pp 305–366

Chatfield LK, Williams PA (1986) Naturally occurring TOL plasmids in *Pseudomonas* strains carry either two homologous or two nonhomologous catechol 2,3-oxygenase genes. J Bacteriol 168:878–885

Dennis JJ (2005) The evolution of IncP catabolic plasmids. Curr Opin Biotechnol 16:291–298

Dennis JJ, Zylstra GJ (2004) Complete sequence and genetic organization of pDTG1, the 83 kilobase naphthalene degradation plasmid from *Pseudomonas putida* strain NCIB 9816–4. J Mol Biol 341:753–768

Devers M, Rouard N, Martin-Laurent F (2008) Fitness drift of an atrazine-degrading population under atrazine selection pressure. Environ Microbiol 10:676–684

Dunn N, Gunsalus IC (1973) Transmissible plasmid coding early enzymes of naphthalene oxidation in *Pseudomonas putida*. J Bacteriol 114:974–979

Durland RH, Helinski DR (1987) The sequence encoding the 43-kilodalton *trfA* protein is required for efficient replication or maintenance of minimal RK2 replicons in *Pseudomonas aeruginosa*. Plasmid 18:164–169

Frost LS, Leplae R, Summers AO, Toussaint A (2005) Mobile genetic elements: the agents of open source evolution. Nat Rev Micro 3:722–732

Gaillard M, Vallaeys T, Vorhölter FJ, Minoia M, Werlen C, Sentchilo V, Pühler A, van der Meer JR (2006) The *clc* element of *Pseudomonas* sp. strain B13, a genomic island with various catabolic properties. J Bacteriol 188:1999–2013

Gallegos MT, Williams PA, Ramos JL (1997) Transcriptional control of the multiple catabolic pathways encoded on the TOL plasmid pWW53 of *Pseudomonas putida* MT53. J Bacteriol 179:5024–5029

Genka H, Nagata Y, Tsuda M (2002) Site-specific recombination system encoded by toluene catabolic transposon Tn*4651*. J Bacteriol 184:4757–4766

Greated A, Lambertsen L, Williams PA, Thomas CM (2002) Complete sequence of the IncP-9 TOL plasmid pWW0 from *Pseudomonas putida*. Environ Microbiol 4:856–871

Grindley NDF (2002) The movement of Tn*3*-like elements: transposition and cointegrate resolution. In: Craig NL, Craigie R, Gellert M, Lambowitz AM (eds) Mobile DNA II. ASM, Washington, DC, pp 272–302

Hansen JB, Olsen RH (1978) IncP2 group of *Pseudomonas*, a class of uniquely large plasmids. Nature 274:715–717

Heinaru E, Vedler E, Jutkina J, Aava M, Heinaru A (2009) Conjugal transfer and mobilization capacity of the completely sequenced naphthalene plasmid pNAH20 from multiplasmid strain *Pseudomonas fluorescens* PC20. FEMS Microbiol Ecol 70:563–574

Heuer H, Smalla K (2012) Plasmids foster diversification and adaptation of bacterial populations in soil. FEMS Microbiol Rev 36:1083–1104

Hickey WJ, Chen S, Zhao J (2012) The *phn* Island: a new genomic island encoding catabolism of polynuclear aromatic hydrocarbons. Front Microbiol 3:125

Hill KE, Weightman AJ, Fry JC (1992) Isolation and screening of plasmids from the epilithon which mobilize recombinant plasmid pD10. Appl Environ Microbiol 58:1292–1300

Hõrak R, Kivisaar M (1998) Expression of the transposase gene *tnpA* of Tn*4652* is positively affected by integration host factor. J Bacteriol 180:2822–2829

Hõrak R, Kivisaar M (1999) Regulation of the transposase of Tn*4652* by the transposon-encoded protein TnpC. J Bacteriol 181:6312–6318

Ilves H, Hõrak R, Kivisaar M (2001) Involvement of σ^S in starvation-induced transposition of *Pseudomonas putida* transposon Tn*4652*. J Bacteriol 183:5445–5448

Ilves H, Hõrak R, Teras R, Kivisaar M (2004) IHF is the limiting host factor in transposition of *Pseudomonas putida* transposon Tn*4652* in stationary phase. Mol Microbiol 51:1773–1785

Inoue K, Miyazaki R, Ohtsubo Y, Nagata Y, Tsuda M (2013) Inhibitory effect of *Pseudomonas putida* nitrogen-related phosphotransferase system on conjugative transfer of IncP-9 plasmid from *Escherichia coli*. FEMS Microbiol Lett 345:102–109

Izmalkova TY, Mavrodi DV, Sokolov SL, Kosheleva IA, Smalla K, Thomas CM, Boronin AM (2006) Molecular classification of IncP-9 naphthalene degradation plasmids. Plasmid 56:1–10

Jeenes DJ, Williams PA (1982) Excision and integration of degradative pathway genes from TOL plasmid pWW0. J Bacteriol 150:188–194

Juhas M, van der Meer JR, Gaillard M, Harding RM, Hood DW, Crook DW (2009) Genomic islands: tools of bacterial horizontal gene transfer and evolution. FEMS Microbiol Rev 33:376–393

Keil H, Keil S, Pickup RW, Williams PA (1985a) Evolutionary conservation of genes coding for *meta* pathway enzymes within TOL plasmids pWW0 and pWW53. J Bacteriol 164:887–895

Keil H, Lebens MR, Williams PA (1985b) TOL plasmid pWW15 contains two nonhomologous, independently regulated catechol 2,3-oxygenase genes. J Bacteriol 163:248–255

Keil H, Keil S, Williams PA (1987) Molecular analysis of regulatory and structural *xyl* genes of the TOL plasmid p WW53–4. J Gen Microbiol 133:1149–1158

Kholodii GY, Yurieva OV, Gorlenko Z, Mindlin SZ, Bass IA, Lomovskaya OL, Kopteva AV, Nikiforov VG (1997) Tn*5041*: a chimeric mercury resistance transposon closely related to the toluene degradative transposon Tn*4651*. Microbiology 143:2549–2556

Kivisaar M (2004) Transposition and other mutational processes in *Pseudomonas*. In: Ramos JL (ed) Pseudomonas, vol 1. Kluwer Academic/Plenum, New York, pp 261–316

Klockgether J, Wurdemann D, Reva O, Wiehlmann L, Tummler B (2007) Diversity of the abundant pKLC102/PAGI-2 family of genomic islands in *Pseudomonas aeruginosa*. J Bacteriol 189:2443–2459

Król JE, Penrod JT, McCaslin H, Rogers LM, Yano H, Stancik AD, Dejonghe W, Brown CJ, Parales RE, Wuertz S, Top EM (2012) Role of IncP-1β plasmids pWDL7::*rfp* and pNB8c in chloroaniline catabolism as determined by genomic and functional analyses. Appl Environ Microbiol 78:828–838

Kunz DA, Chapman PJ (1981) Isolation and characterization of spontaneously occurring TOL plasmid mutants of *Pseudomonas putida* HS1. J Bacteriol 146:952–964

Lehrbach PR, Ward J, Meulien P, Broda P (1982) Physical mapping of TOL plasmids pWWO and pND2 and various R plasmid-TOL derivatives from *Pseudomonas* spp. J Bacteriol 152:1280–1283

Li W, Shi J, Wang X, Han Y, Tong W, Ma L, Liu B, Cai B (2004) Complete nucleotide sequence and organization of the naphthalene catabolic plasmid pND6-1 from *Pseudomonas* sp. strain ND6. Gene 336:231–240

Liebert CA, Hall RM, Summers AO (1999) Transposon Tn*21*, flagship of the floating genome. Microbiol Mol Biol Rev 63:507–522

Maeda K, Nojiri H, Shintani M, Yoshida T, Habe H, Omori T (2003) Complete nucleotide sequence of carbazole/dioxin-degrading plasmid pCAR1 in *Pseudomonas resinovorans* strain CA10 indicates its mosaicity and the presence of large catabolic transposon Tn*4676*. J Mol Biol 326:21–33

Martinez B, Tomkins J, Wackett LP, Wing R, Sadowsky MJ (2001) Complete nucleotide sequence and organization of the atrazine catabolic plasmid pADP-1 from *Pseudomonas* sp. strain ADP. J Bacteriol 183:5684–5697

Minakhina S, Kholodii G, Mindlin S, Yurieva O, Nikiforov V (1999) Tn*5053* family transposons are *res* site hunters sensing plasmidal *res* sites occupied by cognate resolvases. Mol Microbiol 33:1059–1068

Miyazaki R, Sato Y, Ito M, Ohtsubo Y, Nagata Y, Tsuda M (2006) Complete nucleotide sequence of an exogenously isolated plasmid, pLB1, involved in gamma-hexachlorocyclohexane degradation. Appl Environ Microbiol 72:6923–6933

Miyazaki R, Ohtsubo Y, Nagata Y, Tsuda M (2008) Characterization of the *traD* operon of naphthalene-catabolic plasmid NAH7: a host-range modifier in conjugative transfer. J Bacteriol 190:6281–6289

Miyazaki R, Minoia M, Pradervand N, Sentchilo V, Sulser S, Reinhard F, van der Meer JR (2012) The *clc* element and related genomic islands in *Proteobacteria*. In: Roberts AP, Mullany P (eds) Bacterial integrative mobile genetic elements. Landes Bioscience, Austin, pp 261–272

Nakazawa T (2002) Travels of a *Pseudomonas*, from Japan around the world. Environ Microbiol 4:782–786

Nishi A, Tominaga K, Furukawa K (2000) A 90-kilobase conjugative chromosomal element coding for biphenyl and salicylate catabolism in *Pseudomonas putida* KF715. J Bacteriol 182:1949–1955

Nojiri H, Shintani M, Omori T (2004) Divergence of mobile genetic elements involved in the distribution of xenobiotic-catabolic capacity. Appl Microbiol Biotechnol 64:154–174

Norman A, Hansen LH, Sørensen SJ (2009) Conjugative plasmids: vessels of the communal gene pool. Philos Trans R Soc Lond B Biol Sci 364:2275–2289

Ohtsubo Y, Ishibashi Y, Naganawa H, Hirokawa S, Atobe S, Nagata Y, Tsuda M (2012a) Conjugal transfer of polychlorinated biphenyl/biphenyl degradation genes in *Acidovorax* sp. strain KKS102, which are located on an integrative and conjugative element. J Bacteriol 194:4237–4248

Ohtsubo Y, Maruyama F, Mitsui H, Nagata Y, Tsuda M (2012b) Complete genome sequence of *Acidovorax* sp. KKS102, a polychlorinated biphenyl-degrading strain. J Bacteriol 194: 6970–6971

Ono A, Miyazaki R, Sota M, Ohtsubo Y, Nagata Y, Tsuda M (2007) Isolation and characterization of naphthalene-catabolic genes and plasmids from oil-contaminated soil by using two cultivation-independent approaches. Appl Microbiol Biotechnol 74:501–510

Petrovski S, Stanisich VA (2011) Embedded elements in the IncP1-β plasmids R772 and R906 can be mobilized and can serve as a source of diverse and novel elements. Microbiology 157: 1714–1725

Qiu X, Gurkar AU, Lory S (2006) Interstrain transfer of the large pathogenicity island (PAPI-1) of *Pseudomonas aeruginosa*. Proc Natl Acad Sci USA 103:19830–19835

Rajeev L, Malanowska K, Gardner JF (2009) Challenging a paradigm: the role of DNA homology in tyrosine recombinase reactions. Microbiol Mol Biol Rev 73:300–309

Ramos JL, Marqués S, Timmis KN (1997) Transcriptional control of the *Pseudomonas* TOL plasmid catabolic operons is achieved through an interplay of host factors and plasmid-encoded regulators. Annu Rev Microbiol 51:341–373

Ravatn R, Zehnder AJ, van der Meer JR (1998) Low-frequency horizontal transfer of an element containing the chlorocatechol degradation genes from *Pseudomonas* sp. strain B13 to *Pseudomonas putida* F1 and to indigenous bacteria in laboratory-scale activated-sludge microcosms. Appl Environ Microbiol 64:2126–2132

Riuz R, Aranda-Olmedo MI, Domíniquez-Cuevas P, Ramos-Gonzalez MI, Marqués S (2004) Transcriptional regulation of the toluene catabolic pathways. In: Ramos JL (ed) Pseudomonas, vol 2. Kluwer Academic/Plenum, New York, pp 509–537

Roberts AP, Chandler M, Courvalin P, Guedon G, Mullany P, Pembroke T, Rood JI, Smith CJ, Summers AO, Tsuda M, Berg DE (2008) Revised nomenclature for transposable genetic elements. Plasmid 60:167–173

Ryan MP, Pembroke JT, Adley CC (2009) Novel Tn*4371*-ICE like element in *Ralstonia pickettii* and genome mining for comparative elements. BMC Microbiol 9:242

Sen D, Van der Auwera GA, Rogers LM, Thomas CM, Brown CJ, Top EM (2011) Broad-host-range plasmids from agricultural soils have IncP-1 backbones with diverse accessory genes. Appl Environ Microbiol 77:7975–7983

Sen D, Brown CJ, Top EM, Sullivan J (2013) Inferring the evolutionary history of IncP-1 plasmids despite incongruence among backbone gene trees. Mol Biol Evol 30:154–166

Sentchilo VS, Perebituk AN, Zehnder AJ, van der Meer JR (2000) Molecular diversity of plasmids bearing genes that encode toluene and xylene metabolism in *Pseudomonas* strains isolated from different contaminated sites in Belarus. Appl Environ Microbiol 66:2842–2852

Sevastsyanovich YR, Krasowiak R, Bingle LE, Haines AS, Sokolov SL, Kosheleva IA, Leuchuk AA, Titok MA, Smalla K, Thomas CM (2008) Diversity of IncP-9 plasmids of *Pseudomonas*. Microbiology 154:2929–2941

Shaw LE, Williams PA (1988) Physical and functional mapping of two cointegrate plasmids derived from RP4 and TOL plasmid pDK1. J Gen Microbiol 134:2463–2474

Shintani M, Nojiri H (2013) Mobile genetic elements (MGEs) carrying catabolic genes. In: Malik A, Grohmann E, Alves M (eds) Management of microbial resources in the environment. Springer, Netherlands, pp 167–214

Sørensen SJ, Bailey M, Hansen LH, Kroer N, Wuertz S (2005) Studying plasmid horizontal transfer in situ: a critical review. Nat Rev Micro 3:700–710

Sota M, Endo M, Nitta K, Kawasaki H, Tsuda M (2002) Characterization of a class II defective transposon carrying two haloacetate dehalogenase genes from *Delftia acidovorans* plasmid pUO1. Appl Environ Microbiol 68:2307–2315

Sota M, Kawasaki H, Tsuda M (2003) Structure of haloacetate-catabolic IncP-1β plasmid pUO1 and genetic mobility of its residing haloacetate-catabolic transposon. J Bacteriol 185:6741–6745

Sota M, Yano H, Nagata Y, Ohtsubo Y, Genka H, Anbutsu H, Kawasaki H, Tsuda M (2006a) Functional analysis of unique class II insertion sequence IS*1071*. Appl Environ Microbiol 72:291–297

Sota M, Yano H, Ono A, Miyazaki R, Ishii H, Genka H, Top EM, Tsuda M (2006b) Genomic and functional analysis of the IncP-9 naphthalene-catabolic plasmid NAH7 and its transposon Tn*4655* suggests catabolic gene spread by a tyrosine recombinase. J Bacteriol 188:4057–4067

Sota M, Tsuda M, Yano H, Suzuki H, Forney LJ, Top EM (2007) Region-specific insertion of transposons in combination with selection for high plasmid transferability and stability accounts for the structural similarity of IncP-1 plasmids. J Bacteriol 189:3091–3098

Sota M, Yano H, Tsuda M (2008) Bacterial class II catabolic transposons. In: Yoshida Y, Aoki M (eds) DNA transposable elements research. Nova Science, New York, pp 23–67

Springael D, Kreps S, Mergeay M (1993) Identification of a catabolic transposon, Tn*4371*, carrying biphenyl and 4-chlorobiphenyl degradation genes in *Alcaligenes eutrophus* A5. J Bacteriol 175:1674–1681

Suenaga H, Koyama Y, Miyakoshi M, Miyazaki R, Yano H, Sota M, Ohtsubo Y, Tsuda M, Miyazaki K (2009) Novel organization of aromatic degradation pathway genes in a microbial community as revealed by metagenomic analysis. ISME J 3:1335–1348

Teras R, Hõrak R, Kivisaar M (2000) Transcription from fusion promoters generated during transposition of transposon Tn*4652* is positively affected by integration host factor in *Pseudomonas putida*. J Bacteriol 182:589–598

Teras R, Jakovleva J, Kivisaar M (2009) Fis negatively affects binding of Tn*4652* transposase by out-competing IHF from the left end of Tn*4652*. Microbiology 155:1203–1214

Thomas CM, Haines A (2004) Plasmids of the genus *Pseudomonas*. In: Ramos JL (ed) Pseudomonas, vol 1. Kluwer Academic/Plenum, New York, pp 197–231

Thomas CM, Nielsen KM (2005) Mechanisms of, and barriers to, horizontal gene transfer between bacteria. Nat Rev Micro 3:711–721

Toleman MA, Walsh TR (2011) Combinatorial events of insertion sequences and ICE in Gram-negative bacteria. FEMS Microbiol Rev 35:912–935

Top E, de Smet I, Verstraete W, Dijkmans R. Mergeay M (1994) Exogenous isolation of mobilizing plasmids from polluted soils and sludges. Appl Environ Microbiol 60:831–839

Top EM, Springael D (2003) The role of mobile genetic elements in bacterial adaptation to xenobiotic organic compounds. Curr Opin Biotechnol 14:262–269

Toussaint A, Merlin C, Monchy S, Benotmane MA, Leplae R, Mergeay M, Springael D (2003) The biphenyl- and 4-chlorobiphenyl-catabolic transposon Tn*4371*, a member of a new family of genomic islands related to IncP and Ti plasmids. Appl Environ Microbiol 69:4837–4845

Trefault N, De la Iglesia R, Molina AM, Manzano M, Ledger T, Pérez-Pantoja D, Sánchez MA, Stuardo M, González B (2004) Genetic organization of the catabolic plasmid pJP4 from *Ralstonia eutropha* JMP134 (pJP4) reveals mechanisms of adaptation to chloroaromatic pollutants and evolution of specialized chloroaromatic degradation pathways. Environ Microbiol 6:655–668

Tsuda M, Genka H (2001) Identification and characterization of Tn*4656*, a novel class II transposon carrying a set of toluene-degrading genes from TOL plasmid pWW53. J Bacteriol 183:6215–6224

Tsuda M, Iino T (1987) Genetic analysis of a transposon carrying toluene degrading genes on a TOL plasmid pWW0. Mol Gen Genet 210:270–276

Tsuda M, Iino T (1988) Identification and characterization of Tn*4653*, a transposon covering the toluene transposon Tn*4651* on TOL plasmid pWW0. Mol Gen Genet 213:72–77

Tsuda M, Iino T (1990) Naphthalene degrading genes on plasmid NAH7 are on a defective transposon. Mol Gen Genet 223:33–39

Tsuda M, Minegishi K, Iino T (1989) Toluene transposons Tn*4651* and Tn*4653* are class II transposons. J Bacteriol 171:1386–1393

Tsuda M, Tan HM, Nishi A, Furukawa K (1999) Mobile catabolic genes in bacteria. J Biosci Bioeng 87:401–410

van der Meer JR, Sentchilo V (2003) Genomic islands and the evolution of catabolic pathways in bacteria. Curr Opin Biotechnol 14:248–254

van Houdt R, Toussaint A, Ryan MP, Pembroke JT, Meargy M, Adley CC (2012) The Tn*4731* ICE family of bacterial mobile genetic elements. In: Roberts AP, Mullany P (eds) Bacterial integrative mobile genetic elements. Landes Bioscience, Austin, pp 179–200

Vanhooff V, Galloy C, Agaisse H, Lereclus D, Revet B, Hallet B (2006) Self-control in DNA site-specific recombination mediated by the tyrosine recombinase TnpI. Mol Microbiol 60: 617–629

Williams PA, Murray K (1974) Metabolism of benzoate and the methylbenzoates by *Pseudomonas putida* (*arvilla*) mt-2: evidence for the existence of a TOL plasmid. J Bacteriol 120:416–423

Williams PA, Worsey MJ (1976) Ubiquity of plasmids in coding for toluene and xylene metabolism in soil bacteria: evidence for the existence of new TOL plasmids. J Bacteriol 125:818–828

Williams PA, Rheinallt MJ, Zylstra GJ (2004) Genomics of catabolic plasmids. In: Ramos JL (ed) Pseudomonas, vol 1. Kluwer Academic/Plenum, New York

Yano H, Deckert GE, Rogers LM, Top EM (2012) Roles of long and short replication initiation proteins in the fate of IncP-1 plasmids. J Bacteriol 194:1533–1543

Yano H, Garruto CE, Sota M, Ohtsubo Y, Nagata Y, Zylstra GJ, Williams PA, Tsuda M (2007) Complete sequence determination combined with analysis of transposition/site-specific recombination events to explain genetic organization of IncP-7 TOL plasmid pWW53 and related mobile genetic elements. J Mol Biol 369:11–26

Yano H, Miyakoshi M, Ohshima K, Tabata M, Nagata Y, Hattori M, Tsuda M (2010) Complete nucleotide sequence of TOL plasmid pDK1 provides evidence for evolutionary history of IncP-7 catabolic plasmids. J Bacteriol 192:4337–4347

Yano H, Genka H, Ohtsubo Y, Nagata Y, Top EM, Tsuda M (2013) Cointegrate-resolution of toluene-catabolic transposon Tn*4651*: determination of crossover site and the segment required for full resolution activity. Plasmid 69:24–35

Chapter 6
Adaptation to Xenobiotics and Toxic Compounds by *Cupriavidus* and *Ralstonia* with Special Reference to *Cupriavidus metallidurans* CH34 and Mobile Genetic Elements

Max Mergeay and Rob Van Houdt

Abstract The purpose of this book chapter is to give a view of the relationships between mobile genetic elements and genes related to environmental adaptations, and the development of new catabolic properties but also of other properties linked to the survival in anthropogenic (mostly industrial) environments. The insights and knowledge in the genetics and genomics of the β-proteobacteria belonging to the closely related *Cupriavidus* and *Ralstonia* genera, with some focus on the genome of *Cupriavidus metallidurans* CH34, will be central.

Keywords Acetone • Betaproteobacteria • Biphenyl • Chromid • *Cupriavidus* • Heavy metals • ICE • Integrative conjugative elements • PCB • Tn*4371* • Toluene

6.1 *Cupriavidus* and *Ralstonia* Genera: A Branch of Burkholderiaceae in β-Proteobacteria: Historical and Taxonomic Introduction

6.1.1 From Pseudomonas and Alcaligenes to Cupriavidus and Ralstonia

Bacteria resistant to heavy metals and displaying both oxidative metabolism and facultative hydrogenotrophy were first discovered in sediments of a decantation basin in a nonferrous metallurgical plant (Mergeay et al. 1978) located

M. Mergeay (✉) • R. Van Houdt
Unit of Microbiology, Belgian Nuclear Research Centre (SCK•CEN), Boeretang 200, 2400 Mol, Belgium
e-mail: mmergeay@sckcen.be

H. Nojiri et al. (eds.), *Biodegradative Bacteria: How Bacteria Degrade, Survive, Adapt, and Evolve*, DOI 10.1007/978-4-431-54520-0_6, © Springer Japan 2014

in the province of Liège (Belgium). This region was a major player during the industrial revolution in the early nineteenth century (development of the zinc metallurgy under the process discovered in 1810 by Jean-Jacques Dony (1759–1819)). Similar strains were also regularly found in soils of other metallurgical plants in northeastern Belgium as well as in Germany and Congo (Diels and Mergeay 1990). These isolates, which displayed multiple resistances to heavy metals (especially cadmium, zinc, cobalt, and nickel and later to copper, chromate, mercury, lead and silver) and with strain CH34 as the best studied representative, were first assigned to the genus *Pseudomonas* at a moment where the taxonomy of this genus was already in full and radical reassessment (Palleroni 2003; Stanier et al. 1966). Next, they were seen as close relatives to *Alcaligenes eutrophus*, mainly due to the presence of two hydrogenases (one soluble and one particulate) (Lejeune et al. 1983; Mergeay et al. 1978, 1985). Later, the taxonomic proximity of strain CH34 with plant pathogens as *Ralstonia solanacearum* (Yabuuchi et al. 1995) was also recognized as well as with some strains able to efficiently degrade xenobiotic organics (Don and Pemberton 1981, 1985; Springael et al. 1993, 1994). This occurred with the isolation of other metal-resistant strains from a variety of industrial sites or biotopes that shared typical genetic features with strain CH34 (Brim et al. 1999; Diels and Mergeay 1990; Goris et al. 2001; Mergeay 2000; Schmidt and Schlegel 1994; Schmidt et al. 1991; Stoppel et al. 1995; Stoppel and Schlegel 1995) and especially the genetic determinants for high resistance to heavy metals as nickel, cobalt, cadmium, zinc, lead, copper, mercury cations, and chromate carried by plasmids or genomic islands (Borremans et al. 2001; Diels et al. 1985; Juhnke et al. 2002; Liesegang et al. 1993; Monchy et al. 2007; Nies et al. 1987, 1989; Nies 1995; Peitzsch et al. 1998; Ryan et al. 2009; Sensfuss and Schlegel 1988). Yet the variety of new strains from different geographical origins that displayed uncommon phenotypes also led to further and rather drastic taxonomic reevaluations and additions (as far as strain CH34 is concerned, from *Alcaligenes* via *Ralstonia* and *Wautersia* to *Cupriavidus* in a couple of years) that seem to be more or less stabilized now (Chen et al. 2001; Sato et al. 2006; Vandamme and Coenye 2004; Vaneechoutte et al. 2004). As far as the resistance to cadmium, cobalt, and zinc mediated by the *czc* genes of pMOL30 is concerned, other *Cupriavidus* strains carrying these genes and belonging to the species *Cupriavidus campinensis* (especially strain DS185: plasmid pMOL85 of *C. campinensis* DS185 (from a "zinc desert") is transferred at high frequency to plasmid-free derivatives of *Cupriavidus metallidurans* CH34 by selecting for metal resistance) and *Cupriavidus basilensis* ER121 were isolated from various nonferrous metallurgical sites in northeastern Belgium (Brim et al. 1999; Diels et al. 1989, 1995a; Diels and Mergeay 1990; Goris et al. 2001), but their genome is not yet sequenced.

6.2 Available Genomes: A Material to Study Natural Genetic Engineering and (Recent) Evolution

This study is mainly built around the genome of the β-proteobacterium *C. metallidurans* CH34 in close comparison with other genomes of strains belonging to the genera *Cupriavidus* and *Ralstonia*. The available *Cupriavidus* genomes include that of the catabolic strain *C. pinatubonensis* (*R. eutropha*) JMP134 encoding around 300 genes predicted to be directly involved in the catabolism of aromatic compounds and typified by plasmid pJP4 governing the degradation of herbicides as 2,4-dichlorophenoxyacetic acid (Don and Pemberton 1981, 1985; Lykidis et al. 2010; Sato et al. 2006), the hydrogenotrophs *C. eutrophus* (*C. necator*, *R. eutropha*) H16 and *C. necator* N-1 (Poehlein et al. 2012), and the β-rhizobium *C. taiwanensis*, a nitrogen-fixing symbiont of tropical legumes (Amadou et al. 2008). The available *Ralstonia* genomes include that of the plant pathogen *R. solanacearum* G1000 (Salanoubat et al. 2002) and other *R. solanacearum* sensu lato genomes (including the plant pathogen *R. haywardii* subspecies *celebensis* R229 formerly called blood disease bacteria R229 due to the red color appearing on infected banana plants (Remenant et al. 2011), which is not a nosocomial infection despite a misleading appellation) and the environmental strains *R. pickettii* 12J and 12D.

New strains of *C. metallidurans* were recently isolated from space-related environments such as clean rooms for the construction of satellites (strain NE12) and drinking water systems of the International Space Station (ISS) (strains NA1, NA2, and NA4), and from clinical settings (H1130). The heavy metal resistance determinants and mobile genetic element pool of these strains were already compared to CH34 by comparative genome hybridization (Mijnendonckx et al. 2013; Van Houdt et al. 2012b). Furthermore, the genomes of NA4 and H1130 are sequenced. *C. metallidurans* strains could be compared to other atypical isolates of *Cupriavidus* (*eutrophus*, and *taiwanensis*) and *Ralstonia* that were found in clinical settings (Coenye et al. 2005). The first *R. pickettii* strains were found in clinical settings (Ryan et al. 2006; Yabuuchi et al. 1995); however, the only available sequenced genomes are from the environmental strains 12J and 12D that share many genomic islands and characteristic metal resistance genes with *C. metallidurans* CH34 (Ryan et al. 2007, 2009; Van Houdt et al. 2009). All the sequenced *Ralstonia* and *Cupriavidus* genomes contain two chromosomes; the smaller one is now referred to as a chromid, due to the fact that the replication function is more similar to that of a plasmid (megaplasmid) than to that of a chromosome. *Burkholderia* strains from the closely related *Burkholderia* genus have also chromids in their genome and are even more ubiquitous than strains from the *Ralstonia* and *Cupriavidus* genera, especially in soil, environmental, and clinical settings. Other β-proteobacterial genomes that will be taken into consideration are those of the environmental strains *Delftia acidovorans* SPH-1, *Comamonas testosteroni* KF-1, and *Bordetella petrii* DSM12804 for which mobile genetic elements very similar and sometimes identical to some of *C. metallidurans* CH34 have been found (Ryan et al. 2009; Van Houdt et al. 2009). These environmental β-proteobacteria have only one chromosome of ~6 Mb, in contrast with the various replicons found in *Cupriavidus* strains (Table 6.1), although totalizing a comparable amount of genetic material.

6.3 Various Types of Replicons: Chromosomes, Chromids, Megaplasmids, and Plasmids

Besides the chromosome and megaplasmids [for a discussion of this word see (Schwartz 2009)] such as pHG1 (Schwartz et al. 2003) in *C. eutrophus* H16, pRALTA in *C. taiwanensis* LMG19424 (Amadou et al. 2008), and pMOL28 and pMOL30 in *C. metallidurans* CH34 (Mergeay et al. 2009; Monchy et al. 2007), and an IncP1 catabolic pJP4 plasmid in *C. pinatubonensis* JMP134 (Don and Pemberton 1985), all *Cupriavidus* and *Ralstonia* genomes are typified by the presence of a second large replicon with a size around 2 Mb.

These large replicons, which are secondary to the main chromosome, have been termed both "second chromosomes" if they carry essential genes and are indispensable for cell viability and "megaplasmids" if they do not use chromosome-type but plasmid-type replication systems. Recently, the term "chromid" was introduced to distinguish this replicon as it is neither a chromosome nor a plasmid (Harrison et al. 2010). Three criteria were defined: (1) chromids have plasmid-type maintenance and replication systems, (2) chromids have a nucleotide composition close to that of the chromosome, and (3) chromids carry core genes that are found on the chromosome in other species (Harrison et al. 2010). In the *Cupriavidus* and *Ralstonia* chromids the gene cluster *csp repA csp parA parB* involved in replication and partitioning is strongly conserved. Some *Burkholderia* genomes, which are closely related β-proteobacteria, even carry more than one chromid and have an extraordinary total number of replicons, e.g., *B. xenovorans* LB400 has one chromosome and two large chromids, while *B. vietnamiensis* G4 has five additional plasmids. Many features detected in *Cupriavidus* strains are also shared by these two strains (see Table 6.1 for an inventory of the replicons in the different strains). As illustrated below, chromids are an important repository of typical "accessory genes" including virulence, catabolism, and metal resistance genes (Van Houdt and Mergeay 2012).

The prevailing hypothesis for the origin of the chromid in the multipartite genomes of *Cupriavidus* and *Burkholderia* implies that it evolved from ancestral plasmids. To determine whether these putative ancestral plasmids were the same in the *Cupriavidus/Ralstonia* and *Burkholderia* lineages, the similarity and phylogenetic relationships of the ParA and ParB proteins encoded by chromids as well as of DnaA proteins encoded by the chromosome were investigated in 19 β-proteobacteria from those three genera (Lykidis et al. 2010). The reported data suggest that two distinct

Table 6.1 Genome size and organization of representative strains (discussed in this chapter)

Strain	Biotope or main phenotype of attention	Genome size (kb)	Chromosome size (kb)	Chromid(s) size (kb)	Megaplasmid(s) size (kb)	Plasmid(s) size (kb)
Cupriavidus eutrophus H16	Chemolithotrophy	7,416	4,052	2,912	452 (pHG1)	–
Cupriavidus metallidurans CH34	Environmental strain, heavy metals	6,913	3,928	2,580	171 (pMOL28) 234 (pMOL30)	–
Cupriavidus necator N-1		8,481	3,873	2,685 1,499	424	–
Cupriavidus pinatubonensis JMP134	Degradation of xenobiotics	7,255	3,807	2,726	635	88 (pJP4)
Cupriavidus taiwanensis LMG19424	Nitrogen fixation, β-rhizobium	6,476	3,417	2,502	557 (pRALTA)	–
Ralstonia pickettii 12J	Environmental strain, heavy metals	5,326	3,943	1,302	–	81
Ralstonia pickettii 12D	Environmental strain, heavy metals	5,685	3,647	1,323	390 (pRp12D01) 273 (pRp12D02)	51 (pRp12D03)
Ralstonia solanacearum GMI1000	Plant pathogen	5,811	3,716	2,095	–	–
Bordetella petrii DSM12804	Environmental strain, heavy metals	5,288	5,288	–	–	–
Delftia acidovorans SPH-1	Environmental strain, heavy metals	6,768	6,768	–	–	–
Comamonas testosteroni KF-1	Environmental strain, heavy metals	6,027	6,027	–	–	–
Burkholderia vietnamiensis G4	Degrader of trichloroethylene, benzene, and toluene	8,391	3,653	2,412 1,241	398 (pBVIE01) 266 (pBVIE02) 227 (pBVIE03)	107 (pBVIE04) 88 (pBVIE05)
Burkholderia xenovorans LB400	Degrader of biphenyl and polychloro-biphenyls (PCBs)	9,731	4,896	3,364 1,472	–	–

plasmids (one for *Cupriavidus/Ralstonia* and one for *Burkholderia*) may have been the origin of the chromids present in the genera *Cupriavidus/Ralstonia* and *Burkholderia* while DnaA is well conserved in the three genera (Lykidis et al. 2010).

6.4 *C. metallidurans* CH34, a Bacterium Adapted to Heavy Metals and Industrial Environments, a Short Update

Besides taxonomy, extensive information about the biology of *C. metallidurans* CH34 is already available (see (Mergeay 2000). Other comprehensive reports addressed the catalogue of heavy metal resistance genes (Mergeay et al. 2003; Nies 1999, 2000, 2003; von Rozycki and Nies 2009), the metallotranscriptome (all the genes that are regulated after metallic challenges tested with 16 different metals) (Monsieurs et al. 2011), the genome (Janssen et al. 2010), and the content of mobile genetic elements (Mergeay et al. 2009; Mijnendonckx et al. 2011; Monchy et al. 2007; Van Houdt et al. 2009), most of these features are discussed below.

The applications and perspectives in environmental biotechnology may be recalled as well (Diels et al. 1993, 1995b, 1996, 2009; Magrisso et al. 2008). Taking into account the diversity of mobile genetic elements in *C. metallidurans* and their implication in horizontal gene transfer, it looks relevant to quote again the capacity of *C. metallidurans* CH34 to express foreign genes as shown by heterologous complementation either by electroporation (Taghavi et al. 1994) or by R-prime plasmids carrying genes from *Pseudomonas fluorescens*, *Salmonella enterica* serovar Typhimurium and other bacteria (Lejeune et al. 1983; Van Gijsegem and Toussaint 1982). A recent report shed a new light on this property to easily accept and express foreign genetic material and suggests attractive developments in a modern perspective by showing that *C. metallidurans* CH34 is fully suitable for expressing genes from the soil metagenome and subsequently synthetizing natural products such as pigments and antibiotics after transformation by broad host range cosmid vectors carrying metagenomic DNA, which appeared to be not expressed in an *Escherichia coli* host (Craig et al. 2009).

6.5 A Specificity of the *C. metallidurans* CH34 Chromosome: A Source of Genomic Islands and Mobile Genetic Elements

The full genome sequencing and complete closure of the *C. metallidurans* CH34 replicons combined with an extensive manual annotation unraveled a comprehensive catalogue of mobile genetic elements (Table 6.2). Both aspects contributed to the completeness of this catalogue, as draft genomic assemblies are very often interrupted by transposable elements and genomic islands, thereby limiting the ability to fully assess and understand the genomic context of important bacterial

Table 6.2 Catalogue of mobile genetic elements and genomic islands identified in *C. metallidurans* CH34[a]

Element	Location (size kb)	Features
Genomic islands		
CMGI-1 (109.6)	CHR	Identical to island PAGI-2C of *Pseudomonas aeruginosa* clone C, except inactivation of integrase (TBSSR[b]) by Tn*6049*
CMGI-2 (101.6)	CHR	Tn4371 ICE family (ICE_{Tn4371}6054), involved in hydrogenotrophy and degradation of aromatic compounds
CMGI-3 (97.0)	CHR	Tn4371 ICE family (ICE_{Tn4371}6055), involved in hydrogenotrophy and CO_2 fixation
CMGI-4 (56.5)	CHR	Tn4371 ICE family (ΔICE_{Tn4371}6055), partial element, complete element found in *D. acidovorans* SPH1
CMGI-5 (25.4)	CHR	Remnant of integrated plasmid
CMGI-6 (17.6)	CHR	TBSSR, no defined function for accessory genes
CMGI-7 (15.4)	CHR	Arsenic resistance, TBSSR is part of a four-gene module (PRQ)
CMGI-8 (12.3)	CHR	TBSSR, no defined function for accessory genes
CMGI-9 (20.6)	CHR	TBSSR, no defined function for accessory genes
CMGI-10 (21.0)	CHR	No defined function for accessory genes
CMGI-11 (10.8)	CHR	Putative fimbrial operon, flanked by two IS elements (IS*Rme7*)
CMGI-12 (9.1)	CHR	No defined function for accessory genes
CMGI-13 (15.9)	CHR	Genes involved in polysaccharide biosynthesis
CMGI-A (87.1)	Chromid	No defined function for accessory genes
CMGI-B ± D (19.5 + 141.2)	Chromid	Multiple genes coding for phage-related proteins (associated to phages of *R. solanacearum*)
CMGI-C (7.1)	Chromid	TBSSR fragment, gene coding for mannose-binding lectin
CMGI-E (120.2)	Chromid	Tn*7*-related genes at one extremity, genes putatively involved in degradation of aromatic compounds
CMGI-28a (45.9)	pMOL28	Heavy metal resistance genes (*mer*, *cnr*, and *chr*), flanked by IS*1071* elements
CMGI-28b (23.0)	pMOL28	TBSSR is part of a four-gene module (PRQ), three *rhs*-like genes
CMGI-28c (15.0)	pMOL28	TBSSR, no defined function for accessory genes
CMGI-30a (74.4)	pMOL30	TBSSR (heavy metal resistance genes (*mer*, *czc*, and *pbr*)), flanked by Tn*4380* elements (one intact and one truncated copy)

(continued)

Table 6.2 (continued)

Element	Location (size kb)	Features
CMGI-30b (88.0)	pMOL30	TBSSR, heavy metal resistance genes (*sil*, *ncc-nre*, and *cop*)
Element (size kb)	Location (copies)	
Transposons		
Tn*4378* (8.2)	pMOL28 (1)	Tn*501* family, mercury resistance
Tn*4380* (8.0)	pMOL30 (1)	Tn*501* family, mercury resistance
Tn*6048* (10.4)	CHR (1), chromid (2)	Contains eight metal-inducible genes (peptidase, phosphatase, permeases, and two-component regulatory system)
Tn*6049* (3.5)	CHR (7), chromid (3), pMOL28 (1), pMOL30 (1)	Often associated with genomic islands
Tn*6050* (6.8)	Chromid (2)	Tn*501* family, accessory genes encode a sulfate permease, a universal stress protein (UspA), and a DksA-like protein

[a]The *C. metallidurans* CH34 genome also bears 21 distinct insertion sequence (IS) elements from 10 different IS families and reaching a total of 57 intact IS copies in CH34 (Mijnendonckx et al. 2011)
[b]TBSSR stands for tyrosine-based site-specific recombinase, a more appropriate name for "integrase"

adaptations (Ricker et al. 2012). In addition, manual annotation combined with comparative syntenic analyses allowed verifying and curating the high-throughput automatic annotations (Janssen et al. 2010; Vallenet et al. 2006).

6.6 Integrated Mobile Genetic Elements and Transposons

6.6.1 The Tn4371 Family with a Modular Structure, Not Only Catabolic Genes

The *C. metallidurans* CH34 chromosome carries four large genomic islands. One is identical to island PAGI-2C of *Pseudomonas aeruginosa* clone C (Klockgether et al. 2007; Larbig et al. 2002), while the other three belong to the Tn*4371* ICE (integrative conjugative element) family. This family consists of mobile genetic elements with four modules containing genes involved in integration (via a tyrosine-based site-specific recombinase (TBSSR)), maintenance/stability, accessory genes conferring a special phenotypic trait to the host, and genes involved in conjugational transfer (Fig. 6.1).

The conjugative module displays similarity with conjugative genes of Ti plasmids and IncP broad-host-range plasmids. It is now recognized how frequent conjugation genes are integrated in the bacterial chromosomes mainly in ICE and how they may be essential actors in horizontal gene transfer (Guglielmini et al. 2011).

Tn*4371* (Springael et al. 1993, 2001), the first reported element of this family, sometimes nicknamed as a "biphenyl transposon," was isolated from *Cupriavidus oxalaticus* A5 and transferred to *C. metallidurans* CH34 (Merlin et al. 1997, 1999; Springael et al. 1993, 2001). Another "biphenyl transposon" from *Acidovorax* sp. KKS102 was the first member of the whole family where conjugational mediated

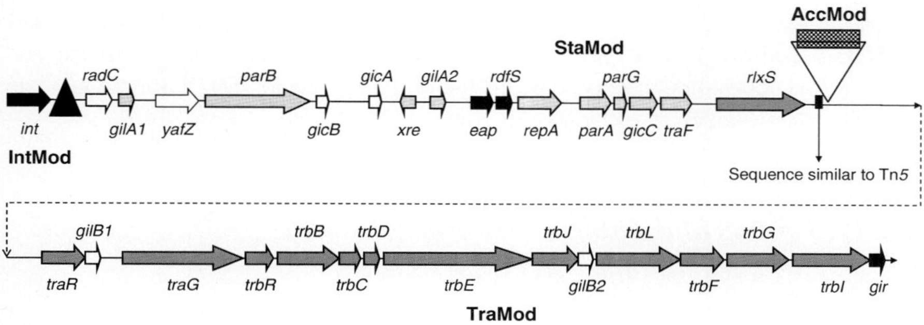

Fig. 6.1 Characteristic genes of the Tn*4371*-ICE family. *Black arrows*: involved in integration and excision; *light gray arrows*: involved in the stabilization of the element; *dark gray arrows*: involved in the transfer of the element; *white arrows*: various genes whose functions are not known (not always present in every element); *checkered box*: accessory gene module; *black triangle*: accessory genes can also be present (adapted from Van Houdt et al. 2011)

transfer was reported and quantified (Ohtsubo et al. 2012). Currently, this ICE_{Tn4371} family harbors around 40 elements with sizes ranging from 38 to 101 kb. Elements carrying accessory genes, which are mostly flanked by the conjugative genes *rlxS* (*virD2*) and *traR*, reside in β- and γ-proteobacteria, while elements in α-proteobacteria apparently lack accessory genes and display a more divergent pattern (Ohtsubo et al. 2012; Van Houdt et al. 2011). Accessory genes have not only very diverse functions including catabolism of xenobiotic compounds, resistance to heavy metals and to antibiotics, chemolithotrophic metabolism but also some unknown functions that should attract more interest in the future. Strains from man-made environments (sewages, industrial wastes, and clinical settings) are predominant among the bearers of Tn*4371* ICEs next to plant pathogens, which are also well represented (Van Houdt et al. 2011).

6.6.2 *Tn*6049*: 12 Copies in* C. metallidurans *CH34*

Transposon Tn*6049* (3.5 kb) is often associated with genomic islands and is found in 12 copies. The element carries a transposase, two conserved hypothetical protein-coding genes, and an ATPase-encoding gene, putatively named *gspA* because its product shows strong similarities to putative general secretion pathway ATPases (Mergeay et al. 2009). Tn*6049* appears to play a peculiar role in "stabilizing" genomic islands in strain CH34. It inactivated the terminal TBSSR gene of three genomic islands, a key conjugation gene (relaxase) of CMGI-5 and IS*1071*. Furthermore, it transposed into IS*1071* at the extremity of the CMGI-28a island rich in metal resistance genes (Ryan et al. 2009; Van Houdt et al. 2009). IS*1071* is a very active IS element often observed in catabolic plasmids (Providenti et al. 2006; Wyndham et al. 1994) and considered to be involved in their generation (Sen et al. 2011). A similar transposon, named Tn*6053*, was observed in *B. xenovorans* LB400, and it shares around 60 % protein identity with the corresponding Tn*6049* orthologs, except for *tnmC* where the 72-amino-acid gene product of Tn*6053* shares only 24 % identity with TnmC of CH34. Similar transposons were also identified in *B. phymatum* STMB15 (one copy), *B. phytofirmans* PsJN (six copies), and *B. petrii*. The identities between these *Burkholderia* and *Bordetella* Tn*6049*-like transposons are much higher (up to 100 % for *tnmC*) than with Tn*6049* (Ryan et al. 2009; Van Houdt et al. 2009).

6.6.3 *Tn*6048 *and Tn*6050*: Transposons with "Unconventional" Accessory Genes*

Transposon Tn*6048* (Tn*5053*-like family) contains next to a module of four genes involved in mobility eight metal-inducible genes. The *mmfABC1DC2RSM* cluster codes for a peptidase (*mmfA*), a phosphatase (*mmfB*), two major facilitator superfamily (MFS) permeases (*mmfC1* and *mmfC2*), a drug/metabolite transporter (DMT) super family permease (*mmfM*), a two-component regulatory system (*mmfRS*), and a putative secreted protein (*mmfD*). Highly similar clusters were found on plasmid pBVIE02 of *B. vietnamiensis* G4 (on Tn*6048*-like transposon), in *Acidovorax delafieldii* 2AN (draft sequence), and in *Dechloromonas aromatica* RCB. In a related cluster in *Thiomonas* sp. 3As, the peptidase, phosphatase, and two-component regulatory system are associated with a resistance-nodulation-cell division (RND) superfamily efflux pump instead of permeases. Together with the *pbr* cluster on plasmid pMOL30 of *C. metallidurans* CH34 (Borremans et al. 2001; Hynninen et al. 2009; Taghavi et al. 2009), they make up three examples where a peptidase and a phosphatase are associated with a transport system (ATPase, RND efflux pump, or MFS permease).

Transposon Tn*6050* (two copies in the chromid of *C. metallidurans* CH34) contains five genes. Two are involved in its mobility and are 100 % identical (protein level) with their counterparts on Tn*4380*, the mercury transposon of pMOL30. Yet, the accessory genes *sulP* (sulfate permease), *uspA11* (universal stress response), and *dksA* (suppressor of *dnaK*) are very different from metal resistance genes. These genes are often syntenic (not always associated with a transposon) in a variety of genomes. The massive sequencing of bacterial genomes is likely to uncover new combinations of accessory genes in mobile genetic elements that are not related to classical resistance, virulence, or catabolic genes, with Tn*6049* and Tn*6050* as examples.

6.7 Insight in Integrases (TBSSRs)

The *C. metallidurans* CH34 genome carries in total a quite high number of TBSSRs (28 intact and 16 fragments), especially compared to closely related bacteria like *C. taiwanensis* LMG19424 (1 intact and 5 fragments) (Amadou et al. 2008). The paucity of integrase genes in the chromosome and chromid of *C. taiwanensis* LMG19424 and also of *C. eutrophus* H16, which both have expanded from a common backbone (Amadou et al. 2008), suggests that putative horizontally transferred islands have become permanently domesticated. Contrary to the *C. eutrophus* H16 chromosome and cosmid, the megaplasmid pHG1 carries an extensive "junkyard" region encompassing 39 remnants of transposases and phage-type integrases (Schwartz et al. 2003). This suggests that the acquisition and rearrangements of pHG1 are more recent similar to the development of pMOL28 and pMOL30 (see above). Two particular cases/modules will be discussed in more detail below.

6.7.1 RIT Elements: A Three-Gene Structure for Site-Specific Recombinases in Trio

The *C. metallidurans* CH34 genome revealed the presence of particular modules with three adjacent and overlapping genes encoding tyrosine-based site-specific recombinases (TBSSRs). These elements were named RIT (recombinases in trio).

One RIT belongs to the integrase module of CMGI-2 (ICE_{Tn4371}6054) and inactivates a *radC*-like gene (putatively involved in DNA repair). A second RIT belongs to the complex accessory module of CMGI-2 where it inactivated a *tnpA* gene (IS*66* family). These RIT elements were later found in other bacteria and taxa (e.g., Firmicutes, α-, β-, and δ-proteobacteria) (Van Houdt et al. 2012a). For instance, identical RIT copies were found in *Burkholderia phytofirmans* PsJN (three copies), *Aromatoleum aromaticum* EbN1 (three copies), *Dinoroseobacter shibae* DFL 12 (two copies), *Heliobacterium modesticaldum* Ice1 (two copies), *Bordetella petrii* DSM 12804 (two copies), *Caulobacter* sp. K31 (three copies), *Mesorhizobium loti* MAFF303099 (two copies), and *Gramella forsetii* KT0803 (two copies). Their distribution among different taxa and multiple copies in a genome provides some hints into the mobility of the RIT elements. Finally, although not particularly well conserved, the three proteins in RITs make distinct families when included in a set of over 1,300 TBSSR protein sequences (Van Houdt et al. 2012a). All three proteins (families) display a possible catalytic motif, suggesting that the three enzymes may be active. Apparently, RITs are more frequent in chromosomes (in 62.3 % of the cases) than in plasmids (in 37.7 % of the cases). At least for 19 chromosomally embedded RITs more information is available on the genomic context, indicating that for this group approximately 68 % is located on a predicted genomic island.

6.7.2 PRQ Modules

Next to RIT elements, the *C. metallidurans* CH34 genome also harbors two modules of four adjacent and overlapping genes encoding three putative TBSSRs and a putative small regulatory protein. The modules were provisionally called PRQ (Para RIT in Quatuor), carrying *prq* genes. The fourth and smallest gene clearly points to two groups (*prqABCP* and *prqABCR*). Two PRQ modules were found in *C. metallidurans* CH34. The small *prqP* gene belongs to a family of conserved hypothetical protein-coding genes while *prqR* is often annotated as *tetR*-like pointing to a regulatory function. One is located at the extremity of CMGI-7, which carries genes involved in the resistance to arsenic. Synteny analysis revealed at least nine conserved modules (*prqA1B1C1prqP1*) in other bacteria. The second module is associated with genomic island CMGI-28b in plasmid pMOL28 (Rmet_6252-6255; *prqA2B2C2R2*) and displayed synteny with around 35 other bacterial genomes. The two *prq* clusters of CH34 share very little identity and looked to have been inherited separately. Both appeared to lie at the extremity of a genomic island (possible difference with RIT) and resembled more to other clusters such as *bimAB* (Van Houdt et al. 2012a) found at the extremity of a pMOL30 metal resistance genomic island that is also found in *R. pickettii* 12J (Van Houdt et al. 2009). Maybe two distinct families of *prq* elements have to be considered separately as the corresponding syntenies do not or barely overlap. Inside the *Cupriavidus* genus, there is one equivalent (with 25 % protein identity) in the *C. necator* N-1 genome (Fig. 6.2) and two copies in *C. metallidurans* NA4 (100 % conserved). This is surprisingly low and may reflect the peculiar ability of *C. metallidurans* CH34

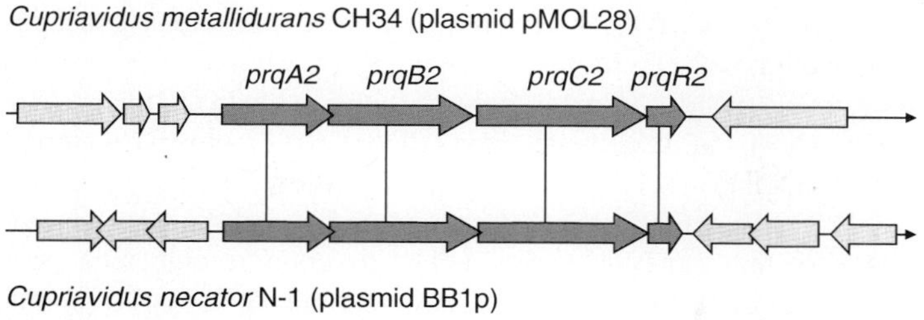

Fig. 6.2 Syntenic comparison of two regions from *C. metallidurans* CH34 and C. *necator* N-1. The PRQ module genes belonging to a synteny block are in *dark gray* and connected

to capture a variety of mobile genetic elements that are not necessarily present in related strains and its talent as a dedicated "collector" from different hosts. These *prq* elements are observed frequently in γ-proteobacteria, especially in vibrios and in enterics (pathogens), in α-proteobacteria, and in some actinobacteria. Many of them are closely linked to *uvrA1* and *ssb* genes as it is the case in CH34 and NA4. This points to some DNA processing that could be linked to the action of site-specific recombinases and also suggests that these elements have defined accessory genes. On the other hand, the duplication in *C. metallidurans* NA4 suggests that the element is mobile. Still, insufficient information is available about the mobility and functions of these elements.

6.8 The Megaplasmids pMOL28 and pMOL30: Specialized in the Maximal Viable Response to Heavy Metals

Megaplasmid pMOL28 carries genes for resistance to chromate, mercury, cobalt, and nickel cations, all clustered in the complex genomic island CMGI-28a (Mergeay et al. 2009; Monchy et al. 2007). The plasmid core with the main functions for replication, maintenance, and transfer is very similar to the megaplasmid cores of pHG1 from *C. eutrophus* H16 (Schwartz et al. 2003) and pRALTA from *C. taiwanensis* LMG19424. These megaplasmids might well be regarded as the decisive factor in fixing their hosts to their ecological niche considering the distinct and specialized functions on pMOL28 (heavy metal resistance in collaboration with pMOL30), pHG1 (hydrogenotrophy and chemolithotrophy), and pRALTA (nitrogen fixation and legume symbiosis). A fourth megaplasmid, this of *C. necator* N-1, belongs to the same group. However, the accessory genes of the latter are very different than those found on pHG1, and there is no evidence for a specialization in chemolithotrophy in the N-1 megaplasmid.

Despite the detailed annotation of megaplasmids pMOL28 and pMOL30, continuous manual curation of the CH34 genome via updated databases, and earlier functional information from proteomic and transcriptomic approaches, many enigmatic genes remain in both plasmids. Recent transcriptomic analyses revealed that five of such genes, located on CMGI-28a, are induced by submillimolar concentrations of chromate (Juhnke et al. 2002; Monsieurs et al. 2011). All five genes are sited immediately downstream of the chromate-regulated operon *chrIA1B1CEF1* (Rmet_6199 through Rmet_6204) and were named *chrO* (Rmet_6198), *chrN* (Rmet_6197), *chrP* (Rmet_6196, encoding a MFS family permease with ten transmembrane α-helical segments), *chrY* (Rmet_6195, encoding a putative membrane protein), and *chrZ* (Rmet_6194). Synteny analysis shows a fair to good resemblance between the pMOL28 *chr* locus of *C. metallidurans* CH34 and the *chr* loci of *R. solanacearum* Y45 (Li et al. 2011), *Methylobacterium* sp. 4–46, *B. vietnamiensis* G4, *B. pseudomallei*, and *Arthrobacter sp*. FB24 (an actinobacterium); however, various gene combinations are observed (Janssen et al. 2010). This puzzling variation in *chr* locus structure may point to differences among these bacteria in terms of selectivity to chromate or another metal oxyanion. The situation in *R. solanacearum* Y45 is interesting because the chromate genes are adjacent to the *cnr* genes as in pMOL28. There are no *bimAB* genes (with *bimA* encoding a TBSSR as in pMOL28) but a different TBSSR encoding gene at the extremity of the *cnr* genes, adjacent to the first *cnr* gene *cnrY*. It suggests again a new combination between mobile modules (here the integrase module and this of accessory genes).

Plasmid pMOL30 encodes mainly resistance to cadmium, zinc, and cobalt (in a ~25 kb genomic island CMGI-30a that is also found with a high identity in the *R. pickettii* 12J chromosome) as well as to copper (Monchy et al. 2006), lead (Borremans et al. 2001; Hynninen et al. 2009; Taghavi et al. 2009), and mercury cations. Especially as far as the *cop* and *czc* genes are concerned, it is striking to see the amount of additional genes that are required besides the major well-known mechanisms of efflux and detoxification. For instance, the copper-inducible genes of pMOL30 constitute a cluster of around 35 genes from which only 11 encode proteins responsible for the copper resistance genes (P-ATPase, periplasmic detoxification *pco/copABCD* and *silCBA*, a RND-based cation/proton antiporter efflux system) (Mergeay et al. 2003). It was hypothesized that these additional genes are necessary for the adaptation to very high concentrations of heavy metals (with concentrations ranging up to 20 mM) in circumstances where the already numerous metal resistance genes on the chromid of *C. metallidurans* CH34 are insufficient (Janssen et al. 2010; Mergeay 2000; Monchy et al. 2007). Some recent reports about such peculiar genes of pMOL30 and their encoded proteins begin to be released: CopK (Bersch et al. 2008), a methionine-rich small periplasmic protein; SilB (Bersch et al. 2011) with a metallochaperone domain; CopH (Sendra et al. 2006), a periplasmic copper-binding protein and its paralog CzcE (Pompidor et al. 2007; Zoropogui et al. 2008); CzcP (Scherer and Nies 2009), a novel PIB4-type ATPase; and CnrX (actually on pMOL28), an anti-sigma factor (Pompidor et al. 2008), but a comprehensive view of their role and synergies is still needed.

A direct relationship of the megaplasmid pMOL30 to other known megaplasmids is less clear than the relationship observed with pMOL28 and reported above. Nonetheless, through synteny analysis, it was noted that the backbone of pMOL30 shares, over a region of approximately 60 kb, many characteristics with the pBVIE01 megaplasmid (398 kb) (Table 6.1) of *B. vietnamiensis* G4 (Mergeay et al. 2009).

6.9 *Cupriavidus* and *Ralstonia* Chromid Genes and Environmental Adaptation

A few characteristic features of the chromids will be reviewed with a focus on genomic islands, the resistance-nodulation-cell division superfamily (RND) efflux systems and catabolic genes.

6.9.1 The Genomic Islands of the C. metallidurans *CH34 Chromid*

The mosaic structure of the C. metallidurans CH34 chromid (and other *Cupriavidus*, *Ralstonia*, and *Burkholderia* strains) made it difficult to identify genomic islands based on (1) lack of synteny, (2) a base composition and/or phylogeny differing from the bulk of the genome, and (3) a higher content of hypothetical genes than neighboring regions. Consequently, no genomic islands were identified on the *C. metallidurans* CH34 chromid in a first approach (Ryan et al. 2009; Van Houdt et al. 2009). However, comparison of 16 different *C. metallidurans* strains to strain CH34 via genomic hybridization revealed the presence of at least five genomic islands on the *C. metallidurans* CH34 chromid (Van Houdt et al. 2012b) (Table 6.2). One island (CMGI-B ± D) carries multiple genes coding for phage-related proteins (associated to phages of *Ralstonia solanacearum*) and other genes in synteny with *R. solanacearum*. Genomic island CMGI-E carries genes putatively involved in the degradation of aromatic compounds (vanillate O-demethylase oxygenase and phthalate 4,5-dioxygenase) (Van Houdt et al. 2012b).

6.9.2 Metal Resistance Genes and Especially RND Genes

The *C. metallidurans* CH34 genome harbors 12 loci coding for RND efflux systems. These are distributed among the megaplasmids pMOL28 (*cnr*) and pMOL30 (*czc*, *sil*, and *ncc*), the chromid (*cus*, *nim*, *czc2*, *hmv*, *zni*, *zne*, *hmy*), and the chromosome (*agr*, *hmz*). Despite considerable efforts, the diversity of the RND mechanisms, especially on the chromid, remains to be fully elucidated. Likewise, many heavy metal resistance loci carry genes accessory to the efflux mechanism that are still poorly studied and only partly understood. The *czc* and *cnr* systems on pMOL30 and pMOL28, respectively, were analyzed in most detail. A possibility is that the chromid allowed a fairly good adaptation to heavy metals in a (volcanic) biotope colonized by the ancestor of strain CH34 before its more recent adaptation to industrial biotopes in which the acquisition and rearrangements of megaplasmids became decisive for the survival of the bacteria. The duplication of some genes and functions could have been followed by the inactivation of redundant genes (that were maybe less adapted to extreme metallic conditions).

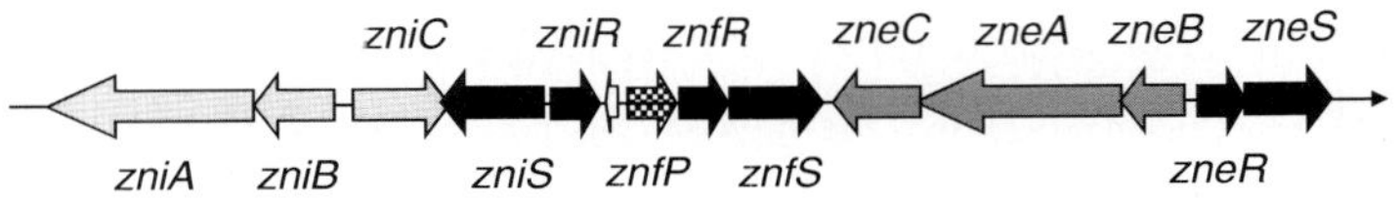

Fig. 6.3 Genomic organization of two RND systems on the *C. metallidurans* CH34 chromid. *Dark* and *light gray*: genes coding for RND systems; *black*: genes coding for two-component regulatory systems; *checkered*: gene coding for a MFS permease

One particular locus on the chromid with two RND systems and a MFS permease attracts the attention (Fig. 6.3). The *zneBAC* operon structure consists of a gene coding for a membrane fusion protein (B), an integral membrane pump protein belonging to the RND superfamily (A), and an outer membrane pore protein (C) instead of the more common CBA structure. A similar arrangement is found in chromids of most *Ralstonia* spp. (*R. syzygii* R24, *R. solanacearum* PS107, CMR15, Y45, and GMI1000, *R. haywardii* ssp. *celebensis* R229). The *zne* cluster is likely linked to zinc resistance since the ZneB protein (membrane fusion protein), which has been purified and crystallized to a resolution of 2.8 Å, was shown to be highly specific for Zn^{2+} (De Angelis et al. 2010). However, the system is not induced by any of the 16 metals (including Zn^{2+}) analyzed in a transcriptomic study (Monsieurs et al. 2011).

The *zniBA zniC zniS zniR* cluster has also a very peculiar structure. Both the genes coding for the efflux pump and the regulatory system are divergently transcribed (Fig. 6.3). A similar arrangement is also found in most *Ralstonia* spp. and in *Bradyrhizobium* BTAi1. Furthermore, *R. pickettii* 12J carries three copies of the *zni* cluster, and all three are associated with another efflux gene coding for a CDF protein (cation diffusion factor) and a transposon with *tnsAtniABQ* genes (somewhat similar to those in Tn*6048* (*tniABQR*)). In fact one of the three transposons is slightly different in identity and contains a second CDF gene.

Finally, the *zne* and *zni* systems appear to mainly belong to the specific patrimony of the genus *Ralstonia* as they are not present in other *Cupriavidus* genomes. These genes were not induced by a monometallic challenge (Monsieurs et al. 2011), suggesting that they are either inactive or that they respond to very different inducers or physiological circumstances. Examples are the induction of *zneBAC* and *zneRS* in magnesium-free growth medium (unpublished results) and the induction of *hmy* when basalt was used as iron source (Olsson-Francis et al. 2010).

6.9.3 Catabolic Genes in Cupriavidus *and* Ralstonia

Chromids are a good observatory to look at catabolic genes and especially those of *C. pinatubonensis* JMP134 (Lykidis et al. 2010) and *R. pickettii* 12J. As an example, a cluster of 25 *R. pickettii* 12J chromid genes (Rpic_4616-4641) involved in the catabolism of aromatic compounds (*tmo*, *tom*, *mhp*) has high identity orthologs in perfect synteny in the *C. pinatubonensis* JMP134 chromid (79 % protein identity on average) but also in the *C. metallidurans* CH34 chromosomally embedded genomic island ICE_{Tn4371}6064 (with some rearrangements and somewhat lower identity). It may be recalled that the three strains are environmental strains typical of anthropogenic/industrial biotopes. Codon usage of the *C. metallidurans* CH34 catabolic genes is clearly divergent from the host supporting recent horizontal acquisition (some genes of this island are also almost 100 % identical with *B. vietnamiensis* G4 orthologs), while the orthologous *R. pickettii* genes share the codon usage of their host.

Other catabolic clusters on the *R. pickettii* chromid that illustrate the variety of acquired catabolic genes are (1) the degradation of phenylacetate (*paa* genes; also observed in all *Ralstonia* chromids and *Cupriavidus* chromosomes), (2) the degradation of aliphatic sulfonates (also observed in *C. metallidurans* chromosome and the chromids of most of the other *Cupriavidus*), (3) the metabolism of phosphonoacetaldehyde (also observed in all *Ralstonia* chromids and the nosocomial *C. metallidurans*), (4) the degradation of catechol via dioxygenase (also observed in all *Cupriavidus* except *C. metallidurans*), (5) homogentisate 1–2 dioxygenase (also observed in all *Cupriavidus*/*Ralstonia* chromids), and (6) 2-nitropropane dehydrogenase (also observed in all *Cupriavidus*/*Ralstonia* chromids).

An interesting catabolic pathway of *C. metallidurans* CH34 is the degradation of acetone and isopropanol (Rosier et al. 2012), which was serendipitously discovered through proteomic studies made with bacteria exposed to a spaceflight to the International Space Station (Leys et al. 2009). The involved *acxRABC* genes are found in the *C. metallidurans* CH34 and NA4 chromids, in the *C. necator* N-1 chromid, and in the megaplasmid of *C. pinatubonensis* JMP134. Although quite rare the cluster can be found in α-, β-, γ-, and ε-proteobacteria as well as in some actinobacteria.

6.10 Conclusion: Some Visibility on the Genomic Strategies Used by Some β-Proteobacteria for Their Adaptation to Harsh Anthropogenic Environments

The study of the accessory genes associated to the various genetic elements and involved in a variety of environmental and metabolic responses showed that:

1. The accessory genes can be associated to different recognizable mobile elements (e.g., autonomous vs integrated MGEs) and yet be quite conserved.
2. They may appear fully "integrated" in the chromosome of their host (most often integrated in the chromid as far as the genera *Ralstonia*, *Cupriavidus*, and *Burkholderia* are concerned). Evidence of the mobile genetic element carrying the accessory genes has been lost. The indications of a horizontal gene transfer are generally restricted to differences in GC content or codon usage as well as lack of synteny.
3. Intermediate situations do also exist as suggested by the inactivation of genes involved in mobility (deletions, insertions in tyrosine-based site-specific

recombinases or divergences between flanking IS elements). Transposon Tn*6049* appears to play a peculiar role in "stabilizing" genomic islands in strain CH34 as reported above. Again, we hypothesize that such inactivation of key mobile elements helps to stabilize the genomic island conferring a specific advantage to its host. This could be one of the very first steps of the further irreversible erosion of mobility genes. Especially for genes related to the response to high concentrations of heavy metals as it is observed in metallurgical sites, natural selection is expected to force their stability by inactivating the mobility of their MGE carriers.

All three categories have been found in the *C. metallidurans* CH34 genome by comparison with other closely related strains and especially with the plant-associated *C. taiwanensis* LMG19424 and *R. solanacearum* for which the adaptation to plants could be considered much more ancient than the adaptation of *C. metallidurans* CH34 to man-made situations (e.g., wastes of chemical or metallurgical industry).

Genes involved in the catabolism of man-made very recalcitrant organic chemicals (also called organic xenobiotics) that were likely synthetized during the late nineteenth or the twentieth century are often associated to mobile genetic elements of infectious character as judged by their presence and maximal conservation in the genomes of taxonomically distant bacteria (broad host range). Some of these catabolic genes may be considered to be recently emerged via genomic rearrangements, i.e., during the twentieth century and captured in some way by active mobile genetic elements. This kind of rationale may also be extended to other functions that are linked to anthropogenic activities as the resistance to heavy metals, to antibiotics, and even some virulence genes. This also makes perceptible the evolution of living cells and their genomes inside a measurable time period with a striking feeling of immediacy and accessibility even at the experimental level. Some steps of the evolutionary mechanisms involved in horizontal gene transfer, the acquisition of new functions, and their stabilization in the genomes of interest look to be accessible to experimentation. In this respect, to speak about natural gene engineering (Shapiro 2011) by reference to "in vitro" genetic engineering and genetically modified organisms (GMOs) is not anymore a metaphor but also a way to address the current controversy about GMOs in a more consensual way by exploring the mechanisms by which bacteria may (1) acquire genes from taxonomic distant sources, (2) modify and rearrange genes as a response to environmental characteristics (such as the availability of potential new carbon sources even from xenobiotic sources), and (3) integrate new genetic combinations into its regulatory and structural network (genome structure). As it appears from the studies on mobile genetic elements and their accessory genes in bacteria, natural genetic engineering sorted out by natural selection can be powerful, finely regulated, and fast in its accomplishments.

A major challenge rising from these studies on genomes of β-proteobacteria adapted to anthropogenic environments and rich in new kinds of mobile genetic elements and accessory functions is to scrutinize the inventories and to decipher the correlated functions.

Acknowledgments This work was funded by Belspo ESA grants MELGEN-1, MISSEX, and COMICS in the frame of projects around life support in space environments and environmental microbiology of ISS (International Space Station).

References

Amadou C, Pascal G, Mangenot S, Glew M, Bontemps C, Capela D, Carrere S, Cruveiller S, Dossat C, Lajus A, Marchetti M, Poinsot V, Rouy Z, Servin B, Saad M, Schenowitz C, Barbe V, Batut J, Medigue C, Masson-Boivin C (2008) Genome sequence of the beta-rhizobium *Cupriavidus taiwanensis* and comparative genomics of rhizobia. Genome Res 18:1472–1483

Bersch B, Favier A, Schanda P, van Aelst S, Vallaeys T, Coves J, Mergeay M, Wattiez R (2008) Molecular structure and metal-binding properties of the periplasmic CopK protein expressed in *Cupriavidus metallidurans* CH34 during copper challenge. J Mol Biol 380:386–403

Bersch B, Derfoufi KM, De Angelis F, Auquier V, Ekende EN, Mergeay M, Ruysschaert JM, Vandenbussche G (2011) Structural and metal binding characterization of the C-terminal metallochaperone domain of membrane fusion protein SilB from *Cupriavidus metallidurans* CH34. Biochemistry (Mosc) 50:2194–2204

Borremans B, Hobman JL, Provoost A, Brown NL, Van der Lelie D (2001) Cloning and functional analysis of the *pbr* lead resistance determinant of *Ralstonia metallidurans* CH34. J Bacteriol 183:5651–5658

Brim H, Heyndrickx M, de Vos P, Wilmotte A, Springael D, Schlegel HG, Mergeay M (1999) Amplified rDNA restriction analysis and further genotypic characterisation of metal-resistant soil bacteria and related facultative hydrogenotrophs. Syst Appl Microbiol 22:258–268

Chen WM, Laevens S, Lee TM, Coenye T, De Vos P, Mergeay M, Vandamme P (2001) *Ralstonia taiwanensis* sp. nov., isolated from root nodules of Mimosa species and sputum of a cystic fibrosis patient. Int J Syst Evol Microbiol 51:1729–1735

Coenye T, Spilker T, Reik R, Vandamme P, Lipuma JJ (2005) Use of PCR analyses to define the distribution of *Ralstonia* species recovered from patients with cystic fibrosis. J Clin Microbiol 43:3463–3466

Craig JW, Chang FY, Brady SF (2009) Natural products from environmental DNA hosted in *Ralstonia metallidurans*. ACS Chem Biol 4:23–28

De Angelis F, Lee JK, O'Connell JD 3rd, Miercke LJ, Verschueren KH, Srinivasan V, Bauvois C, Govaerts C, Robbins RA, Ruysschaert JM, Stroud RM, Vandenbussche G (2010) Metal-induced conformational changes in ZneB suggest an active role of membrane fusion proteins in efflux resistance systems. Proc Natl Acad Sci USA 107:11038–11043

Diels L, Mergeay M (1990) DNA probe-mediated detection of resistant bacteria from soils highly polluted by heavy metals. Appl Environ Microbiol 56:1485–1491

Diels L, Faelen M, Mergeay M, Nies D (1985) Mercury transposons from plasmids governing multiple resistance to heavy metals in *Alcaligenes eutrophus* CH34. Arch Physiol Biochem 93:27–28

Diels L, Sadouk A, Mergeay M (1989) Large plasmids governing multiple resistances to heavy metals - a genetic approach. Toxicol Environ Chem 23:79–89

Diels L, Vanroy S, Taghavi S, Doyen W, Leysen R, Mergeay M (1993) The use of *Alcaligenes eutrophus* immobilized in a tubular membrane reactor for heavy-metal recuperation. In: Torma AE, Apel ML, Brierley CL (eds) International biohydrometallurgy symposium (IBS-93), Jackson Hole, 22–25 August 1993, pp 133–144

Diels L, Dong QH, Vanderlelie D, Baeyens W, Mergeay M (1995a) The *czc* operon of *Alcaligenes eutrophus* CH34: from resistance mechanism to the removal of heavy metals. J Ind Microbiol 14:142–153

Diels L, Van Roy S, Somers K, Willems I, Doyen W, Mergeay M, Springael D, Leysen R (1995b) The use of bacteria immobilized in tubular membrane reactors for heavy metal recovery and degradation of chlorinated aromatics. J Membr Sci 100:249–258

Diels L, Van Roy S, Leysen R, Mergeay M (1996) Heavy metal bioprecipitation by *Alcaligenes Eutrophus* CH34 immobilized in a membrane bioreactor. Int Biodeterioration Biodegredation 37:239

Diels L, Van Roy S, Taghavi S, Van Houdt R (2009) From industrial sites to environmental applications with *Cupriavidus metallidurans*. Antonie Van Leeuwenhoek 96:247–258

Don RH, Pemberton JM (1981) Properties of six pesticide degradation plasmids isolated from *Alcaligenes paradoxus* and *Alcaligenes eutrophus*. J Bacteriol 145:681–686

Don RH, Pemberton JM (1985) Genetic and physical map of the 2,4-dichlorophenoxyacetic acid-degradative plasmid pJP4. J Bacteriol 161:466–468

Goris J, De Vos P, Coenye T, Hoste B, Janssens D, Brim H, Diels L, Mergeay M, Kersters K, Vandamme P (2001) Classification of metal-resistant bacteria from industrial biotopes as *Ralstonia campinensis* sp nov., *Ralstonia metallidurans* sp nov and *Ralstonia basilensis* Steinle et al. 1998 emend. Int J Syst Evol Microbiol 51:1773–1782

Guglielmini J, Quintais L, Garcillan-Barcia MP, de la Cruz F, Rocha EP (2011) The repertoire of ICE in prokaryotes underscores the unity, diversity, and ubiquity of conjugation. PLoS Genet 7:e1002222

Harrison PW, Lower RP, Kim NK, Young JP (2010) Introducing the bacterial 'chromid': not a chromosome, not a plasmid. Trends Microbiol 18:141–148

Hynninen A, Touze T, Pitkanen L, Mengin-Lecreulx D, Virta M (2009) An efflux transporter PbrA and a phosphatase PbrB cooperate in a lead-resistance mechanism in bacteria. Mol Microbiol 74:384–394

Janssen PJ, Van Houdt R, Moors H, Monsieurs P, Morin N, Michaux A, Benotmane MA, Leys N, Vallaeys T, Lapidus A, Monchy S, Medigue C, Taghavi S, McCorkle S, Dunn J, van der Lelie D, Mergeay M (2010) The complete genome sequence of *Cupriavidus metallidurans* strain CH34, a master survivalist in harsh and anthropogenic environments. PLoS One 5:e10433

Juhnke S, Peitzsch N, Hubener N, Grosse C, Nies DH (2002) New genes involved in chromate resistance in *Ralstonia metallidurans* strain CH34. Arch Microbiol 179:15–25

Klockgether J, Wurdemann D, Reva O, Wiehlmann L, Tummler B (2007) Diversity of the abundant pKLC102/PAGI-2 family of genomic islands in *Pseudomonas aeruginosa*. J Bacteriol 189:2443–2459

Larbig KD, Christmann A, Johann A, Klockgether J, Hartsch T, Merkl R, Wiehlmann L, Fritz HJ, Tummler B (2002) Gene islands integrated into tRNA(Gly) genes confer genome diversity on a *Pseudomonas aeruginosa* clone. J Bacteriol 184:6665–6680

Lejeune P, Mergeay M, Vangijsegem F, Faelen M, Gerits J, Toussaint A (1983) Chromosome transfer and R-prime plasmid formation mediated by plasmid pULB113 (RP4-mini-Mu) in *Alcaligenes eutrophus* CH34 and *Pseudomonas fluorescens* 6.2. J Bacteriol 155:1015–1026

Leys N, Baatout S, Rosier C, Dams A, s'Heeren C, Wattiez R, Mergeay M (2009) The response of *Cupriavidus metallidurans* CH34 to spaceflight in the international space station. Antonie Van Leeuwenhoek 96:227–245

Li Z, Wu S, Bai X, Liu Y, Lu J, Xiao B, Lu X, Fan L (2011) Genome sequence of the tobacco bacterial wilt pathogen *Ralstonia solanacearum*. J Bacteriol 193:6088–6089

Liesegang H, Lemke K, Siddiqui RA, Schlegel HG (1993) Characterization of the inducible nickel and cobalt resistance determinant *cnr* from pMOL28 of *Alcaligenes eutrophus* ch34. J Bacteriol 175:767–778

Lykidis A, Perez-Pantoja D, Ledger T, Mavromatis K, Anderson IJ, Ivanova NN, Hooper SD, Lapidus A, Lucas S, Gonzalez B, Kyrpides NC (2010) The complete multipartite genome sequence of *Cupriavidus necator* JMP134, a versatile pollutant degrader. PLoS One 5:e9729

Magrisso S, Erel Y, Belkin S (2008) Microbial reporters of metal bioavailability. Microb Biotechnol 1:320–330

Mergeay M (2000) Bacteria adapted to industrial biotopes: the metal resistant *Ralstonia*. In: Storz G, Hengge-Aronis R (eds) Bacterial stress responses. ASM, Washington D.C., pp 403–414

Mergeay M, Houba C, Gerits J (1978) Extrachromosomal inheritance controlling resistance to cadmium, cobalt, copper and zinc ions: evidence from curing a *Pseudomonas*. Archives Internationales De Physiologie De Biochimie Et De Biophysique 86:440–442

Mergeay M, Nies D, Schlegel HG, Gerits J, Charles P, Van Gijsegem F (1985) *Alcaligenes eutrophus* CH34 is a facultative chemolithotroph with plasmid-bound resistance to heavy metals. J Bacteriol 162:328–334

Mergeay M, Monchy S, Vallaeys T, Auquier V, Benotmane A, Bertin P, Taghavi S, Dunn J, van der Lelie D, Wattiez R (2003) *Ralstonia metallidurans*, a bacterium specifically adapted to toxic metals: towards a catalogue of metal-responsive genes. FEMS Microbiol Rev 27:385–410

Mergeay M, Monchy S, Janssen P, Van Houdt R, Leys N (2009) Megaplasmids in *Cupriavidus* genus and metal resistance. In: Schwartz E (ed) Microbial megaplasmids, vol 11, Microbiology Monographs. Springer, Berlin, pp 209–238

Merlin C, Springael D, Mergeay M, Toussaint A (1997) Organisation of the bph gene cluster of transposon Tn*4371*, encoding enzymes for the degradation of biphenyl and 4-chlorobiphenyl compounds. Mol Gen Genet 253:499–506

Merlin C, Springael D, Toussaint A (1999) Tn*4371*: A modular structure encoding a phage-like integrase, a Pseudomonas-like catabolic pathway, and RP4/Ti-like transfer functions. Plasmid 41:40–54

Mijnendonckx K, Provoost A, Monsieurs P, Leys N, Mergeay M, Mahillon J, Van Houdt R (2011) Insertion sequence elements in *Cupriavidus metallidurans* CH34: distribution and role in adaptation. Plasmid 65:193–203

Mijnendonckx K, Provoost A, Ott CM, Venkateswaran A, Mahillon J, Leys N, Van Houdt R (2013) Characterization of the survival ability of *Cupriavidus metallidurans* and *Ralstonia pickettii* from space-related environments. Microb Ecol 65:347–360

Monchy S, Benotmane MA, Wattiez R, van Aelst S, Auquier V, Borremans B, Mergeay M, Taghavi S, van der Lelie D, Vallaeys T (2006) Transcriptomic and proteomic analyses of the pMOL30-encoded copper resistance in *Cupriavidus metallidurans* strain CH34. Microbiology 152: 1765–1776

Monchy S, Benotmane MA, Janssen P, Vallaeys T, Taghavi S, van der Lelie D, Mergeay M (2007) Plasmids pMOL28 and pMOL30 of *Cupriavidus metallidurans* are specialized in the maximal viable response to heavy metals. J Bacteriol 189:7417–7425

Monsieurs P, Moors H, Van Houdt R, Janssen PJ, Janssen A, Coninx I, Mergeay M, Leys N (2011) Heavy metal resistance in *Cupriavidus metallidurans* CH34 is governed by an intricate transcriptional network. Biometals 24:1133–1151

Nies DH (1995) The cobalt, zinc, and cadmium efflux system CzcABC from *Alcaligenes eutrophus* functions as a cation-proton antiporter in *Escherichia coli*. J Bacteriol 177:2707–2712

Nies DH (1999) Microbial heavy-metal resistance. Appl Microbiol Biotechnol 51:730–750

Nies DH (2000) Heavy metal-resistant bacteria as extremophiles: molecular physiology and biotechnological use of *Ralstonia* sp CH34. Extremophiles 4:77–82

Nies DH (2003) Efflux-mediated heavy metal resistance in prokaryotes. FEMS Microbiol Rev 27:313–339

Nies D, Mergeay M, Friedrich B, Schlegel HG (1987) Cloning of plasmid genes encoding resistance to cadmium, zinc, and cobalt in *Alcaligenes eutrophus* CH34. J Bacteriol 169:4865–4868

Nies A, Nies DH, Silver S (1989) Cloning and expression of plasmid genes encoding resistances to chromate and cobalt in *Alcaligenes eutrophus*. J Bacteriol 171:5065–5070

Ohtsubo Y, Ishibashi Y, Naganawa H, Hirokawa S, Atobe S, Nagata Y, Tsuda M (2012) Conjugal transfer of polychlorinated biphenyl/biphenyl degradation genes in *Acidovorax* sp. strain KKS102, which are located on an integrative and conjugative element. J Bacteriol 194: 4237–4248

Olsson-Francis K, VH R, Mergeay M, Leys N, Cockell CS (2010) Microarray analysis of a microbe-mineral interaction. Geobiology 8:446–456

Palleroni NJ (2003) Prokaryote taxonomy of the 20th century and the impact of studies on the genus *Pseudomonas*: a personal view. Microbiology 149:1–7

Peitzsch N, Eberz G, Nies DH (1998) *Alcaligenes eutrophus* as a bacterial chromate sensor. Appl Environ Microbiol 64:453–458

Poehlein A, Kusian B, Friedrich B, Daniel R, Bowien B (2012) Complete genome sequence of the type strain *Cupriavidus necator* N-1. J Bacteriol 193:5017

Pompidor G, Zoropogui A, Kahn R, Coves J (2007) Overproduction, purification and preliminary X-ray diffraction analysis of CzcE from *Cupriavidus metallidurans* CH34. Acta Crystallogr Sect F Struct Biol Cryst Commun 63:884–886

Pompidor G, Maillard AP, Girard E, Gambarelli S, Kahn R, Coves J (2008) X-ray structure of the metal-sensor CnrX in both the apo- and copper-bound forms. FEBS Lett 582:3954–3958

Providenti MA, Shaye RE, Lynes KD, McKenna NT, O'Brien JM, Rosolen S, Wyndham RC, Lambert IB (2006) The locus coding for the 3-nitrobenzoate dioxygenase of *Comamonas* sp. strain JS46 is flanked by IS*1071* elements and is subject to deletion and inversion events. Appl Environ Microbiol 72:2651–2660

Remenant B, de Cambiaire JC, Cellier G, Jacobs JM, Mangenot S, Barbe V, Lajus A, Vallenet D, Medigue C, Fegan M, Allen C, Prior P (2011) *Ralstonia syzygii*, the blood disease bacterium and some Asian *R. solanacearum* strains form a single genomic species despite divergent lifestyles. PLoS One 6:e24356

Ricker N, Qian H, Fulthorpe RR (2012) The limitations of draft assemblies for understanding prokaryotic adaptation and evolution. Genomics 100:167–175

Rosier C, Leys N, Henoumont C, Mergeay M, Wattiez R (2012) Purification and characterization of the acetone carboxylase of *Cupriavidus metallidurans* strain CH34. Appl Environ Microbiol 78:4516–4518

Ryan MP, Pembroke JT, Adley CC (2006) *Ralstonia pickettii*: a persistent gram-negative nosocomial infectious organism. J Hosp Infect 62:278–284

Ryan MP, Pembroke JT, Adley CC (2007) *Ralstonia pickettii* in environmental biotechnology: potential and applications. J Appl Microbiol 103:754–764

Ryan MP, Pembroke JT, Adley CC (2009) Novel Tn*4371*-ICE like element in *Ralstonia pickettii* and genome mining for comparative elements. BMC Microbiol 9:242

Salanoubat M, Genin S, Artiguenave F, Gouzy J, Mangenot S, Arlat M, Billault A, Brottier P, Camus JC, Cattolico L, Chandler M, Choisne N, Claudel-Renard C, Cunnac S, Demange N, Gaspin C, Lavie M, Moisan A, Robert C, Saurin W, Schiex T, Siguier P, Thebault P, Whalen M, Wincker P, Levy M, Weissenbach J, Boucher CA (2002) Genome sequence of the plant pathogen *Ralstonia solanacearum*. Nature 415:497–502

Sato Y, Nishihara H, Yoshida M, Watanabe M, Rondal JD, Concepcion RN, Ohta H (2006) *Cupriavidus pinatubonensis* sp. nov. and *Cupriavidus laharis* sp. nov., novel hydrogen-oxidizing, facultatively chemolithotrophic bacteria isolated from volcanic mudflow deposits from Mt. Pinatubo in the Philippines. Int J Syst Evol Microbiol 56:973–978

Scherer J, Nies DH (2009) CzcP is a novel efflux system contributing to transition metal resistance in *Cupriavidus metallidurans* CH34. Mol Microbiol 73:601–621

Schmidt T, Schlegel HG (1994) Combined nickel-cobalt-cadmium resistance encoded by the *ncc* locus of *Alcaligenes xylosoxidans* 31A. J Bacteriol 176:7045–7054

Schmidt T, Stoppel RD, Schlegel HG (1991) High-level nickel resistance in *Alcaligenes xylosoxydans* 31A and *Alcaligenes eutrophus* KTO2. Appl Environ Microbiol 57:3301–3309

Schwartz E (2009) Microbial megaplasmids, Microbiology Monographs. Springer, Berlin

Schwartz E, Henne A, Cramm R, Eitinger T, Friedrich B, Gottschalk G (2003) Complete nucleotide sequence of pHG1: a *Ralstonia eutropha* H16 megaplasmid encoding key enzymes of H(2)-based ithoautotrophy and anaerobiosis. J Mol Biol 332:369–383

Sen D, Van der Auwera GA, Rogers LM, Thomas CM, Brown CJ, Top EM (2011) Broad-host-range plasmids from agricultural soils have IncP-1 backbones with diverse accessory genes. Appl Environ Microbiol 77:7975–7983

Sendra V, Cannella D, Bersch B, Fieschi F, Menage S, Lascoux D, Coves J (2006) CopH from Cupriavidus metallidurans CH34. A novel periplasmic copper-binding protein. Biochemistry (Mosc) 45:5557–5566

Sensfuss C, Schlegel HG (1988) Plasmid pMOL28-encoded resistance to nickel is due to specific efflux. FEMS Microbiol Lett 55:295–298

Shapiro J (2011) Evolution, a view from 21st century. FT Press Science, Upper Saddle River

Springael D, Kreps S, Mergeay M (1993) Identification of a catabolic transposon, Tn*4371*, carrying biphenyl and 4-chlorobiphenyl degradation genes in *Alcaligenes eutrophus* A5. J Bacteriol 175:1674–1681

Springael D, Diels L, Mergeay M (1994) Transfer and expression of PCB-degradative genes into heavy metal resistant *Alcaligenes eutrophus* strains. Biodegradation 5:343–357

Springael D, Ryngaert A, Merlin C, Toussaint A, Mergeay M (2001) Occurrence of Tn*4371*-related mobile elements and sequences in (chloro)biphenyl-degrading bacteria. Appl Environ Microbiol 67:42–50

Stanier RY, Palleroni NJ, Doudoroff M (1966) The aerobic pseudomonads: a taxonomic study. J Gen Microbiol 43:159–271

Stoppel RD, Schlegel HG (1995) Nickel-resistant bacteria from anthropogenically nickel-polluted and naturally nickel-percolated ecosystems. Appl Environ Microbiol 61:2276–2285

Stoppel RD, Meyer M, Schlegel HG (1995) The nickel resistance determinant cloned from the Enterobacterium *Klebsiella oxytoca* - conjugational transfer, expression, regulation and DNA homologies to various nickel-resistant bacteria. Biometals 8:70–79

Taghavi S, Vanderlelie D, Mergeay M (1994) Electroporation of *Alcaligenes eutrophus* with (mega) plasmids and genomic DNA fragments. Appl Environ Microbiol 60:3585–3591

Taghavi S, Lesaulnier C, Monchy S, Wattiez R, Mergeay M, van der Lelie D (2009) Lead(II) resistance in *Cupriavidus metallidurans* CH34: interplay between plasmid and chromosomally-located functions. Antonie Van Leeuwenhoek 96:171–182

Vallenet D, Labarre L, Rouy Z, Barbe V, Bocs S, Cruveiller S, Lajus A, Pascal G, Scarpelli C, Medigue C (2006) MaGe: a microbial genome annotation system supported by synteny results. Nucleic Acids Res 34:53–65

Van Gijsegem F, Toussaint A (1982) Chromosome transfer and R-prime formation by an RP4::mini-Mu derivative in *Escherichia coli*, *Salmonella typhimurium*, *Klebsiella pneumoniae*, and *Proteus mirabilis*. Plasmid 7:30–44

Van Houdt R, Mergeay M (2012) Plasmids as secondary chromosomes. In: Bell E, Bond J, Klinman J, Masters B, Wells R (eds) Molecular life sciences: an encyclopedic reference, Springer reference (www.springerreference.com). doi:10.1007/SpringerReference_333639

Van Houdt R, Monchy S, Leys N, Mergeay M (2009) New mobile genetic elements in *Cupriavidus metallidurans* CH34, their possible roles and occurrence in other bacteria. Antonie Van Leeuwenhoek 96:205–226

Van Houdt R, Toussaint A, Ryan MP, Pembroke JT, Mergeay M, Adley CC (2011) The Tn*4371* ICE family of bacterial mobile genetic elements. In: Roberts AP, Mullany P (eds) Bacterial integrative mobile genetic elements. Landes Bioscience, Austin

Van Houdt R, Leplae R, Lima-Mendez G, Mergeay M, Toussaint A (2012a) Towards a more accurate annotation of tyrosine-based site-specific recombinases in bacterial genomes. Mob DNA 3:6

Van Houdt R, Monsieurs P, Mijnendonckx K, Provoost A, Janssen A, Mergeay M, Leys N (2012b) Variation in genomic islands contribute to genome plasticity in *Cupriavidus metallidurans*. BMC Genomics 13:111

Vandamme P, Coenye T (2004) Taxonomy of the genus *Cupriavidus*: a tale of lost and found. Int J Syst Evol Microbiol 54:2285–2289

Vaneechoutte M, Kampfer P, De Baere T, Falsen E, Verschraegen G (2004) *Wautersia* gen. nov., a novel genus accommodating the phylogenetic lineage including *Ralstonia eutropha* and related species, and proposal of *Ralstonia* [*Pseudomonas*] *syzygii* (Roberts et al. 1990) comb. nov. Int J Syst Evol Microbiol 54:317–327

von Rozycki T, Nies DH (2009) *Cupriavidus metallidurans*: evolution of a metal-resistant bacterium. Antonie Van Leeuwenhoek Int J Gen Mol Microbiol 96:115–139

Wyndham RC, Nakatsu C, Peel M, Cashore A, Ng J, Szilagyi F (1994) Distribution of the catabolic transposon Tn*5271* in a groundwater bioremediation system. Appl Environ Microbiol 60:86–93

Yabuuchi E, Kosako Y, Yano I, Hotta H, Nishiuchi Y (1995) Transfer of two *Burkholderia* and an *Alcaligenes* species to *Ralstonia* gen. Nov.: proposal of *Ralstonia pickettii* (Ralston, Palleroni and Doudoroff 1973) comb. Nov., *Ralstonia solanacearum* (Smith 1896) comb. Nov. and *Ralstonia eutropha* (Davis 1969) comb. Nov. Microbiol Immunol 39:897–904

Zoropogui A, Gambarelli S, Coves J (2008) CzcE from *Cupriavidus metallidurans* CH34 is a copper-binding protein. Biochem Biophys Res Commun 365:735–739

Chapter 7
Conjugative Elements: Host Chromosome Function Modifiers

Masaki Shintani, Yurika Takahashi, and Hideaki Nojiri

Abstract Conjugative transfer of plasmids and integrative and conjugative elements (ICEs) play a key role in rapid bacterial evolution and adaptation. These elements give a host several new phenotypes that can be advantageous for host survival in some cases, but disadvantageous in others. Comparisons of growth rate and biomass yield as well as competition assays have shown that carriage of plasmids frequently imposes a fitness cost on the host cells. New phenotypes are caused by cross talk between plasmid- and chromosome-encoded factors during implantation of regulatory circuits into the host native transcriptional network. Nucleoid-associated proteins (NAPs) are important factors for adaptation of the ICEs to the new host cell environment. Plasmids and ICEs, containing genes involved in the biodegradation of various compounds, are excellent models to study the impact on the host. A chlorocatechol-degradative ICE, ICE$_{clc}$, can express its degradative genes without changing the fitness of the new host. The host of toluene/xylene-degradative plasmid pWW0 builds cooperative regulatory systems to express

M. Shintani (✉)
Japan Collection of Microorganisms, RIKEN BioResource Center,
3-1-1 Kouyadai, Tsukuba, Ibaraki 305-0074, Japan

Department of Applied Chemistry and Biochemical Engineering,
Graduate School of Engineering, Shizuoka University, 3-5-1 Johoku, Naka-ku,
Hamamatsu 432-8561, Shizuoka, Japan
e-mail: tmshint@ipc.shizuoka.ac.jp

Y. Takahashi
Biotechnology Research Center, The University of Tokyo,
1-1-1 Yayoi, Bukyo-ku, Tokyo 113-8657, Japan

H. Nojiri
Biotechnology Research Center, The University of Tokyo,
1-1-1 Yayoi, Bukyo-ku, Tokyo 113-8657, Japan

Agricultural Bioinformatics Research Unit, The University of Tokyo,
1-1-1 Yayoi, Bukyo-ku, Tokyo 113-8657, Japan

H. Nojiri et al. (eds.), *Biodegradative Bacteria: How Bacteria Degrade, Survive, Adapt, and Evolve*, DOI 10.1007/978-4-431-54520-0_7, © Springer Japan 2014

metabolic activity. The influence of naphthalene-degradative plasmid NAH7 on the host cells is modest, and the carriage of NAH7 alleviates host stress when exposed to naphthalene. The effects of carbazole-degradative pCAR1 carriage on the host cells depend on the type of host, and the NAPs encoded on pCAR1 are important for controlling the transcriptional network of the host. The extent of the impact on the host cells changes the genetic stability of these elements in the host, which is important for spreading these genetic elements among different bacteria.

Keywords Fitness • Horizontal gene transfer • Integrative conjugative elements • Nucleoid-associated protein • Plasmid

7.1 Introduction

Horizontal gene transfer (HGT) enables bacteria to acquire various new traits. Among the processes of HGT, conjugation is the most effective means to spreading genes (Guglielmini et al. 2011) and is therefore one of the most important phenomena for rapid bacterial evolution and adaptation. Bacterial plasmids are mobile genetic elements (MGEs) and extrachromosomal genetic units that can self-replicate and be transferred among various bacteria by conjugation (Sota and Top 2008; Frost et al. 2005). Integrative and conjugative elements (ICEs) are also transferable by conjugation and are similar to plasmids but are usually integrated into the host chromosome (Wozniak and Waldor 2010). Plasmids and ICEs are able to transfer very large DNA regions of genomes, containing many genes that endow various host phenotypes such as antibiotic and heavy metal resistance, pathogenicity, and biodegradation ability for various compounds, including persistent organic pollutants. In some cases, expression of new phenotypes is advantageous for host survival, but in other cases, expression of new phenotypes is disadvantageous. Expression of phenotypes beneficial for host survival requires the successful implantation of gene regulatory circuits on MGEs into the host native transcriptional network. Occasionally, the implantation causes a fitness cost to the host cell, resulting in decreased host survival. These effects occur by various interactions between plasmid- and chromosome-encoded factors. In this chapter, we review recent reports of how plasmids and ICEs alter host phenotypes as well as how these alterations are caused by MGEs.

7.2 Classification of Plasmids and ICEs

A large number of plasmids have been identified, and the nucleotide sequences of 3,173 plasmids have been determined (as of Sept. 2012, ftp://ftp.ncbi.nih.gov/genomes/Plasmids/). Bacterial plasmids have been classified into incompatibility (Inc) groups based on their replication and partition systems. When two different plasmids cannot be maintained in a single bacterial cell line, these two plasmids are deemed incompatible and belong to the same Inc group. There are 27 Inc groups for the *Enterobacteriaceae* (Carattoli 2009), at least 14 groups for pseudomonads (Thomas and Haines 2004), and approximately 18 groups for staphylococci (Frost et al. 2005; Sota and Top 2008), although these groups do not comprise all plasmids. Notably, six mobility (MOB) classes (MOB_F, MOB_H, MOB_Q, MOB_C, MOB_P, and MOB_V) and four mating pair formation (MPF) types (MPF_F, MPF_I, MPF_G, and MPF_T) have been proposed on the basis of amino acid sequences of the proteins involved in plasmid MOB or the type IV secretion system during MPF. These groupings enabled us to classify more plasmids in plasmid databases (Smillie et al. 2010; Garcillán-Barcia et al. 2009; 2011).

Fewer ICEs have been experimentally identified than plasmids (approximately 10 kinds of ICEs have been described) because ICEs are ordinarily linked to the host chromosome (Wozniak and Waldor 2010). Guglielmini et al. found 682 putative conjugative systems through in silico analysis of more than 1,000 prokaryotic genome sequences (Guglielmini et al. 2011). Thus, ICEs are likely more abundant MGEs than plasmids in bacteria, although a detailed classification of ICEs is not yet available.

7.3 Effects of Carriage of New Genetic Elements on Host Cells

Since the 1970s, plasmids have been used as a vector for genes to produce various proteins (including enzymes), and many studies on how to improve protein production in *Escherichia coli* using ColE1-type vectors and their derivatives have been described. In these studies, various comparisons have been performed to identify the effect of the plasmid (vector) carriage on host cells, including growth rate, fitness cost, biomass yield, glucose consumption, oxygen uptake rate, and survivability of the host (Lee and Edlin 1985; Bouma and Lenski 1988; Birnbaum and Bailey 1991; Rhee et al. 1994; Andersson et al. 1996; Rozkov et al. 2004; Buch et al. 2010; Sharma et al. 2011). Similarly, the above comparisons have been performed with naturally occurring plasmids, which are conjugative or non-conjugative plasmids isolated from nature (Table 7.1). Several studies have shown that plasmid carriage induces biofilm formation (D'Alvise et al. 2010; Engberg et al. 1975; Ghigo 2001; Reisner et al. 2003; Yang et al. 2008). The results of these studies suggest that plasmid carriage is frequently a burden to the host.

Over the past decade, comprehensive analyses of the burden caused by plasmids have been performed by comparing host transcriptomic and proteomic data with information on plasmid and host chromosome DNA sequences. Notably, ColE1-type plasmid carriage commonly reduces the glycolysis and TCA cycle activity and increases the production of acetate (Diaz Ricci and Hernández 2000; Wang et al. 2006; Ow et al. 2006).

Currently, only two studies have reported the impact of ICEs on host cells: one report on ICE_{clc} (Sect. 7.6) and the second report on Tn*1549* (Foucault et al. 2010). Tn*1549* was identified in *Enterococcus* spp. and carries vancomycin resistance genes (Garnier et al. 2000). The fitness costs to the hosts carrying these ICEs were not significant; however, artificial constitutive expression of vancomycin resistance

Table 7.1 Effects of plasmid carriage

Host	Plasmid	Analysis	Results	Reference
E. coli K-12	R1*drd19* (IncFII, 100 kb) and variants with different copy numbers	• Growth rate • Scanning electron microscopy	As the plasmid copy number increased, the growth rate of the host was reduced The cell length increased by the carriage of the plasmid and correlated with the plasmid copy number	Engberg et al. (1975)
E. coli 114	pH121(= PB165, IncQ, non-conjugative)	• Growth curve • Competition assay	No differences were observed in the respective growth curve for plasmid-free and plasmid-containing hosts. In the mixed culture of plasmid-free and plasmid-containing hosts, the ratio of plasmid-containing hosts gradually decreased	Dale and Smith (1979)
E. coli W3110	RP1 (=RP4, IncP-1, 60 kb)	• Minimal nutritional requirements • Growth rate	RP1-containing hosts have greater requirement for Mg^{2+}, K^+, Fe^{2+}, and PO_4^{3-} than RP1-free hosts The growth rate of plasmid-free and plasmid-containing hosts was identical in a simple salt medium; however, the growth rate of plasmid-containing hosts was reduced in the presence of low concentrations of phosphate	Klemperer et al. (1979)
E. coli K12 921	Drug resistance plasmids (101 plasmids) (the original hosts were *Escherichia*, *Salmonella*, *Enterobacter*, *Klebsiella*)	• Growth rate	The majority of the larger (>80 kb) R plasmids increased the doubling time of the host, whereas the majority of the smaller (<80 kb) plasmids did not increase the doubling time of the host	Zünd and Lebek (1980)
E. coli K12	R1 (IncFII, 100 kb) RP4 (IncP-1, 60 kb)	• Competition assay • Long-term culturing	No plasmid loss occurred. The fitness cost of the host was reduced through genetic changes in both the plasmids and host chromosome	Dahlberg and Chao (2003)
E. coli K12 MG1655	R1 (IncFII, ~100 kb)	• Competition assay • Long-term culturing	An evolved plasmid appeared in 2 of 5 experiments, which conferred a fitness advantage to the host. The evolved plasmid increased the relative fitness of *S. enterica*	Dionisio et al. (2005)
E. coli K12-derived strain XK1502	R1 (IncFII, ~100 kb) and its derivative	• Competition assay • *In silico* simulation	The fitness cost to hosts with the R1 plasmid carrying the fertility inhibition (*fin*) system was smaller than the fitness cost to hosts with the *fin*-deficient R1-derivative	Haft et al. (2009)

P. fluorescens SBW25	pQBR103 (425 kb)	• Survival of plasmid-free and plasmid-containing strains on sugar beet	The population of the plasmid-containing strains decreased in comparison with plasmid-free strains after seed inoculation. One hundred days after the inoculation, as the sugar beets matured, no differences were detected in the populations of plasmid-free and plasmid-containing strains	Lilley and Bailey (1997)
P. chlororaphis SPR044	pJP4 (IncP-1, 80-kb)	• Survival of *P. chlororaphis* (and its derivatives) with pJP4 in the *Arabidopsis thaliana* rhizosphere was assessed.	The competitiveness of hosts carrying pJP4 was reduced in the rhizosphere Both abiotic and biotic stresses resulted in disadvantages of plasmid-containing strains	Schmidt-Eisenlohr and Baron (2003)
S. typhimurium	pSf-R27 (IncHI, no antibiotic resistance) pSf-R27Δ*sfh* (the Sfh protein is closely related to the global regulator H-NS)	• Transcriptome • Competition assay • Virulence • Motility	The influence of the host carrying pSF-R27Δ*sfh* on the transcriptome and fitness cost was larger than the influence of the host carrying pSF-R27 pSf-R27Δ*sfh* imposed a mild H-NS-deficient phenotype, with fitness and motility reduction	Doyle et al. (2007)
P. fluorescens SBW25	pQBR103 (425 kb)	• Single-cell Raman spectroscopy (whole-cell chemical composition, reflecting metabolic status)	Single-cell Raman spectral profiles were significantly changed by carriage of the plasmid The changes were due to alteration of nucleic acids or their component amount in the cells, likely because of DNA replication stress	Ude et al. (2007)
P. putida KT2440 *P. aeruginosa* PAO1 *P. fluorescens* Pf0-1	pCAR1 (IncP-7, 200 kb)	• Growth rate • Transcriptome • Pyoverdine yield	The growth rates of plasmid-free and plasmid-containing strains were identical Transcriptome alterations were observed (see main text)	Shintani et al. (2010a)

genes resulted in a large reduction of host fitness (Foucault et al. 2010). This fact, together with the studies of ICE$_{clc}$ (Sect. 7.6), ICEs may impose a smaller impact on their host, although the number of reports of ICEs is still insufficient.

7.4 Roles of Nucleoid-Associated Proteins (NAPs) That Carry Conjugative Elements

Bacterial chromosomal DNA is folded by NAPs to a compact structure, the nucleoid (Dorman and Kane 2009; Dillon and Dorman 2010). Some NAPs are well-known DNA-binding proteins, such as Fis (factor for inversion stimulation), H-NS (histone-like protein H1 from *E. coli* strain U93), IHF (integration host factor), and Lrp (leucine-responsive regulatory protein), but there are less well-studied NAPs, such as NdpA, which are distributed among Gram-negative bacteria (Takeda et al. 2011). These NAPs are not only involved in forming the nucleoid, but are also global regulators of gene expression (Dillon and Dorman 2010). Notably, H-NS, which binds to A+T-rich DNA regions, provides a host defense system against acquired foreign DNA with low G+C content by binding to low G+C regions and silencing the expression of (detrimental) genes in the xenogenetic region (Navarre et al. 2006).

H-NS and other NAPs are not only encoded on the bacterial chromosome but also on plasmids (Takeda et al. 2011). One such plasmid, pSf-R27 in *Salmonella*, has a gene encoding an H-NS-like protein, Sfh (Doyle et al. 2007). Carriage of *sfh-deleted* pSf-R27 results in alteration of the transcription of more genes, in addition to the transcription of the native pSF-R27 (Doyle et al. 2007). Thus, a protein homologous to H-NS, encoded on the plasmid, likely has a stealth function that enables the plasmid to transfer its genetic material into the host cell without interfering with the host transcriptional network, by silencing host xenogenetic genes and maintaining host fitness. Takeda et al. found relationships between plasmid features (size, G+C content, transferability) and the presence of NAP-encoding genes on the plasmid, based on a comparison of more than 2,000 plasmids: (1) a large plasmid frequently carries NAPs genes, (2) a (putative) transferable plasmid tends to carry NAP genes, and (3) the G+C content of a plasmid with an H-NS-encoding gene is relatively low (Takeda et al. 2011). Therefore, NAPs may have important roles in plasmid distribution. In addition, H-NS encoded on the plasmid may allow the host to stably maintain the plasmid via the stealth function, even in hosts with higher G+C contents. Recently, Castang and Dove reported that the genes for two H-NS family proteins, MvaT and MvaU, encoded on the chromosome cannot be removed from the host *P. aeruginosa* PAO1 because these proteins are crucial for repressing the expression of Pf4 prophage (an acquired foreign DNA), which kills cells or inhibits cell growth (Castang and Dove 2012). NAPs, including H-NS family proteins, may have key functions in the coexistence of the host and MGEs.

7.5 Degradative Plasmids and ICEs

In the last three decades, a large number of plasmids have been identified that carry genes involved in the biodegradation of various compounds, known as degradative plasmids (Nojiri et al. 2004; Shintani and Nojiri 2013). Among these plasmids, IncP-1 (MOB_P, MPF_T), P-7 (MOB_H, MPF_F), and P-9 (MOB_F, MPF_T) group plasmids have been thoroughly analyzed in regard to their basic features, such as replication, partition, and conjugative transfer. It should be noted that many other plasmids, such as carbazole-degradative plasmid pCAR3 (MOB_F, MPF_F) (Shintani et al. 2007), biphenyl-degradative plasmid pNL1 (MOB_F, MPF_F) (Romine et al. 1999), and dibenzo-*p*-dioxin-degradative plasmid pSWIT02 (MOB_P, MPF_T) (Miller et al. 2010), which probably belong to unidentified Inc groups, have not been studied in detail, although their sequences have been determined. One of the characteristic features of the IncP-1, IncP-7, and IncP-9 groups is the host range of these plasmids. The host range is determined by replication and conjugative transfer systems and can be broad (IncP-1), narrow (IncP-7), or intermediate (IncP-9) (Shintani et al. 2010b). Several ICEs carry degradative genes, such as biphenyl-degradative genes on the *bph-sal* element (Nishi et al. 2000) or Tn*4371*-like element (or $ICE_{Tn\mathit{4371}}$) (Toussaint et al. 2003; Ryan et al. 2009) as well as the $ICE_{KKS102}\mathit{4677}$ (Ohtsubo et al. 2003, 2006, 2012) and 3-chlorocatechol-degradative genes on the *clc* element (or ICE_{clc}) (Ravatn et al. 1998a).

The depletion of the xenobiotics is dependent on the replication and (stable) maintenance of these MGEs, expression of degradative genes on the MGEs, efficient degradation of intermediates during the degradative reactions (i.e., no accumulation of intermediates) by cooperative regulation(s) of enzyme expression. In other words, the degradative ability can change depending on the host strain, and some hosts are not able to degrade xenobiotics because (1) the degradative ability is not expressed in the host (Goris et al. 2002), (2) the genes on the host chromosome are not efficiently transcribed because they are not induced by the intermediate compounds of the substrate degradation (or accumulation of an intermediate compound) (Takahashi et al. 2009a; Shintani et al. 2011a), and (3) genetic instability of degradative genes in the host cells (Williams et al. 1988; Duetz and van Andel 1991; Leddy et al. 1995). Therefore, degradative MGEs represent an excellent model system for understanding the influence of genetic elements on host cells by analyzing cooperative regulation of transcriptional networks.

The following sections describe the effect of degradative MGEs on host cells in more detail. The discussion is focused on ICE_{clc}, two IncP-9 plasmids, pWW0 and NAH7, and the IncP-7 plasmid pCAR1.

7.6 3- and 4-Chlorocatechol and 2-Aminophenol-Degradative ICE_{clc}

ICE_{clc} in *Pseudomonas knackmussii* B13, a 3-chlorobenzoate (3-CBA) metabolizing strain, normally carries ICE_{clc} at two chromosomal locations (Ravatn et al. 1998a). The nucleotide sequence of the 103 kb ICE_{clc} element was determined, and the

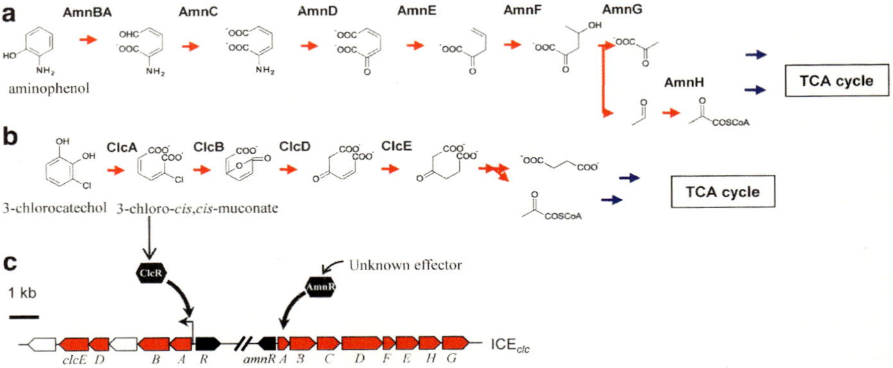

Fig. 7.1 Metabolic pathway of aminophenol (**a**) and 3-chlorocatechol (**b**) in *P. knackmussii* B13 carrying ICE$_{clc}$. *Red arrows* indicate conversions catalyzed by the Amn (**a**) and Clc (**b**) enzymes encoded on ICE$_{clc}$, while *blue arrows* indicate other enzymes encoded on the host chromosome. (**c**) Genetic organization of the *clc* and *amn* genes on ICE$_{clc}$ of the *P. knackmussii* B13 chromosome. *Black*, *red*, and *white pentagons* indicate regulatory genes, catabolic genes, and other unrelated ORFs, respectively. *Arrows* from 3-chloro-*cis*,*cis*-muconate in panel (**b**) indicate that the compound is an effector of the transcriptional regulator ClcR. Transcription upstream from the promoter on *clcA* is induced by the effector. The effector of AmnR has not yet been identified

element was found to carry not only genes essential for 3-CBA degradation, *clcRABDE*, which encodes the enzymes for 3- and 4-chlorocatechol degradation, but also 2-aminophenol metabolism, *amn* (Fig. 7.1; Gaillard et al. 2006). ICE*clc*-like elements are distributed among various bacteria. One biphenyl-degrading bacterium, *Burkholderia xenovorans* LB400, carries a nearly identical ICE$_{clc}$ with additional *o*-halobenzoate-degradative genes (Gaillard et al. 2006; Chain et al. 2006). *Ralstonia* sp. strain JS705, isolated from groundwater, also carries an ICE*clc*-like element with another insertion of chlorobenzene-degradative genes (Müller et al. 2003). The element was initially recognized as an IncJ plasmid able to transfer among *Beta*- and *Gammaproteobacteria* (Ravatn et al. 1998c); however, this transfer is not due to a conjugative plasmid but an ICE (Ravatn et al. 1998a, b). Interestingly, the transfer system of ICE$_{clc}$ has two functional origins of transfer (*oriT*s), which is rather different from other transferable genetic elements such as plasmids (Miyazaki and van der Meer 2011a, b).

The *clc* genes are induced in the presence of 2-chloro-*cis*, *cis*-muconate, which is an intermediate compound of the 3-CBA and 4-chlorobenzoate-degradation pathway and is acting as an inducer of ClcR (Fig. 7.1; Coco et al. 1994). These genes are also induced by anthranilate, benzoate, and salicylate (Gaillard et al. 2006), which are predicted to be converted to catechol and *cis*, *cis*-muconate, an analog of 2-chloro-*cis*, *cis*-muconate. ClcR is, however, not efficiently induced by *cis*, *cis-muconate* (Park and Kim 2001). Therefore, activation of the *clcA* promoter by anthranilate, benzoate, and salicylate likely occurs via cross-activation by a non-ICEclc, chromosome-encoded transcriptional activator such as CatR (Parsek et al. 1994).

The ICE is ordinarily integrated at the 3′-end of *tRNA*^{Gly}-encoding genes on the host chromosome (Gaillard et al. 2006). The element is occasionally excised from the chromosome and forms a circular shape, primarily in stationary phase cells, and then transferred to a new host cell and site-specifically integrates into the *tRNAGly-encoding* gene in the chromosome of the new host (Sentchilo et al. 2009). Fifteen transcripts of ICE$_{clc}$ have been identified by northern blot and microarray analyses, and eight of these transcripts are upregulated during the stationary phase (Gaillard et al. 2010). These transcripts are found only after pre-growth of the cells on 3-CBA or fructose, on which the transfer of this element is promoted (Gaillard et al. 2010).

Gaillard et al. assessed the impact of acquiring ICE$_{clc}$ in the *P. aeruginosa* PAO1 chromosome, a nonnative carrier of this element, using transcriptome profiling, phenotype arrays, competition assays for fitness cost, and biofilm formation studies (Gaillard et al. 2008). Only a few differences between ICE$_{clc}$-free PAO1 and ICE$_{clc}$-containing PAO1 were detected (Gaillard et al. 2008), contrary to most studies on plasmids, as previously mentioned. Repression of three potential operons on the PAO1 chromosome is observed by carriage of ICE$_{clc}$. One of the operons is likely involved in transport of small molecules (PA3232–PA3235), and the two other operons are involved in acetoin (*aco*) and glycolate catabolism (*glcDEF*) (Gaillard et al. 2008). No fitness cost was detected in ICE$_{clc}$-containing PAO1 by the competition assay, suggesting that the repression of the operons did not affect the host fitness (Gaillard et al. 2008). Comprehensive comparisons of host phenotypes using phenotypic microarrays showed that no significant changes occurred by carriage of ICE$_{clc}$ (Gaillard et al. 2008). The carriage of ICE$_{clc}$ on the PAO1 chromosome resulted in the reduction of biofilm formation during growth on succinate, glucose, or acetate, whereas the growth rate and biomass yield were not significantly affected by these substrates (Gaillard et al. 2008). In contrast, the catabolic genes on ICE$_{clc}$ were not silent. The ICE$_{clc}$-containing PAO1 was able to grow on 3-chlorobenzoate and 2-aminophenol as sole carbon sources (Gaillard et al. 2008). ICE$_{clc}$ could be transferred from PAO1 to *P. putida* UWC1, indicating that the transfer machinery of ICE$_{clc}$ was also correctly expressed in PAO1 (Gaillard et al. 2008). The presence of xenogenetic silencing by MvaT and MvaU (H-NS family proteins in PAO1) for the ICE$_{clc}$ in PAO1 is unclear; however, gene transcription on the element in PAO1 may not be silenced but comparable to the original host, B13 (Gaillard et al. 2006). These results indicate that the newly acquired MGE not always reduces host fitness. Although the fitness cost to other hosts than PAO1 is still unclear, the low fitness cost of ICE$_{clc}$ may result in successful implantation of the element into the host, enlarging the host range.

7.7 IncP-9-Degradative Plasmids

IncP-9 is a major group of degradative plasmids (Shintani et al. 2010b). The replication and partition systems of IncP-9-degradative plasmids have not been well investigated; however, the systems of the IncP-9 plasmid pM3 have been characterized

(Sevastsyanovich et al. 2005). The *oriV-rep* system of IncP-9 cannot function in *E. coli* without a part of the *par* region *in cis* (Sevastsyanovich et al. 2005). The IncP-9 plasmids are known to transfer to *Pseudomonas* species as well as *E. coli*, *Erwinia chrysanthemi*, *Hydrogenophaga palleronii*, *Serratia* species, and *Burkholderia* species (Benson and Shapiro 1978; Nakazawa 1978; Nancharaiah et al. 2003; Ramos-Gonzalez et al. 1991; Sarand et al. 2000). Notably, Miyazaki et al. reported that the *traD* operon of NAH7 is a host-range modifier in conjugative transfer (Miyazaki et al. 2008).

7.7.1 Toluene/Xylene-Degradative Plasmid pWW0

pWW0 is an IncP-9 plasmid identified in *P. putida* mt-2, carrying toluene/xylene-degradative (*xyl*) genes contained by two kinds of class II transposons, Tn*4651* and Tn*4653* (Tsuda and Iino 1987, 1988; Tsuda et al. 1989). The 117-kb nucleotide sequence of pWW0 has been determined (Greated et al. 2002). The transcriptional and translational regulatory mechanisms of pWW0 have been well characterized (reviewed in Ramos et al. 1997; Ruiz et al. 2004). Toluene is converted to benzoate by gene products encoded on an upper operon, and benzoate is converted to acetaldehyde and pyruvate via *meta*-cleavage by gene products on the *meta* operon (Fig. 7.2; Greated et al. 2002; Ramos et al. 1997). *xylR* and *xylS*, located downstream of the *meta* operon, encode specific transcriptional regulators that activate the expression of pathway genes. Toluene induces transcription of the upper operon from promoter *Pu* by acting as an effector of XylR. Benzoate is a transcriptional inducer of the *meta* operon from promoter *Pm* and an effector of XylS. Toluene also induces the expression of the *meta* operon without being converted to benzoate (Inouye et al. 1987; Ramos et al. 1987). The regulatory circuit of these operons on pWW0 in the host *P. putida* is built by interplay of factors encoded on both the plasmid and the host chromosome. At least four host sigma factors are required for expressing the entire set of *xyl* genes. *Pu* is a σ^{54}-dependent promoter, under positive control of the NtrC family protein XylR (de Lorenzo et al. 1991; Pérez-Martín and de Lorenzo 1996). The *Pu* activity is also influenced by IHF and the nitrogen-related phosphoenolpyruvate:carbohydrate phosphotransferase system (PtsN) of *P. putida* (Cases et al. 1999; Aranda-Olmedo et al. 2006). Another NAP, HU, is predicted to be involved in *Ps* activity (Ramos et al. 1997). Activated XylR also positively regulates another σ^{54}-dependent promoter of *xylS*, *Ps1*, and negatively regulates its own promoter *Pr*, containing the tandem promoters *Pr1* and *Pr2* (Marqués et al. 1998; Bertoni et al. 1997). In addition to σ^{54}, other sigma factors are involved in the transcriptional regulation; σ^{38} and σ^{32} in the transcription from *Pm*, and the housekeeping σ^{70} in the transcription from *Pr1*/*Pr2* and from *Ps2*, the other *xylS* promoter present in the *Ps* region (Fig. 7.2; Marqués et al. 1995, 1999). TurA, an H-NS family protein is an additional host factor that binds to *Pu* and likely represses transcription at low temperatures (Rescalli et al. 2004). PprA, a LytTR-type two-component response regulator, is another host factor that inhibits *Pu* transcription by a mutual exclusion mechanism between PprA and XylR that bind to upstream

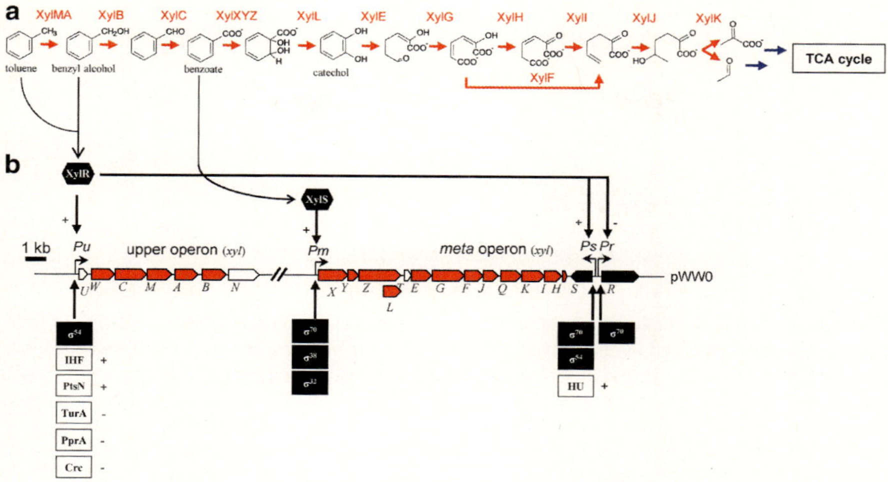

Fig. 7.2 (**a**) Catabolic pathway of toluene in *P. putida* mt-2, carrying pWW0. *Red arrows* indicate conversions catalyzed by the Xyl enzymes encoded on pWW0, and *blue arrows* indicate other enzymes encoded on the host chromosome. (**b**) Genetic organization and regulatory circuit of the *xyl* genes on pWW0 in mt-2 (reviewed by Ramos et al. 1997 and Ruiz et al. 2004). *Black*, *red*, and *white pentagons* indicate regulatory genes, catabolic genes, and other related genes, respectively. *Pr* and *Ps* contain tandem promoters *Pr1* and *Pr2* as well as *Ps1* and *Ps2*, respectively. *Arrows* from toluene, benzyl alcohol, and benzoate in panel (**a**) indicate effectors of the transcriptional regulators XylR and XylS. Host factors involved in the promoter activity are shown in *black* (sigma factors) and *white* (other host factors) *boxes*. The symbols "+" and "–" indicate induction and repression of transcription from the promoter, respectively

activating sequences (Vitale et al. 2008). The Crc protein, a global regulator that controls carbon flow in the host cell, has an important role in inhibition via catabolite repression of preferred carbon sources. Crc is a translational repressor that binds to the translation initiation regions of the mRNAs coding for XylR and XylS (Moreno et al. 2010). Crc also binds to the translation initiation regions on the mRNAs of eight *xyl* genes (*xylU*, *xylX*, *xylA*, *xylB*, *xylN*, *xylL*, and *xylQ*). Notably, five of the seven genes on the upper operon contained a strong binding site for Crc, suggesting that Crc tightly regulated the genes involved in the upper pathway of toluene-to-benzoate conversion (Moreno et al. 2010). In summary, implantation of the regulatory circuit of the degradation pathway of pWW0 into the host cell requires cross talk between plasmid- and chromosome-encoded factors.

7.7.2 Naphthalene-Degradative Plasmid NAH7

NAH7 is a self-transmissible plasmid of the IncP-9 group, identified in the naphthalene-degrading bacterium *P. putida* G7 (Dunn and Gunsalus 1973). The 83-kb nucleotide sequence of this plasmid has been determined (Sota et al. 2006).

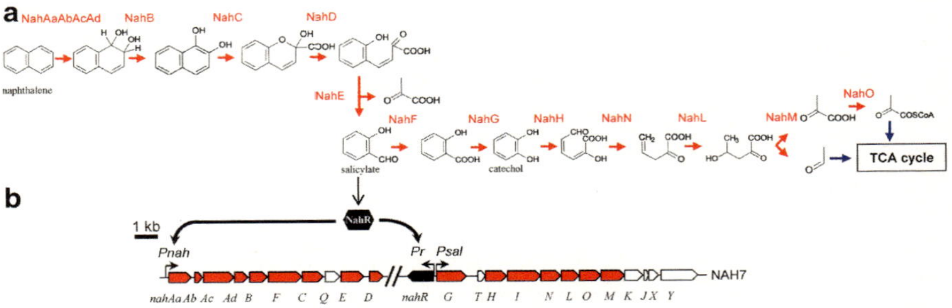

Fig. 7.3 (**a**) Metabolic pathway of naphthalene in *P. putida* G7 carrying NAH7. *Red arrows* indicate conversions catalyzed by the Nah enzymes encoded on NAH7, and *blue arrows* indicate other enzymes encoded on the host chromosome. (**b**) Genetic organization of the *nah* genes on NAH7. *Black*, *red*, and *white pentagons* indicate regulatory genes, catabolic genes, and other related genes, respectively. An *arrow* from salicylate in panel (**a**) indicates that salicylate is an effector of the transcriptional regulator NahR, which regulates the transcriptional activity from the two promoters *Pnah* and *Psal*. Transcription from each promoter is induced by the effector

The naphthalene catabolic (*nah*) genes on the plasmid have been investigated in detail; the enzymes encoded by the upper *nah* operon (*nahAa* to *narE*) convert naphthalene to salicylate, and those encoded by the lower *nah* operon (*nahG* to *nahJ*) convert salicylate to pyruvate and acetaldehyde via the *meta*-cleavage pathway (Fig. 7.3; Yen and Serdar 1988). The promoters of these operons, *Pnah* and *Psal*, are regulated by NahR, a LysR-type transcriptional regulator, and salicylate acts as an effector (Schell and Poser 1989). Both the upper and lower operons on NAH7 are located on a class II transposon, Tn*4655* (Tsuda and Iino 1990). Although this transposon lacks the *tnpA* gene involved in self-regulated transposition (Sota et al. 2006), *Tn4655* can form a cointegrate when the *tnpA* gene of Tn*4653* is supplied *in trans* (Tsuda and Iino 1990; Sota et al. 2006).

Recently, Fernández et al. reported the effect of carriage of NAH7 on *P. putida* KT2440 (Fernández et al. 2012). The growth rate of KT2440 was influenced by the presence of naphthalene, although no bactericidal effect was observed. Only a slight difference in the growth rate was found by carriage of NAH7, whereas the host containing NAH7 could degrade naphthalene by efficient expression of the Nah catabolic pathway. The colonization ability of the plant rhizosphere was not affected by the carriage of NAH7 (Fernández et al. 2012). Transcriptome comparisons of the plasmid-free and plasmid-containing KT2440 cells were performed. Transcription of only 19 genes altered by carriage of NAH7, indicating that NAH7 carriage modestly affected the gene expression in the host cell (Fernández et al. 2012). Based on transcriptome comparisons of NAH7-free and NAH7-containing hosts (in the absence and presence of naphthalene), cellular stress through exposure to naphthalene was alleviated by the carriage of NAH7 (Fernández et al. 2012).

7.8 IncP-7 Carbazole-Degradative Plasmid pCAR1

pCAR1 is found in a carbazole (nitrogen-containing aromatic heterocyclic compound)-degrading bacterium, *P. resinovorans* CA10 (Ouchiyama et al. 1993; Nojiri et al. 2001). The 200 kb sequence of pCAR1 has been determined (Maeda et al. 2003; Takahashi et al. 2009b), and the plasmid belongs to the IncP-7 group (Shintani et al. 2006). The basic features of pCAR1 replication, partition, and conjugative transfer have been characterized (Shintani et al. 2005a, 2006). The carbazole-degradation pathway of CA10 is shown in Fig. 7.4. pCAR1 carries carbazole-degradative (*car*) genes and anthranilate-degradative (*ant*) gene that allows the bacterium to convert the carbazole to anthranilate and the anthranilate to catechol (Nojiri 2012). The chromosome of CA10 possesses catechol-degradative (*cat*) genes (Nojiri 2012), as do most *Pseudomonas*. The catechol is host-dependently metabolized by CatABC and enzymes encoded further downstream on the chromosome (Harwood and Parales 1996). The *car* and *ant* genes are induced by AntR, and anthranilate is an effector of AntR (Fig. 7.4; Urata et al. 2004). Transcription of *antR* is regulated by the σ^{54}-dependent promoter P_{antR} (Miyakoshi et al. 2007), and

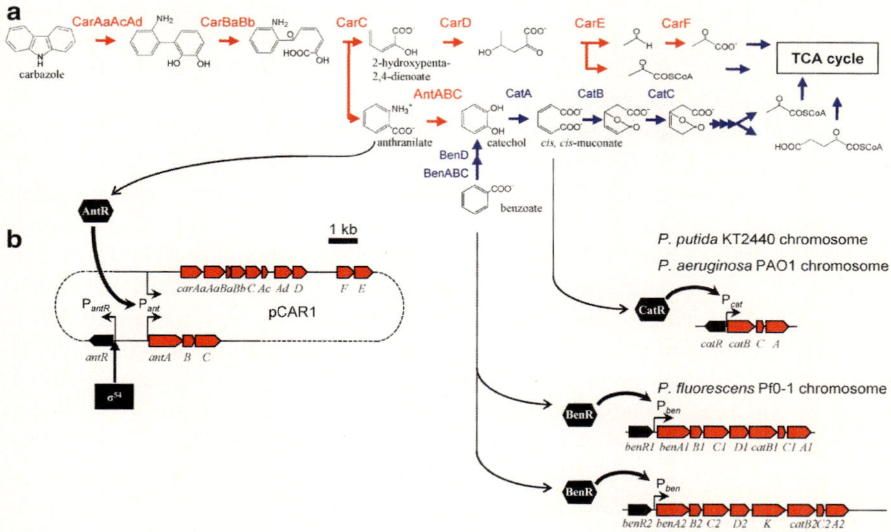

Fig. 7.4 (**a**) Catabolic pathway of carbazole in *Pseudomonas* spp. carrying pCAR1. *Red arrows* indicate conversions catalyzed by the Car and Ant enzymes encoded on pCAR1, and *blue arrows* indicate Cat and other enzymes encoded on the host chromosome. (**b**) Genetic organization of the *car* and *ant* genes on pCAR1, the *cat* gene cluster on the *P. putida* KT2440 and *P. aeruginosa* PAO1 chromosomes, and the *ben-cat* gene clusters on the *P. fluorescens* Pf0-1 chromosome. *Black* and *red pentagons* indicate regulatory genes and catabolic genes, respectively. *Arrows* from anthranilate, benzoate, and *cis*, *cis*-muconate in panel (**a**) indicate effectors of the transcriptional regulators AntR, CatR, and BenR, respectively. Transcription from each promoter is induced by the effector. P_{antR} is regulated by σ^{54}

thus transcriptional regulation of the *car* and *ant* genes is mediated by factors encoded on the plasmid, and the host chromosome, as is the case with *xyl* genes on pWW0 in *P. putida* mt-2. The *cat* genes in most *Pseudomonas* are induced by CatR, the effecter of which is *cis*, *cis*-muconate (Harwood and Parales 1996). This plasmid carries a functional 73-kb transposon, Tn*4676*, containing the *car* and *ant* genes (Maeda et al. 2003; Shintani et al. 2005b; 2011b). Conjugative transfer of pCAR1 is found in the *Pseudomonas*, and the frequency of pCAR1 transfer can be altered by changing the host donor (Shintani et al. 2005a; 2006). pCAR1 is also transferred to *Stenotrophomonas* spp., whose bacterial order (*Xanthomonadales*) is different from that of *Pseudomonas* (*Pseudomonadales*), in bacterial populations extracted from natural river water (Shintani et al. 2008a). Note that the transfer from *P. putida* to the pCAR1-cured *Stenotrophomonas* has not been detected, likely due to the different conjugation conditions in the laboratory and in the environment (Shintani et al. unpublished data).

In our recent work, comparisons of phenotypes and transcriptome profiles of different hosts of pCAR1 other than CA10 (the original host of pCAR1) have been performed, e.g., using *P. putida* KT2440, *P. aeruginosa* PAO1, *P. fluorescens* Pf0-1, *P. chlororaphis* subsp. *chlororaphis* JCM 2778 (previously *P. chlororaphis* IAM 1511), and *P. putida* HS01 (Miyakoshi et al. 2007; 2009; Shintani et al. 2008b, 2010a, c, 2011c; Takahashi et al. 2009a). Among these, Pf0-1 is an unsuitable host of pCAR1 for carbazole degradation (Takahashi et al. 2009a) because the transcriptional regulation of the *cat* genes in Pf0-1 is different from in other hosts (Takahashi et al. 2009a). CatR is the transcriptional regulator of the *cat* genes in hosts such as KT2440 and PAO1, whereas BenR is the transcriptional regulator in Pf0-1 (Fig. 7.4). *cis*, *cis*-muconate, an intermediate in the catechol-degradative pathway, acts as an effector of CatR and inducer of the *cat* genes (Fig. 7.4). Conversely, the *cat* genes of Pf0-1 are not induced by catechol but by benzoate, which is not an intermediate in carbazole catabolism (Fig. 7.4). In fact, the initiation of carbazole degradation in pCAR1-containing Pf0-1 is significantly slower than in other hosts (Takahashi et al. 2009a). Catechol accumulation is always deleterious to the host cells, and thus expression of the *ant* genes in the presence of carbazole can be deleterious to Pf0-1. This is an example of an implantation failure of a regulatory circuit in the degradation pathway encoded on pCAR1, in the context of the native transcriptional network of the host. Notably, deletion of DNA regions containing *ant* genes is frequently detected in Pf0-1 in the presence of carbazole (Takahashi et al. 2009a; Shintani et al. 2011a, b). The host of *ant*-deleted pCAR1 can still use carbazole as the sole carbon source because the host carries the *carDEF* genes, encoding enzymes that catalyze the conversion of 2-hydroxy-penta-2,4-dienoate to acetyl-CoA (Fig. 7.4; Shintani et al. 2011a). The deletion results in the improvement of survival in the presence of carbazole because excessive accumulation of catechol is avoided (Takahashi et al. 2009a; Shintani et al. 2011a, b). This is one example of plasmid and host adaptation when the host encounters a new carbon source.

The influence of pCAR1 carriage on different host strains was compared for three *Pseudomonas* hosts: *P. putida* KT2440, *P. aeruginosa* PAO1, and *P. fluorescens* Pf0-1. Competition assays with the plasmid-free and plasmid-containing hosts showed that pCAR1 carriage imposes a fitness cost; this cost is the smallest in Pf0-1 and the largest in KT2440 (Takahashi et al. submitted). Transcriptome comparisons with tiling arrays have been performed to evaluate the effect of pCAR1 carriage on the host, both during early-log phase (Miyakoshi et al. 2007; Shintani et al. 2010a) and from the exponential phase to the early stationary phase (Takahashi et al. submitted). Phenotypic comparisons have also been performed between pCAR1-free and pCAR1-containing host cells using phenotype microarrays (Takahashi et al. submitted). The number of upregulated and downregulated genes by pCAR1 carriage is the largest in KT2440 and the smallest in Pf0-1. Genes on putative prophages in each host chromosome are commonly upregulated by carriage of pCAR1 (Shintani et al. 2010a; Takahashi et al. submitted). pCAR1 carriage alters the host gene expression of the iron acquisition system, especially in KT2440 and PAO1 (Shintani et al. 2010a). A greater amount of a major siderophore for iron acquisition, pyoverdine, was found in pCAR1-containing hosts as compared with pCAR1-free hosts, partly due to the expression of the *car* genes on pCAR1 (Shintani et al. 2010a). Comparisons of the transcriptome profiles of the pCAR1-free and pCAR1-containing hosts showed that pCAR1 carriage affected the host more greatly during the transition and stationary phases than during the log phase (Takahashi et al. submitted). The putative functional categories of the affected genes are more similar between KT2440 and PAO1 than between other host combinations. These categories contained genes encoding the RNA polymerase core (*rpo*), ribosome protein subunits (*rps*, *rpl*), ATP synthetase (*atp*), and succinate dehydrogenase (*sdh*) (Takahashi et al. submitted). Not only the phenotype but also the primary metabolic capacity of the TCA cycle, and several steps away from the TCA cycle, is commonly affected in the host cells (Takahashi et al. submitted).

The transcriptions of several genes in KT2440 are specifically affected by pCAR1 carriage (Shintani et al. 2010a). PP_3700 is significantly upregulated by pCAR1 carriage, and the expression of PP_3700 is almost silent in the absence of pCAR1 (Miyakoshi et al. 2007; Shintani et al. 2010a). PP_3700, designated *parI*, is located on a genomic island in the KT2440 chromosome of three hosts; thus, the gene is also specifically present in KT2440 (Miyakoshi et al. 2007; 2012). ParI is a putative type Ia ParA family member, involved in chromosomal and plasmid DNA partitioning (Miyakoshi et al. 2007; 2012). ParI interferes with the IncP-7 plasmid partitioning system, suggesting that *parI* on the genomic island can restrict the host range of the plasmids (Miyakoshi et al. 2012). Therefore, ParI and the partitioning system of IncP-7 plasmids can mediate incompatibility between the plasmid and the chromosomal genomic island.

MexEFOprN, a host efflux pump, is also upregulated by pCAR1 carriage in KT2440 (Shintani et al. 2010a). Although genes orthologous to *mexEFoprN* are also present on two other host chromosomes, the transcriptional levels are not affected by pCAR1 carriage (Shintani et al. 2010a). The higher expression of the pump by pCAR1 carriage results in enhancement of KT2440 chloramphenicol resistance; hence, the resistance is not mediated by antibiotic-resistance genes on the plasmid (Shintani et al. 2010a). Although the physiological explanation why these genes are upregulated by pCAR1 carriage is presently unclear, the higher expression of the efflux pump can be costly to the host. Overexpression of the efflux pump may thus have resulted in a large fitness cost of the host KT2440.

pCAR1 has three NAP genes, *pmr*, *phu*, and *pnd* (Yun et al. 2010; Nojiri 2012). Disruption of *pmr* in KT2440 (pCAR1) affected the host transcriptome more than pCAR1 carriage (Yun et al. 2010). The transcriptional levels of several genes that increased upon pCAR1 carriage, such as the expression of the *parI* or *mexEFoprN* genes, reverted with the *pmr* disruption (Yun et al. 2010). Notably, the transcriptional levels of putative horizontally acquired genes are altered by the disruption, whereas these levels are not affected by pCAR1 carriage. This suggests that Pmr has a stealth function (similar to H-NS) on other plasmids, e.g., pSf-R27 (Doyle et al. 2007). Indeed, Pmr binds to many DNA regions on pCAR1 and the KT2440 chromosome, especially to A + T-rich regions (Yun et al. 2010). Recently, double disruptions of two of the three above-described NAP genes resulted in genetic instability of pCAR1, suggesting that the NAP genes on the plasmid may be involved in maintaining genetic stability in the host cell (Hirotani et al. unpublished). The transcription of genes on pCAR1 was also compared by tiling arrays of six different *Pseudomonas* hosts, including the three above-described hosts (Miyakoshi et al. 2009; Shintani et al. 2011c). The transcription levels of the accessory genes (*car* and *ant* genes) and genes on the IncP-7 plasmid backbone involved in plasmid replication, partition, and conjugative transfer are different depending on the host (Miyakoshi et al. 2009; Shintani et al. 2011c). Notably, the transcription levels of the three NAP genes also depended on the host, suggesting that these genes may control the transcriptional network interplay between the plasmid and host chromosome.

7.9 Conclusions and Perspective

MGEs alter the host phenotype by modulating the gene expression of the MGEs and the host chromosomes. The influence of MGE carriage clearly changes depending on the host cell environment. Plasmid carriage often imposes fitness costs to the host cells, although the extent of these costs depends on the type of host. The effects are influenced by (1) the size and copy number of the plasmid, (2) expression of deleterious genes on the plasmid, and (3) regulation conflicts of gene expression between the plasmid and host chromosome, processes in which NAPs are crucial. Plasmids and hosts appear to modulate the transcriptional network for coexistence.

Plasmids can adapt to new host environments by introducing mutations, such as pCAR1 in Pf0-1 (Shintani et al. 2011a, b). Sota et al. reported that the IncP-1 plasmid improves its stability in hosts where the plasmid is unstable by introducing mutations to the replication initiation protein (Sota et al. 2010). Naturally, host adaptation mechanisms may also exist, e.g., point mutations, chromosomal rearrangements, and transcriptional adjustment by NAPs; however, detailed analyses of host adaptation are lacking.

The extent of the influence of the carriage of ICEs is still unclear. Considering that ICEs have been conserved during bacterial evolution, these elements may have developed mechanisms to adapt to new hosts. Gaillard et al. proposed several possibilities for the stealth function of ICE_{clc}. The site-specific insertion of ICE_{clc} into *tRNA*^{Gly} may depress high expression of the integrase (*intB13*) gene, which is deleterious to the cell. Possibly, no other deleterious gene functions are present on ICE_{clc}, and no regulation conflicts arise between the regulators on ICE_{clc} and the host chromosome (Gaillard et al. 2008). Guglielmini et al. showed that plasmids and ICEs use similar conjugative systems and that ICEs and conjugative plasmids should be regarded as the same. Differences in the means of existence of extrachromosomal or intrachromosomal ICEs and conjugative plasmids in cells are likely due to different requirements for stable maintenance (Guglielmini et al. 2011).

Previous studies of the effects of MGEs have focused on host cell populations rather than on single cells. However, gene expression and regulation is often stochastic events that occur in individual cells (Rosenfeld et al. 2005; Pedraza and van Oudenaarden 2005). Indeed, Silva-Rocha and de Lorenzo reported a high-degree stochastic transcription of *xyl* genes on pWW0 in *P. putida* mt-2 (the original host) at the single-cell level, when the cells were exposed to *m*-xylene (Silva-Rocha and de Lorenzo 2012). They proposed that this phenomenon is essential for host survival when encountering mixtures of nutrients in the environment (Silva-Rocha and de Lorenzo 2012). Thus, single-cell genomics, transcriptomics, proteomics, and metabolomics will be necessary to understand how the MGEs and their hosts have evolved and adapted in nature.

Acknowledgements The writing of this book chapter was supported by the Special Postdoctoral Researcher Program of Riken and by Grant-in-Aid for Young Scientists (Grant Number 24780087) to M.S.

References

Andersson L, Yang S, Neubauer P, Enfors SO (1996) Impact of plasmid presence and induction on cellular responses in fed batch cultures of *Escherichia coli*. J Biotechnol 46:255–263

Aranda-Olmedo I, Marín P, Ramos JL, Marqués S (2006) Role of the *ptsN* gene product in catabolite repression of the *Pseudomonas putida* TOL toluene degradation pathway in chemostat cultures. Appl Environ Microbiol 72:7418–7421

Benson S, Shapiro J (1978) TOL is a broad-host-range plasmid. J Bacteriol 135:278–280

Bertoni G, Pérez-Martín J, de Lorenzo V (1997) Genetic evidence of separate repressor and activator activities of the XylR regulator of the TOL plasmid, pWW0, of *Pseudomonas putida*. Mol Microbiol 23:1221–1227

Birnbaum S, Bailey JE (1991) Plasmid presence changes the relative levels of many host cell proteins and ribosome components in recombinant *Escherichia coli*. Biotechnol Bioeng 37:736–745

Bouma JE, Lenski RE (1988) Evolution of a bacteria/plasmid association. Nature 335:351–352

Buch AD, Archana G, Naresh Kumar G (2010) Broad-host-range plasmid-mediated metabolic perturbations in *Pseudomonas fluorescens* 13525. Appl Microbiol Biotechnol 88:209–218

Carattoli A (2009) Resistance plasmid families in *Enterobacteriaceae*. Antimicrob Agents Chemother 53:2227–2238

Cases I, Pérez-Martín J, de Lorenzo V (1999) The IIANtr (PtsN) protein of *Pseudomonas putida* mediates the C source inhibition of the sigma54-dependent *Pu* promoter of the TOL plasmid. J Biol Chem 274:15562–15568

Castang S, Dove SL (2012) Basis for the essentiality of H-NS family members in *Pseudomonas aeruginosa*. J Bacteriol 194:5101–5109

Chain PS, Denef VJ, Konstantinidis KT, Vergez LM, Agulló L, Reyes VL, Hauser L, Córdova M, Gómez L, González M, Land M, Lao V, Larimer F, LiPuma JJ, Mahenthiralingam E, Malfatti SA, Marx CJ, Parnell JJ, Ramette A, Richardson P, Seeger M, Smith D, Spilker T, Sul WJ, Tsoi TV, Ulrich LE, Zhulin IB, Tiedje JM (2006) *Burkholderia xenovorans* LB400 harbors a multi-replicon, 9.73-Mbp genome shaped for versatility. Proc Natl Acad Sci USA 103: 15280–15287

Coco WM, Parsek MR, Chakrabarty AM (1994) Purification of the LysR family regulator, ClcR, and its interaction with the *Pseudomonas putida clcABD* chlorocatechol operon promoter. J Bacteriol 176:5530–5533

Dahlberg C, Chao L (2003) Amelioration of the cost of conjugative plasmid carriage in *Escherichia coli* K12. Genetics 165:1641–1649

Dale JW, Smith JT (1979) The effect of a plasmid on growth and survival of *E. coli*. Antonie Van Leeuwenhoek 45:103–111

D'Alvise PW, Sjøholm OR, Yankelevich T, Jin Y, Wuertz S, Smets BF (2010) TOL plasmid carriage enhances biofilm formation and increases extracellular DNA content in *Pseudomonas putida* KT2440. FEMS Microbiol Lett 312:84–92

de Lorenzo V, Herrero M, Metzke M, Timmis KN (1991) An upstream XylR- and IHF-induced nucleoprotein complex regulates the sigma 54-dependent *Pu* promoter of TOL plasmid. EMBO J 10:1159–1167

Diaz Ricci JC, Hernández ME (2000) Plasmid effects on *Escherichia coli* metabolism. Crit Rev Biotechnol 20:79–108

Dillon SC, Dorman CJ (2010) Bacterial nucleoid-associated proteins, nucleoid structure and gene expression. Nat Rev Microbiol 8:185–195

Dionisio F, Conceição IC, Marques AC, Fernandes L, Gordo I (2005) The evolution of a conjugative plasmid and its ability to increase bacterial fitness. Biol Lett 1:250–252

Dorman CJ, Kane KA (2009) DNA bridging and antibridging: a role for bacterial nucleoid-associated proteins in regulating the expression of laterally acquired genes. FEMS Microbiol Rev 33:587–592

Doyle M, Fookes M, Ivens A, Mangan MW, Wain J, Dorman CJ (2007) An H-NS-like stealth protein aids horizontal DNA transmission in bacteria. Science 315:251–252

Duetz WA, van Andel JG (1991) Stability of TOL plasmid pWW0 in *Pseudomonas putida* mt-2 under non-selective conditions in continuous culture. J Gen Microbiol 137:1369–1374

Dunn NW, Gunsalus IC (1973) Transmissible plasmid coding early enzymes of naphthalene oxidation in *Pseudomonas putida*. J Bacteriol 114:974–979

Engberg B, Hjalmarsson K, Nordström K (1975) Inhibition of cell division in *Escherichia coli* K-12 by the R-factor R1 and copy mutants of R1. J Bacteriol 124:633–640

Fernández M, Niqui-Arroyo JL, Conde S, Ramos JL, Duque E (2012) Enhanced tolerance to naphthalene and enhanced rhizoremediation performance for *Pseudomonas putida* KT2440 via the NAH7 catabolic plasmid. Appl Environ Microbiol 78:5104–5110

Foucault ML, Depardieu F, Courvalin P, Grillot-Courvalin C (2010) Inducible expression eliminates the fitness cost of vancomycin resistance in enterococci. Proc Natl Acad Sci USA 107:16964–16969

Frost LS, Leplae R, Summers AO, Toussaint A (2005) Mobile genetic elements: the agents of open source evolution. Nat Rev Microbiol 3:722–732

Gaillard M, Vallaeys T, Vorhölter FJ, Minoia M, Werlen C, Sentchilo V, Pühler A, van der Meer JR (2006) The *clc* element of *Pseudomonas* sp. strain B13, a genomic island with various catabolic properties. J Bacteriol 188:1999–2013

Gaillard M, Pernet N, Vogne C, Hagenbüchle O, van der Meer JR (2008) Host and invader impact of transfer of the *clc* genomic island into *Pseudomonas aeruginosa* PAO1. Proc Natl Acad Sci USA 105:7058–7063

Gaillard M, Pradervand N, Minoia M, Sentchilo V, Johnson D, van der Meer J (2010) Transcriptome analysis of the mobile genome ICE$_{clc}$ in *Pseudomonas knackmussii* B13. BMC Microbiol 10:153

Garcillán-Barcia MP, Francia MV, de la Cruz F (2009) The diversity of conjugative relaxases and its application in plasmid classification. FEMS Microbiol Rev 33:657–687

Garcillán-Barcia MP, Alvarado A, de la Cruz F (2011) Identification of bacterial plasmids based on mobility and plasmid population biology. FEMS Microbiol Rev 35:936–956

Garnier F, Taourit S, Glaser P, Courvalin P, Galimand M (2000) Characterization of transposon Tn*1549*, conferring VanB-type resistance in *Enterococcus* spp. Microbiology 146:1481–1489

Ghigo JM (2001) Natural conjugative plasmids induce bacterial biofilm development. Nature 412:442–445

Goris J, Dejonghe W, Falsen E, De Clerck E, Geeraerts B, Willems A, Top E, Vandamme P, De Vos P (2002) Diversity of transconjugants that acquired plasmid pJP4 or pEMT1 after inoculation of a donor strain in the A- and B-horizon of an agricultural soil and description of *Burkholderia hospita* sp. nov. and *Burkholderia terricola* sp. nov. Syst Appl Microbiol 25:340–352

Greated A, Lambertsen L, Williams PA, Thomas CM (2002) Complete sequence of the IncP-9 TOL plasmid pWW0 from *Pseudomonas putida*. Environ Microbiol 4:856–871

Guglielmini J, Quintais L, Garcillán-Barcia MP, de la Cruz F, Rocha EP (2011) The repertoire of ICE in prokaryotes underscores the unity, diversity, and ubiquity of conjugation. PLoS Genet 7:e1002222

Haft RJ, Mittler JE, Traxler B (2009) Competition favours reduced cost of plasmids to host bacteria. ISME J 3:761–769

Harwood CS, Parales RE (1996) The beta-ketoadipate pathway and the biology of self-identity. Annu Rev Microbiol 50:553–590

Inouye S, Nakazawa A, Nakazawa T (1987) Expression of the regulatory gene *xylS* on the TOL plasmid is positively controlled by the *xylR* gene product. Proc Natl Acad Sci USA 84:5182–5186

Klemperer RM, Ismail NT, Brown MR (1979) Effect of R plasmid RP1 on the nutritional requirements of *Escherichia coli* in batch culture. J Gen Microbiol 115:325–331

Leddy MB, Phipps DW, Ridgway HF (1995) Catabolite-mediated mutations in alternate toluene degradative pathways in *Pseudomonas putida*. J Bacteriol 177:4713–4720

Lee SW, Edlin G (1985) Expression of tetracycline resistance in pBR322 derivatives reduces the reproductive fitness of plasmid-containing *Escherichia coli*. Gene 39:173–180

Lilley AK, Bailey MJ (1997) Impact of plasmid pQBR103 acquisition and carriage on the phytosphere fitness of *Pseudomonas fluorescens* SBW25: burden and benefit. Appl Environ Microbiol 63:1584–1587

Maeda K, Nojiri H, Shintani M, Yoshida T, Habe H, Omori T (2003) Complete nucleotide sequence of carbazole/dioxin-degrading plasmid pCAR1 in *Pseudomonas resinovorans* strain CA10 indicates its mosaicity and the presence of large catabolic transposon Tn*4676*. J Mol Biol 326:21–33

Marqués S, Gallegos MT, Ramos JL (1995) Role of sigma S in transcription from the positively controlled *Pm* promoter of the TOL plasmid of *Pseudomonas putida*. Mol Microbiol 18:851–857

Marqués S, Gallegos MT, Manzanera M, Holtel A, Timmis KN, Ramos JL (1998) Activation and repression of transcription at the double tandem divergent promoters for the *xylR* and *xylS* genes of the TOL plasmid of *Pseudomonas putida*. J Bacteriol 180:2889–2894

Marqués S, Manzanera M, González-Pérez MM, Gallegos MT, Ramos JL (1999) The XylS-dependent *Pm* promoter is transcribed in vivo by RNA polymerase with sigma 32 or sigma 38 depending on the growth phase. Mol Microbiol 31:1105–1113

Miller TR, Delcher AL, Salzberg SL, Saunders E, Detter JC, Halden RU (2010) Genome sequence of the dioxin-mineralizing bacterium *Sphingomonas wittichii* RW1. J Bacteriol 192:6101–6102

Miyakoshi M, Shintani M, Terabayashi T, Kai S, Yamane H, Nojiri H (2007) Transcriptome analysis of *Pseudomonas putida* KT2440 harboring the completely sequenced IncP-7 plasmid pCAR1. J Bacteriol 189:6849–6860

Miyakoshi M, Nishida H, Shintani M, Yamane H, Nojiri H (2009) High-resolution mapping of plasmid transcriptomes in different host bacteria. BMC Genomics 10

Miyakoshi M, Shintani M, Inoue K, Terabayashi T, Sai F, Ohkuma M, Nojiri H, Nagata Y, Tsuda M (2012) ParI, an orphan ParA family protein from *Pseudomonas putida* KT2440-specific genomic island, interferes with the partition system of IncP-7 plasmids. Environ Microbiol 14:2946–2959

Miyazaki R, van der Meer JR (2011a) A dual functional origin of transfer in the ICE_{clc} genomic island of *Pseudomonas knackmussii* B13. Mol Microbiol 79:743–758

Miyazaki R, van der Meer JR (2011b) How can a dual *oriT* system contribute to efficient transfer of an integrative and conjugative element? Mob Genet Elem 1:82–84

Miyazaki R, Ohtsubo Y, Nagata Y, Tsuda M (2008) Characterization of the *traD* operon of naphthalene-catabolic plasmid NAH7: a host-range modifier in conjugative transfer. J Bacteriol 190:6281–6289

Moreno R, Fonseca P, Rojo F (2010) The Crc global regulator inhibits the *Pseudomonas putida* pWW0 toluene/xylene assimilation pathway by repressing the translation of regulatory and structural genes. J Biol Chem 285:24412–24419

Müller TA, Werlen C, Spain J, Van Der Meer JR (2003) Evolution of a chlorobenzene degradative pathway among bacteria in a contaminated groundwater mediated by a genomic island in *Ralstonia*. Environ Microbiol 5:163–173

Nakazawa T (1978) TOL plasmid in *Pseudomonas aeruginosa* PAO: thermosensitivity of self-maintenance and inhibition of host cell growth. J Bacteriol 133:527–535

Nancharaiah Y, Wattiau P, Wuertz S, Bathe S, Mohan S, Wilderer P, Hausner M (2003) Dual labeling of *Pseudomonas putida* with fluorescent proteins for in situ monitoring of conjugal transfer of the TOL plasmid. Appl Environ Microbiol 69:4846–4852

Navarre WW, Porwollik S, Wang Y, McClelland M, Rosen H, Libby SJ, Fang FC (2006) Selective silencing of foreign DNA with low GC content by the H-NS protein in *Salmonella*. Science 313:236–238

Nishi A, Tominaga K, Furukawa K (2000) A 90-kilobase conjugative chromosomal element coding for biphenyl and salicylate catabolism in *Pseudomonas putida* KF715. J Bacteriol 182:1949–1955

Nojiri H (2012) Structural and molecular genetic analyses of the bacterial carbazole degradation system. Biosci Biotechnol Biochem 76:1–18

Nojiri H, Sekiguchi H, Maeda K, Urata M, Nakai S, Yoshida T, Habe H, Omori T (2001) Genetic characterization and evolutionary implications of a *car* gene cluster in the carbazole degrader *Pseudomonas* sp. strain CA10. J Bacteriol 183:3663–3679

Nojiri H, Shintani M, Omori T (2004) Divergence of mobile genetic elements involved in the distribution of xenobiotic-catabolic capacity. Appl Microbiol Biotechnol 64:154–174

Ohtsubo Y, Shimura M, Delawary M, Kimbara K, Takagi M, Kudo T, Ohta A, Nagata Y (2003) Novel approach to the improvement of biphenyl and polychlorinated biphenyl degradation activity: promoter implantation by homologous recombination. Appl Environ Microbiol 69:146–153

Ohtsubo Y, Goto H, Nagata Y, Kudo T, Tsuda M (2006) Identification of a response regulator gene for catabolite control from a PCB-degrading beta-proteobacteria, *Acidovorax* sp. KKS102. Mol Microbiol 60:1563–1575

Ohtsubo Y, Ishibashi Y, Naganawa H, Hirokawa S, Atobe S, Nagata Y, Tsuda M (2012) Conjugal transfer of PCB/biphenyl degradation genes in *Acidovorax* sp. KKS102 located on integrative and conjugative element. J Bacteriol 194:4237–4248

Ouchiyama N, Zhang Y, Omori T, Kodama T (1993) Biodegradation of carbazole by *Pseudomonas* spp. CA06 and CA10. Biosci Biotechnol Biochem 57:455–460

Ow D, Nissom P, Philp R, Oh S, Yap M (2006) Global transcriptional analysis of metabolic burden due to plasmid maintenance in *Escherichia coli* DH5 alpha during batch fermentation. Enz Microb Technol 39:391–398

Park HS, Kim HS (2001) Genetic and structural organization of the aminophenol catabolic operon and its implication for evolutionary process. J Bacteriol 183:5074–5081

Parsek MR, McFall SM, Shinabarger DL, Chakrabarty AM (1994) Interaction of two LysR-type regulatory proteins CatR and ClcR with heterologous promoters: functional and evolutionary implications. Proc Natl Acad Sci USA 91:12393–12397

Pedraza JM, van Oudenaarden A (2005) Noise propagation in gene networks. Science 307:1965–1969

Pérez-Martín J, de Lorenzo V (1996) ATP binding to the sigma 54-dependent activator XylR triggers a protein multimerization cycle catalyzed by UAS DNA. Cell 86:331–339

Ramos J, Mermod N, Timmis K (1987) Regulatory circuits controlling transcription of TOL plasmid operon encoding meta-cleavage pathway for degradation of alkylbenzoates by *Pseudomonas*. Mol Microbiol 1:293–300

Ramos J, Marqués S, Timmis K (1997) Transcriptional control of the *Pseudomonas* TOL plasmid catabolic operons is achieved through an interplay of host factors and plasmid-encoded regulators. Annu Rev Microbiol 51:341–373

Ramos-Gonzalez MI, Duque E, Ramos JL (1991) Conjugational transfer of recombinant DNA in cultures and in soils: host range of *Pseudomonas putida* TOL plasmids. Appl Environ Microbiol 57:3020–3027

Ravatn R, Studer S, Springael D, Zehnder AJ, van der Meer JR (1998a) Chromosomal integration, tandem amplification, and deamplification in *Pseudomonas putida* F1 of a 105-kilobase genetic element containing the chlorocatechol degradative genes from *Pseudomonas* sp. strain B13. J Bacteriol 180:4360–4369

Ravatn R, Studer S, Zehnder AJ, van der Meer JR (1998b) Int-B13, an unusual site-specific recombinase of the bacteriophage P4 integrase family, is responsible for chromosomal insertion of the 105-kilobase *clc* element of *Pseudomonas* sp. strain B13. J Bacteriol 180:5505–5514

Ravatn R, Zehnder AJ, van der Meer JR (1998c) Low-frequency horizontal transfer of an element containing the chlorocatechol degradation genes from *Pseudomonas* sp. strain B13 to *Pseudomonas putida* F1 and to indigenous bacteria in laboratory-scale activated-sludge microcosms. Appl Environ Microbiol 64:2126–2132

Reisner A, Haagensen JA, Schembri MA, Zechner EL, Molin S (2003) Development and maturation of *Escherichia coli* K-12 biofilms. Mol Microbiol 48:933–946

Rescalli E, Saini S, Bartocci C, Rychlewski L, De Lorenzo V, Bertoni G (2004) Novel physiological modulation of the *Pu* promoter of TOL plasmid: negative regulatory role of the TurA protein of *Pseudomonas putida* in the response to suboptimal growth temperatures. J Biol Chem 279:7777–7784

Rhee J, Diazricci J, Bode J, Schugerl K (1994) Metabolic enhancement due to plasmid maintenance. Biotechnol Lett 16:881–884

Romine MF, Stillwell LC, Wong KK, Thurston SJ, Sisk EC, Sensen C, Gaasterland T, Fredrickson JK, Saffer JD (1999) Complete sequence of a 184-kilobase catabolic plasmid from *Sphingomonas aromaticivorans* F199. J Bacteriol 181:1585–1602

Rosenfeld N, Young JW, Alon U, Swain PS, Elowitz MB (2005) Gene regulation at the single-cell level. Science 307:1962–1965

Rozkov A, Avignone-Rossa CA, Ertl PF, Jones P, O'Kennedy RD, Smith JJ, Dale JW, Bushell ME (2004) Characterization of the metabolic burden on *Escherichia coli* DH1 cells imposed by the presence of a plasmid containing a gene therapy sequence. Biotechnol Bioeng 88:909–915

Ruiz R, Aranda-Olmedo MI, Dominguez-Cuevas P, Ramos-Gonzalez MI, Marqués S (2004) Transcriptional regulation of the toluene catabolic pathways. In: Ramos JL (ed) Pseudomonas Vol. 2 –virulence and gene regulation. Kluwer Academic/Plenum, New York

Ryan MP, Pembroke JT, Adley CC (2009) Novel Tn*4371*-ICE like element in *Ralstonia pickettii* and genome mining for comparative elements. BMC Microbiol 9:242

Sarand I, Haario H, Jørgensen KS, Romantschuk M (2000) Effect of inoculation of a TOL plasmid containing mycorrhizosphere bacterium on development of Scots pine seedlings, their mycorrhizosphere and the microbial flora in m-toluate-amended soil. FEMS Microbiol Ecol 31:127–141

Schell MA, Poser EF (1989) Demonstration, characterization, and mutational analysis of NahR protein binding to *nah* and *sal* promoters. J Bacteriol 171:837–846
Schmidt-Eisenlohr H, Baron C (2003) The competitiveness of *Pseudomonas chlororaphis* carrying pJP4 is reduced in the *Arabidopsis thaliana* rhizosphere. Appl Environ Microbiol 69:1827–1831
Sentchilo V, Czechowska K, Pradervand N, Minoia M, Miyazaki R, van der Meer J (2009) Intracellular excision and reintegration dynamics of the ICE_{clc} genomic island of *Pseudomonas knackmussii* sp strain B13. Mol Microbiol 72:1293–1306
Sevastsyanovich Y, Titok M, Krasowiak R, Bingle L, Thomas C (2005) Ability of IncP-9 plasmid pM3 to replicate in *Escherichia coli* is dependent on both *rep* and *par* functions. Mol Microbiol 57:819–833
Sharma V, Archana G, Naresh Kumar G (2011) Plasmid load adversely affects growth and gluconic acid secretion ability of mineral phosphate-solubilizing rhizospheric bacterium *Enterobacter asburiae* PSI3 under P limited conditions. Microbiol Res 166:36–46
Shintani M, Nojiri H (2013) Mobile genetic elements (MGEs) carrying catabolic genes. In: Malik A, Grohmann E, Alves M (eds) Management of microbial resources in the environment. Springer, Netherlands
Shintani M, Habe H, Tsuda M, Omori T, Yamane H, Nojiri H (2005a) Recipient range of IncP-7 conjugative plasmid pCAR2 from *Pseudomonas putida* HS01 is broader than from other *Pseudomonas* strains. Biotechnol Lett 27:1847–1853
Shintani M, Yoshida T, Habe H, Omori T, Nojiri H (2005b) Large plasmid pCAR2 and class II transposon Tn*4676* are functional mobile genetic elements to distribute the carbazole/dioxin-degradative *car* gene cluster in different bacteria. Appl Microbiol Biotechnol 67:370–382
Shintani M, Yano H, Habe H, Omori T, Yamane H, Tsuda M, Nojiri H (2006) Characterization of the replication, maintenance, and transfer features of the IncP-7 plasmid pCAR1, which carries genes involved in carbazole and dioxin degradation. Appl Environ Microbiol 72:3206–3216
Shintani M, Urata M, Inoue K, Eto K, Habe H, Omori T, Yamane H, Nojiri H (2007) The *Sphingomonas* plasmid pCAR3 is involved in complete mineralization of carbazole. J Bacteriol 189:2007–2020
Shintani M, Fukushima N, Tezuka M, Yamane H, Nojiri H (2008a) Conjugative transfer of the IncP-7 carbazole degradative plasmid, pCAR1, in river water samples. Biotechnol Lett 30:117–122
Shintani M, Matsui K, Takemura T, Yamane H, Nojiri H (2008b) Behavior of the IncP-7 carbazole-degradative plasmid pCAR1 in artificial environmental samples. Appl Microbiol Biotechnol 80:485–497
Shintani M, Takahashi Y, Tokumaru H, Kadota K, Hara H, Miyakoshi M, Naito K, Yamane H, Nishida H, Nojiri H (2010a) Response of the *Pseudomonas* host chromosomal transcriptome to carriage of the IncP-7 plasmid pCAR1. Environ Microbiol 12:1413–1426
Shintani M, Takahashi Y, Yamane H, Nojiri H (2010b) The behavior and significance of degradative plasmids belonging to Inc groups in *Pseudomonas* within natural environments and microcosms. Microb Environ 25:253–265
Shintani M, Yamane H, Nojiri H (2010c) Behavior of various hosts of the IncP-7 carbazole-degradative plasmid pCAR1 in artificial microcosms. Biosci Biotechnol Biochem 74:343–349
Shintani M, Horisaki T, Yamane H, Ohkuma M, Nojiri H (2011a) Evolution of the IncP-7 carbazole-degradative plasmid pCAR1 improves survival of its host *Pseudomonas fluorescens* Pf0-1 in artificial water microcosms. Microbiology 157:2276–2286
Shintani M, Matsumoto T, Yoshikawa H, Yamane H, Ohkuma M, Nojiri H (2011b) DNA rearrangement has occurred in the carbazole-degradative plasmid pCAR1 and the chromosome of its unsuitable host, *Pseudomonas fluorescens* Pf0-1. Microbiology 157:3405–3416
Shintani M, Tokumaru H, Takahashi Y, Miyakoshi M, Yamane H, Nishida H, Nojiri H (2011c) Alterations of RNA maps of IncP-7 plasmid pCAR1 in various *Pseudomonas* bacteria. Plasmid 66:85–92

Silva-Rocha R, de Lorenzo V (2012) Stochasticity of TOL plasmid catabolic promoters sets a bimodal expression regime in *Pseudomonas putida* mt-2 exposed to *m*-xylene. Mol Microbiol 86:199–211

Smillie C, Garcillán-Barcia MP, Francia MV, Rocha EP, de la Cruz F (2010) Mobility of plasmids. Microbiol Mol Biol Rev 74:434–452

Sota M, Top EM (2008) Horizontal gene transfer mediated by plasmids. In: Lipps G (ed) Plasmids: current research and future trends. Horizon Scientific, Norfolk

Sota M, Yano H, Ono A, Miyazaki R, Ishii H, Genka H, Top E, Tsuda M (2006) Genomic and functional analysis of the IncP-9 naphthalene-catabolic plasmid NAH7 and its transposon Tn*4655* suggests catabolic gene spread by a tyrosine recombinase. J Bacteriol 188:4057–4067

Sota M, Yano H, Hughes JM, Daughdrill GW, Abdo Z, Forney LJ, Top EM (2010) Shifts in the host range of a promiscuous plasmid through parallel evolution of its replication initiation protein. ISME J 4:1568–1580

Takahashi Y, Shintani M, Li L, Yamane H, Nojiri H (2009a) Carbazole-degradative IncP-7 plasmid pCAR1.2 is dtructurally unstable in *Pseudomonas fluorescens* Pf0-1, which accumulates catechol, the intermediate of the carbazole degradation pathway. Appl Environ Microbiol 75:3920–3929

Takahashi Y, Shintani M, Yamane H, Nojiri H (2009b) The complete nucleotide sequence of pCAR2: pCAR2 and pCAR1 were structurally identical IncP-7 carbazole degradative plasmids. Biosci Biotechnol Biochem 73:744–746

Takeda T, Yun CS, Shintani M, Yamane H, Nojiri H (2011) Distribution of genes encoding nucleoid-associated protein homologs in plasmids. Int J Evol Biol 2011:685015

Thomas CM, Haines AS (2004) Plasmids of the genus *Pseudomonas*. In: Ramos J-L (ed) *Pseudomonas*. Kluwer Academic/Plenum Publishers, New York

Toussaint A, Merlin C, Monchy S, Benotmane MA, Leplae R, Mergeay M, Springael D (2003) The biphenyl- and 4-chlorobiphenyl-catabolic transposon Tn*4371*, a member of a new family of genomic islands related to IncP and Ti plasmids. Appl Environ Microbiol 69:4837–4845

Tsuda M, Iino T (1987) Genetic analysis of a transposon carrying toluene degrading genes on a TOL plasmid pWW0. Mol Gen Genet 210:270–276

Tsuda M, Iino T (1988) Identification and characterization of Tn*4653*, a transposon covering the toluene transposon Tn*4651* on TOL plasmid pWW0. Mol Gen Genet 213:72–77

Tsuda M, Iino T (1990) Naphthalene degrading genes on plasmid NAH7 are on a defective transposon. Mol Gen Genet 223:33–39

Tsuda M, Minegishi K, Iino T (1989) Toluene transposons Tn*4651* and Tn*4653* are class II transposons. J Bacteriol 171:1386–1393

Ude S, Bailey MJ, Huang WE, Spiers AJ (2007) The environmental plasmid pQBR103 alters the single-cell Raman spectral profile of *Pseudomonas fluorescens* SBW25. Microb Ecol 53:494–497

Urata M, Miyakoshi M, Kai S, Maeda K, Habe H, Omori T, Yamane H, Nojiri H (2004) Transcriptional regulation of the ant operon, encoding two-component anthranilate 1,2-dioxygenase, on the carbazole-degradative plasmid pCAR1 of *Pseudomonas resinovorans* strain CA10. J Bacteriol 186:6815–6823

Vitale E, Milani A, Renzi F, Galli E, Rescalli E, de Lorenzo V, Bertoni G (2008) Transcriptional wiring of the TOL plasmid regulatory network to its host involves the submission of the sigma54-promoter *Pu* to the response regulator PprA. Mol Microbiol 69:698–713

Wang Z, Xiang L, Shao J, Wegrzyn A, Wegrzyn G (2006) Effects of the presence of ColE1 plasmid DNA in *Escherichia coli* on the host cell metabolism. Microb Cell Fact 5:34

Williams P, Taylor S, Gibb L (1988) Loss of the toluene-xylene catabolic genes of TOL plasmid pWW0 during growth of *Pseudomonas putida* on benzoate is due to a selective growth advantage of 'cured' segregants. J Gen Microbiol 134:2039–2048

Wozniak RA, Waldor MK (2010) Integrative and conjugative elements: mosaic mobile genetic elements enabling dynamic lateral gene flow. Nat Rev Microbiol 8:552–563

Yang X, Ma Q, Wood TK (2008) The R1 conjugative plasmid increases Escherichia coli biofilm formation through an envelope stress response. Appl Environ Microbiol 74:2690–2699

Yen K, Serdar C (1988) Genetics of naphthalene catabolism in pseudomonads. Crit Rev Microbiol 15:247–268

Yun C-S, Suzuki C, Naito K, Takeda T, Takahashi Y, Sai F, Terabayashi T, Miyakoshi M, Shintani M, Nishida H, Yamane H, Nojiri H (2010) Pmr, a Histonelike protein H1 (H-NS) family protein encoded by the IncP-7 plasmid pCAR1, is a key global regulator that alters host function. J Bacteriol 192:4720–4731

Zünd P, Lebek G (1980) Generation time-prolonging R plasmids: correlation between increases in the generation time of *Escherichia coli* caused by R plasmids and their molecular size. Plasmid 3:65–69

Part II
Enzyme Systems

Chapter 8
Online Monitoring of Biodegradation Processes Using Enzymatic Biosensors

Sarka Bidmanova, Eva Hrdlickova, Tana Koudelakova, Jiri Damborsky, and Zbynek Prokop

Abstract Biodegradation of organic compounds by natural attenuation or bioaugmentation is widely used for removal of unwanted chemicals from the environment. The essential component of remediation technologies is monitoring of a contaminant levels. Although the analytical methods of gas and liquid chromatography are available, they are time-, labour- and resource-demanding, placing limitations on the number of samples that can be analysed at a time. Furthermore, these methods cannot be easily adapted for in situ measurements. Biosensors can be used as an alternative or complement to these conventional techniques. Biosensors are based on a biological component coupled to a transducer, which translates the interaction between an analyte and a biocomponent into a signal that can be processed. Application of biosensors in monitoring of environmental contaminants is promising owing to their sensitivity, low costs, user-friendliness and adaptability for in situ measurements. In this contribution, we describe development of haloalkane dehalogenase-based biosensors and their application for detection of halogenated hydrocarbons in the environmental samples.

S. Bidmanova • T. Koudelakova
Loschmidt Laboratories, Department of Experimental Biology
and Research Centre for Toxic Compounds in the Environment,
Masaryk University, Kamenice 5/A13, 62500 Brno, Czech Republic

E. Hrdlickova
Enantis, Ltd., Palackeho trida 1802/129, 61200 Brno, Czech Republic

J. Damborsky • Z. Prokop (✉)
Loschmidt Laboratories, Department of Experimental Biology
and Research Centre for Toxic Compounds in the Environment,
Masaryk University, Kamenice 5/A13, 62500 Brno, Czech Republic

Enantis, Ltd., Palackeho trida 1802/129, 61200 Brno, Czech Republic
e-mail: zbynek@chemi.muni.cz

H. Nojiri et al. (eds.), *Biodegradative Bacteria: How Bacteria Degrade, Survive, Adapt, and Evolve*, DOI 10.1007/978-4-431-54520-0_8, © Springer Japan 2014

Keywords Biodegradation • Enzymatic biosensors • Haloalkane dehalogenases • Halogenated hydrocarbons • Immobilization • Online monitoring

Abbreviations

CB	Chlorobutane
CCMP	3-Chloro-2-(chloromethyl)-1-propene
DBA	1,2-Dibromoethane
DbjA	Haloalkane dehalogenase from *Bradyrhizobium japonicum* USDA110
DCA	1,2-Dichloroethane
DhaA	Haloalkane dehalogenase from *Rhodococcus rhodochrous* NCIMB 13064
DhlA	Haloalkane dehalogenase from *Xanthobacter autotrophicus* GJ10
EC	Enzyme commission number
LED	Light-emitting diode
LinB	Haloalkane dehalogenase from *Sphingobium japonicum* UT26
MW	Molecular weight
pI	Isoelectric point
SD	Standard deviation
TCP	1,2,3-Trichloropropane

8.1 Introduction

Biodegradation of organic compounds by natural attenuation or bioaugmentation is widely used for removal of unwanted chemicals from the environment. The essential component of remediation technologies is monitoring of the contaminant levels in the treated soil or groundwater. Analysis of the multiple samples collected at a locality and transported to a laboratory has to be conducted to secure data describing the real situation in the heterogenous environment. Besides heterogeneity of the soil and groundwater samples, also changes in pollutant concentration within time complicate monitoring of the progress in remediation. Rigorous monitoring of contamination at a locality using conventional analytical methods (Fig. 8.1) is therefore very time-, labour- and resource-demanding. One possible solution to this problem is development of online monitoring systems for continuous analysis of the contaminants. Such a system should be robust, selective, sensitive, easily operable and cheap, to allow sensing of the contaminants at many different locations for longer periods of time.

The tendency to carry out field monitoring has driven the development of biosensors as the new analytical tools. They can be an alternative or a complement to conventional techniques. Their application in environmental monitoring is very promising owing to their sensitivity, low costs, user-friendliness and adaptability for on-site field measurements. Biosensors are highly useful for monitoring of long-term

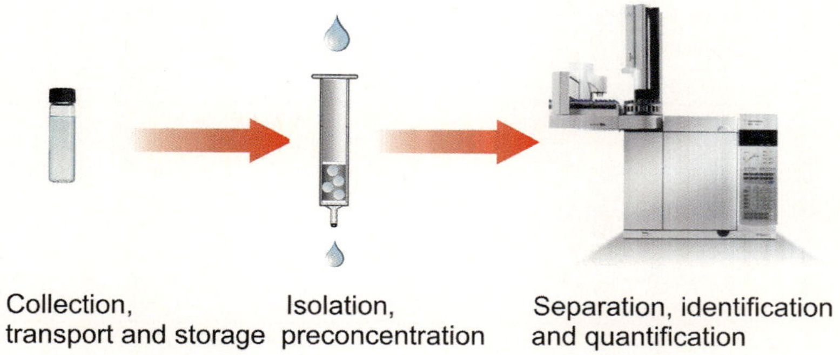

Fig. 8.1 Schematic representation of environmental analysis using conventional analytical methods (adapted from Jakubowska et al. 2009)

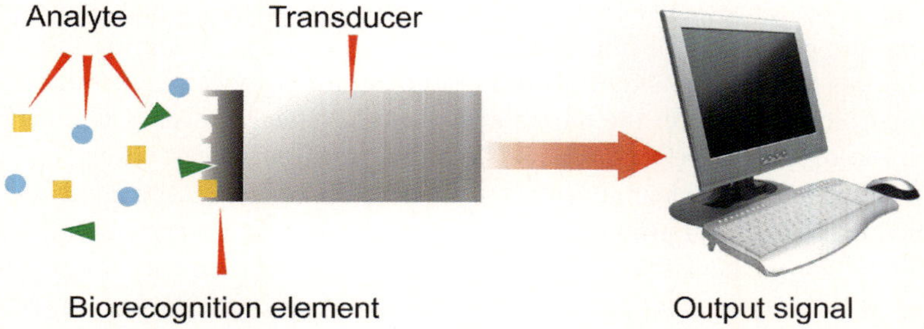

Fig. 8.2 Schematic setup of a biosensor (adapted from Leonard et al. 2003)

changes in contaminant concentrations or preliminary screening of contamination from landfills with inappropriate protective barriers, old sewers, sewage or industrial effluents before applying more costly techniques (Allan et al. 2006). A biosensor is a self-contained integrated device (Fig. 8.2) which is capable of providing specific quantitative or semi-quantitative analytical information using a biorecognition element which is retained in direct spatial contact with a transducer (Thévenot et al. 1999). The chemical compounds to be detected (target analytes) interact selectively with the biorecognition element. This biocatalytic or bioaffinity reaction between the analyte molecules and sensitive layer causes a change in the physicochemical properties of the layer, e.g., changes in mass or optical properties. The altered properties are converted by the transducer to an electronic output signal, which is proportional to the analyte concentration. The signal is amplified, processed and displayed as a measurable signal, which is then evaluated using a computer system. Biosensors should be distinguished from bioassays, where the transducer is not an integral part of the analytical system (Leonard et al. 2003; Rodriguez-Mozaz et al. 2005; Thévenot et al. 1999; Ziegler and Göpel 1998).

Biosensors have to be sensitive and selective for the analyte of interest (Table 8.1). They also need to be cheap, simple to use and able to measure pollutants in complex matrices with minimal sample preparation (Andreescu and Sadik 2004; Eltzov and

Table 8.1 Comparison of conventional analytical methods and biosensors for detection of environmental pollutants

Characteristics	Conventional methods	Biosensors
Sample pretreatment	−	+
Sensitivity	+	±
Selectivity	+	±
Time consumption	−	+
Costs	−	+
Measurement mode	±	+
Requirement on personnel	−	+

+ advantage, − disadvantage, ± limitation (Dennison and Turner 1995; Rodriguez-Mozaz et al. 2007)

Marks 2011). Other advantages offered by biosensors over conventional analytical techniques are fast response time, continuous signal providing a real-time monitoring data, the possibility of miniaturization and portability; permitting their use as field devices working on-site. In addition to determining specific chemicals, some particular biosensors offer the measurement of biological effects of pollutants due to their biological base (Dennison and Turner 1995; Rodriguez-Mozaz et al. 2006, 2007). Despite of their clear advantages, the development of a successful biosensor has encountered several problems, e.g., limited lifetime and reproducibility, which are affected by the sensitive nature of the biorecognition element (Andreescu and Sadik 2004; Karube and Nomura 2000; Wanekaya et al. 2008). Further drawback is also the limited selectivity in differentiating among compounds of similar classes (Rodriguez-Mozaz et al. 2006).

8.2 Enzymatic Biosensors

One of the largest group of potentially harmful pollutants are halogenated hydrocarbons, mainly as a result of their extensive use in agriculture and industry. These compounds belong to prevalent groundwater contaminants and are significant components of hazardous wastes and landfill leachates. Since halogenated hydrocarbons are very resistant to degradation and have adverse health effects, their accurate monitoring in the environment is necessary.

Several biosensing systems for the detection of halogenated hydrocarbons have been reported. These biosensors are catalytic, utilizing microorganisms or purified enzymes as the biorecognition elements (Table 8.2). The *whole-cell biosensors* with microorganisms producing enzymes offer advantages of simple preparation, presence of all co-factors and enzymes involved in a multistep reaction in a single cell. Whole cells are also more tolerant to a significant change in pH, temperature or ionic concentration than purified enzymes. The major limitations for the use of whole cells are long response times due to mass transfer of an analyte and a product and lower specificity than enzymatic biosensors (D'Souza 2001; Orellana and

Table 8.2 Comparison of whole cells and purified enzymes as the biorecognition elements in the biosensors (D'Souza 2001; Mehrvar et al. 2000; Orellana and Haigh 2008; Rogers 2006; Wang and Liu 2010)

Properties	Whole cells	Purified enzymes
Preparation	Simple culturing and harvesting	Tedious, time-consuming and costly purification
Changes of catalytic properties	Modification by genetic engineering, adaptation to adverse conditions	Modification by genetic engineering
Requirement of co-factors	Presence of all needed co-factors in single cell	Need of extra co-factors
Multianalyte detection	Presence of all enzymes in single cell	Immobilization of enzyme mixture on transducer
Sensitivity	Low due to lower concentration of enzyme	High
Selectivity	Low due to unwanted side reactions	High
Response time	Long due to need for cell membrane crossing	Fast
Lifetime	Long due to stabilization of enzymes inside cell	Limited due to low stability of enzymes
Regeneration	Allowing re-growth of cells	Immobilization of fresh enzymes

Haigh 2008; Rogers 2006). They also need more time to return to the baseline level after the use (Wang and Liu 2010). The *enzymatic biosensors* possess several advantages, e.g., high selectivity and ability to modify catalytic properties or substrate specificity by means of genetic engineering (Rogers 1995, 2006). However, enzyme purification is time-consuming and expensive process, which is the main limitation of this type of biosensors (D'Souza 2001; Lei et al. 2006; Orellana and Haigh 2008). Other disadvantages can be lower stability and requirement for soluble co-factors (Lei et al. 2006; Mehrvar et al. 2000).

Dehalogenating enzymes with the potential for application in the biosensors for halogenated pollutants are able to cleave carbon-halogen bonds by using distinct catalytic mechanisms. These enzymes may or may not use co-factors for their activity. During dehalogenation, the halogen substituent is often replaced by a hydrogen atom or a hydroxyl group. Seven principal mechanisms have been described for enzymatic cleavage of the carbon-halogen bonds: (1) reductive dehalogenation, (2) oxidative dehalogenation, (3) hydrolytic dehalogenation, (4) thiolytic dehalogenation, (5) intramolecular nucleophilic substitution, (6) dehydrodehalogenation and (7) hydration (Fetzner 1998; Fetzner and Lingens 1994; Janssen et al. 1994, 2005). Halogenated aliphatic hydrocarbons can be converted by haloalkane dehalogenases (EC 3.8.1.5), haloacid and chloroacrylic acid dehalogenases (EC 3.8.1.2 and 3.8.1.-) and three lyases: haloalcohol dehalogenases (EC 4.5.1.-), dechlorinases (EC 4.5.1.-) and glutathione-dependent dichloromethane dehalogenases (EC 4.5.1.3).

Several bioassays and biosensors for the detection of halogenated aliphatic hydrocarbons have been described in the scientific literature. They are based on whole microbial cells containing haloalkane dehalogenases (Campbell et al. 2006;

Hutter et al. 1995; Peter et al. 1996, 1997; Reardon et al. 2009) and haloacid dehalogenases (Bachas-Daunert et al. 2009; Maliszewska and Wilk 2008). The haloalkane dehalogenases represent an attractive biorecognition element due to co-factor-free hydrolytic mechanism, leading to products easily detectable electrochemically or optically. In addition, haloalkane dehalogenases exhibit a broad substrate specificity, enabling detection of many significant environmental pollutants, e.g., metabolites of γ-hexachlorocyclohexane and/or β-hexachlorocyclohexane, 1,2-dichloroethane, 1,2-dibromoethane and 1,2,3-trichloropropane. Properties of haloalkane dehalogenases and methods for their immobilization are described in the Sects. 8.2.1 and 8.2.2. Sensing principle for the detection of dehalogenation products and analytical performance of current haloalkane dehalogenase-based biosensors are introduced in the Sects. 8.2.3 and 8.2.4.

8.2.1 Haloalkane Dehalogenases

Haloalkane dehalogenases are hydrolytic enzymes able to catalyse cleavage of a carbon-halogen bond in a wide range of halogenated aliphatic hydrocarbons via mechanism of nucleophilic substitution, followed by the nucleophilic addition of water. Products of the reaction are a corresponding alcohol, a halide and a proton (Janssen 2004). Except of biosensing of halogenated hydrocarbons (Bidmanova et al. 2010a, b; Campbell et al. 2006; Hutter et al. 1995; Peter et al. 1996, 1997; Reardon et al. 2009), variety of environmental applications have been described for haloalkane dehalogenases, including by-product recycling and bioremediation of contaminated sites (Koudelakova et al. 2013; Lal et al. 2010; Mena-Benitez et al. 2008; Stucki and Thüer 1995; Swanson 1999). Haloalkane dehalogenases were further applied for detoxification of chemical warfare agents (Blum and Richardt 2008; Prokop et al. 2005, 2006), synthesis of chiral intermediates for pharmaceutical use (Prokop et al. 2004, 2010; Szymański et al. 2011; Westerbeek et al. 2011) and molecular imaging (Hong et al. 2011; Los et al. 2008; Taniguchi and Kawakami 2010; Zhang et al. 2006). Some of these applications have been commercialized by the companies BMG Engineering (www.bmgeng.ch), Enantis (www.enantis.com), OptiEnz (www.optienz.com), Promega (www.promega.com) and Photon System Instruments (www.psi.cz).

During the past two decades, 16 different haloalkane dehalogenases have been identified and biochemically characterized (Drienovska et al. 2012; Hasan et al. 2011; Jesenska et al. 2000, 2005, 2009; Keuning et al. 1985; Nagata et al. 1997; Poelarends et al. 1998; Sallis et al. 1990; Sato et al. 2005; Scholtz et al. 1987; Yokota et al. 1987). They have been originally isolated from bacterial strains living in a soil contaminated with halogenated compounds (Keuning et al. 1985; Nagata et al. 1997; Poelarends et al. 1998; Sallis et al. 1990; Scholtz et al. 1987; Yokota et al. 1987), later also from pathogenic (Hasan et al. 2011; Jesenska et al. 2000, 2005) and symbiotic bacteria (Sato et al. 2005). Recently, the haloalkane dehalogenase originating from extremophilic bacteria has been described and experimentally characterized (Drienovska et al. 2012).

Substrate specificity, catalytic efficiency and stability of haloalkane dehalogenases are important parameters to be considered for their practical applications, including design of catalytic biosensors. These enzymes possess broad *substrate specificity*. They are able to convert chlorinated, brominated and iodinated aliphatic hydrocarbons including alkanes, cycloalkanes, alkenes, alcohols, esters, carboxylic acids, ethers, epoxides, nitriles and amides (Damborsky et al. 2001; Koudelakova et al. 2011). Compounds carrying more than one halogen on a single carbon, fluorinated or aromatic are not converted (Damborsky et al. 2001). Halogenated compounds 1-bromobutane, 1-iodopropane, 1-iodobutane, 1,2-dibromoethane and 4-bromobutanenitrile are the most general substrates, which are converted by overwhelming majority of haloalkane dehalogenases. On the other hand, 1,2-dichloropropane, 1,2,3-trichloropropane, chlorocyclohexane and (bromomethyl)cyclohexane were identified as poor substrates of haloalkane dehalogenases (Koudelakova et al. 2011). Nevertheless, the substrate range as well as the rate of conversion of individual substrates differs among haloalkane dehalogenases (Damborsky et al. 2001; Kmunicek et al. 2005; Sato et al. 2005).

Catalytic efficiencies of haloalkane dehalogenases are in the range from 10^4 to 10^5 M^{-1} s^{-1} with their best substrates (Bosma et al. 2003; Chaloupkova et al. 2003; Pavlova et al. 2009; Schanstra et al. 1996). Even though these numbers are far from the theoretically maximal value 10^8 to 10^9 M^{-1} s^{-1} of diffusion-limited enzymes, they are close to the recently estimated median catalytic efficiency of an "average enzyme" (Bar-Even et al. 2011). However, the catalytic efficiency with the anthropogenic chlorinated substrates such as 1,2,3-trichloropropane is significantly lower, which might be critical for the biosensor performance with these analytes.

Stability of haloalkane dehalogenases is a key factor for their application in biosensing in different environmental conditions. The pH optimum of most haloalkane dehalogenases lies between 8.0 and 9.2, the temperature optimum between 37 and 50 °C (Koudelakova et al. 2013). Activity of haloalkane dehalogenases is reduced approximately five times at 20 °C, in comparison with the activity at 37 °C (Bidmanova et al. unpublished data). Haloalkane dehalogenases, showing good activity at lower temperatures, are attractive for biosensing and biodegradation, since the performance of these enzymes is strongly dependent on conditions at an application site. Therefore, isolation of haloalkane dehalogenases with different temperature and pH preferences or their improvement by protein engineering and immobilization is of interest for biosensing applications.

The haloalkane dehalogenases LinB (Nagata et al. 1993), DhlA (Keuning et al. 1985), DhaA (Kulakova et al. 1997) and DbjA (Sato et al. 2005) exhibiting the highest level of activity and ability to convert otherwise resistant compounds represent useful biocatalysts and will be described here into detail (Table 8.3).

LinB enzyme was isolated from the bacterium *Sphingobium japonicum* UT26 (formerly *Sphingomonas paucimobilis*) that utilizes pesticide γ-hexachlorocyclohexane as the sole carbon and energy source under aerobic conditions. The natural substrate of LinB is 1,3,4,6-tetrachloro-1,4-cyclohexadiene, the intermediate in biodegradation of γ-hexachlorocyclohexane (Nagata et al. 1993). LinB exhibits the broadest substrate specificity of all known haloalkane dehalogenases, preferring

Table 8.3 Characteristics of four haloalkane dehalogenases exhibiting the highest application potential (adapted from Koudelakova et al. 2013)

Enzyme	LinB	DhlA	DhaA	DbjA
Organism	*Sphingobium japonicum* UT26	*Xanthobacter autotrophicus* GJ10	*Rhodococcus rhodochrous* NCIMB 13064	*Bradyrhizobium japonicum* USDA110
Heterologous host	*Escherichia coli*	*Escherichia coli*	*Escherichia coli*	*Escherichia coli*
Non-optimized yield (mg L^{-1})	50	70	120	50
Predicted MW (kDa)	33	35	33	34
Quaternary structure[a]	1	1	1	1, 2 or 4[b]
pI	–[c]	5.4	4.5 or 4.7[d]	–[c]
pH optimum	8.2 or 8.8[d]	8.2	8.0–9.5[d]	9.7
Temperature optimum (°C)	40	37	30–37[d]	50
Preferred substrates[e]	Ethyl 2-bromobutyrate 1,2-Dibromethane Methyl 2-bromopropionate	1,2-Dibromomethane 1,2-Dichloroethane 1,3-Dichloropropene	1,2-Dibromoethane 3-Chloropropene 1,3-Dichloropropene	Ethyl 2-bromopropionate 2-Bromo-1-chloropropane 1,2-Dibromoethane
Attractive substrates[f]	1,3,4,6-Tetrachloro-1,4-cyclohexadiene Sulphur mustard	1,2-Dichloroethane 1,2-Dibromomethane	1,2,3-Trichloropropane Sulphur mustard	1,2-Dibromoethane
Applications	Biosensing Bioremediation Decontamination	Biosensing Bioremediation	Biosensing Bioremediation Biocatalysis Decontamination Molecular imaging	Biosensing Biocatalysis

MW molecular weight, *pI* isoelectric point
[a] Number of subunits
[b] Structure depends on conditions
[c] Not determined
[d] Different sources of data
[e] Substrates converted with the highest rates
[f] Substrates interesting for biosensing application

long-chain and β-substituted haloalkanes (Nagata et al. 1997). Activity of LinB towards chemical warfare agent sulphur mustard, bis(2-chloroethyl) sulphide, has been recently reported (Prokop et al. 2005).

One of the best studied haloalkane dehalogenases to date is *DhlA* enzyme produced by the bacterium *Xanthobacter autotrophicus* GJ10 (Keuning et al. 1985). This enzyme is involved in degradation pathway of 1,2-dichloroethane (Janssen et al. 1989) which is toxic compound used in synthesis of vinyl chloride. DhlA generally dehalogenates only small terminally halogenated aliphatic hydrocarbons (Damborsky et al. 2001).

DhaA enzyme was isolated from the soil bacterium *Rhodococcus rhodochrous* NCIMB 13064, utilizing 1-chloroalkanes as a sole carbon and energy source (Kulakova et al. 1997). DhaA has been extensively studied due to its activity towards 1,2,3-trichloropropane, which is a harmful by-product in the manufacture of epichlorohydrin (Bosma et al. 2003; Gray et al. 2001; Pavlova et al. 2009). DhaA shows activity towards the chemical weapon sulphur mustard (Prokop et al. 2005).

More recently described *DbjA* enzyme originates from the symbiotic bacterium *Bradyrhizobium japonicum* USDA110. DbjA can dehalogenate compounds with chain lengths of up to ten carbon atoms with the optimum activity for compounds with six or seven carbon atoms (Sato et al. 2005). Additionally, this enzyme exhibits high catalytic activity for β-methylated haloalkanes (Chovancova et al. 2007).

Other haloalkane dehalogenases exhibit rather low dehalogenating activity (Koudelakova et al. 2013) and therefore enhancement of their catalytic properties by protein engineering or immobilization strategies is essential for practical applications. Additionally, identification and isolation of novel enzymes from extremophilic bacteria can provide biocatalysts possessing useful features for their applications, including biosensing.

8.2.2 Immobilization of Haloalkane Dehalogenases

Immobilization of biorecognition elements on the transducer surface is a necessary and critical step in the design of biosensors, strongly affecting their analytical performances (Andreescu and Sadik 2004; Sassolas et al. 2012). Therefore, an intensive effort has been devoted to development of successful immobilization strategies for haloalkane dehalogenases. A good immobilization method should be simple, fast and gentle (Biran et al. 2008). Immobilized elements have to remain tightly bound to the surface and should not be desorbed during the use. There are five common methods of immobilization (Fig. 8.3, Table 8.4): (1) entrapment, (2) adsorption, (3) covalent binding, (4) cross-linking and (5) affinity binding (Sassolas et al. 2012).

Immobilization by *entrapment* is a simple method based on the fixation of biocatalysts, most frequently the microbial cells, within three-dimensional matrices without direct binding to the support (Bickerstaff 1997). The biocatalysts and additives can be simultaneously deposited in the same sensing layer. There is no need for modification of the biological element so that the function and activity are

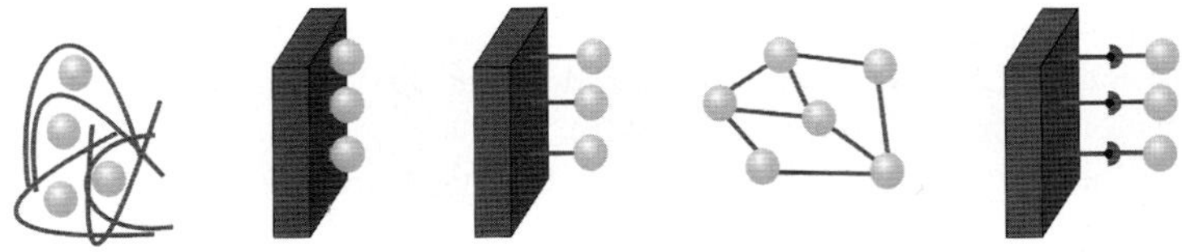

Fig. 8.3 Schematic representation of the basic immobilization methods (Sassolas et al. 2012)

preserved. Biosensors based on entrapped enzymes or cells are often characterized by increased operational and storage stability. Major limitations of entrapment are diffusion restrictions caused by the support. They can be minimized by increasing the porosity of the matrix. On the other hand, entrapment often results in leaching of the enzymes or the cells during continuous use, which can be remedied by cross-linking (Mehrvar et al. 2000; Sassolas et al. 2012).

Adsorption onto solid supports is the simplest technique of immobilization which involves reversible surface interactions between enzymes or cells and support (Sharma et al. 2003). The adsorption mechanism is governed by weak bonds such as Van der Waal's forces, electrostatic and/or hydrophobic interactions between biomaterials and support (Choi 2004), so no chemical activation is required. This procedure is generally non-destructive to the enzymes or the cells (Sassolas et al. 2012). Adsorption is cheap, fast and reversible, allowing regeneration with fresh enzymes or cells (Bickerstaff 1997; Lei et al. 2006). The most significant drawback of this technique is desorption of the element resulting from changes in temperature, pH and ionic strength (Choi 2004; Sassolas et al. 2012). Thus, biosensors based on adsorbed biocatalysts suffer from poor operational and storage stability (Sassolas et al. 2012). Another disadvantage is the non-specific adsorption of other proteins or substances (Bickerstaff 1997; Choi 2004).

Covalent binding is popular immobilization method for the enzymes, while it is not suitable for the microbial cells due to affecting cell viability by harsh conditions during the immobilization (D'Souza 2001). This method is based on binding of the biocatalysts to the surface through functional groups that they contain and that are not essential for their catalytic activity. The binding of enzymes to the solid support is generally carried out by initial activation of the surface using multifunctional reagents, followed by enzyme coupling to the activated support, then the removal of excess and unbound molecules (Sassolas et al. 2012). This technique is more permanent than the adsorption and avoids problems of protein leaching from the support. One drawback of this technique is that covalent binding might cause changes in the enzyme conformation resulting in lower activity (Lei et al. 2006).

Cross-linking is another well-known approach to develop biosensors which utilizes intermolecular cross-linking of proteins using glutaraldehyde or other cross-linking reagents (Mateo et al. 2004). The enzyme molecules can be either cross-linked with each other or in the presence of a functionally inert protein, e.g., bovine serum albumin. This method is attractive due to its simplicity and achieved

Table 8.4 Advantages and limitations of basic immobilization methods (Park et al. 2010; Sassolas et al. 2012)

Method	Binding principle	Immobilized element	Advantages	Disadvantages
Entrapment	Incorporation of biocatalyst within gel or polymer	Cells Enzymes[a]	Activity retained, different biocatalysts in same matrix	High diffusion barrier, leaching
Adsorption	Weak bonds	Cells Enzymes	Simple, no modification of biocatalyst	Desorption, non-specific adsorption
Covalent binding	Covalent bonds between biocatalyst and support	Enzymes	Low diffusion barrier, stable, short response time	Specific groups required, loss of activity, often toxic cross-linker
Cross-linking	Covalent bonds between biocatalyst and cross-linker	Enzymes	Simple	Loss of activity, low mechanical stability
Affinity binding	Affinity bonds between support and biocatalyst	Cells Enzymes	Controlled and oriented immobilization	Specific groups required

[a]Often combined with other immobilization method

strong chemical binding between the biomolecules. The main disadvantage is the possibility of activity losses due to the distortion of the active enzyme conformation and the chemical alterations of the active site during cross-linking (Sassolas et al. 2012). Cross-linking is rarely used as the only means of immobilization because the absence of mechanical properties and the poor stability are severe limitations. This technique is most often used to enhance other immobilization methods, mainly adsorption and entrapment (Bickerstaff 1997).

Affinity binding can provide oriented and site-specific immobilization of enzymes in order to avoid enzyme deactivation by the active site blocking (Campàs et al. 2009; Sassolas et al. 2012). Affinity bonds are created between an activated support, e.g., with avidin or metal chelates, and a specific group of protein sequence, e.g., biotin or histidine (Sassolas et al. 2012). An enzyme can contain affinity tags in its sequence or the affinity tag needs to be attached to the protein sequence by genetic engineering methods (Campàs et al. 2009; Andreescu and Marty 2006). This method might be also useful for immobilization of cells.

All aforementioned strategies were investigated for immobilization of haloalkane dehalogenases in the cells or in a purified form (Bidmanova et al. 2010a; Campbell et al. 2006; Dravis et al. 2001; Hutter et al. 1995; Johnson et al. 2008; Peter et al. 1996, 1997; Reardon et al. 2009; Samorski 2008). Some of them were also applied to the development of biosensors for detection of halogenated aliphatic hydrocarbons. Two previous reports described covalent binding of DhaA onto Eupergit (Samorski 2008) and alumina support impregnated by polyethyleneimine (Dravis et al. 2001). Combination of affinity and covalent binding was used for immobilization of DhlA on iron oxide nanoparticles (Johnson et al. 2008). Whole microbial cells, containing haloalkane dehalogenases, were entrapped in alginate and chitosan gel and used in a form of beads or layer coating the transducer (Campbell et al. 2006; Hutter et al. 1995; Peter et al. 1996, 1997; Reardon et al. 2009). Purified haloalkane dehalogenase used in the optical biosensor was pre-immobilized by adsorption and then cross-linked with glutaraldehyde (Bidmanova et al. 2010a). In this case, cross-linking of LinB was favourable due to the highest activity retention and negligible leaching, unlike the entrapment and the combination of entrapment and cross-linking.

8.2.3 Sensing Principle of Biosensors with Haloalkane Dehalogenases

Upon addition of a halogenated compound to haloalkane dehalogenases immobilized in a purified form or inside the cells, the interaction between enzyme and target analyte can be detected by a number of analytical techniques using appropriate transducers. The measured signal should be correlated with the concentration of the analytes (Su et al. 2011). Among various sensing methods, electrochemical and optical techniques are most widely used in the development of catalytic biosensors (Rodriguez-Mozaz et al. 2004; Su et al. 2011) and were also reported for the detection of products of dehalogenase reaction (Bidmanova et al. 2010a; Campbell et al. 2006; Hutter et al. 1995; Peter et al. 1996, 1997; Reardon et al. 2009).

Detection of halogenated aliphatic hydrocarbons based on potentiometry as one of several *electrochemical approaches* was described by Hutter et al. (1995) and Peter et al. (1996, 1997). The authors used chloride and bromide electrodes with an ion-selective AgX^{-}/AgS membrane in combination with the Ag/AgCl reference electrode. The electrodes were interfaced with the pH-ionometer (Hutter et al. 1995). Firstly, they developed a bioassay based on the aforementioned ion-selective electrode and polymer beads with entrapped microbial cells containing haloalkane dehalogenases (Hutter et al. 1995). This sensing design was adapted in the compact biosensor by coating the working electrode with an immobilized microbial layer. Whole cells containing haloalkane dehalogenases had minimal influence on halide ions in measurement solution, however, the enzymatically formed chloride or bromide ions generated the change in the potential resulting from ion accumulation at the ion-selective membrane interface. Potential difference between the working electrode and the reference electrode exhibited concentration-dependent behaviour (Peter et al. 1996).

Although the potentiometric biosensor employing haloalkane dehalogenase (Peter et al. 1996) was only partially characterized and effect of some factors, e.g., buffering capacity or presence of other chemicals, was not studied, application of potentiometry can be advantageous for environmental analysis. In general, potentiometric systems exhibit a wide detection range, high sensitivity and selectivity due to the ion-selective working electrode. However, a highly stable and accurate reference electrode is always required (Griffiths and Hall 1993; Su et al. 2011). Significantly, an ion-selective potentiometric biosensor only requires measurement of the electrode potential at near zero current, thereby simple and inexpensive instrumentation can be applied for potential field monitoring. Desirable features of potentiometric biosensors for environmental analysis are also possibility of portability and miniaturization (Dennison and Turner 1995; Thévenot et al. 1999). However, this electroanalytical technique is confronted with a wide range of potential problems coming from complex environmental samples: (1) electrode fouling resulting in erroneous response characteristics, (2) electrode drift leading to poor reproducibility, (3) electrode dissolution causing a high surface excess of analyte that degrades the detection limit of the ion-selective electrode and (4) electrode instability resulting in the response characteristics which are erratic and non-representative (De Marco et al. 2007).

Biosensors for *optical detection* of halogenated aliphatic hydrocarbons were firstly described by Campbell et al. (2006) and Reardon et al. (2009). Their sensing concept utilized immobilized whole microbial cells containing haloalkane dehalogenase on the optical transducer. The first biosensor exploiting purified haloalkane dehalogenase has been recently reported by Bidmanova et al. (2010a). All mentioned systems (Bidmanova et al. 2010a; Campbell et al. 2006; Reardon et al. 2009) were based on the measurement of optical signal changes produced in dependence on the concentration of protons as products of dehalogenase reaction.

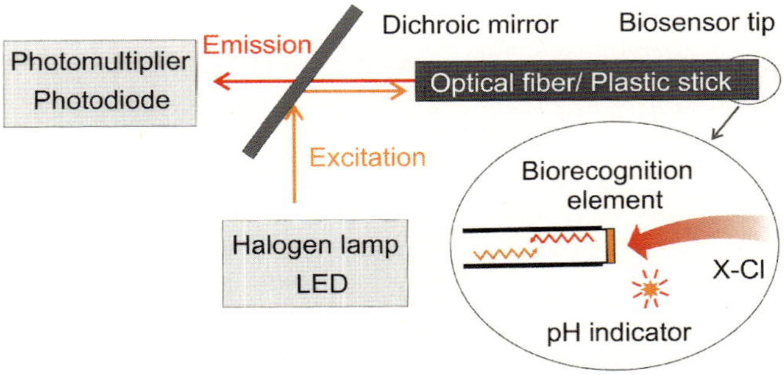

Fig. 8.4 Schematic of optical biosensor system including enlarged view of the biosensor tip. *LED* light-emitting diode (adapted from Campbell et al. 2006)

Reported biosensors consisted of an *pH optode* and *optoelectronic instrumentation*: light source, light detector and signal processing (Fig. 8.4). The optode was constructed by immobilization of a fluorescent pH indicator on an optical fibre, which was used to couple light between the indicator and the measuremet instrumentation. The light from the halogen lamp or light-emitting diode (LED) was filtered to select narrow band of wavelengths at excitation maximum of the indicator. After reflection on a dichroic mirror, the light was introduced into polymethylmethacrylate optical fibre to excite the immobilized indicator on a tip. Emitted light was collected and transmitted by a fibre to a dichroic mirror which reflected the long-wavelength light and passed shorter wavelengths. The light directed into a photomultiplier was then transformed into an electrical signal (Bidmanova et al. 2010a; Campbell et al. 2006; Reardon et al. 2009).

These first prototypes of the biosensor systems provided space for optimization. The most recent biosensor called EnviroPen was developed in collaboration among Loschmidt Laboratories at Masaryk University (www.muni.cz), Enantis (www.enantis.com) and Photon System Instruments (www.psi.cz). The EnviroPen follows the original sensing concept. However, its optoelectronic instrumentation part was fine-tuned to be more sensitive and simultaneously simple, miniature and inexpensive (Fig. 8.5). Photomultiplier in this instrument was replaced by photodiode and optical fibre was replaced by the plastic stick.

Selection of an indicator dye for the pH optode is a critical step in biosensor development. Although many pH indicators are known, only a few meet the requirements for the use in a pH optode. They should possess an appropriate pK_a in the pH range of interest, strong absorption within the wavelength range from 400 to 700 nm, photo and chemical stability, non-toxicity, availability of functional groups suitable for immobilization and high quantum yield (Leiner and Hartmann 1993). *Indicator dyes* which were applied in the biosensors with haloalkane dehalogenases (Fig. 8.6) were (1) fluoresceinamine, (2) 5(6)-carboxyfluorescein and (3) 5(6)-carboxynaphthofluorescein (Bidmanova et al. 2010a; Campbell et al. 2006; Reardon et al. 2009). These molecules undergo reversible changes in their fluorescent intensity upon reaction with protons (Leiner and Hartmann 1993). Although they all belong to the

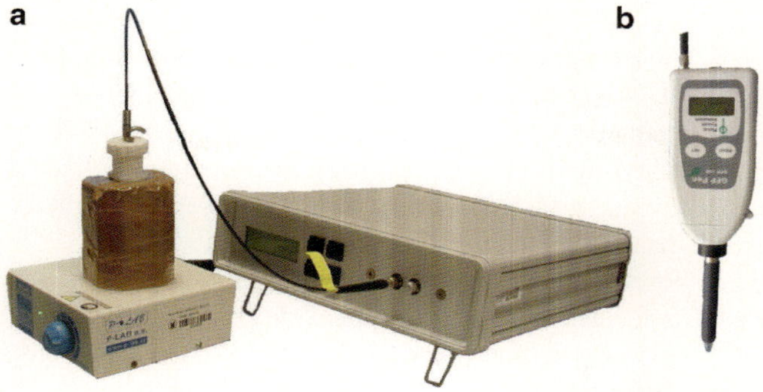

Fig. 8.5 Optical biosensors employing purified haloalkane dehalogenase: (**a**) instrumentation of the first biosensor prototype (25 × 7 × 25 cm, Bidmanova et al. 2010a) and (**b**) the fine-tuned and miniaturized latest version of the biosensor EnviroPen (6 × 3 × 19 cm, www.enantis.com)

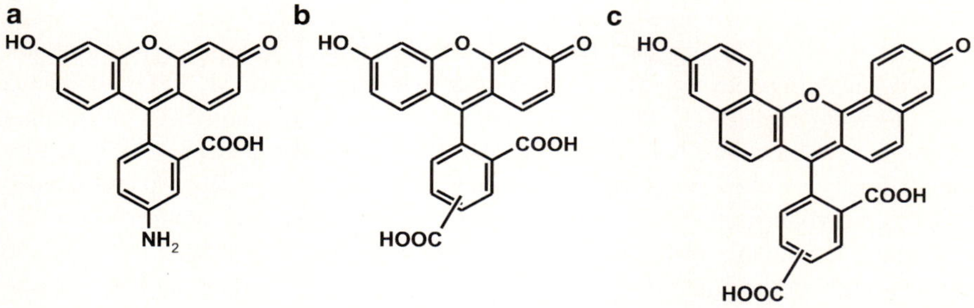

Fig. 8.6 Structure of fluorescent pH indicators used in optical biosensors employing haloalkane dehalogenases: (**a**) fluoresceinamine, (**b**) 5(6)-carboxyfluorescein and (**c**) 5(6)-carboxynaphthofluorescein

derivatives of fluorescein, they differ in their properties, e.g., pK_a value, excitation and emission wavelength, the Stokes shift and quantum yield (Bidmanova et al. 2012; Duong et al. 2006). They also possess different functional groups for the immobilization to ensure close proximity with the optical transducer.

Immobilization method has significant impact on mechanical integrity and response characteristics of the constructed biosensor. Generally, the same immobilization strategies as described for the biorecognition element are applicable for the indicator molecules. Fluoresceinamine in biosensors developed by Campbell et al. (2006) and Reardon et al. (2009) was immobilized on the tip of optical fibre by cross-linking with glutaraldehyde. The microbial cells containing haloalkane dehalogenases were immobilized subsequently on the prepared sensing layer. More convenient approach is co-immobilization of biorecognition element and indicator dye by a single immobilization method, which is suitable for both sensing elements (Biran et al. 2008). Cross-linking with glutaraldehyde was used by Bidmanova et al. (2010a) to prepare a biosensor tip formed by one sensitive layer of the purified haloalkane dehalogenase and the conjugate of fluorescent pH indicator with bovine serum albumin. The microenvironment resulting from the sensing layer usually affects the properties of a pH indicator. In particular, shifts of several units in pK_a value can be observed (Duong et al. 2006; Lobnik et al. 1998; Schulman et al. 1995; Wencel et al. 2009). Furthermore, response time or photostability of immobilized indicator may be altered due to different diffusion rates and hydrophobicities (Lobnik et al. 1998; Wolfbeis et al. 1992). Therefore, the immobilized conjugates of 5(6)-carboxyfluorescein and 5(6)-carboxynaphthofluorescein with bovine serum albumin were characterized into detail (Bidmanova et al. 2012).

The optical biosensors employing haloalkane dehalogenases offer numerous benefits for environmental monitoring. Optical systems can be easily miniaturized (Farré et al. 2009) as it was demonstrated for the latest version of biosensor EnviroPen (Fig. 8.5). These devices are cost-effective and thus can be fabricated as disposable biosensors. Moreover, fibre-optic biosensors can be used for remote or hazardous applications because light is transmitted over long distances in the fibre and the detector can be located at the distance from the sensor tip (Biran et al. 2008; Lin 2000; Marazuela and Moreno-Bondi 2002). These biosensors do not require additional reference element, unlike potentiometric biosensors (Biran et al. 2008; Marazuela and Moreno-Bondi 2002). Further advantage is transmission of multiple signals simultaneously, enabling detection of several analytes at a time (Farré et al. 2009; Marazuela and Moreno-Bondi 2002). Such multichannel device for detection of environmental contaminants including halogenated hydrocarbons has been recently developed by the company OptiEnz (www.optienz.com). On the contrary, optical biosensors also implicate certain limitations. They have a narrow dynamic range compared to the electrodes (Lin 2000; Wolfbeis 2005), can be interfered by the sample turbidity, the ambient light or the scattered excitation light. They offer limited operational lifetime due to possible photobleaching or leaching of the indicator dye out of the biosensor (Biran et al. 2008; Lin 2000; Marazuela and Moreno-Bondi 2002).

8.2.4 Performance of Biosensors with Haloalkane Dehalogenases

The developed biosensors employing haloalkane dehalogenases were characterized by evaluation of their analytical parameters such as limit of detection, selectivity, stability and reproducibility. Selected biosensors were also tested for functionality under operating conditions. Characteristics of both potentiometric and optical systems are summarized in the Table 8.5 (Koudelakova et al. 2013).

Reported biosensors differ in the *detection limit* ranging from 1 to 25,000 μg L^{-1} (Bidmanova et al. 2010a; Campbell et al. 2006; Peter et al. 1996; Reardon et al. 2009). Although such detection limits are sufficient for monitoring of contaminated sites, they would have to be lowered for the analysis of drinking water (Reardon et al. 2009). The maximum contaminant limits of regulated chemicals 1,2-dibromoethane, 1,2-dichloroethane and γ-hexachlorocyclohexane in drinking water set by US EPA range from 0.05 to 5 μg L^{-1} (United States Environmental Protection Agency 2011). Improvement of the detection limit was successfully demonstrated with the optical biosensor based on the purified enzyme LinB from

Table 8.5 Characteristics of the biosensors employing haloalkane dehalogenases (Koudelakova et al. 2013)

Biorecognition element	Transducer	Detection limit ($\mu g\ L^{-1}$)	SD (%)	Time[a] (min)	pH[b]	Temperature[b] (°C)	Stability[b]	Reference
Whole cells of *Rhodococcus* sp. DSM 6344 entrapped in alginate	Potentiometric electrode	220 for CB 40 for DBA	7.8	5	5.0–9.0	20.0–30.0	4 °C, 7 days	Peter et al. (1996)
Whole cells of *Xanthobacter autotrophicus* GJ10 entrapped in alginate	Fiber-optic pH optode	11,000 for DCA	9.0	8–10	7.3	Room temperature	4 °C, 9 days	Campbell et al. (2006)
Whole cells of *Rhodococcus* sp. GJ70 entrapped in alginate	Fiber-optic pH optode	1 for DBA	9.0	7–110	5.5–6.5	Room temperature	4 °C, 9 days	Reardon et al. (2009)
Enzyme LinB from *Sphingobium japonicum* UT26 cross-linked with glutaraldehyde	Fibre-optic pH optode	24,985 for DBA 1,750 for CCMP	–[c]	30	8.2	21.0	4 °C, 90 days[d]	Bidmanova et al. (2010a)
Enzyme LinB from *Sphingobium japonicum* UT26 cross-linked with glutaraldehyde	pH optode	2,630 for DBA 750 for CCMP	8.6	1	6.0–10.0	18.5–22.0	–[c]	–[d]
Mutant enzyme DhaA31 from *Rhodococcus* sp. cross-linked with glutaraldehyde	pH optode	1,327 for TCP	–[c]	1	6.5–7.5; 8.5–9.0	16.0–22.0	–[c]	–[d]

CB chlorobutane, *CCMP* 3-chloro-2-(chloromethyl)-1-propene, *DBA* 1,2-dibromoethane, *DCA* 1,2-dichloroethane, *SD* standard deviation, *TCP* 1,2,3-trichloropropane

[a] Measurement time used for monitoring of enzymatic reaction and calculation of the detection limit

[b] Retention of >80 % biosensor response

[c] Not tested

[d] Not published (Hrdlickova et al. unpublished data), EnviroPen biosensor commercialized by the company Enantis (www.enantis.com)

Table 8.6 Comparison of the optical biosensors based on purified haloalkane dehalogenases demonstrating a progress made from the first prototype to the fine-tuned and miniaturized biosensor

Properties	Prototype	EnviroPen
Chemical compatibility	Aqueous solution[a]	Aqueous solution[a]
Detection limits (μg L^{-1})	24,985 for DBA	2,630 for DBA
	1,750 for CCMP	750 for CCMP
pH range	–[b]	4–10[c]
Temperature range (°C)	–[b]	4–40[d]
Measurement time (min)[e]	30	1
Storage	Dry or wet conditions	Dry or wet conditions
Size (cm)	25×7×25	6×3×19
Estimated manufacturing costs (€)	6,000	1,600

CCMP 3-chloro-2-(chloromethyl)-1-propene, *DBA* 1,2-dibromoethane
[a] Potential for gas biosensing
[b] Not tested
[c] Retention of >60 % biosensor response
[d] Retention of >20 % biosensor response
[e] Measurement time used for monitoring of enzymatic reaction and calculation of detection limit

Sphingobium japonicum UT26. The first prototype of the biosensor (Fig. 8.5a) exhibited detection limit of 25,000 μg L^{-1} for 1,2-dibromoethane (Bidmanova et al. 2010a). Systematic research and development led to construction of innovative biosensor EnviroPen (Fig. 8.5b, Table 8.6) based on different indicator dye and fine-tuned optoelectronics, lowering the detection limit by one order of magnitude, reaching 2,600 μg L^{-1} (Hrdlickova et al. unpublished data). Further decrease of the detection limit will require application of protein engineering to improve the catalytic performance of employed enzymes (Campàs et al. 2009).

Selectivity of a biosensor is determined by employed biorecognition element. Individual haloalkane dehalogenases are not specific to the particular substrate due to their wide substrate scope (Koudelakova et al. 2011). Limited selectivity of haloalkane dehalogenase-based biosensor can be improved through the protein engineering or by using sensor arrays carrying the enzymes with different selectivities (Reardon et al. 2009). Such strategy is applicable for the detection of 1,2-dichloroethane, which is converted only by DhlA (Janssen et al. 2005). Another approach, enabling monitoring of a broad range of halogenated compounds, can be based on combination of several haloalkane dehalogenases with different substrate specificities in the multichannel device. This system might be very useful for preliminary screening of potentially contaminated sites with analysis of the positive samples by confirmatory analytical techniques.

Critical parameters of biosensors influencing their applicability are *reproducibility*, *measurement time* and *storage stability*. All reported biosensors employing both whole cells and purified enzyme exhibit standard deviation lower than 10 %, even though this value is calculated from a different number of measurements (Bidmanova et al. 2010a; Campbell et al. 2006; Peter et al. 1996; Reardon et al. 2009). The time necessary for monitoring of biosensor response is up to 30 min in the most of the reported biosensors. This is acceptable for the long-term monitoring; however, mapping of contamination favours collection of the data more rapidly (Allan et al. 2006). Improvement of the instrumentation and modification of data analysis led to remarkable reduction of the measurement time from 30 to 1 min (Table 8.6). Important issue is also the storage stability of the biosensors. Current biosensors are stored at 4 °C in buffer solution (Campbell et al. 2006; Reardon et al. 2009) or in a dry form (Peter et al. 1996); however, their stability is limited. Storage stability can be improved by lyophilization (Cao 2005).

Biosensor performance, mainly enzymatic reaction in biosensor, is strongly affected by the properties of the application site, particularly by *temperature and pH*. Average temperature of a groundwater in Central Europe is approximately 9.5 °C (Pitter 2009). Temperature of surface water is strongly dependent on the climatic conditions and ranges from 0 °C under ice to 40 °C in the hot springs. The average annual temperature of Central European rivers is approximately 10 °C; however, the temperature strongly varies throughout the year (Pitter 2009). The most common pH is 6.5 to 8.5 for surface water (Hem 1985) and 6.0 to 8.5 for groundwater systems (Hem 1985; Jordana and Piera 2004). However, acidification caused by industry can lead to pH values lower than 5.5 (Scheidleder et al. 1999). Majority of reported biosensors are able to cover the common pH range of waters (Hrdlickova et al. unpublished data, Peter et al. 1996; Reardon et al. 2009); however, broad temperature diversity might be an obstacle. The biosensors perform well in the temperature range from 20 to 30 °C, retaining minimally 80 % of their response (Bidmanova et al. 2010a; Campbell et al. 2006; Peter et al. 1996; Reardon et al. 2009). The optical biosensors utilizing purified haloalkane dehalogenases LinB and DhaA31 are also able to detect halogenated chemicals down to 5 °C. Application of novel haloalkane dehalogenases from extreme environments, active at lower temperature and pH, would be an advantage for preparation of the biosensors suitable for variety of environments.

While monitoring contaminants in the environment, the evaluation of potential *matrix effects* on biosensor performance is of great importance. Environmental samples are usually comprised of extremely complex and variable mixtures of proteins, carbohydrates, lipids, small molecules and salts, which may lead to false positive or false negative results (Long et al. 2009). The potential matrix effect of environmental water samples (Fig. 8.7) was evaluated with the biosensor employing purified haloalkane dehalogenase LinB (Fig. 8.8). The detection limit for the water samples spiked with 1,2-dibromoethane was 64,000 μg L^{-1}, which was slightly higher, compared to the detection limit in the buffer solution (25,000 μg L^{-1}).

8.3 Conclusions

The biosensors employing haloalkane dehalogenases have been demonstrated as fast, reliable, simple and cost-effective tools for detection of halogenated aliphatic hydrocarbons under laboratory conditions. The field trials under

Fig. 8.7 Environmental water samples collected from the landfill site Hajek in Czech Republic. Samples were spiked with 1,2-dibromoethane and analysed using the haloalkane dehalogenase-based biosensor

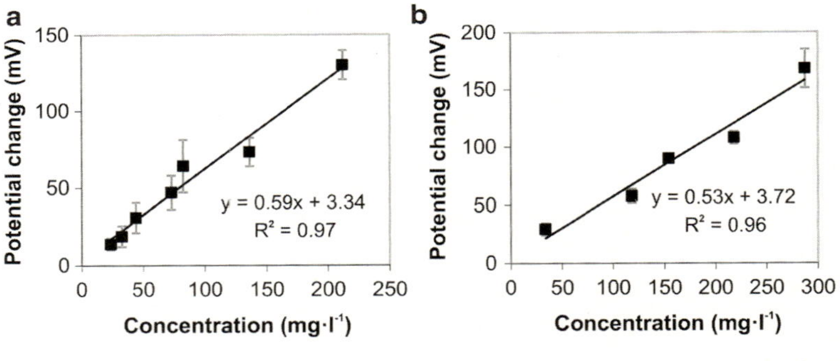

Fig. 8.8 Calibration plot of the dehalogenase-based biosensor for 1,2-dibromoethane (**a**) in buffer solution and (**b**) in spiked water sample collected from landfill site. Error bars represent three to five independent measurements (Bidmanova et al. 2010a, b, Hrdlickova et al. unpublished data)

environmental settings are currently in progress. On the basis of reported features, good perspectives are given to the optimized systems in both field screening and field monitoring. Applications of the biosensors for screening based on high-throughput analyses may include spatial and temporal characterization of specific halogenated contaminants, e.g., identification of hot spots in the contaminated site or concentration changes of a contaminant on a short timescale. Developed systems are also particularly suited for continuous in situ monitoring, where sample removal is inconvenient, difficult or dangerous. Specific examples might include in situ monitoring of a process control stream to determine the efficiency of hazardous waste site remediation, monitoring an agricultural run-off during a peak application periods or continuous monitoring of wells to determine whether concentrations of an analyte of interest during remediation procedure or after the site closure are in compliance.

Acknowledgements The authors would like to express their sincere thanks to Prof. Ken Reardon (Colorado State University, USA) for introducing us to the field of optical biosensors, Prof. Thomas Scheper (University Hannover, Germany) and Prof. Jaromir Hubalek (Brno University of Technology, Czech Republic) for construction of the first prototypes of our optical biosensors and to Dr. Martin Trtilek (Photon System Instruments, Czech Republic) for their miniaturization. The work was supported by the Grant Agency of the Czech Republic (P207/12/0775, P503/12/0572), the Grant Agency of the Czech Academy of Sciences (IAA401630901) and the European Regional Development Fund (CZ.1.05/2.1.00/01.0001).

References

Allan IJ, Vrana B, Greenwood R, Mills GA, Knutsson J, Holmberg A, Guigues N, Fouillac A-M, Laschi S (2006) Strategic monitoring for the European Water Framework Directive. Trends Anal Chem 25:704–715

Andreescu S, Marty J-L (2006) Twenty years research in cholinesterase biosensors: from basic research to practical applications. Biomol Eng 23:1–15

Andreescu S, Sadik OA (2004) Trends and challenges in biochemical sensors for clinical and environmental monitoring. Pure Appl Chem 76:861–878

Bachas-Daunert PG, Sellers ZP, Wei Y (2009) Detection of halogenated organic compounds using immobilized thermophilic dehalogenase. Anal Bioanal Chem 395:1173–1178

Bar-Even A, Noor E, Savir Y, Liebermeister W, Davidi D, Tawfik DS, Milo R (2011) The moderately efficient enzyme: evolutionary and physicochemical trends shaping enzyme parameters. Biochem 50:4402–4410

Bickerstaff GF (1997) Immobilization of enzymes and cells. In: Bickerstaff GF (ed) Methods in biotechnology, 1st edn. Humana Press, New Jersey

Bidmanova S, Chaloupkova R, Damborsky J, Prokop Z (2010a) Development of an enzymatic fiber-optic biosensor for detection of halogenated hydrocarbons. Anal Bioanal Chem 398:1891–1898

Bidmanova S, Pohanka M, Cabal J, Prokop Z, Damborsky J (2010b) Early warning biosensors for detection of chemical warfare agents (in Czech). Chem Listy 104:302–308

Bidmanova S, Hlavacek A, Damborsky J, Prokop Z (2012) Conjugation of 5(6)-carboxyfluorescein and 5(6)-carboxynaphthofluorescein with bovine serum albumin and their immobilization for optical pH sensing. Sens Actuators: B Chem 161:93–99

Biran I, Yu X, Walt DR (2008) Optrode-based fiber optic biosensors (bio-optrode). In: Ligler FS, Taitt CR (eds) Optical biosensors: Today and tomorrow, 2nd edn. Elsevier, Amsterdam

Blum M-M, Richardt A (2008) Hydrolytic enzymes for chemical warfare agent decontamination. In: Richardt A, Blum M-M (eds) Decontamination of warfare agent, 1st edn. Wiley-VCH, Weinheim

Bosma T, Pikkemaat MG, Kingma J, Dijk J, Janssen DB (2003) Steady-state and pre-steady-state kinetic analysis of halopropane conversion by a *Rhodococcus* haloalkane dehalogenase. Biochem 42:8047–8053

Campàs M, Prieto-Simón B, Marty J-L (2009) A review of the use of genetically engineered enzymes in electrochemical biosensors. Semin Cell Dev Biol 20:3–9

Campbell DW, Müller C, Reardon KF (2006) Development of a fiber optic enzymatic biosensor for 1,2-dichloroethane. Biotechnol Lett 28:883–887

Cao L (2005) Unconventional enzyme immobilization. In: Cao L (ed) Carrier-bound immobilized enzymes, 1st edn. Wiley-VCH, Weinheim
Chaloupkova R, Sykorova J, Prokop Z, Jesenska A, Monincova M, Pavlova M, Tsuda M, Nagata Y, Damborsky J (2003) Modification of activity and specificity of haloalkane dehalogenase from *Sphingomonas paucimobilis* UT26 by engineering of its entrance tunnel. J Biol Chem 278:52622–52628
Choi MMF (2004) Progress in enzyme-based biosensors using optical transducers. Microchim Acta 148:107–132
Chovancova E, Kosinski J, Bujnicki JM, Damborsky J (2007) Phylogenetic analysis of haloalkane dehalogenases. Proteins 67:305–316
D'Souza SF (2001) Microbial biosensors. Biosens Bioelectron 16:337–353
Damborsky J, Rorije E, Jesenska A, Nagata Y, Klopman G, Peijnenburg WJGM (2001) Structure-specificity relationships for haloalkane dehalogenases. Environ Toxicol Chem 20:2681–2689
De Marco R, Clarke G, Pejcic B (2007) Ion-selective electrode potentiometry in environmental analysis. Electroanal 19:1987–2001
Dennison MJ, Turner AP (1995) Biosensors for environmental monitoring. Biotech Adv 13:1–12
Dravis BC, Swanson PE, Russell AJ (2001) Haloalkane hydrolysis with an immobilized haloalkane dehalogenase. Biotechnol Bioeng 75:416–423
Drienovska I, Chovancova E, Koudelakova T, Damborsky J, Chaloupkova R (2012) Biochemical characterization of the first haloalkane dehalogenase from the cold-adapted bacterium. Appl Environ Microbiol 78:4995–4998
Duong HD, Sohn O-J, Lam HT, Rhee JI (2006) An optical pH sensor with extended detection range based on fluoresceinamine covalently bound to sol–gel support. Microchem J 84:50–55
Eltzov E, Marks RS (2011) Whole-cell aquatic biosensors. Anal Bioanal Chem 400:895–913
Farré M, Kantiani L, Pérez S, Barceló D (2009) Sensors and biosensors in support of EU Directives. Trends Anal Chem 28:170–185
Fetzner S (1998) Bacterial dehalogenation. Appl Microbiol Biotechnol 50:633–657
Fetzner S, Lingens F (1994) Bacterial dehalogenases: Biochemistry, genetics, and biotechnological applications. Microbiol Mol Biol Rev 58:641–685
Gray KA, Richardson TH, Kretz KK, Short JM, Bartnek F, Knowles R, Kan L, Swanson PE, Robertson DE (2001) Rapid evolution of reversible denaturation and elevated melting temperature in a microbial haloalkane dehalogenase. Adv Synth Catal 343:607–617
Griffiths D, Hall G (1993) Biosensors - what real progress is being made? Trends Biotechnol 11:122–130
Hasan K, Fortova A, Koudelakova T, Chaloupkova R, Ishitsuka M, Nagata Y, Damborsky J, Prokop Z (2011) Biochemical characteristics of the novel haloalkane dehalogenase DatA, isolated from the plant pathogen *Agrobacterium tumefaciens* C58. Appl Environ Microbiol 77:1881–1884
Hem JD (1985) Study and interpretation of the chemical characteristics of natural water. United States Geological Survey, Virginia
Hong H, Benink HA, Zhang Y, Yang Y, Uyeda HT, Engle JW, Severin GW, McDougall MG, Barnhart TE, Klaubert DH, Nickles RJ, Fan F, Cai W (2011) HaloTag: a novel reporter gene for positron emission tomography. Am J Transl Res 3:392–403
Hutter W, Peter J, Swoboda H, Hampel W, Rosenberg E, Krämer D, Kellner R (1995) Development of a microbial bioassay for chlorinated and brominated hydrocarbons. Anal Chim Acta 306:237–241
Jakubowska N, Zygmunt B, Polkowska Ż, Zabiegała B, Namieśnik J (2009) Sample preparation for gas chromatographic determination of halogenated volatile organic compounds in environmental and biological samples. J Chromatogr A 1216:422–441
Janssen DB (2004) Evolving haloalkane dehalogenases. Curr Opin Chem Biol 8:150–159
Janssen DB, Pries F, van der Ploeg J, Kazemier B, Terpstra P, Witholt B (1989) Cloning of 1,2-dichloroethane degradation genes of *Xanthobacter autotrophicus* GJ10 and expression and sequencing of the *dhlA* gene. J Bacteriol 171:6791–6799
Janssen DB, Pries F, van der Ploeg JR (1994) Genetics and biochemistry of dehalogenating enzymes. Annu Rev Microbiol 48:163–191
Janssen DB, Dinkla IJT, Poelarends GJ, Terpstra P (2005) Bacterial degradation of xenobiotic compounds: Evolution and distribution of novel enzyme activities. Environ Microbiol 7:1868–1882

Jesenska A, Sedlacek I, Damborsky J (2000) Dehalogenation of haloalkanes by *Mycobacterium tuberculosis* H37Rv and other mycobacteria. Appl Environ Microbiol 66:219–222

Jesenska A, Pavlova M, Strouhal M, Chaloupkova R, Tesinska I, Monincova M, Prokop Z, Bartos M, Pavlik I, Rychlik I, Möbius P, Nagata Y, Damborsky J (2005) Cloning, biochemical properties, and distribution of mycobacterial haloalkane dehalogenases. Appl Environ Microbiol 71:6736–6745

Jesenska A, Monincova M, Koudelakova T, Hasan K, Chaloupkova R, Prokop Z, Geerlof A, Damborsky J (2009) Biochemical characterization of haloalkane dehalogenases DrbA and DmbC, representatives of a novel subfamily. Appl Environ Microbiol 75:5157–5160

Johnson AK, Zawadzka AM, Deobald LA, Crawford RL, Paszczynski AJ (2008) Novel method for immobilization of enzymes to magnetic nanoparticles. J Nanopart Res 10:1009–1025

Jordana S, Piera EB (2004) Natural groundwater quality and health. Geol Acta 2:175–188

Karube I, Nomura Y (2000) Enzyme sensors for environmental analysis. J Mol Catal B Enzym 10:177–181

Keuning S, Janssen DB, Witholt B (1985) Purification and characterization of hydrolytic haloalkane dehalogenase from *Xanthobacter autotrophicus* GJ10. J Bacteriol 163:635–639

Kmunicek J, Hynkova K, Jedlicka T, Nagata Y, Negri A, Gago F, Wade RC, Damborsky J (2005) Quantitative analysis of substrate specificity of haloalkane dehalogenase LinB from *Sphingomonas paucimobilis* UT26. Biochem 44:3390–3401

Koudelakova T, Chovancova E, Brezovsky J, Monincova M, Fortova A, Jarkovsky J, Damborsky J (2011) Substrate specificity of haloalkane dehalogenases. Biochem J 435:345–354

Koudelakova T, Bidmanova S, Dvorak P, Pavelka A, Chaloupkova R, Prokop Z, Damborsky J (2013) Haloalkane dehalogenases: Biotechnological applications. Biotechnol J 8:32–45

Kulakova AN, Larkin MJ, Kulakov LA (1997) The plasmid-located haloalkane dehalogenase gene from *Rhodococcus rhodochrous* NCIMB 13064. Microbiol 143:109–115

Lal R, Pandey G, Sharma P, Kumari K, Malhotra S, Pandey R, Raina V, Kohler HP, Holliger C, Jackson C, Oakeshott JG (2010) Biochemistry of microbial degradation of hexachlorocyclohexane and prospects for bioremediation. Microbiol Mol Biol Rev 74:58–80

Lei Y, Chen W, Mulchandani A (2006) Microbial biosensors. Anal Chim Acta 568:200–210

Leiner MJP, Hartmann P (1993) Theory and practice in optical pH sensing. Sens Actuators B: Chem 11:281–289

Leonard P, Hearty S, Brennan J, Dunne L, Quinn J, Chakraborty T, O'Kennedy R (2003) Advances in biosensors for detection of pathogens in food and water. Enzym Microb Technol 32:3–13

Lin J (2000) Recent development and applications of optical and fiber-optic pH sensors. Trends Anal Chem 19:541–552

Lobnik A, Oehme I, Murkovic I, Wolfbeis OS (1998) pH optical sensors based on sol-gels: Chemical doping versus covalent immobilization. Anal Chim Acta 367:159–165

Long F, Zhu A, Sheng JW, He M, Shi H-C (2009) Matrix effects on the microcystin-LR fluorescent immunoassay based on optical biosensor. Sens 9:3000–3010

Los GV, Encell LP, McDougall MG, Hartzell DD, Karassina N, Zimprich C, Wood MG, Learish R, Ohana RF, Urh M, Simpson D, Mendez J, Zimmerman K, Otto P, Vidugiris G, Zhu J, Darzins A, Klaubert DH, Bulleit RF, Wood KV (2008) HaloTag: a novel protein labeling technology for cell imaging and protein analysis. ACS Chem Biol 3:373–382

Maliszewska I, Wilk KA (2008) Detection of some chloroorganic compounds by a microbial sensor system. Mat Sci-Pol 26:451–458

Marazuela MD, Moreno-Bondi MC (2002) Fiber-optic biosensors – an overview. Anal Bioanal Chem 372:664–682

Mateo C, Palomo JM, van Langen LM, van Rantwijk F, Sheldon RA (2004) A new, mild cross-linking methodology to prepare cross-linked enzyme aggregates. Biotechnol Bioeng 86:273–276

Mehrvar M, Bis C, Scharer JM, Moo-Young M, Luong JH (2000) Fiber-optic biosensors – trends and advances. Anal Sci 16:677–692

Mena-Benitez GL, Gandia-Herrero F, Graham S, Larson TR, McQueen-Mason SJ, French CE, Rylott EL, Bruce NC (2008) Engineering a catabolic pathway in plants for the degradation of 1,2-dichloroethane. Plant Physiol 147:1192–1198

Nagata Y, Nariya T, Ohtomo R, Fukuda M, Yano K, Takagi M (1993) Cloning and sequencing of a dehalogenase gene encoding an enzyme with hydrolase activity involved in the degradation of γ-hexachlorocyclohexane in *Pseudomonas paucimobilis*. J Bacteriol 175:6403–6410

Nagata Y, Miyauchi K, Damborsky J, Manova K, Ansorgova A, Takagi M (1997) Purification and characterization of a haloalkane dehalogenase of a new substrate class from a γ-hexachlorocyclohexane-degrading bacterium, *Sphingomonas paucimobilis* UT26. Appl Environ Microbiol 63:3707–3710

Orellana G, Haigh D (2008) New trends in fiber-optic chemical and biological sensors. Curr Anal Chem 4:273–295

Park B-W, Yoon D-Y, Kim D-S (2010) Recent progress in bio-sensing techniques with encapsulated enzymes. Biosens Bioelectron 26:1–10

Pavlova M, Klvana M, Prokop Z, Chaloupkova R, Banas P, Otyepka M, Wade RC, Tsuda M, Nagata Y, Damborsky J (2009) Redesigning dehalogenase access tunnels as a strategy for degrading an anthropogenic substrate. Nat Chem Biol 5:727–733

Peter J, Hutter W, Stöllnberger W, Hampel W (1996) Detection of chlorinated and brominated hydrocarbons by an ion sensitive whole cell biosensor. Biosens Bioelectron 11:1215–1219

Peter J, Hutter W, Stöllnberger W, Karner F, Hampel W (1997) Semicontinuous detection of 1,2-dichloroethane in water samples using *Xanthobacter autotrophicus* GJ10 encapsulated in chitosan beads. Anal Chem 69:2077–2079

Pitter P (2009) Physicochemical properties of water. In: Pitter P (ed) Hydrochemistry (in Czech), 4th edn. VŠCHT Publisher, Praha

Poelarends GJ, Wilkens M, Larkin MJ, van Elsas JD, Janssen DB (1998) Degradation of 1,3-dichloropropene by *Pseudomonas cichorii* 170. Appl Environ Microbiol 64:2931–2936

Prokop Z, Damborsky J, Nagata Y, Janssen DB (2004) Method of production of optically active hydrocarbons and alcohols using hydrolytic dehalogenation catalysed by haloalkane dehalogenases. Patent WO 2006079295 (A2)

Prokop Z, Damborsky J, Oplustil F, Jesenska A, Nagata Y (2005) Method of detoxification of yperite by using haloalkane dehalogenases. Patent CZ 298287; EA 011311; NZ 564218; EP 1899022; AU 2006254625 B2; US 7,888,103; JP 4793947; WO 2006128390 (A1)

Prokop Z, Oplustil F, DeFrank J, Damborsky J (2006) Enzymes fight chemical weapons. Biotechnol J 1:1370–1380

Prokop Z, Sato Y, Brezovsky J, Mozga T, Chaloupkova R, Koudelakova T, Jerabek P, Stepankova V, Natsume R, van Leeuwen JG, Janssen DB, Florian J, Nagata Y, Senda T, Damborsky J (2010) Enantioselectivity of haloalkane dehalogenases and its modulation by surface loop engineering. Angew Chem Int Ed Engl 49:6111–6115

Reardon KF, Campbell DW, Müller C (2009) Optical fiber enzymatic biosensor for reagentless measurement of ethylene dibromide. Eng Life Sci 9:291–297

Rodriguez-Mozaz S, Marco M-P, López de Alda MJ, Barceló D (2004) Biosensors for environmental applications: Future development trends. Pure Appl Chem 76:723–752

Rodriguez-Mozaz S, López de Alda MJ, Marco M-P, Barceló D (2005) Biosensors for environmental monitoring. A global perspective. Talanta 65:291–297

Rodriguez-Mozaz S, López de Alda MJ, Barceló D (2006) Biosensors as useful tools for environmental analysis and monitoring. Anal Bioanal Chem 386:1025–1041

Rodriguez-Mozaz S, López de Alda MJ, Barceló D (2007) Advantages and limitations of on-line solid phase extraction coupled to liquid chromatography-mass spectrometry technologies versus biosensors for monitoring of emerging contaminants in water. J Chromatogr A 1152:97–115

Rogers KR (1995) Biosensors for environmental applications. Biosens Bioelectron 10:533–541

Rogers KR (2006) Recent advances in biosensor techniques for environmental monitoring. Anal Chim Acta 568:222–231

Sallis PJ, Armfield SJ, Bull AT, Hardman DJ (1990) Isolation and characterization of a haloalkane halidohydrolase from *Rhodococcus erythropolis* Y2. J Gen Microbiol 136:115–120

Samorski M (2008) Immobilisierung von Haloalkan-Dehalogenasen und Prozessentwicklung der enzymatischen Produktion von optisch aktivem 2,3-Dichlor-1-propanol. University Stuttgart, Dissertation

Sassolas A, Blum LJ, Leca-Bouvier BD (2012) Immobilization strategies to develop enzymatic biosensors. Biotechnol Adv 30:489–511

Sato Y, Monincova M, Chaloupkova R, Prokop Z, Ohtsubo Y, Minamisawa K, Tsuda M, Damborsky J, Nagata Y (2005) Two rhizobial strains, *Mesorhizobium loti* MAFF303099 and *Bradyrhizobium japonicum* USDA110, encode haloalkane dehalogenases with novel structures and substrate specificities. Appl Environ Microbiol 71:4372–4379

Schanstra JP, Kingma J, Janssen DB (1996) Specificity and kinetics of haloalkane dehalogenase. J Biol Chem 271:14747–14753

Scheidleder A, Grath J, Winkler G, Stärk U, Koreimann C, Gmeiner C, Nixon S, Casillas J, Gravesen P, Leonard J, Elvira M (1999) Groundwater quality and quantity in Europe. Office for Official Publications of the European Communities, Copenhagen

Scholtz R, Leisinger T, Suter F, Cook AM (1987) Characterization of 1-chlorohexane halidohydrolase, a dehalogenase of wide substrate range from an *Arthrobacter* sp. J Bacteriol 169:5016–5021

Schulman SG, Chen S, Bai F, Leiner MJP, Weis L, Wolfbeis OS (1995) Dependence of the fluorescence of immobilized 1-hydroxypyrene-3,6,8-trisulfonate on solution pH: Extension of the range of applicability of a pH fluorosensor. Anal Chim Acta 304:165–170

Sharma SK, Sehgal N, Kumar A (2003) Biomolecules for development of biosensors and their applications. Curr Appl Phys 3:307–316

Stucki G, Thüer M (1995) Experiences of a large-scale application of 1,2-dichloroethane degrading microorganisms for groundwater treatment. Environ Sci Technol 29:2339–2345

Su L, Jia W, Hou C, Lei Y (2011) Microbial biosensors: A review. Biosens Bioelectron 26:1788–1799

Swanson PE (1999) Dehalogenases applied to industrial-scale biocatalysis. Curr Opin Biotechnol 10:365–369

Szymański W, Westerbeek A, Janssen DB, Feringa BL (2011) A simple enantioconvergent and chemoenzymatic synthesis of optically active α-substituted amides. Angew Chem Int Ed Engl 50:10712–10715

Taniguchi Y, Kawakami M (2010) Application of HaloTag protein to covalent immobilization of recombinant proteins for single molecule force spectroscopy. Langmuir 26:10433–10436

Thévenot DR, Toth K, Durst RA, Wilson GS (1999) Electrochemical biosensors: recommended definitions and classification. Pure Appl Chem 71:2333–2348

United States Environmental Protection Agency (2011) Edition of the drinking water standards and health advisories. Office of Water United States Environmental Protection Agency. Washington, DC

Wanekaya AK, Chen W, Mulchandani A (2008) Recent biosensing developments in environmental security. J Environ Monit 10:703–712

Wang P, Liu Q (2010) Introduction. In: Wang P, Liu Q (eds) Cell-based biosensors: Principles and applications, 1st edn. Artech House, Norwood

Wencel D, MacCraith BD, McDonagh C (2009) High performance optical ratiometric sol–gel-based pH sensor. Sens Actuators B: Chem 139:208–213

Westerbeek A, Szymański W, Wijma HJ, Marrink SJ, Feringa BL, Janssen DB (2011) Kinetic resolution of α-bromoamides: Experimental and theoretical investigation of highly enantioselective reactions catalyzed by haloalkane dehalogenases. Adv Synth Catal 353:931–944

Wolfbeis OS (2005) Materials for fluorescence-based optical chemical sensors. J Mater Chem 15:2657–2669

Wolfbeis OS, Rodriguez NV, Werner T (1992) LED-compatible fluorosensor for measurement of near-neutral pH values. Mikrochim Acta 108:133–141

Yokota T, Omori T, Kodama T (1987) Purification and properties of haloalkane dehalogenase from *Corynebacterium* sp. strain m15-3. J Bacteriol 169:4049–4054

Zhang Y, So M-K, Loening AM, Yao H, Gambhir SS, Rao J (2006) HaloTag protein-mediated site-specific conjugation of bioluminescent proteins to quantum dots. Angew Chem Int Ed Engl 45:4936–4940

Ziegler C, Göpel W (1998) Biosensor development. Curr Opin Chem Biol 2:585–591

Chapter 9
Structure and Function of Aromatic-Ring Hydroxylating Dioxygenase System

Kengo Inoue and Hideaki Nojiri

Abstract A wide variety of aromatic compounds are aerobically degraded by bacteria. Aromatic-ring hydroxylation is one of the most common initial degradation step and thus is an essential catalytic reaction for aromatic-ring degradation by bacteria in nature. Most of the cases, the hydroxylation is catalyzed by an oxygenase family known as Rieske nonheme iron (di)oxygenase or Rieske oxygenase (RO). ROs catalyze a broad range of aromatic-ring compounds including mono- and polycyclic aromatic and hetero-aromatic compounds. ROs are composed of terminal oxygenase and electron transfer components. The terminal oxygenase component has Rieske [2Fe-2S] cluster as a redox center for receive electrons from the electron transfer component(s) and mononuclear iron as a catalytic site for dioxygen activation. In addition to genetic and biochemical studies, their crystal structures have been extensively studied recently. To date (11/7/2012), the structures of 15 different terminal oxygenases and 11 electron transfer components have been determined and deposited to Protein Data Bank (76 structures including their variants in total).

This chapter will address structures and functions of aromatic-ring hydroxylating dioxygenases. The overviews of ROs (Sect. 9.1) and overall structure of ROs containing their electron transfer components (Sect. 9.2), inter- and intramolecular electron transfer in ROs (Sect. 9.3), catalytic mechanism (Sect. 9.4), and substrate specificity and enzyme engineering (Sect. 9.5) are reviewed.

K. Inoue (✉)
Interdisciplinary Research Organization, University of Miyazaki,
5200 Kihara, Kiyotake, Miyazaki 889-1692, Japan
e-mail: kinoue@cc.miyazaki-u.ac.jp

H. Nojiri
Biotechnology Research Center, The University of Tokyo,
1-1-1 Yayoi, Bunkyo-ku, Tokyo 113-8657, Japan

H. Nojiri et al. (eds.), *Biodegradative Bacteria: How Bacteria Degrade, Survive, Adapt, and Evolve*, DOI 10.1007/978-4-431-54520-0_9, © Springer Japan 2014

Keywords Aromatic ring • Crystal structure • Dioxygenation • Electron transfer • Hydroxylation • Nonheme iron • Protein–protein interaction • Rieske oxygenase • Substrate specificity

9.1 Aromatic-Ring Hydroxylating Dioxygenases

Aromatic compounds are widely distributed and abundant organic compounds in nature. They are derived from aromatic amino acids, lignin components, flavonoids, quinones, crude oils, etc. Some of the toxic xenobiotics, such as dioxins, polychlorinated biphenyls, and nitroaromatics, also contain aromatic ring(s). Aromatic compounds are utilized by microorganisms as the carbon source and/or electron donor for respiration. Bacterial aerobic metabolism of aromatic compounds has been extensively studied for their applications of bioremediation and chemical and pharmaceutical industries. Initial reaction of aromatic-ring degradation is generally dihydroxylation by Rieske nonheme iron dioxygenase (RO) to yield *cis-dihydrodiols* and derivatives with strict stereo- and regio-specificity (Boyd and Sheldrake 1998; Gibson and Parales 2000; Ullrich and Hofrichter 2007). ROs catalyze not only dioxygenation but also monooxygenation, sulfoxidation, denitrification, and demethylation (Fig. 9.1) (Rosche et al. 1995; Herman et al. 2005; Nojiri 2012). In contrast to bacterial ROs, less numbers of these family enzymes in higher organisms are found, but known ROs in plants and insects play essential roles in biological processes, such as chlorophyll metabolism, cell death, development, and cholesterol metabolism (Gray et al. 1997; Tanaka et al. 1998; Yoshiyama et al. 2006; Yoshiyama-Yanagawa et al. 2011). Among the ROs, relatively simple compounds, such as naphthalene and biphenyl, are the substrates for the best-studied RO members. In these ROs, enzymatic mechanisms based three-dimensional structures are understood in detail.

9.2 RO Structures

9.2.1 *Classification*

ROs are composed of terminal oxygenase and electron transfer components. Oxygenation reaction is catalyzed by the terminal oxygenase component and two electrons are required for the reaction (Wolfe et al. 2001). Electron transfer components transfer the electrons from NAD(P)H to terminal oxygenase component. Electron transfer components are only reductase or ferredoxin and ferredoxin reductase (Table 9.1). All the components contain redox center(s) for storage of electrons. According to Batie's classification system, ROs are divided into five groups by their component numbers and prosthetic groups (Table 9.1) (Batie et al. 1991). In class I ROs, the electron transfer component is only reductase which

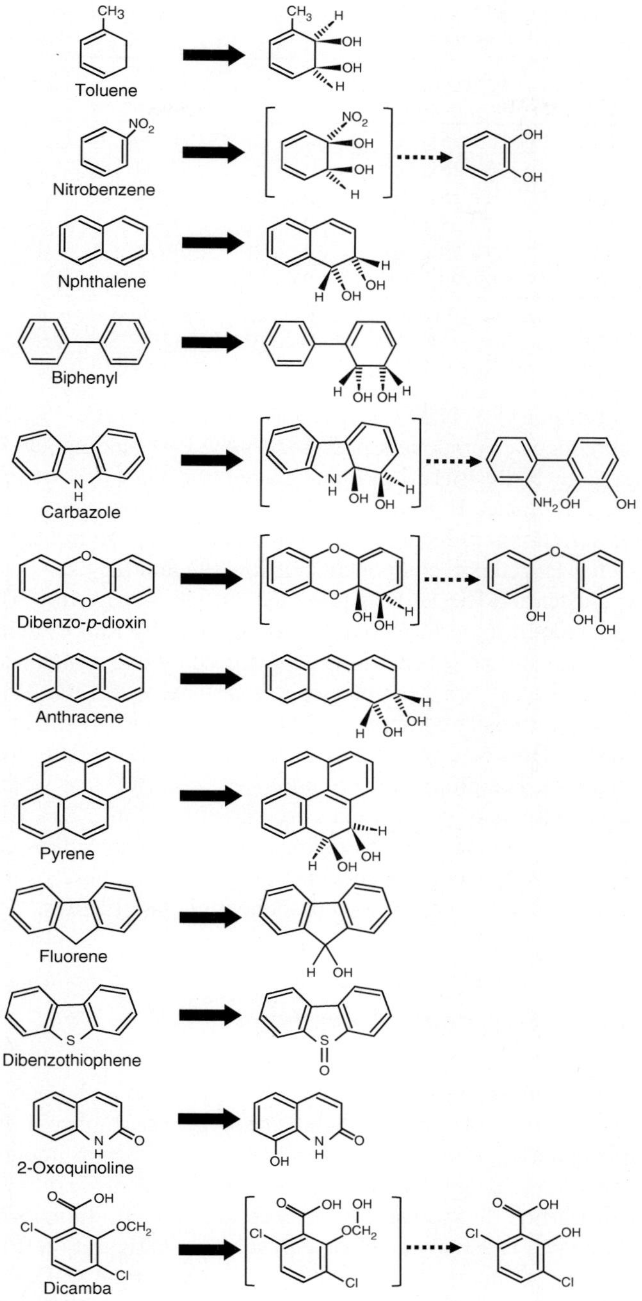

Fig. 9.1 Oxidation of aromatic compounds catalyzed by Rieske oxygenases. Compounds in parenthesis are unstable and spontaneous reaction (*dotted arrow*) to form the diol compounds is shown, except for the dicamba monooxygenation reaction (in the monooxygenated compound of dicamba in parenthesis is proposed intermediate of the monooxygenation reaction and the 3,6-dichlorosalicylic acid as the final product is shown)

Table 9.1 ROs classification by electron transfer components [modified classification by Batie et al. (1991)]

	Prosthetic group(s)[a]		
Class	Oxygenase	Ferredoxin	Reductase
IA	$[2Fe\text{-}2S]_R$ Fe^{2+}	–	FMN $[2Fe\text{-}2S]_P$
IB	$[2Fe\text{-}2S]_R$ Fe^{2+}	–	FAD $[2Fe\text{-}2S]_P$
IIA	$[2Fe\text{-}2S]_R$ Fe^{2+}	$[2Fe\text{-}2S]_{Pu}$	FAD
IIB	$[2Fe\text{-}2S]_R$ Fe^{2+}	$[2Fe\text{-}2S]_R$	FAD
IIC	$[2Fe\text{-}2S]_R$ Fe^{2+}	[3Fe-4S]	FAD
III	$[2Fe\text{-}2S]_R$ Fe^{2+}	$[2Fe\text{-}2S]_R$	FDA $[2Fe\text{-}2S]_P$

[a]*$[2Fe\text{-}2S]_R$* Rieske-type [2Fe-2S] cluster, *$[2Fe{:}2S]_P$* plant-type [2Fe-2S] cluster, *$[2Fe\text{-}2S]_{Pu}$* putidaredoxin-type [2Fe-2S] cluster

contains FMN (class IA) or FAD (class IB) and plant-type [2Fe-2S] cluster as the cofactors. The reductase component possesses NAD(P)H binding site for transferring electrons from NAD(P)H to terminal oxygenase via FMN or FAD and the plant-type [2Fe-2S] cluster. ROs in class II have two electron transfer components, ferredoxin and ferredoxin reductase. The ferredoxin reductase in class II has only FAD as a cofactor. Ferredoxin components in class IIA and IIB have putidaredoxin-type [2Fe-2S] cluster and Rieske-type [2Fe-2S] cluster, respectively. The [2Fe-2S] cluster of putidaredoxin-type ferredoxin is coordinated by four cysteine residues, whereas that of Rieske-type ferredoxin is coordinated by two cysteine and two histidine residues. There are few examples of RO systems with [3Fe-4S]-type ferredoxin as the electron transfer component (Martin and Mohn 1999; Kim et al. 2006; Takagi et al. 2005). However, so far, they have not been classified into any class in Batie's classification. Therefore, here we propose "class IIC" into which ROs with [3Fe-4S]-type ferredoxin as the examples are classified (Table 9.1). Class III ROs have both ferredoxin reductase and ferredoxin component as well as class II ROs, but the ferredoxin reductase component contains plant-type [2Fe-2S] cluster and the FAD cofactor similarly to the reductase component of class IB ROs.

9.2.2 Terminal Oxygenase Component

Terminal oxygenase in RO has Rieske-type [2Fe-2S] cluster and mononuclear non-heme iron. Crystal structures of naphthalene dioxygenase, biphenyl dioxygenase, carbazole dioxygenase, nitrobenzene dioxygenase, dicamba monooxygenase, 2-oxoquinoline 8-monooxygenase, toluene dioxygenase, cumene dioxygenase, and polycyclic hydrocarbon dioxygenase have been determined (Table 9.2). Terminal oxygenase component is usually an α_3 or $\alpha_3\beta_3$ subunit configuration (Fig. 9.2). The terminal oxygenase component of naphthalene dioxygenase (NDO) is the first example of terminal oxygenase structure of ROs (Kauppi et al. 1998). The terminal oxygenase of NDO is $\alpha_3\beta_3$ configuration and resembles shape of mushroom. The height of the mushroom is 75 Å and the diameters of the cap and stem are 102 Å and

Table 9.2 A list of PDB entry codes of terminal oxygenase and electron components of ROs

Enzyme[a]	Complex with	Organism	Class	PDB ID	Reference
Terminal oxygenase					
Naphthalene dioxygenase	–	*Pseudomonas putida* NCIB 9816-4	III	1NDO	Kauppi et al. (1998)
Naphthalene dioxygenase	Indole	*Pseudomonas putida* NCIB 9816-4	III	1EG9	Carredano et al. (2000)
Naphthalene dioxygenase	Naphthalene	*Pseudomonas putida* NCIB 9816-4	III	1O7G	Karlsson et al. (2003)
Naphthalene dioxygenase (oxidized Rieske center)	–	*Pseudomonas putida* NCIB 9816-4	III	1O7H	Karlsson et al. (2003)
Naphthalene dioxygenase	O_2	*Pseudomonas putida* NCIB 9816-4	III	1O7M	Karlsson et al. (2003)
Naphthalene dioxygenase	Indole and O_2	*Pseudomonas putida* NCIB 9816-4	III	1O7N	Karlsson et al. (2003)
Naphthalene dioxygenase	(1R,2S)-*cis* 1,2 dihydroxy-1,2-dihydronaphthalene	*Pseudomonas putida* NCIB 9816-4	III	1O7P	Karlsson et al. (2003)
Naphthalene dioxygenase (fully reduced)	–	*Pseudomonas putida* NCIB 9816-4	III	1O7W	Karlsson et al. (2003)
Naphthalene dioxygenase	NO	*Pseudomonas putida* NCIB 9816-4	III	1UUW	Karlsson et al. (2005)
Naphthalene dioxygenase	NO and indole	*Pseudomonas putida* NCIB 9816-4	III	1UUV	Karlsson et al. (2005)
Naphthalene dioxygenase	–	*Rhodococcus* sp. NCIMB12038	III	2B1X	Gakhar et al. (2005)
Naphthalene dioxygenase	Indole	*Rhodococcus* sp. NCIMB12038	III	2B24	Gakhar et al. (2005)
Naphthalene dioxygenase	Phenanthrene	*Pseudomonas putida* NCIB 9816-4	III	2HMK	Ferraro et al. (2006)
Naphthalene dioxygenase Phe325Val mutant	–	*Pseudomonas putida* NCIB 9816-4	III	2HMJ	Ferraro et al. (2006)
Naphthalene dioxygenase Phe325Val mutant	Phenanthrene	*Pseudomonas putida* NCIB 9816-4	III	2HML	Ferraro et al. (2006)
Naphthalene dioxygenase	Anthracene	*Pseudomonas putida* NCIB 9816-4	III	2HMM	Ferraro et al. (2006)
Naphthalene dioxygenase Phe325Val mutant	Anthracene	*Pseudomonas putida* NCIB 9816-4	III	2HMN	Ferraro et al. (2006)
Naphthalene dioxygenase	3-Nitrotoluene	*Pseudomonas putida* NCIB 9816-4	III	2HMO	Ferraro et al. (2006)
Biphenyl dioxygenase	–	*Rhodococcus jostii* RHA1	IIB	1ULI	Furusawa et al. (2004)
Biphenyl dioxygenase	Biphenyl	*Rhodococcus jostii* RHA1	IIB	1ULJ	Furusawa et al. (2004)
Biphenyl dioxygenase	–	*Sphingobium yanoikuyae* B1	IIB	2GBW	Ferraro et al. (2007)

(continued)

Table 9.2 (continued)

Enzyme[a]	Complex with	Organism	Class	PDB ID	Reference
Biphenyl dioxygenase	Biphenyl	*Sphingobium yanoikuyae* B1	IIB	2GBX	Ferraro et al. (2007)
Biphenyl dioxygenase	–	*Burkholderia xenovorans* LB400	IIB	2XR8	Kumar et al. (2011)
Biphenyl dioxygenase	Biphenyl	*Burkholderia xenovorans* LB400	IIB	2XRX	Kumar et al. (2011)
Biphenyl dioxygenase variant P4	–	*Burkholderia xenovorans* LB400	IIB	2XSO	Kumar et al. (2011)
Biphenyl dioxygenase variant P4	2,6-Dichlorobiphenyl	*Burkholderia xenovorans* LB400	IIB	2XSH	Kumar et al. (2011)
Biphenyl dioxygenase variant RR41	–	*Burkholderia xenovorans* LB400	IIB	2YFI	Mohammadi et al. (2011)
Biphenyl dioxygenase variant RR41	Dibenzofuran	*Burkholderia xenovorans* LB400	IIB	2YFJ	Mohammadi et al. (2011)
Biphenyl dioxygenase variant RR41	2-Chloro dibenzofuran	*Burkholderia xenovorans* LB400	IIB	2YFL	Kumar et al. (2012)
Biphenyl dioxygenase	–	*Pandoraea pnomenusa* B-356	IIB	3GZY	Kumar et al. (unpublished)
Biphenyl dioxygenase	Biphenyl	*Pandoraea pnomenusa* B-356	IIB	3GZX	Kumar et al. (unpublished)
Biphenyl dioxygenase	Phenylcyclohexadienediol	*Pandoraea pnomenusa* B-356	IIB	3GZZ	Kumar et al. (unpublished)
Carbazole dioxygenase	–	*Janthinobacterium* sp. J3	III	1WW9	Nojiri et al. (2005)
Carbazole dioxygenase	–	*Nocardioides aromaticivorans* IC177	IIB	3GCF	Inoue et al. (2009)
Carbazole dioxygenase	–	*Novosphingobium* sp. KA1	IIA	3GKQ	Umeda et al. (unpublished)
Nitrobenzene dioxygenase	–	*Comamonas* sp. JS765	III	2BMO	Friemann et al. (2005)
Nitrobenzene dioxygenase	Nitrobenzene	*Comamonas* sp. JS765	III	2BMQ	Friemann et al. (2005)
Nitrobenzene dioxygenase	3-Nitrotoluene	*Comamonas* sp. JS765	III	2BMR	Friemann et al. (2005)
Dicamba monooxygenase	–	*Stenotrophomonas maltophilia* DI-6	IIA	3GTE	D'Ordine et al. (2009)
Dicamba monooxygenase	Dicamba	*Stenotrophomonas maltophilia* DI-6	IIA	3GTS	D'Ordine et al. (2009)
Dicamba monooxygenase	Co and dicamba	*Stenotrophomonas maltophilia* DI-6	IIA	3GB4	D'Ordine et al. (2009)
Dicamba monooxygenase	Co and 3,6-dichlorosalicylate	*Stenotrophomonas maltophilia* DI-6	IIA	3GOB	D'Ordine et al. (2009)

Dicamba monooxygenase	–	*Stenotrophomonas maltophilia* DI-6	IIA	3GKE	Dumitru et al. (2009)
Dicamba monooxygenase	Dicamba	*Stenotrophomonas maltophilia* DI-6	IIA	3GL2	Dumitru et al. (2009)
Dicamba monooxygenase	3,6-Dichlorosalicylate	*Stenotrophomonas maltophilia* DI-6	IIA	3GL0	Dumitru et al. (2009)
2-Oxoquinoline 8-monooxygenase (oxidized)	–	*Pseudomonas putida* 86	IB	1Z01	Martins et al. (2005)
2-Oxoquinoline 8-monooxygenase (reduced)	–	*Pseudomonas putida* 86	IB	1Z02	Martins et al. (2005)
2-Oxoquinoline 8-monooxygenase (oxidized)	2-Oxoquinolin	*Pseudomonas putida* 86	IB	1Z03	Martins et al. (2005)
Toluene dioxygenase	–	*Pseudomonas putida* F1	IIB	3EN1	Friemann et al. (2009)
Toluene dioxygenase (without mononuclear iron)	–	*Pseudomonas putida* F1	IIB	3EQQ	Friemann et al. (2009)
Cumene dioxygenase	O_2	*Pseudomonas fluorescens*	IIB	1WQL	Dong et al. (2005)
Polycyclic aromatic hydrocarbon dioxygenase	–	*Sphingomonas* sp. CHY-1	IIB	2CKF	Jakoncic et al. (2007)
Reductas					
Phthalate dioxygenase	–	*Burkholderia cepacia* DB01	IA	2PIA	Correll et al. (1992)
Benzoate dioxygenase	–	*Acinetobacter baylyi* ADP1	IB	1KRH	Karlsson et al. (2002)
Ferredoxin					
Naphthalene dioxygenase	–	*Pseudomonas putida* NCIB 9816-4	III	2QPZ	Brown et al. (2008)
Biphenyl dioxygenase	–	*Burkholderia xenovorans* LB400	IIB	1FQT	Colbert et al. (2000)
Biphenyl dioxygenase	–	*Sphingobium yanoikuyae* B1	IIB	2I7F	Ferraro et al. (2007)
Biphenyl dioxygenase (oxidized)	–	*Acidovorax* sp. KKS102	IIB	2E4P	Senda et al. (2007)
Biphenyl dioxygenase (reduced)	–	*Acidovorax* sp. KKS102	IIB	2E4Q	Senda et al. (2007)
Carbazole dioxygenase	–	*Pseudomonas resinovorans* CA10	III	1VCK	Nam et al. (2005)
Carbazole dioxygenase	–	*Nocardioides aromaticivorans* IC177	IIB	3GCE	Inoue et al. (2009)
Carbazole dioxygenase	–	*Novosphingobium* sp. KA1	IIA	–	Umeda et al. (unpublished)
Toluene dioxygenase	–	*Pseudomonas putida* F1	IIB	3DQY	Friemann et al. (2009)

(continued)

Table 9.2 (continued)

Enzyme[a]	Complex with	Organism	Class	PDB ID	Reference
Ferredoxin reductase					
Biphenyl dioxygenase	–	*Acidovorax* sp. KKS102	IIB	1D7Y	Senda et al. (2000)
Biphenyl dioxygenase	NADH	*Acidovorax* sp. KKS102	IIB	1F3P	Senda et al. (2000)
Biphenyl dioxygenase (oxidized)	–	*Acidovorax* sp. KKS102	IIB	2GR2, 2GR3, 2GQW	Senda et al. (2007)
Biphenyl dioxygenase (hydroquinone)	–	*Acidovorax* sp. KKS102	IIB	2GR1, 2YVF	Senda et al. (2007)
Biphenyl dioxygenase (semiquinone)	–	*Acidovorax* sp. KKS102	IIB	2YVG	Senda et al. (2007)
Biphenyl dioxygenase (reoxidized)	NAD^+	*Acidovorax* sp. KKS102	IIB	2GR0	Senda et al. (2007)
Carbazole dioxygenase	–	*Janthinobacterium* sp. J3	III	–	Ashikawa et al. (unpublished)
Carbazole dioxygenase (without FAD)	–	*Janthinobacterium* sp. J3	III	–	Ashikawa et al. (unpublished)
Carbazole dioxygenase	–	*Novosphingobium* sp. KA1	IIA	–	Umeda et al. (unpublished)
Toluene dioxygenase	–	*Pseudomonas putida* F1	IIB	3EF6	Friemann et al. (2009)
Ferredoxin and ferredoxin reductase complex					
Biphenyl dioxygenase	–	*Acidovorax* sp. KKS102	IIB	2YVJ	Senda et al. (2007)
Terminal oxygenase and ferredoxin complex					
Carbazole dioxygenase	–	*Janthinobacterium* sp. J3 (terminal oxygenase) and *Pseudomonas resinovorans* CA10 (ferredoxin)	III	2DE5	Ashikawa et al. (2006)
Carbazole dioxygenase (reduced)	–	*Janthinobacterium* sp. J3 (terminal oxygenase) and *Pseudomonas resinovorans* CA10 (ferredoxin)	III	2DE6	Ashikawa et al. (2006)
Carbazole dioxygenase	Carbazole	*Janthinobacterium* sp. J3 (terminal oxygenase) and *Pseudomonas resinovorans* CA10 (ferredoxin)	III	2DE7	Ashikawa et al. (2006)

Carbazole dioxygenase (reduced)	Carbazole	*Janthinobacterium* sp. J3 (terminal oxygenase) and *Pseudomonas resinovorans* CA10 (ferredoxin)	III	3VMG	Ashikawa et al. (2012)
Carbazole dioxygenase (reoxidized)	O_2	*Janthinobacterium* sp. J3 (terminal oxygenase) and *Pseudomonas resinovorans* CA10 (ferredoxin)	III	3VMH	Ashikawa et al. (2012)
Carbazole dioxygenase	Carbazole and O_2	*Janthinobacterium* sp. J3 (terminal oxygenase) and *Pseudomonas resinovorans* CA10 (ferredoxin)	III	3VMI	Ashikawa et al. (2012)
Carbazole dioxygenase Ile262Leu mutant	–	*Janthinobacterium* sp. J3 (terminal oxygenase) and *Pseudomonas resinovorans* CA10 (ferredoxin)	III	–	Inoue et al. (unpublished)
Carbazole dioxygenase Ile262Val mutant	–	*Janthinobacterium* sp. J3 (terminal oxygenase) and *Pseudomonas resinovorans* CA10 (ferredoxin)	III	–	Inoue et al. (unpublished)
Carbazole dioxygenase Ile262Val mutant	Carbazole	*Janthinobacterium* sp. J3 (terminal oxygenase) and *Pseudomonas resinovorans* CA10 (ferredoxin)	III	–	Inoue et al. (unpublished)
Carbazole dioxygenase Ile262Val mutant	Carbazole and O_2	*Janthinobacterium* sp. J3 (terminal oxygenase) and *Pseudomonas resinovorans* CA10 (ferredoxin)	III	–	Inoue et al. (unpublished)
Carbazole dioxygenase Phe275Trp mutant	–	*Janthinobacterium* sp. J3 (terminal oxygenase) and *Pseudomonas resinovorans* CA10 (ferredoxin)	III	–	Inoue et al. (unpublished)
Carbazole dioxygenase Phe275Trp mutant	Carbazole	*Janthinobacterium* sp. J3 (terminal oxygenase) and *Pseudomonas resinovorans* CA10 (ferredoxin)	III	–	Inoue et al. (unpublished)
Carbazole dioxygenase Phe275Trp mutant	Fluorene	*Janthinobacterium* sp. J3 (terminal oxygenase) and *Pseudomonas resinovorans* CA10 (ferredoxin)	III	–	Inoue et al. (unpublished)

(continued)

Table 9.2 (continued)

Enzyme[a]	Complex with	Organism	Class	PDB ID	Reference
Carbazole dioxygenase Gln282Asn mutant	–	*Janthinobacterium* sp. J3 (terminal oxygenase) and *Pseudomonas resinovorans* CA10 (ferredoxin)	III	–	Inoue et al. (unpublished)
Carbazole dioxygenase Gln282Tyr mutant	–	*Janthinobacterium* sp. J3 (terminal oxygenase) and *Pseudomonas resinovorans* CA10 (ferredoxin)	III	–	Inoue et al. (unpublished)
Carbazole dioxygenase Gln282Tyr mutant	Carbazole	*Janthinobacterium* sp. J3 (terminal oxygenase) and *Pseudomonas resinovorans* CA10 (ferredoxin)	III	–	Inoue et al. (unpublished)

[a]Mutations were introduced into the terminal oxygenase component in all the mutant enzymes

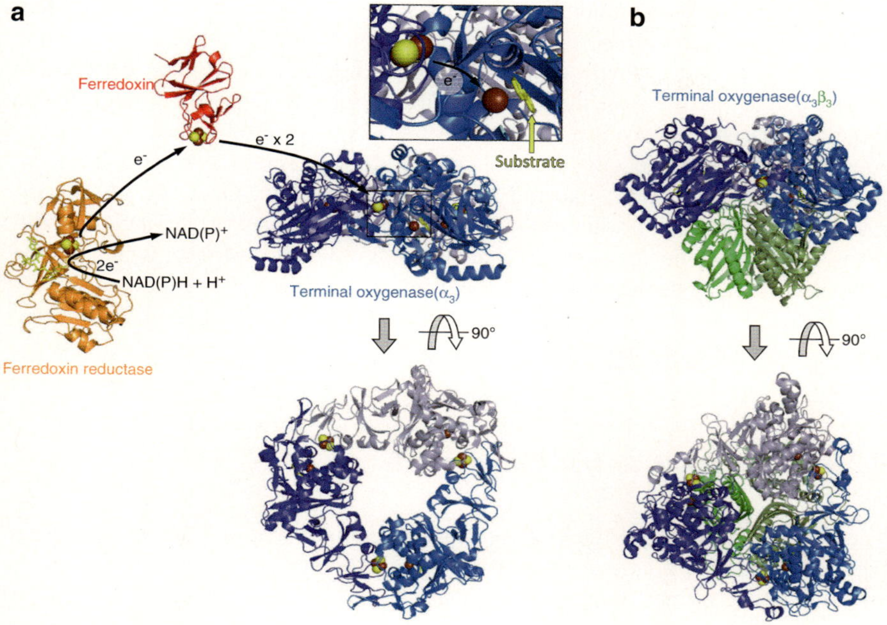

Fig. 9.2 Overall structures of Rieske oxygenase components. The iron atom, sulfur atom, FAD, and carbazole are shown in *brown sphere, yellow sphere, yellow sticks*, and *yellow and blue sticks* (imino nitrogen), respectively. (**a**) The terminal oxygenase component with α_3 configuration (*blue*; PDB entry code: 2DE7), ferredoxin (*red*; PDB entry code: 1VCK), and ferredoxin reductase (*orange*; unpublished) of carbazole 1,9a-dioxygenase. Schematic electron transfer is also presented. Magnified view of the substrate-binding pocket is shown in the in set. (**b**) The terminal oxygenase of naphthalene dioxygenase with $\alpha_3\beta_3$ configuration (PDB entry code: 1NDO). The α and β subunits of the terminal oxygenase component are colored in *blue* and *green*, respectively

50 Å, respectively. The terminal oxygenase component of carbazole dioxygenase (CARDO) and 2-oxoquinoline 8-monooxygenase are the first examples of RO which have α_3 configuration (Nojiri et al. 2005; Martins et al. 2005). The terminal oxygenase of CARDO is a doughnut shape with 100 Å in diameter, a central hole in 30 Å, and 45 Å thickness. Both of the redox centers (Rieske [2Fe-2S] cluster and mononuclear nonheme iron) and substrate-binding site are in the α-subunit. The β-subunit is thought to have a structural role in the holoenzyme but, in biphenyl dioxygenase (BDO) from *Pandoraea pnomenusa* (formerly *Comamonas testosteroni*) strain B-356, the β-subunit influences substrate specificity (Hurtubise et al. 1998). The α-subunit is consist of two domains, Rieske and catalytic domains. One iron (Fe_1) of the Rieske cluster is coordinated by two cysteine residues and the other iron (Fe_2) is coordinated by two histidine residues. Two inorganic sulfide ions bridge the two iron ions, forming a flat rhombic arrangement. This domain of ROs is similar to the Rieske iron-sulfur protein of the cytochrome bc_1 complex in the electron transport chain for ATP generation (Iwata et al. 1996) and chloroplast b_6f complex (Carrell et al. 1997). The Rieske domain of terminal oxygenase component is likely

to be derived from a common ancestral Rieske protein with these Rieske proteins because of their similarities. The distance between the Rieske center and the mononuclear iron within a single α subunit is further from the distance between the Rieske center and the iron in a neighboring α subunit. Therefore, electrons are thought to transfer from a Rieske center to a mononuclear iron in the neighboring α subunit (Fig. 9.2). The catalytic domain of RO is dominated by a stranded antiparallel β-sheet which extends into the middle of the domain. One side of the large β-sheet is covered with α-helices that contain ligands, an aspartate, and two histidines, to the mononuclear iron. A channel from molecular surface, which enables substrates to access to the active site, is formed by the sheet and the helices. The region close to the active site iron is primarily composed of hydrophobic residues so that the terminal oxygenase of ROs can accommodate hydrophobic substrates.

9.2.3 Electron Transfer Components

The crystal structures of the electron transfer components of ROs have not been determined as many as the terminal oxygenase components (Table 9.2), but the structures have offered important insights into structure-function relationships of electron carrier proteins.

9.2.3.1 Reductase

The first three-dimensional structure of class I reductase component is phthalate dioxygenase from *Burkholderia cepacia* strain DB01 (Table 9.2) (Correll et al. 1992). This protein has three individual domains: FMN, NAD, and [2Fe-2S] domains (N-terminal to C-terminal). The FMN-binding, [2Fe-2S] cluster, and NADH-binding domains are brought together near a central cleft in the molecule with only 4.9 Å separating the 8-methyl group of FMN and a cysteine sulfur ligated to iron. The phthalate dioxygenase belongs to class IA but another example of reductase structure of class I, benzoate dioxygenase, belongs to class IB. The reductase component of benzoate dioxygenase from *Acinetobacter baylyi* strain ADP1 also has three domains, [2Fe-2S] cluster, FAD-binding, and NADH-binding domains (N-terminal to C-terminal), with different order from that of phthalate dioxygenase (Karlsson et al. 2002).

9.2.3.2 Ferredoxin

ROs in class II have a ferredoxin component which transfers electrons from ferredoxin reductase component to terminal oxygenase. Crystal structure of ferredoxin component of BDO (BphF) from *Burkholderia xenovorans* strain LB400 was determined as the first example of RO ferredoxin (Colbert et al. 2000). This small protein composed of 109 amino acid residues has a Rieske [2Fe-2S] cluster which is located at the apex of this molecule with its two histidine ligands exposed to solvent. Redox potential of BphF was approximately −150 mV, whereas Rieske proteins from the bc_1 or b_6f complexes have redox potentials near +300 mV. Some other ferredoxin structures of ROs; two different CARDOs, NDO, toluene dioxygenase; and other two different BDOs have been determined (Table 9.2). They are primarily similar to BphF protein at structural level except for ferredoxin of CARDO from *Novosphingobium* sp. strain KA1, which is not Rieske-type ferredoxin but putidaredoxin type. This protein structure is round shape rather than wedge shape as the Rieske-type ferredoxins (Umeda et al. unpublished). This difference probably influences interaction with terminal oxygenase and ferredoxin reductase component.

9.2.3.3 Ferredoxin Reductase

Crystal structure of ferredoxin reductase (BphA4) in class IIB from *Acidovorax* sp. strain KKS102 with and without NADH has been determined as the first example of ferredoxin reductase in this class (Senda et al. 2000). This protein has significant similarity to not only ferredoxin reductases of ROs in other bacteria but also putidaredoxin reductase of P450cam system of *Pseudomonas putida*, and it has essentially the same fold as the enzymes of the glutathione reductase family. This protein has three domains, FAD-binding, NADH-binding, and C-terminal domain. The FAD-binding and C-terminal domains have interacting surface areas for its counterpart, ferredoxin component (BphA3) (Senda et al. 2007). Crystal structures of ferredoxin reductase components of CARDOs from *Janthinobacterium* sp. strain J3 and *Novosphingobium* sp. strain KA1 and toluene dioxygenase from *Pseudomonas putida* strain F1 have been also determined (Table 9.2). They are essentially similar to BphA4 structure.

9.3 Electron Transfer in Rieske Oxygenase

Both intra- and intermolecular electron transfers in ROs are essential for catalytic activity. Redox potential of each redox center is an important factor for electron transfer but limited numbers of reports which demonstrated experimentally measured redox potentials (Table 9.3). The Rieske proteins have reduction potentials ranging from −150 to +400 mV which is proposed to be due to electrostatic environment of the cluster and/or structural differences, whereas the redox centers in RO are relatively lower (Colbert et al. 2000; Brown et al. 2008). The redox potential of Rieske proteins is known to be dependent on pH probably because of the two iron-coordinating histidines. The wide range redox potential of Rieske proteins is first proposed to be due to solvent accessibility or protein structure. However, determined structures of Rieske proteins are significantly similar and do not have large variations in solvent accessibility. Thus, the redox potential is likely to be determined

Table 9.3 Redox potentials of Rieske oxygenase components[a]

Class	Enzyme	Reductase (mV)[b]	Ferredoxin (mV)	Terminal oxygenase (mV)	Reference
IA	Phthalate dioxygenase	−287 ($FMN_{sq/hq}$) −174 ($FMN_{ox/sq}$) −174 ([2Fe-2S])	–	−150	Gassner et al. (1995)
	Hydrobenzoate dioxygenase	−200	–	−125	Riedel et al. (1995)
IB	Anthranilate dioxygenase	N.D.	–	−86	Beharry et al. (2003)
	2-Oxoquinoline-8-monooxygenase	−180	–	−100	Rosche et al. (1995)
IIA	Dioxin dioxygenase	–	−245, −247	N.D.	Armengaud and Timmis (1997) Armengaud et al. (2000)
	Dicamba demethylase	N.D.	−171	−21	Chakraborty et al. (2005)
IIB	Biphenyl dioxygenase	N.D.	−157	N.D.	Couture et al. (2001)
	Benzene dioxygenase	N.D.	−155	−112	Geary et al. (1984)
	Carbazole dioxygenase	N.D.	−185	N.D.	Inoue et al. (2009)
III	Toluene dioxygenase	N.D.	−109	N.D.	Subramanian et al. (1985)
	Carbazole dioxygenase	N.D.	−169	N.D.	Nam et al. (2005)

[a]*N.D.* not determined

[b]sq, hq, and ox represent semiquinone, reduced, oxidized from of FMN, respectively. [2Fe-2S] represents plant-type [2Fe-2S] cluster

by structural difference rather than solvent accessibility. This is supported by biochemical studies using mutated enzymes (Klingen and Ullmann 2004).

One redox center must be in close proximity to the other so that electron transfer occurs. In class II or III ROs, the ferredoxin component shuttles between the ferredoxin reductase and terminal oxygenase components and thus the ferredoxin noncovalently binds to the other components. To date, two electron transfer complexes have been reported: ferredoxin reductase (BphA4) and ferredoxin (BphA3) of BDO from *Acidovorax* sp. strain KKS102 and ferredoxin and terminal oxygenase of CARDO (from *Pseudomonas resinovorans* strain CA10 and *Janthinobacterium* sp. strain J3, respectively). These crystal structures have revealed interacting site, binding force, and conformational changes by complex formation. The crystal structure of BphA4 and BphA3 complex demonstrated BphA3 binds to middle of BphA4 by hydrophobic interaction (four tryptophan residues in BphA4 and two proline residues in BphA3), electrostatic-like interaction (negatively charged surface of BphA4 interacts with positively charged BphA3 surface), and hydrogen bonds (Fig. 9.3a) (Senda et al. 2007). Electrons are likely to be transferred from FAD in BphA4 to the [2Fe-2S] cluster in BphA3 via Trp320 of BphA4 and His66 of BphA3. Structural comparison between the free and complex forms of BphA3 and BphA4 revealed conformational changes, clockwise rotation of NADH and C-terminal domain of BphA4, side chain rotation of His66 (a ligand in [2Fe-2S] cluster) of BphA3, and the peptide bond between Gly46 and Glu47 of BphA3. This report also presents the butterfly-like motion of the isoalloxazine ring of FAD along with redox status. The other electron transfer complex of ferredoxin and terminal oxygenase of CARDO revealed interacting amino acid residues between the components (Fig. 9.3b, c) (Ashikawa et al. 2006). Ferredoxin binds to boundary of the terminal oxygenase subunits by electrostatic and hydrophobic interactions. Free forms of ferredoxin and terminal oxygenase component of CARDO have also been determined. Detailed comparison revealed two significant conformational changes of main chain, Lys12-Trp15 and Asp347-Asn352 of the terminal oxygenase component. These amino acid residues are located at the component boundary to form hydrophobic interactions and the hydrophobic residues: Trp15 and Val351 of Oxy interact with Phe67 and Pro83 of Fd, respectively. On the interacting molecular surfaces, electrostatic interactions (Lys13, Arg118, Arg210, and Glu353 of the terminal oxygenase interact with Glu64, Glu43, Glu55, and His68 of the ferredoxin component, respectively) were also observed. According to the complex structure, electron transfer pathway from the terminal oxygenase and the ferredoxin was proposed (Fig. 9.3c). The distance between the Rieske cluster of ferredoxin and mononuclear iron of the terminal oxygenase is approximately 22 Å. This is too far for the direct electron transfer without Rieske cluster of the terminal oxygenase. However, the average distance between the ligand atoms of the Rieske clusters of terminal oxygenase and ferredoxin was approximately 12–13 Å which is shorter than 14 Å threshold, the limit of electron tunneling in a protein medium (Page et al. 1999).

The distance from Rieske [2Fe-2S] cluster of one terminal oxygenase α subunit and mononuclear iron of the neighboring α subunit is shorter than 14 Å as well. A conserved aspartate residue (D180 in CARDO; Fig. 9.3c) binds to a histidine ligand to the Rieske cluster and a histidine ligand to the mononuclear iron. A mutational study of the

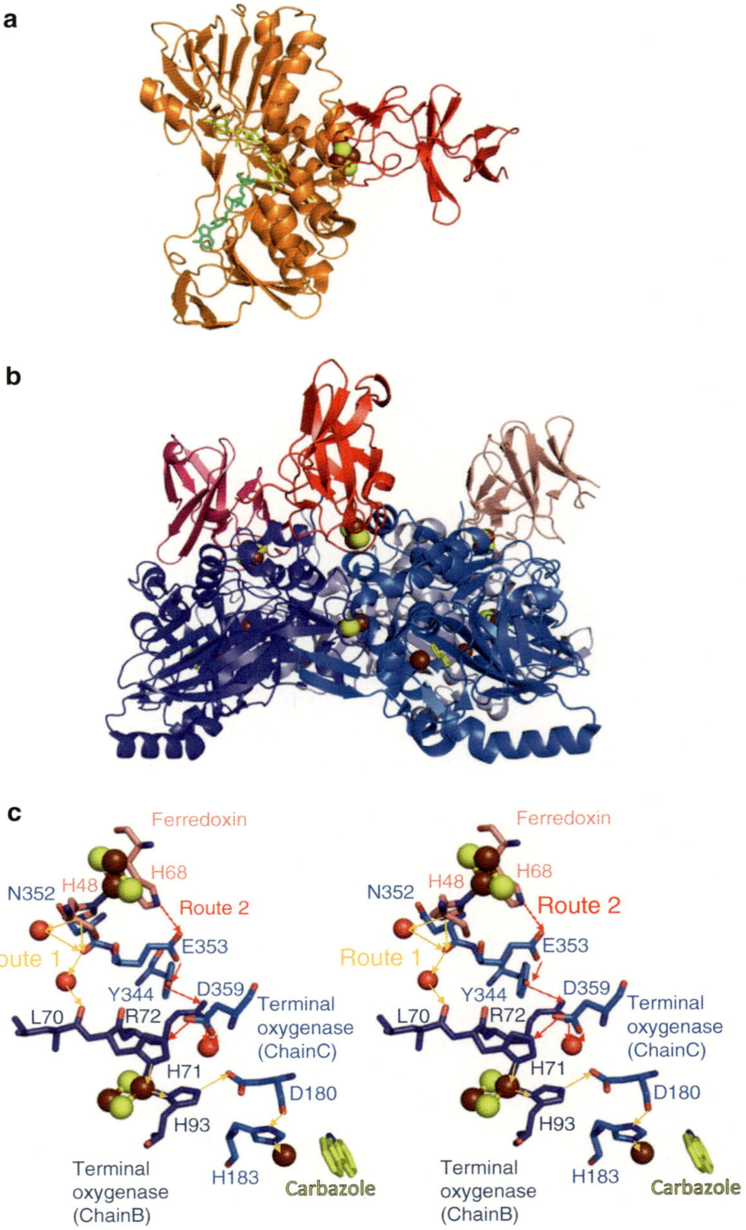

Fig. 9.3 The crystal structures of the electron transfer complexes. (**a**) The ferredoxin and ferredoxin reductase complex of biphenyl dioxygenase from *Acidovorax* sp. strain KKS102 (PDB entry code: 2YVJ). The ferredoxin and ferredoxin reductase are colored in *red* and *orange*, respectively. The [2Fe-2S] cluster, FAD, and NAD^+ molecules are shown as the *brown and yellow balls, yellow sticks*, and *green sticks*, respectively. The Rieske [2Fe-2S] cluster of the ferredoxin binds closely to the FAD and NAD+ molecules in the middle of ferredoxin reductase component.

corresponding residue to the aspartate residue in toluene dioxygenase from *Pseudomonas putida* strain F1 (Asp219) demonstrated this residue is essential for enzyme activity (Jiang et al. 1996; Friemann et al. 2009), indicating the aspartate residue construct a main path for electron transfer from the Rieske cluster to mononuclear iron.

9.4 Catalytic Mechanism

In ROs, dioxygen is reductively activated as well as other oxidase and oxygenase enzymes. Biochemical studies of the ROs, especially for single turnover studies by regulating the enzymes to be oxidized and reduced, have provided important insights into reaction mechanisms. A scheme of proposed catalytic mechanism of ROs based on the previous studies (Ashikawa et al. 2012; Karlsson et al. 2003; Neibergall et al. 2007; Wolfe and Lipscomb 2003; Wolfe et al. 2001, 2002; Kovaleva and Lipscomb 2008) is presented in Fig. 9.4. Single turnover studies of NDO have revealed two electrons (reduction of one Fe of Rieske [2Fe-2S] cluster and the mononuclear iron), and substrate binding is required for oxygen reactivity (Wolfe et al. 2001). After substrate binding and oxygen reactivation, an Fe(III)-OOH intermediate is likely to undergo O–O bond homolytic fission to yield an Fe(V)-oxo-hydroxo intermediate. The *cis*-hydroxylated product is released when one or two electron(s) come from the electron transfer component. ROs with the Rieske cluster oxidized and the mononuclear iron reduced are under resting state. Addition of H_2O_2 to NDO under resting state allows reaction with the substrate to yield the product (Fig. 9.4) (Wolfe and Lipscomb 2003). This is so-called "peroxide shunt" reaction that does not require a reduced Rieske cluster. In benzoate 1,2-dioxygenase, even fully oxidized form with the Rieske cluster oxidized and mononuclear iron in the Fe(III) state can utilized H_2O_2 as a source of reduced oxygen atom to form the appropriate product from benzoate (Neibergall et al. 2007). Several crystal structures with substrate and oxygen and reaction intermediates suggested further detailed reaction mechanisms. Karlsson and coworkers revealed that a molecular oxygen species bound to the mononuclear iron in a side-on fashion by determining crystal structures of NDO with substrate and dioxygen (Karlsson et al. 2003). The side-on binding of dioxygen enables each oxygen to attach neighboring carbons from the same face of the aromatic ring to produce the *cis*-dihydrodiols. However, when NO is introduced instead of O_2, NO is bound end-on to the mononuclear iron, whereas NO is commonly used as an analogue for dioxygen

Fig. 9.3 (continued) (**b**) The terminal oxygenase and ferredoxin complex of carbazole dioxygenase from *Janthinobacterium* sp. strain J3 and *Pseudomonas resinovorans* strain CA10 (PDB entry code: 2DE7). Chains *A*, *B*, *C* (terminal oxygenase) and chains *D*, *E*, *F* (ferredoxin) are colored in *light blue, marine, deep blue, light pink, red*, and *dark pink*, respectively. The iron atom, sulfur atom, and carbazole are shown in *brown sphere, yellow sphere*, and *yellow* and *blue* (imino nitrogen) sticks, respectively. The ferredoxin components bind to the terminal oxygenase one for one in close proximity between each Rieske [2Fe-2S] cluster. (**c**) Stereoview of proposed electron transfer pathways between the ferredoxin and terminal oxygenase of carbazole dioxygenase. Two possible routes are presented as *orange* and *red arrows*, according to a report by Ashikawa et al. (2006)

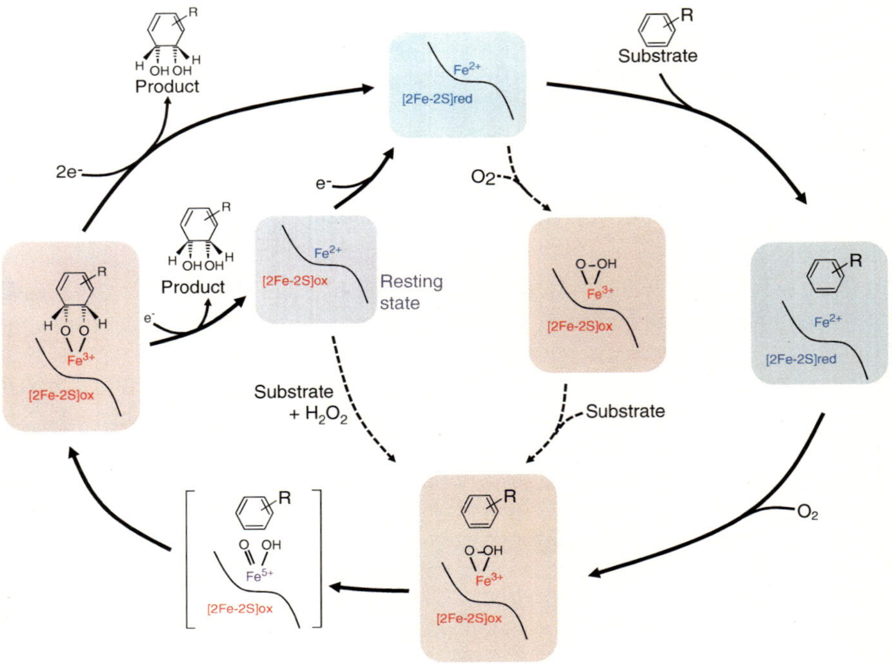

Fig. 9.4 Scheme of proposed catalytic cycles of Rieske oxygenases. Oxidized Rieske [2Fe-2S] cluster, reduced Rieske [2Fe-2S] cluster, oxidized nonheme iron, and reduced nonheme iron are shown as [2Fe-2S]ox, [2Fe-2S]red, Fe^{3+}, and Fe^{2+}, respectively. Oxidized, reduced, and resting state of the redox centers in the terminal oxygenase are shown in *red, blue*, and *purple* background. *Solid arrows* indicate the proposed main pathway taking account into the biochemically and structurally well-characterized ROs. *Broken arrow* shows the possible pathway based on biochemical studies and the complex structures determined

(Karlsson et al. 2005). Crystal structures of CARDO with substrate, dioxygen, and both carbazole and dioxygen revealed that a room for oxygen binding in the substrate-binding pocket is created by substrate binding, suggesting the substrate binds to the active site before dioxygen binding (Ashikawa et al. 2012). On the other hand, CARDO with the substrate revealed conformational changes with closure of the entranceway of the substrate by substrate binding (Fig. 9.5b, c) (Ashikawa et al. 2006). This conformational change seems to trap the substrate at the substrate-binding site and to exclude water from the active site during turnover in order to minimize the risk of leakage of partially processed substrates.

9.5 Substrate Specificity

ROs are known for their broad range of substrate (not only aromatic compounds but cholesterols and chlorophylls as described above). ROs also catalyze monooxygenation, sulfoxidation, *O*- and *N*-dealkylation. Substrate specificity analyses and recent structural studies revealed what determines the substrate specificity and the relationship between structure and substrate specificity, which provides critical clues to the modification of the substrate-binding pocket to obtain an enhanced catalyst for bioremediation (degradation) of toxic contaminants and enzymatic synthesis of chiral precursors for the production of specialty compounds. Several examples of studies on mutant enzymes demonstrated that regio- and enantio-selectivity can be altered by the mutation as follows (Boyd and Sheldrake 1998).

BDO is one of the best-characterized RO by its mutant and chimeric enzymes (Furukawa et al. 2004). Randomly recombined terminal oxygenase components of BDO from *Pseudomonas pseudoalcaligenes* strain KF707 and *Burkholderia xenovorans* strain LB400 by DNA shuffling showed expanded substrate specificities not only for biphenyl derivatives and polychlorinated biphenyl (PCB) but for single aromatic compounds (Kumamaru et al. 1998). Subsequent random priming and site-directed mutants of BDO from strain KF707 showed expanded substrate range and different regio-specificities for various polychlorinated biphenyls from the wild-type enzyme (Suenaga et al. 2001, 2002). Many of the mutants which showed different substrate specificity from the wild-type enzyme had mutation at hydrophobic amino acid residues, such as Phe227, Ile335, and Phe377. According to structural studies of ROs, the substrate-binding pocket is commonly composed of hydrophobic residues consistent with RO's preference for hydrophobic substrates (Fig. 9.5a). These mutational studies indicate that hydrophobic residues also play an important role in determining substrate-binding orientation. BDO from *Burkholderia xenovorans* strain LB400 is also well-studied enzyme by mutations. Mutated terminal oxygenase $BphAE_{p4}$ (Thr335Ala/Phe336Met) and its derivative $BphAE_{RR41}$ metabolize dibenzofuran two and three times faster than wild-type enzyme, respectively (Mohammadi et al. 2011). The crystal structures of $BphAE_{p4}$ and $BphAE_{RR41}$ revealed that replacement altered the constraints (hydrogen bonds) or plasticity of the substrate-binding pocket and increased available space for bulkier compounds than biphenyl, such as dibenzofuran and the doubly *ortho*-chlorinated congener 2,6-dichlorobiphenyl (Kumar et al. 2011; Mohammadi et al. 2011).

NDO is also one of the extensively studied ROs by mutational studies. Mutant enzymes of NDO from *Pseudomonas putida* strain NCIB9816-4 have demonstrated alteration of a series of amino acid residues composing the substrate-binding pocket can affect the regio-selectivity of product formation (Parales et al. 2000a, b; Yu et al. 2001). Detailed comparison between wild-type NDO and its amino acid replacement variants in which Phe352 is substituted by valine demonstrated region- and stereo-selectivity was principally dependent on the orientation of the substrate binding at the active site (Ferraro et al. 2006).

Another example of mutational study based on the crystal structure is CARDO from *Janthinobacterium* sp. strain J3. A series of site-directed mutagenesis variants in which the mutated amino acid residues are located at substrate-binding pocket wall showed different oxygenation site to hetero-aromatic compounds (Uchimura et al. 2008). Among the mutant enzymes, Ile262Val and Gln282Tyr converted carbazole to 1-hydroxycarbazole which is probably a dehydrated compound of lateral dioxygenation product (*cis*-1,2-dihydroxy-1,2-dihydrocarbazole) with higher yield than the wild-type enzyme. Another mutated variant Phe275Trp converted fluorene to 4-hydroxyfluorene (probable dehydration product of *cis*-3,4-dihydroxy-3,4-dihydrofluorene), whereas

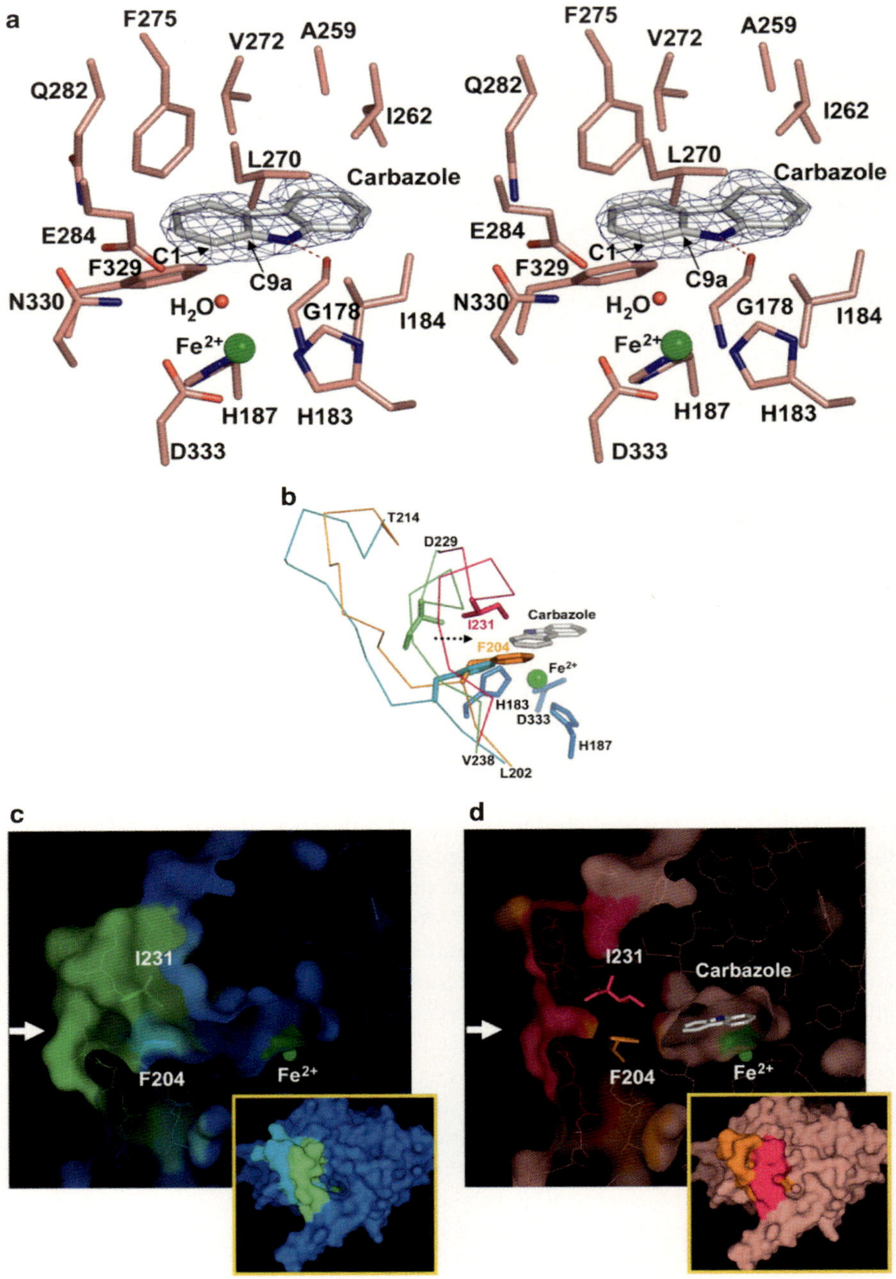

Fig. 9.5 Substrate-binding site of RO with a substrate (crystal structure of carbazole 1,9 a-dioxygenase and carbazole complex by Ashikawa et al. 2006). (**a**) Stereoview of carbazole bound to the substrate-binding pocket of CARDO. The amino acid residues which contact the substrate are shown in *light pink sticks* and carbazole is shown in a *white* and *blue* (imino nitrogen) *sticks*.

wild-type enzyme cannot catalyze this substrate. This variant also could oxidize fluoranthene to yield unidentified *cis*-dihydrodiol. Subsequent structural studies of mutated CARDO variants revealed difference in substrate-binding orientation between the variants and wild-type CARDO (Inoue et al. unpublished). In Ile262Val and Gln282Tyr variant structures with carbazole, the substrate bound to the active site rotated approximately 15° and 25°, respectively, from the substrate position of the wild type. In CARDO, imino nitrogen of carbazole forms hydrogen bonds with carbonyl oxygen of Gly178 at the active site (Fig. 9.5a). This hydrogen bond is likely to stabilize the substrate binding [in 2-oxoquinoline-8-monooxygenase from *Pseudomonas putida*, strain 86 also has similar hydrogen bond between NH group of the substrate, 2-oxoquinoline, and carbonyl oxygen of Gly216 (Martins et al. 2005)]. In the mutant enzymes, there is no "direct" hydrogen bonds between carbazole and the carbonyl oxygen of Gly178 but, instead, two water molecules were introduced into the space which created by the substrate rotation, and they formed different hydrogen bond networks with substrate and carbonyl oxygen of Gly178. The structural analysis of the mutant enzymes of CARDO suggested that different substrate orientation stabilized by the hydrogen bond networks increased formation of the lateral dioxygenation product.

Protein engineering technology combined with three-dimensional structure information has enabled alteration of substrate specificity of ROs. However, limited numbers of modification of the substrate specificity of ROs were reported so far. Further structural and biochemical studies of ROs will develop novel catalytic reactions and applicable enzymes.

9.6 Outlook

In this chapter structural and functional properties of aromatic-ring hydroxylating dioxygenases which is known as Rieske oxygenases were presented. In addition to the genetic and biochemical studies, recent structural studies of ROs have demonstrated interactions between the components, the substrate recognition, and catalytic mechanisms at the atomic level. Considering that ROs have been found to be extensively diverse and widespread in nature (Capyk and Eltis 2012; Iwai et al. 2011), basic structural and functional studies on each enzyme are becoming more and more important. The structural information and mutational studies with genetic and biochemical approach help each other to generate enzyme which is attractive to application for bioremediation and industrial biocatalysis.

Fig. 9.5 (continued) (**b**) Superposition of the entrance of the substrate-binding pocket of CARDO. The carbazole-free and carbazole-binding structures are shown in *blue green* and *green* and *orange* and *pink*, respectively. (**c** and **d**) The surface plots of the carbazole-free (**c**) and carbazole-binding structures (**d**). *White arrows* indicate the entrance of the substrate-binding pocket. The insets show the molecular surfaces viewed from the orientation directed by the white arrows (Reprinted from Ashikawa et al. (2006), with permission of Elsevier)

References

Armengaud J, Timmis KN (1997) Molecular characterization of Fdx1, a putidaredoxin-type [2Fe-2S] ferredoxin able to transfer electrons to the dioxin dioxygenase of *Sphingomonas* sp. RW1. Eur J Biochem 247:833–842

Armengaud J, Gaillard J, Timmis KN (2000) A second [2Fe-2S] ferredoxin from *Sphingomonas* sp. Strain RW1 can function as an electron donor for the dioxin dioxygenase. J Bacteriol 182:2238–2244

Ashikawa Y, Fujimoto Z, Noguchi H, Habe H, Omori T, Yamane H, Nojiri H (2006) Electron transfer complex formation between oxygenase and ferredoxin components in Rieske nonheme iron oxygenase system. Structure 14:1779–1789

Ashikawa Y, Fujimoto Z, Usami Y, Inoue K, Noguchi H, Yamane H, Nojiri H (2012) Structural insight into the substrate- and dioxygen-binding manner in the catalytic cycle of rieske non-heme iron oxygenase system, carbazole 1,9a-dioxygenase. BMC Struct Biol 12:15

Batie CJ, Ballou DP, Correll CC (1991) In: Muller F (ed) Chemistry and biochemistry of flavoenzymes, vol 3. CRC Press, Boca Raton, pp 543–556

Beharry ZM, Eby DM, Coulter ED, Viswanathan R, Neidle EL, Phillips RS, Kurtz DM Jr (2003) Histidine ligand protonation and redox potential in the Rieske dioxygenases: role of a conserved aspartate in anthranilate 1,2-dioxygenase. Biochemistry 42:13625–13636

Boyd DR, Sheldrake GN (1998) The Dioxygenase-catalysed formation of vicinal *cis*-diols. Nat Prod Rep 15:309–324

Brown EN, Friemann R, Karlsson A, Parales JV, Couture MM, Eltis LD, Ramaswamy S (2008) Determining Rieske cluster reduction potentials. J Biol Inorg Chem 13:1301–1313

Capyk JK, Eltis LD (2012) Phylogenetic analysis reveals the surprising diversity of an oxygenase class. J Biol Inorg Chem 17:425–436

Carredano E, Karlsson A, Kauppi B, Choudhury D, Parales RE, Parales JV, Lee K, Gibson DT, Eklund H, Ramaswamy S (2000) Substrate binding site of naphthalene 1,2-dioxygenase: functional implications of indole binding. J Mol Biol 296:701–712

Carrell CJ, Zhang H, Cramer WA, Smith JL (1997) Biological identity and diversity in photosynthesis and respiration: structure of the lumen-side domain of the chloroplast Rieske protein. Structure 5:1613–1625

Chakraborty S, Behrens M, Herman PL, Arendsen AF, Hagen WR, Carlson DL, Wang XZ, Weeks DP (2005) A three-component dicamba O-demethylase from *Pseudomonas maltophilia*, strain DI-6: purification and characterization. Arch Biochem Biophys 437:20–28

Colbert CL, Couture MM, Eltis LD, Bolin JT (2000) A cluster exposed: structure of the Rieske ferredoxin from biphenyl dioxygenase and the redox properties of Rieske Fe-S proteins. Structure 8:1267–1278

Correll CC, Batie CJ, Ballou DP, Ludwig ML (1992) Phthalate dioxygenase reductase: a modular structure for electron transfer from pyridine nucleotides to [2Fe-2S]. Science 258:1604–1610

Couture MM, Colbert CL, Babini E, Rosell FI, Mauk AG, Bolin JT, Eltis LD (2001) Characterization of BphF, a Rieske-type ferredoxin with a low reduction potential. Biochemistry 40:84–92

Dong X, Fushinobu S, Fukuda E, Terada T, Nakamura S, Shimizu K, Nojiri H, Omori T, Shoun H, Wakagi T (2005) Crystal structure of the terminal oxygenase component of cumene dioxygenase from *Pseudomonas fluorescens* IP01. J Bacteriol 187:2483–2490

D'Ordine RL, Rydel TJ, Storek MJ, Sturman EJ, Moshiri F, Bartlett RK, Brown GR, Eilers RJ, Dart C, Qi Y, Flasinski S, Franklin SJ (2009) Dicamba monooxygenase: structural insights into a dynamic Rieske oxygenase that catalyzes an exocyclic monooxygenation. J Mol Biol 392:481–497

Dumitru R, Jiang WZ, Weeks DP, Wilson MA (2009) Crystal structure of dicamba monooxygenase: a Rieske nonheme oxygenase that catalyzes oxidative demethylation. J Mol Biol 392:498–510

Ferraro DJ, Okerlund AL, Mowers JC, Ramaswamy S (2006) Structural basis for regioselectivity and stereoselectivity of product formation by naphthalene 1,2-dioxygenase. J Bacteriol 188:6986–6994

Ferraro DJ, Brown EN, Yu CL, Parales RE, Gibson DT, Ramaswamy S (2007) Structural investigations of the ferredoxin and terminal oxygenase components of the biphenyl 2,3-dioxygenase from *Sphingobium yanoikuyae* B1. BMC Struct Biol 7:10

Friemann R, Ivkovic-Jensen MM, Lessner DJ, Yu C-L, Gibson DT, Parales RE, Eklund H, Ramaswamy S (2005) Structural insight into the dioxygenation of nitroarene compounds: the crystal structure of nitrobenzene dioxygenase. J Mol Biol 348:1139–1151

Friemann R, Lee K, Brown EN, Gibson DT, Eklunda H, Ramaswamy S (2009) Structures of the multicomponent Rieske non-heme iron toluene 2,3-dioxygenase enzyme system. Acta Crystallogr Sect D 65:24–33

Furukawa K, Suenaga H, Goto M (2004) Biphenyl dioxygenases: functional versatilities and directed evolution. J Bacteriol 186:5189–5196

Furusawa Y, Nagarajan V, Tanokura M, Masai E, Fukuda M, Senda T (2004) Crystal structure of the terminal oxygenase component of biphenyl dioxygenase derived from *Rhodococcus* sp. strain RHA1. J Mol Biol 342:1041–1052

Gakhar L, Malik ZA, Allen CC, Lipscomb DA, Larkin MJ, Ramaswamy S (2005) Structure and increased thermostability of *Rhodococcus* sp. naphthalene 1,2-dioxygenase. J Bacteriol 187:7222–7231

Gassner GT, Ludwig ML, Gatti DL, Correll CC, Ballou DP (1995) Structure and mechanism of the iron-sulfur flavoprotein phthalate dioxygenase reductase. FASEB J 9:1411–1418

Geary PJ, Saboowalla F, Patil D, Cammack R (1984) An investigation of the iron-sulphur proteins of benzene dioxygenase from *Pseudomonas putida* by electron-spin-resonance spectroscopy. Biochem J 217:667–673

Gibson DT, Parales RE (2000) Aromatic hydrocarbon dioxygenases in environmental biotechnology. Curr Opin Biotechnol 11:236–243

Gray J, Close PS, Briggs SP, Johal GS (1997) A novel suppressor of cell death in plants encoded by the Lls1 gene of maize. Cell 89:25–31

Herman PL, Behrens M, Chakraborty S, Chrastil BM, Barycki J, Weeks DP (2005) A Three-component dicamba *O*-demethylase from *Pseudomonas maltophilia*, strain DI-6. J Biol Chem 280:24759–24767

Hurtubise Y, Barriault D, Sylvestre M (1998) Involvement of the terminal oxygenase beta subunit in the biphenyl dioxygenase reactivity pattern toward chlorobiphenyls. J Bacteriol 180:5828–5835

Inoue K, Ashikawa Y, Umeda T, Abo M, Katsuki J, Usami Y, Noguchi H, Fujimoto Z, Terada T, Yamane H, Nojiri H (2009) Specific interactions between the ferredoxin and terminal oxygenase components of a class IIB Rieske nonheme iron oxygenase, carbazole 1,9a-dioxygenase. J Mol Biol 392:436–451

Iwai S, Johnson TA, Chai B, Hashsham SA, Tiedje JM (2011) Comparison of the specificities and efficacies of primers for aromatic dioxygenase gene analysis of environmental samples. Appl Environ Microbiol 77:3551–3557

Iwata S, Saynovits M, Link TA, Michel H (1996) Structure of a water soluble fragment of the 'Rieske' iron-sulfur protein of the bovine heart mitochondrial cytochrome *bc1* complex determined by MAD phasing at 1.5 Å resolution. Structure 4:567–579

Jakoncic J, Jouanneau Y, Meyer C, Stojanoff V (2007) The crystal structure of the ring-hydroxylating dioxygenase from *Sphingomonas* CHY-1. FEBS J 274:2470–2481

Jiang H, Parales RE, Lynch NA, Gibson DT (1996) Site-directed mutagenesis of conserved amino acids in the alpha subunit of toluene dioxygenase: potential mononuclear non-heme iron coordination sites. J Bacteriol 178:3133–3139

Karlsson A, Beharry ZM, Matthew Eby D, Coulter ED, Neidle EL, Kurtz DM Jr, Eklund H, Ramaswamy S (2002) X-ray crystal structure of benzoate 1,2-dioxygenase reductase from *Acinetobacter* sp. strain ADP1. J Mol Biol 318:261–272

Karlsson A, Parales JV, Parales RE, Gibson DT, Eklund H, Ramaswamy S (2003) Crystal structure of naphthalene dioxygenase: side-on binding of dioxygen to iron. Science 299:1039–1042

Karlsson A, Parales JV, Parales RE, Gibson DT, Eklund H, Ramaswamy S (2005) NO binding to naphthalene dioxygenase. J Biol Inorg Chem 10:483–489

Kauppi B, Lee K, Carredano E, Parales RE, Gibson DT, Eklund H, Ramaswamy S (1998) Structure of an aromatic-ring-hydroxylating dioxygenase-naphthalene 1,2-dioxygenase. Structure 6:571–586

Kim SJ, Kweon O, Freeman JP, Jones RC, Adjei MD, Jhoo JW, Edmondson RD, Cerniglia CE (2006) Molecular cloning and expression of genes encoding a novel dioxygenase involved in low- and high-molecular-weight polycyclic aromatic hydrocarbon degradation in *Mycobacterium vanbaalenii* PYR-1. Appl Environ Microbiol 72:1045–1054

Klingen AR, Ullmann GM (2004) Negatively charged residues and hydrogen bonds tune the ligand histidine pKa values of Rieske iron-sulfur proteins. Biochemistry 43:12383–12389

Kovaleva EG, Lipscomb JD (2008) Versatility of biological non-heme Fe(II) centers in oxygen activation reactions. Nat Chem Biol 4:186–193

Kumamaru T, Suenaga H, Mitsuoka M, Watanabe T, Furukawa K (1998) Enhanced degradation of polychlorinated biphenyls by directed evolution of biphenyl dioxygenase. Nat Biotechnol 16:663–666

Kumar P, Mohammadi M, Viger JF, Barriault D, Gomez-Gil L, Eltis LD, Bolin JT, Sylvestre M (2011) Structural insight into the expanded PCB-degrading abilities of a biphenyl dioxygenase obtained by directed evolution. J Mol Biol 405:531–547

Kumar P, Mohammadi M, Dhindwal S, Pham TT, Bolin JT, Sylvestre M (2012) Structural insights into the metabolism of 2-chlorodibenzofuran by an evolved biphenyl dioxygenase. Biochem Biophys Res Commun 421:757–762

Martin VJ, Mohn WW (1999) A novel aromatic-ring-hydroxylating dioxygenase from the diterpenoid-degrading bacterium *Pseudomonas abietaniphila* BKME-9. J Bacteriol 181: 2675–2682

Martins BM, Svetlitchnaia T, Dobbek H (2005) 2-Oxoquinoline 8-monooxygenase oxygenase component: active site modulation by Rieske-[2Fe-2S] center oxidation/reduction. Structure 13:817–824

Mohammadi M, Viger JF, Kumar P, Barriault D, Bolin JT, Sylvestre M (2011) Retuning Rieske-type oxygenases to expand substrate range. J Biol Chem 286:27612–27621

Nam J-W, Noguchi H, Fujimoto Z, Mizuno H, Ashikawa Y, Abo M, Fushinobu S, Kobashi N, Wakagi T, Iwata K, Yoshida T, Habe H, Yamane H, Omori T, Nojiri H (2005) Crystal structure of the ferredoxin component of carbazole 1,9a-dioxygenase of *Pseudomonas resinovorans* strain CA10, a novel Rieske non-heme iron oxygenase system. Proteins 58:779–789

Neibergall MB, Stubna A, Mekmouche Y, Munck E, Lipscomb JD (2007) Hydrogen peroxide dependent cis-dihydroxylation of benzoate by fully oxidized benzoate 1,2-dioxygenase. Biochemistry 46:8004–8016

Nojiri H (2012) Structural and molecular genetic analyses of the bacterial carbazole degradation system. Biosci Biotechnol Biochem 76:1–18

Nojiri H, Ashikawa Y, Noguchi H, Nam JW, Urata M, Fujimoto Z, Uchimura H, Terada T, Nakamura S, Shimizu K, Yoshida T, Habe H, Omori T (2005) Structure of the terminal oxygenase component of angular dioxygenase, carbazole 1,9a-dioxygenase. J Mol Biol 351:355–370

Page CC, Moser CC, Chen X, Dutton PL (1999) Natural engineering principles of electron tunnelling in biological oxidation-reduction. Nature 402:47–52

Parales RE, Lee K, Resnick SM, Jiang H, Lessner DJ, Gibson DT (2000a) Substrate specificity of naphthalene dioxygenase: effect of specific amino acids at the active site of the enzyme. J Bacteriol 182:1641–1649

Parales RE, Resnick SM, Yu CL, Boyd DR, Sharma ND, Gibson DT (2000b) Regioselectivity and enantioselectivity of naphthalene dioxygenase during arene *cis*-dihydroxylation: control by phenylalanine 352 in the alpha subunit. J Bacteriol 182:5495–5504

Riedel A, Fetzner S, Rampp M, Lingens F, Liebl U, Zimmermann JL, Nitschke W (1995) EPR, electron spin echo envelope modulation, and electron nuclear double resonance studies of the 2Fe2S centers of the 2-halobenzoate 1,2-dioxygenase from *Burkholderia* (*Pseudomonas*) *cepacia* 2CBS. J Biol Chem 270:30869–30873

Rosche B, Fetzner S, Lingens F, Nitschke W, Riedel A (1995) The 2Fe2S centres of the 2-oxo-1, 2-dihydroquinoline 8-monooxygenase from *Pseudomonas putida* 86 studied by EPR spectroscopy. Biochim Biophys Acta 1252:177–179

Senda T, Yamada T, Sakurai N, Kubota M, Nishizaki T, Masai E, Fukuda M, Mitsuidagger Y (2000) Crystal structure of NADH-dependent ferredoxin reductase component in biphenyl dioxygenase. J Mol Biol 304:397–410

Senda M, Kishigami S, Kimura S, Fukuda M, Ishida T, Senda T (2007) Molecular mechanism of the redox-dependent interaction between NADH-dependent ferredoxin reductase and Rieske-type [2Fe-2S] ferredoxin. J Mol Biol 373:382–400

Subramanian V, Liu TN, Yeh WK, Serdar CM, Wackett LP, Gibson DT (1985) Purification and properties of ferredoxin TOL. A component of toluene dioxygenase from *Pseudomonas putida* F1. J Biol Chem 260:2355–2363

Suenaga H, Goto M, Furukawa K (2001) Emergence of multifunctional oxygenase activities by random priming recombination. J Biol Chem 276:22500–22506

Suenaga H, Watanabe T, Sato M, Ngadiman FK (2002) Alteration of regiospecificity in biphenyl dioxygenase by active-site engineering. J Bacteriol 184:3682–3688

Takagi T, Habe H, Yoshida T, Yamane H, Omori T, Nojiri H (2005) Characterization of [3Fe-4S] ferredoxin DbfA3, which functions in the angular dioxygenase system of Terrabacter sp. strain DBF63. Appl Microbiol Biotechnol 68:336–345

Tanaka A, Ito H, Tanaka R, Tanaka NK, Yoshida K, Okada K (1998) Chlorophyll a oxygenase (CAO) is involved in chlorophyll *b* formation from chlorophyll *a*. Proc Natl Acad Sci USA 95:12719–12723

Uchimura H, Horisaki T, Umeda T, Noguchi H, Usami Y, Li L, Terada T, Nakamura S, Shimizu K, Takemura T, Habe H, Furihata K, Omori T, Yamane H, Nojiri H (2008) Alteration of the substrate specificity of the angular dioxygenase carbazole 1,9a-dioxygenase. Biosci Biotechnol Biochem 72:3237–3248

Ullrich R, Hofrichter M (2007) Enzymatic hydroxylation of aromatic compounds. Cell Mol Life Sci 64:271–293

Wolfe MD, Lipscomb JD (2003) Hydrogen peroxide-coupled *cis*-diol formation catalyzed by naphthalene 1,2-dioxygenase. J Biol Chem 278:829–835

Wolfe MD, Parales JV, Gibson DT, Lipscomb JD (2001) Single turnover chemistry and regulation of O_2 activation by the oxygenase component of naphthalene 1,2-dioxygenase. J Biol Chem 276:1945–1953

Wolfe MD, Altier DJ, Stubna A, Popescu CV, Münck E, Lipscomb JD (2002) Benzoate 1,2-dioxygenase from *Pseudomonas putida*: single turnover kinetics and regulation of a two-component Rieske dioxygenase. Biochemistry 41:9611–9626

Yoshiyama T, Namiki T, Mita K, Kataoka H, Niwa R (2006) Neverland is an evolutionally conserved Rieske-domain protein that is essential for ecdysone synthesis and insect growth. Development 133:2565–2574

Yoshiyama-Yanagawa T, Enya S, Shimada-Niwa Y, Yaguchi S, Haramoto Y, Matsuya T, Shiomi K, Sasakura Y, Takahashi S, Asashima M, Kataoka H, Niwa R (2011) The conserved Rieske oxygenase DAF-36/Neverland is a novel cholesterol-metabolizing enzyme. J Biol Chem 286:25756–25762

Yu CL, Parales RE, Gibson DT (2001) Multiple mutations at the active site of naphthalene dioxygenase affect regioselectivity and enantioselectivity. J Ind Microbiol Biotechnol 27:94–103

Chapter 10
The Protocatechuate 4,5-Cleavage Pathway: Overview and New Findings

Naofumi Kamimura and Eiji Masai

Abstract The protocatechuate (PCA) 4,5-cleavage pathway is one of the key catabolic routes for the degradation of various aromatic compounds, such as lignin-derived low-molecular-weight aromatic compounds, phthalate isomers, and fluorene. In this pathway, the aromatic ring of PCA is initially cleaved by PCA 4,5-dioxygenase, and the resultant product is degraded to pyruvate and oxaloacetate, via the five-enzymatic steps. The gene organization of the PCA 4,5-cleavage pathway genes is divided into two types, which constitute several transcriptional units (*Sphingobium* type) or a single operon (*Comamonas* type). This chapter summarizes properties of each enzyme involved in this pathway and focuses on the recently characterized transcriptional regulation of the pathway genes. We also described a newly found enzyme involved in the pathway, 4-oxalomesaconate tautomerase. The role of the PCA 4,5-cleavage pathway in the catabolism of lignin-derived aromatic compounds is discussed at the end of this chapter.

Keywords Biodegradation • *Comamonas* • Lignin • Protocatechuate • *Sphingobium*

Abbreviations

CHA	4-Carboxy-4-hydroxy-2-oxoadipate
CHMS	4-Carboxy-2-hydroxymuconate-6-semialdehyde
OMA	4-Oxalomesaconate
PCA	Protocatechuate
PDC	2-Pyrone-4,6-dicarboxylate

N. Kamimura • E. Masai (✉)
Department of Bioengineering, Nagaoka University of Technology,
1603-1 Kamitomioka, Nagaoka, Niigata 940-2188, Japan
e-mail: emasai@vos.nagaokaut.ac.jp

H. Nojiri et al. (eds.), *Biodegradative Bacteria: How Bacteria Degrade, Survive, Adapt, and Evolve*, DOI 10.1007/978-4-431-54520-0_10, © Springer Japan 2014

10.1 Introduction

Protocatechuate (PCA) is one of the key intermediate metabolites in the bacterial degradation of various aromatic compounds, including lignin-derived low-molecular-weight compounds such as ferulate and vanillate (Masai et al. 2007), petrochemical aromatic compounds such as phthalate isomers (Sasoh et al. 2006; Eaton 2001; Wang et al. 1995), chlorobenzoates (Nakatsu et al. 1997; Nakatsu and Wyndham 1993), *p*-toluenesulfonate (Locher et al. 1989), and polycyclic aromatic hydrocarbons from fossil fuels and coal derivatives such as naphthalene, fluorene, and its analogs (Seo et al. 2009; Wattiau et al. 2001; Goyal and Zylstra 1997) (Fig. 10.1). It is commonly known that PCA is degraded via three distinct catabolic pathways: the PCA 2,3-cleavage (Crawford et al. 1979); PCA 3,4-cleavage (Harwood and Parales 1996); and PCA 4,5-cleavage (PCA45) pathways (Dagley et al. 1960). These pathways funnel PCA to the Krebs cycle. In the case of the PCA 2,3-cleavage pathway, the aromatic ring of PCA is cleaved by PCA 2,3-dioxygenase to yield 5-carboxy-2-hydroxymuconate-6-semialdehyde. The subsequent product is

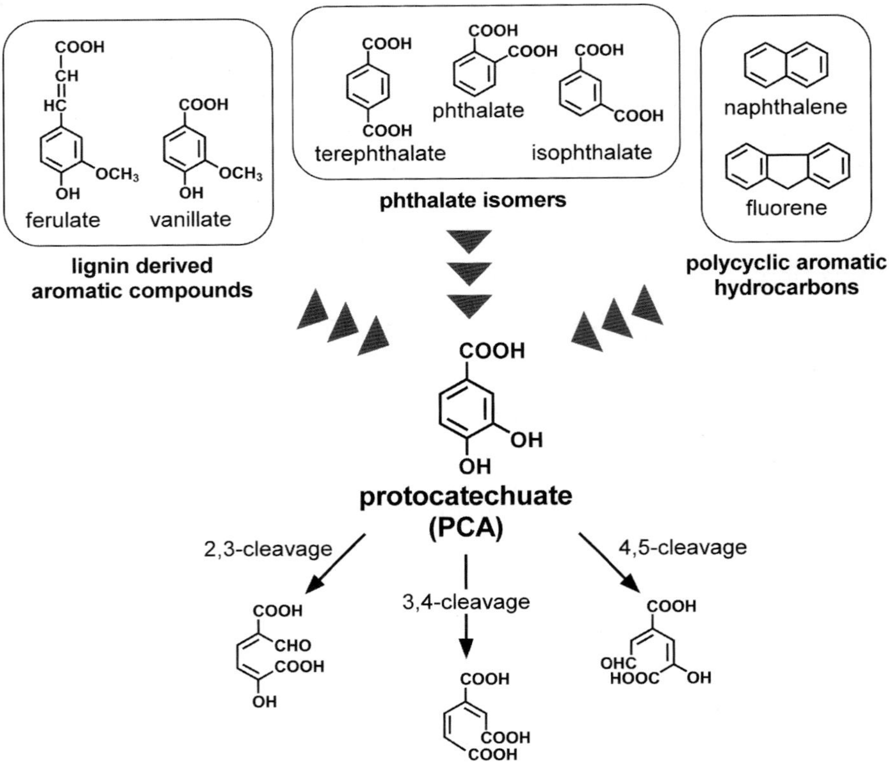

Fig. 10.1 Aromatic compounds funneled to protocatechuate (PCA), which is further metabolized through the three distinct ring-cleavage pathways

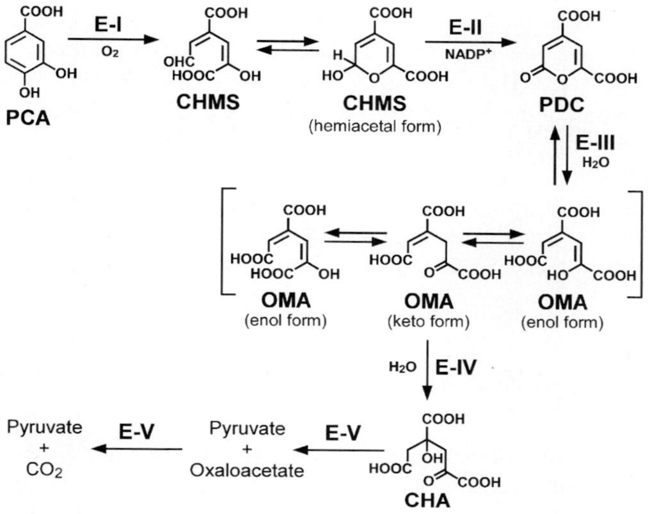

Fig. 10.2 The PCA 4,5-cleavage pathway. Enzymes: E-I, PCA 4,5-dioxygenase; E-II, CHMS dehydrogenase; E-III, PDC hydrolase; E-IV, OMA hydratase; E-V, CHA aldolase/oxaloacetate decarboxylase. Abbreviations: CHMS, 4-carboxy-2-hydroxymuconate-6-semialdehyde; PDC, 2-pyrone-4,6-dicarboxylate; OMA, 4-oxalomesaconate; and CHA, 4-carboxy-4-hydroxy-2-oxoadipate

finally degraded to pyruvate and acetyl coenzyme A (Crawford et al. 1979). The PCA 2,3-cleavage pathway has been previously reported only in *Paenibacillus* sp. (formerly *Bacillus macerans*) strain JJ-1B. This pathway was first described by Crawford (Crawford 1975), and the genes responsible for this pathway were recently identified and characterized (Kasai et al. 2009). In the PCA 3,4-cleavage pathway, PCA is converted into 3-carboxy-*cis*,*cis*-muconate by the reaction catalyzed by PCA 3,4-dioxygenase, and the resulting product is subsequently degraded to succinyl CoA and acetyl CoA via the β-ketoadipate pathway, which is widely distributed among soil bacteria. The varied organization and regulation of the genes have been characterized in detail in *Acinetobacter baylyi* (formerly *Acinetobacter calcoaceticus*) ADP1, *Pseudomonas putida* PRS2000, and *Agrobacterium tumefaciens* A348 (Harwood and Parales 1996).

The PCA45 pathway was reported in 1960 (Dagley et al. 1960), and it was enzymatically characterized by Arciero et al. (1983, 1985), Arciero and Lipscomb (1986), Kersten et al. (1982), and Maruyama and colleagues (Maruyama et al. 1978; Maruyama 1979, 1983, 1985, 1990). In this pathway, PCA is initially transformed to 4-carboxy-2-hydroxymuconate-6-semialdehyde (CHMS) by PCA 4,5-dioxygenase (Fig. 10.2). CHMS is nonenzymatically converted to an intramolecular hemiacetal form and then oxidized by CHMS dehydrogenase. The resultant 2-pyrone-4,6-dicarboxylate (PDC) is hydrolyzed by PDC hydrolase to yield the keto and enol tautomers of 4-oxalomesaconate (OMA), which are in equilibrium. OMA is converted to 4-carboxy-4-hydroxy-2-oxoadipate (CHA) by OMA hydratase. Finally, CHA is cleaved by CHA aldolase to produce pyruvate and oxaloacetate. In 1990, the PCA45 pathway genes coding the above enzymes were first isolated from *Sphingobium* sp. strain (formerly *Sphingomonas paucimobilis*) SYK-6 (Noda et al. 1990), and their functions have been subsequently revealed (Hara et al. 2000, 2003; Masai et al. 1999, 2000). Fifty years after the first report of the PCA45 pathway, the involvement of a novel enzyme, OMA tautomerase, in this pathway was reported in separate studies on *Comamonas* sp. strain E6 (Kamimura et al. 2010a) and *P. putida* KT2440 (Nogales et al. 2011).

In this chapter, we review the PCA45 pathway, its molecular and biochemical profiles, its enzymology, and the transcriptional regulation of its genes. We also discuss the roles of this pathway in the catabolism of lignin-derived aromatic compounds.

10.2 Genes and Enzymes

To date, the PCA45 pathway genes have been isolated from *Sphingomonas* sp. strain LB126 (Wattiau et al. 2001), *Comamonas* sp. strain E6 (Kamimura et al. 2010a), *Comamonas testosteroni* BR6020 (Providenti et al. 2001), *Pseudomonas straminea* (formerly *Pseudomonas ochraceae*) NGJ1 (Maruyama et al. 2004), and *Arthrobacter keyseri* 12B (Eaton 2001) in addition to *Sphingobium* sp. strain SYK-6. The gene organization of the PCA45 pathway genes was suggested to be divided into two types (Fig. 10.3) (Masai et al. 2007). The organization of the SYK-6 *lig* genes for the PCA45 pathway is quite similar to *fld* genes in *Sphingomonas* sp. strain LB126, which appears to constitute several transcriptional units (*Sphingobium* type). On the other hand, the organization and order of the PCA45 pathway genes in *Comamonas* sp. strain E6 (*pmd*), *C. testosteroni* BR6020 (*pmd*), *P. straminea* NGJ1 (*pro*), and *A. keyseri* 12B (*pcm*) are similar, and this type of gene cluster seems to constitute a single operon (*Comamonas* type). Recent studies indicated that the SYK-6 *lig* genes for the PCA45 pathway consist of three transcriptional units, including the *ligK-orf1-ligI-lsdA* operon, the *ligJ-ligA-ligB-ligC* operon, and the monocistronic *ligR* (Kamimura et al. 2010b), whereas the *pmdUKEFDABC* genes of E6 is a single transcriptional unit (Kamimura et al. 2010a). Amino acid sequence identities of PCA45 pathway enzymes from SYK-6 and E6 with those from other bacteria are shown in Table 10.1.

10.2.1 PCA 4,5-Dioxygenase

PCA 4,5-dioxygenase catalyzes the 4,5-cleavage of PCA by incorporating molecular oxygen to produce CHMS. In the 1980s, Arciero et al. investigated the enzymatic characteristics and mechanisms for a ring-cleavage reaction using the purified PCA 4,5-dioxygenase from *C. testosteroni* (formerly *Pseudomonas testosteroni*) (Arciero and Lipscomb 1986; Arciero et al. 1983, 1985). The enzyme genes were isolated from *Sphingobium* sp. strain SYK-6 (*ligAB*) (Noda et al. 1990), *C. testosteroni* BR6020 (*pmdAB*) (Providenti et al. 2001), *P. straminea* NGJ1 (*proOaOb*)

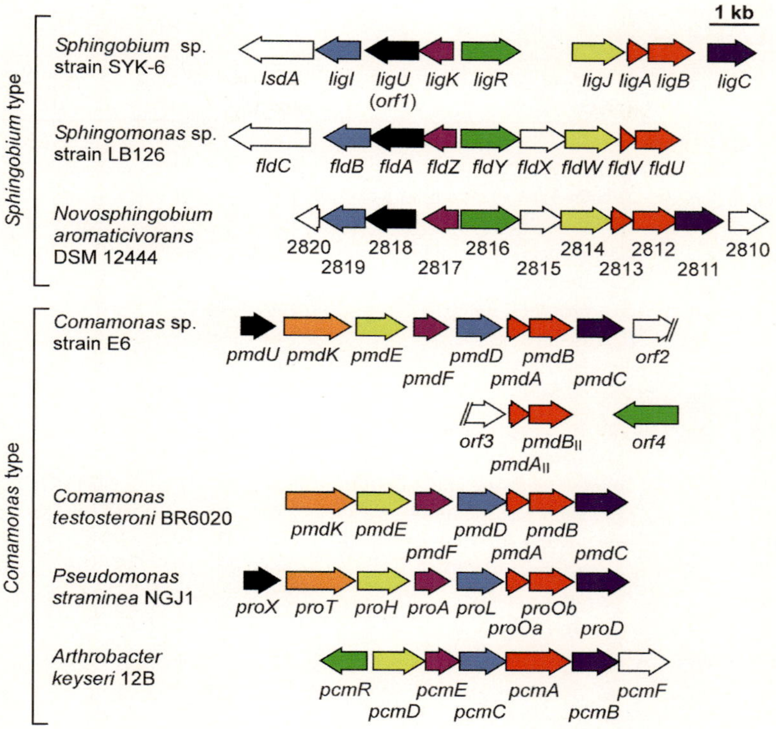

Fig. 10.3 Comparison of the genetic organizations of the PCA45 pathway gene clusters in various bacteria. Genes: α-subunit gene of PCA 4,5-dioxygenase, *ligA*, *fldV*, *pmdA*, *pmdA*$_{II}$, and *proOa*, β-subunit gene of PCA 4,5-dioxygenase; *ligB*, *fldU*, *pmdB*, *pmdB*$_{II}$, and *proOb*; PCA 4,5-dioxygenase gene, *pcmA*; CHMS dehydrogenase gene, *ligC*, *pmdC*, *proD*, and *pcmB*; PDC hydrolase gene, *ligI*, *fldB*, *pmdD*, *proL*, and *pcmC*; OMA hydratase gene, *ligJ*, *fldW*, *pmdE*, *proH*, and *pcmD*; CHA aldolase gene, *ligK*, *fldZ*, *pmdF*, *proA*, and *pcmE*; OMA tautomerase gene, *ligU* (*orf1*), *fldA*, *pmdU*, and *proX*; transcriptional regulatory gene, *ligR*; putative transcriptional regulatory gene, *fldY* and *pcmR*; putative PCA transporter gene, *pmdK* and *proT*; putative lignostilbene α,β-dioxygenase gene, *lsdA*; putative alcohol dehydrogenase gene, *fldC* and *pcmF*

(Maruyama et al. 2004), and *C. testosteroni* T-2 (*pmdAB*) (Mampel et al. 2005), and LigAB and PmdAB of *C. testosteroni* T-2 were purified and characterized. Most PCA 4,5-dioxygenases studied to date consist of small (α) and large (β) subunits and have a $\alpha_2\beta_2$ heterotetrameric structure exemplified by LigAB and *C. testosteroni* enzyme (Sugimoto et al. 1999; Arciero et al. 1990). On the other hand, the *Rhizobium leguminosarum* biovar *viceae* USDA 2370 enzyme consists of only one subunit, and this enzyme is a homodimer (Chen and Lovell 1994). The PCA 4,5-dioxygenase of *A. keyseri* 12B (PcmA) also consists of one subunit, which contains the region corresponding to the α and β subunit of PCA 4,5-dioxygenases at the N- and C-terminals, respectively (Eaton 2001). The enzyme of *C. testosteroni* T-2 shows an absolute requirement for vicinal hydroxyl groups in the 3- and 4-positions of the aromatic

Table 10.1 Amino acid sequence identities of PCA45 pathway enzymes from SYK-6 and E6 with those from other bacteria

Enzymes	Source	Size (number of amino acids)	% Identity[a] with: SYK-6	E6
PCA 4,5-dioxygenase				
α subunit				
LigA	*Sphingobium* sp. strain SYK-6	139	–	47.6
FldV	*Sphingomonas* sp. strain LB126	103	49.6	44.3
PmdA	*Comamonas* sp. strain E6	149	47.6	–
$PmdA_{II}$	*Comamonas* sp. strain E6	132	50.7	69.1
PmdA	*C. testosteroni* BR6020	149	47.6	98.0
ProOa	*P. straminea* NGJ1	149	46.3	97.3
PcmA	*A. keyseri* 12B	433	48.1[b]	56.7[b]
PCA 4,5-dioxygenase				
β subunit				
LigB	*Sphingobium* sp. strain SYK-6	302	–	57.3
FldU	*Sphingomonas* sp. strain LB126	281	66.4	64.6
PmdB	*Comamonas* sp. strain E6	289	57.3	–
$PmdB_{II}$	*Comamonas* sp. strain E6	290	57.3	97.9
PmdB	*C. testosteroni* BR6020	289	57.3	99.3
ProOb	*P. straminea* NGJ1	289	57.0	99.7
PcmA	*A. keyseri* 12B	433	52.8[c]	65.8[c]
CHMS dehydrogenase				
LigC	*Sphingobium* sp. strain SYK-6	315	–	75.0
PmdC	*Comamonas* sp. strain E6	319	75.0	–
PmdC	*C. testosteroni* BR6020	319	74.7	99.1
ProD	*P. straminea* NGJ1	319	73.8	96.9
PcmB	*A. keyseri* 12B	317	69.4	67.2
PDC hydrolase				
LigI	*Sphingobium* sp. strain SYK-6	293	–	53.9
FldB	*Sphingomonas* sp. strain LB126	295	80.1	56.5
PmdD	*Comamonas* sp. strain E6	305	53.9	–
PmdD	*C. testosteroni* BR6020	305	54.5	97.4
ProL	*P. straminea* NGJ1	305	54.5	98.0
PcmC	*A. keyseri* 12B	312	50.5	73.4
OMA tautomerase				
LigU	*Sphingobium* sp. strain SYK-6	351	–	15.4
FldB	*Sphingomonas* sp. strain LB126	356	78.1	17.8
PmdU	*Comamonas* sp. strain E6	230	15.4	–
ProX	*P. straminea* NGJ1	230	14.1	94.8
GalD	*P. putida* KT2440	361	56.9	11.3
PmdU	*C. testosteroni* CNB-2	230	15.2	96.5
OMA hydratase				
LigJ	*Sphingobium* sp. strain SYK-6	341	–	63.1
FldW	*Sphingomonas* sp. strain LB126	340	87.7	63.1
PmdE	*Comamonas* sp. strain E6	342	63.1	–
PmdE	*C. testosteroni* BR6020	342	62.8	99.4
ProH	*P. straminea* NGJ1	342	63.1	100

(continued)

Table 10.1 (continued)

Enzymes	Source	Size (number of amino acids)	% Identity[a] with:	
			SYK-6	E6
PcmD	*A. keyseri* 12B	342	63.2	71.6
LigJ	*Rhodopseudomonas palustris* CGA009	342	61.6	79.5
GalB	*P. putida* KT2440	258	14.7	6.8
CHA aldolase				
LigK	*Sphingobium* sp. strain SYK-6	228	–	65.1
FldZ	*Sphingomonas* sp. strain LB126	224	79.8	63.0
PmdF	*Comamonas* sp. strain E6	227	65.1	–
PmdF	*C. testosteroni* BR6020	227	65.1	99.1
ProA	*P. straminea* NGJ1	227	65.1	98.7
PcmE	*A. keyseri* 12B	230	57.4	60.4
Pput_1361	*P. putida* F1	238	58.2	56.7

[a]Identities were measured using EMBOSS Needle alignment tool (www.ebi.ac.uk/Tools/psa/emboss_needle)
[b]Amino acid residues 1–149 of PcmA were used for the alignment
[c]Amino acid residues 148–433 of PcmA were used for the alignment

ring and for the carboxyl group of PCA, this enzyme additionally catalyzes 5-hydroxy-PCA (gallate) (Mampel et al. 2005). LigAB exhibited a significantly higher activity with PCA than with gallate and 3-*O*-methylgallate. The V_{max}/K_m of LigAB for gallate and 3-*O*-methylgallate were 9.5 % and 0.24 % of the value for PCA, respectively (Kasai et al. 2005).

Vaillancourt et al. proposed that extradiol dioxygenases belong to at least three evolutionarily independent families (Vaillancourt et al. 2006). Type I extradiol dioxygenases belong to the vicinal oxygen chelate superfamily, including 2,3-dihydroxybiphenyl 1,2-dioxygenase of *Burkholderia xenovorans* LB400 and catechol 2,3-dioxygenase of *P. putida* mt-2. Type II extradiol dioxygenases include enzymes consisting of one (e.g., 2,3-dihydroxyphenylpropionate 1,2-dioxygenase and homoprotocatechuate 2,3-dioxygenase of *E. coli*) or two different subunits (PCA 4,5-dioxygenase of *Sphingobium* sp. strain SYK-6, *C. testosteroni*, and *P. straminea* NGJ1, and 2-aminophenol 1,6-dioxygenase of *C. testosteroni* CNB-1 and *Pseudomonas pseudoalcaligenes* JS45). Type III enzymes are part of the cupin superfamily (Fetzner 2012), including gentisate 1,2-dioxygenase of *C. testosteroni*, 3-hydroxyanthranilate 3,4-dioxygenase of *Saccharomyces cerevisiae*, and flavonol 2,4-dioxygenase (quercetinase) of *Aspergillus japonicus*. All the Fe(II)-dependent extradiol dioxygenases share similar active sites, and all have the same iron ligands: the two histidines and one glutamate that constitute the 2-His 1-carboxylate structural motif.

A refined crystal structure of LigAB was determined by Sugimoto et al. (1999). The larger β subunit has 302 residues which form a globular α/β structure composed of 11 β strands, nine α helices, and one 3_{10} helix. The 139 amino acids of the α subunit are composed of ten α-helices, which assemble into a platelike shape that interacts extensively with one face of the β subunit of the same protomer and with the β subunit of the second

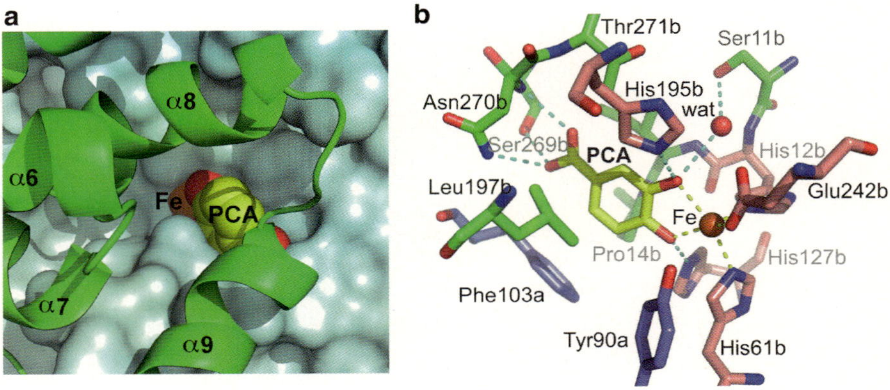

Fig. 10.4 The active site of the LigAB-PCA complex. (**a**) The binding pocket of PCA covered by the α subunit. PCA (represented as CPK) is accommodated in a cavity on the upper part of the β subunit (shown as *grey surface*). Helices α8 and α9 from the α subunit (shown in *green*) form a lid that closes the open end of the pocket. The Fe ion is colored *orange*. (**b**) The iron coordination sphere in the LigAB-PCA complex. The coordination sphere is a distorted tetragonal bipyramid with His12b, Glu242b (suffix "b" indicates amino acid residues of the β subunit), and O4 of PCA as equatorial ligands and His61b and O3 of PCA as axial ligands. Amino acid residues conserved among the type II extradiol dioxygenases (Sugimoto et al. 1999) are shown in *pink*. The residues from the α subunit are shown in *blue* (suffix "a" indicates amino acid residues of the α subunit). Hydrogen bonds between PCA and amino acid residues, and coordinate bonds of the Fe ion are shown as *dotted cyan* and *yellow lines*, respectively. These figures were kindly provided by Dr. Toshiya Senda (National Institute of Advanced Industrial Science and Technology)

protomer. The latter α-β contacts stabilize the $\alpha_2\beta_2$ structure. The active site contains a nonheme iron coordinated by His12b, His61b, and Glu242b (suffix "b" indicates amino acid residues of the β subunit) and a water molecule located in a cleft of the β subunit, which is covered by the α subunit (Fig. 10.4). His195b is thought to act as an active site base to facilitate deprotonation of the hydroxyl group of the substrate.

Comamonas sp. strain E6 has two sets of the PCA 4,5-dioxygenase genes (Fig. 10.3), *pmdAB* and $pmdA_{II}B_{II}$, the products of which share high amino acid sequence identities (PmdA and $\mathrm{PmdA_{II}}$, 69.1 %; PmdB and $\mathrm{PmdB_{II}}$, 97.9 %) (Kamimura et al. 2010a). The *pmdAB* genes are included in the PCA45 pathway genes (*pmdUKEFDABC*), and *pmdB* was found to be essential for the growth of E6 on PCA. On the other hand, there were no other PCA45 pathway genes proximal to $pmdA_{II}B_{II}$. Disruption of $pmdB_{II}$ did not affect the growth on PCA. Our recent study suggested that both genes are necessary for the syringate catabolism in E6 (Kamimura et al., unpublished results).

10.2.2 CHMS Dehydrogenase

Since CHMS is in equilibrium between the open form and the cyclic hemiacetal form (Kersten et al. 1982), it is thought that CHMS dehydrogenase oxidizes the hemiacetal form of CHMS to generate PDC. The enzyme is a homodimer. Both NAD^+ and $NADP^+$

acted as coenzymes, although the apparent K_m value for NAD^+ is greater than that for $NADP^+$. The enzyme activity was not affected by various metal ions, metal chelating reagents, or reducing reagents. Meanwhile, the catalytic activity was strongly inhibited by the addition of sulfhydryl reagents, suggesting that the cysteine residue plays a key role in the enzyme reaction (Maruyama et al. 1978; Masai et al. 2000).

The amino acid sequence of characterized CHMS dehydrogenases, LigC from *Sphingobium* sp. strain SYK-6 and ProD from *P. straminea* NGJ1 showed 15–19 % identity with CbaC family enzymes (Nakatsu et al. 1997). These are the dehydrogenases for carboxylic *cis*-dihydrodiol compounds involved in the degradation of 3-chlorobenzoate or phthalate in *C. testosteroni* (formerly *Alcaligenes* sp. strain) BR60, *P. putida* NMH102-2, and *Burkholderia cepacia* DBO1 (Chang and Zylstra 1998). These dehydrogenases do not have any apparent relationship with group I (long-chain zinc-dependent enzyme), group II (short-chain zinc-dependent enzyme), or group III (iron-activated enzyme) microbial $NAD(P)^+$-dependent alcohol dehydrogenases. Among the enzymes related to the CbaC family, the amino acid sequence H-$(X)_{11}$-K-H-V-L-X-E-K-P-X-A (located in the α helical domain within N-terminal 130 amino acids) is conserved (Nakatsu et al. 1997). LigC and ProD contain the largest part of this conserved sequence at their N-terminals. The CbaC family members also have the putative $NAD(P)^+$-binding motif G-X-X-G-X-G at their N-terminals. Instead, LigC and ProD have G-X-G-X-X-G at their terminals.

10.2.3 PDC Hydrolase

PDC hydrolase catalyzes the reversible conversion of PDC to OMA. This conversion is in equilibrium among two tautomeric enol forms (OMA_{enol}) and a keto form (OMA_{keto}). In the reversible reaction catalyzed by the enzyme, the equilibrium concentration of PDC and OMA_{enol}/OMA_{keto} is dependent on the solution pH. The relative concentrations of PDC and the hydrolysis products are equal to one another at pH 8.25 (Hobbs et al. 2012). The enzyme from *Sphingobium* sp. strain SYK-6 (LigI) is a monomer (Masai et al. 1999), and the kinetic constants, k_{cat} and k_{cat}/K_m, of this enzyme for the hydrolysis of PDC are 340 s^{-1} and 9.8×10^6 M^{-1} s^{-1}, respectively (Hobbs et al. 2012). The kinetic constants are dependent on pH, which suggests that a single active site residue is deprotonated for the hydrolysis of PDC (Hobbs et al. 2012).

The enzyme belongs to the amidohydrolase superfamily (Holm and Sander 1997), members of which catalyze a diverse set of chemical reactions, including the hydrolysis of amide or ester bonds, deamination of nucleic acids, decarboxylation, isomerization, and hydration (Seibert and Raushel 2005). The enzymes in the amidohydrolase superfamily have a distorted $(\beta/\alpha)_8$-TIM barrel structure that typically possesses an active site containing one to three divalent metal ions. In contrast, LigI does not contain bound metal ions, and the addition of metal chelators or divalent metal ions to the assay mixture did not affect the enzyme activity (Hobbs et al. 2012). In the presence of the reaction product, the crystal structures of LigI and the Asp248Ala mutant revealed the roles of conserved residues and a proposed mechanism of enzyme action. The C8 and C11 carboxylates of PDC are coordinated with Arg130 and Arg124, respectively, through ion pair interactions, with an additional electrostatic interaction between Tyr49 and the C11 carboxylate. The lactone carbonyl group of PDC is polarized by electrostatic interactions with His31, His33, and His180. Asp248 deprotonates an active site water molecule for a nucleophilic attack on the C2 carbonyl group of PDC. After formation of the tetrahedral intermediate, the C2-O1 bond is cleaved. All these active site residues are conserved among the enzymes of *Sphingomonas* sp. strain LB126 (FldB), *Comamonas* sp. strain E6 (PmdD), *C. testosteroni* BR6020 (PmdD), *P. straminea* NGJ1 (ProL), and *A. keyseri* 12B (PcmC), suggesting that these proteins seem to catalyze the same reaction as LigI.

10.2.4 OMA Hydratase

OMA hydratase catalyzes the conversion of OMA into CHA. The enzyme is a dimer and does not require the addition of any metal ions for catalytic activity (Hara et al. 2000). X-ray crystallographic data for OMA hydratase from *Rhodopseudomonas palustris* CGA009 (PDB 2GWG) showed that the enzyme contains one Zn ion per subunit. The imidazole group of His223 located near the Zn ion is thought to be involved in enzyme catalysis. The Zn ion was suggested to be strongly bound to the enzyme because the metal chelator reacted with only denatured enzymes (Li et al. 2007). While the enzyme was activated by various reducing reagents such as cysteine and dithiothreitol, the addition of $HgCl_2$ and *p*-chloromercuribenzoate strongly inhibited the activity; therefore, cysteine residue is suggested to be a part of the catalytic site (Maruyama 1985; Hara et al. 2000). Site-directed mutagenesis analyses of the cysteine residues in the *P. straminea* enzyme (ProH) done by Li et al. showed that Cys186 is important for the protection of the Zn ion from chelators. Although the region between Thr181 and Thr190 is missing in the crystallographic data, the authors predicted that this region forms a loop structure and lies just above the Zn ion to protect Zn ion from chelators (Li et al. 2007).

To date, the substrate of OMA hydratase had been considered to be OMA_{keto} (Hara et al. 2000; Kersten et al. 1982; Maruyama 1985). However, Nogales et al. recently reported that GalB, an OMA hydratase involved in the gallate degradation pathway in *P. putida* KT2440, utilizes OMA_{enol} as the substrate for the formation of CHA (Nogales et al. 2011). Although GalB (a hexameric protein) does not show significant amino acid sequence similarities with OMA hydratases of other bacteria (Table 10.1), these proteins have similar enzymatic characteristics. In the report by Nogales et al., *galB* was able to complement the function of *ligJ*, suggesting that the substrate of LigJ and GalB is the same. This will be further discussed in the following section.

10.2.5 CHA Aldolase

CHA aldolase catalyzes the aldol cleavage of CHA to produce pyruvate and oxaloacetate, which can then be channeled to the Krebs cycle. The enzymes of *Sphingobium* sp. strain SYK-6 (LigK) (Hara et al. 2003), *P. straminea* NGJ1 (ProA)

(Maruyama 1990), and *P. putida* (Tack et al. 1972; Wang et al. 2010; Nogales et al. 2011) have been enzymatically characterized. The subunit structure of all the enzymes is a hexamer, and their catalytic activities were dependent on a divalent cation such as Mg^{2+}. Aldolases are typically divided into two groups based on their catalytic mechanism. The class I aldolases utilize a Schiff base mechanism to stabilize a carbanion intermediate. On the other hand, the class II aldolases utilize divalent metal ions to stabilize an enolate intermediate. Since the catalytic activity of CHA aldolase required a divalent metal ion and the enzyme was activated by the presence of phosphate ion, the enzyme is thought to belong to the class II aldolase (Hara et al. 2003). 4-Hydroxy-4-methyl-2-oxoglutarate (HMG)/CHA aldolase of *P. putida* F1 has higher catalytic specificity (k_{cat}/K_m) toward CHA (2.2×10^5) and HMG (7.4×10^4) than other 4-hydroxy-2-ketoacids, α-keto-γ-hydroxyglutarate, and 4-hydroxy-2-oxopentanoate (Wang et al. 2010). The enzyme also exhibits oxaloacetate decarboxylation activity, which involves C–C bond cleavage leading to a formation of pyruvate and CO_2, and pyruvate α-proton exchange activity.

The crystal structure of CHA aldolase of *P. putida* F1 revealed that the enzyme forms a four-layered α-β-β-α sandwich structure with the active site at the interface of two neighboring subunits of a hexamer. This fold is similar to the RNase E inhibitor, RraA (Wang et al. 2010). The catalytic site contains Mg^{2+} ligated by Asp124 and three water molecules (one forms a hydrogen bond to Asp124, one forms a hydrogen bond to Asp102 and Glu199′ (the prime indicates an amino acid from a different protomer), and the third forms a hydrogen bond to Glu199′). The mechanism for CHA aldolase was proposed as follows: deprotonated metal bound water (Wat-1) abstracts a proton from C4–OH of the substrate, leading to the aldol cleavage of the C3–C4 bond and the concomitant formation of the enolate intermediate. The hydrogen bond between Arg123 and the C2 keto oxygen appeared to stabilize the development of a negative charge on this oxygen. A water molecule (Wat-4) binds and donates a proton to pyruvate enolate, forming pyruvate.

10.3 Tautomerization of 4-Oxalomesaconate

In the PCA45 pathway, tautomerization of the enol and keto forms of OMA was previously believed to have been converted nonenzymatically. Recently, Kamimura et al. and Nogales et al. independently discovered the enzyme genes encoding OMA tautomerase, *pmdU* and *galD*, from *Comamonas* sp. strain E6 and *P. putida* KT2440, respectively (Kamimura et al. 2010a; Nogales et al. 2011).

Analyses of a *pmdU* mutant of *Comamonas* sp. strain E6 indicated that *pmdU* is essential for the growth of E6 on PCA (Kamimura et al. 2010a). This mutant was suggested to lose the ability to convert OMA. Since the *pmd* operon contains the OMA hydratase gene, *pmdE* (Fig. 10.3), *pmdU* was assumed to encode tautomerase involved in the conversion of OMA_{enol} into OMA_{keto}. OMA_{keto} had been thought to be the substrate of OMA hydratase (Hara et al. 2000; Kersten et al. 1982; Maruyama 1985). On the other hand, Nogales et al. characterized the *gal* gene cluster including *galD* for the gallate catabolism in *P. putida* KT2440 (Nogales et al. 2011). Gallate dioxygenase, GalA, cleaved the aromatic ring of gallate to produce OMA_{keto}, and the resultant OMA_{keto} was concluded to convert into OMA_{enol} by the further addition of crude GalD. The conversion of gallate into OMA and the tautomerization of OMA were performed continuously at pH 7.0. The chemical structures of the products were then determined by NMR. The most recent work by Ni et al. also concluded that OMA tautomerase of *C. testosteroni* CNB-1, PmdU, converted OMA_{keto} into OMA_{enol} (Ni et al. 2012). The substrate for OMA tautomerase used in this study was prepared using the method of Nogales et al. and by incubation of gallate with recombinant PmdAB at pH 7.0. In 2012, Hobbs et al. estimated the relative amount of OMA_{enol} and OMA_{keto} at various pH values (pH 6–10) using a heteronuclear multiple-bond correlation NMR spectroscopy (Hobbs et al. 2012). At pH 7.0, the relative amounts of the two forms were estimated at 95 % OMA_{enol} and 5 % OMA_{keto} (Hobbs et al. 2012). This fact appeared to suggest that OMA tautomerase catalyzes the conversion of OMA_{enol} into OMA_{keto}.

OMA tautomerases (PmdU) of *Comamonas* sp. strain E6 and *C. testosteroni* CNB-1 show only 11.3 % and 12.0 % amino acid sequence identity to that of GalD of *P. putida* KT2440, respectively, and their sizes are also different (Table 10.1). In spite of the low sequence similarity, these proteins seem to have the same function, since GalD was able to replace the function of PmdU in *C. testosteroni* CNB-1(Ni et al. 2012). The amino acid sequence of GalD shows 36.7 % and 35.7 % of identity, respectively, with 3-methylitaconate-Δ-isomerase, Mii, from *Eubacterium barkeri* DSM1223 (Alhapel et al. 2006) and methylaconitate isomerase, PrpF, from *Shewanella oneidensis* MR-1 (Garvey et al. 2007) that share a common structural fold.

In a previous study, it was reported that an ORF (*orf1*) downstream of *ligI* is essential for the growth of *Sphingobium* sp. strain SYK-6 on vanillate and syringate (Hara et al. 2003). The product of *orf1* (designated here *ligU*) shows 56.9 % of amino acid sequence identity with GalD while it shows only 15 % identity with PmdU (Table 10.1). Our recent studies showed that a *ligU* mutant of SYK-6 lost the ability to convert OMA, and an introduction of *ligU* in a *pmdU* mutant of *Comamonas* sp. strain E6 restored its ability to grow on PCA. In addition, *pmdU* was also able to complement the function of *ligU* in SYK-6 (Kamimura et al., unpublished results). These observations suggest that *ligU* encodes OMA tautomerase. In addition to *ligU*, orthologs of *pmdU* and *galD* were found in the *pro* genes of *P. straminea* NGJ1 (*proX*) and *fld* genes of *Sphingomonas* sp. strain LB126 (*fldA*), respectively (Fig. 10.3). In the *pmd* cluster of *C. testosteroni* BR6020, a *pmdU* ortholog is most likely to be just upstream of *pmdK* because the genetic organization of the *pmd* genes in this strain is quite similar to those in E6 and NGJ1. Interestingly, there is no similar gene of *pmdU* or *galD* in the *A. keyseri* 12B plasmid, pRE1, carrying the *pcm* genes (accession number, AF331043). Correlated to this fact, a *galD* mutant of *P. putida* grew normally on 5 mM gallate but grew poorly at gallate concentrations lower than 2 mM (Nogales et al. 2011). Considering that the relative amount of OMA_{enol} and OMA_{keto} is almost equal at pH 8.0 (Hobbs et al. 2012) and the cytoplasmic pH is maintained within a range of 7.4–7.8, in a neutrophilic bacteria, such as *Escherichia coli* and *Bacillus subtilis* (Slonczewski et al. 2009), OMA tautomerase is not indispensable for certain types of bacteria to catabolize OMA.

10.4 Transcriptional Regulation

The transcriptional regulations of the PCA45 pathway genes have been investigated in *Sphingobium* sp. strain SYK-6 (Fig. 10.5) and *Comamonas* sp. strain E6. The only characterized regulatory protein in the PCA45 pathway is LigR. LigR positively regulates transcription of the *ligK-ligU-ligI-lsdA* operon (*ligK* operon), the *ligJ-ligA-ligB-ligC* operon (*ligJ* operon) in SYK-6 in response to the effector molecules, PCA or gallate, and represses transcription of its own gene (Kamimura et al. 2010b). LigR is a member of LysR-type transcriptional regulators (LTTRs) and shows 20.8 % and 22.1 % of amino acid sequence identities with PcaQ of *Agrobacterium tumefaciens* A348 (Parke 1996) and *Sinorhizobium meliloti* (MacLean et al. 2006, 2008), respectively. These proteins positively regulate the expression of the *pcaDCHGB* operons involved in the PCA 3,4-cleavage pathway.

LTTR is one of the most common types of regulatory proteins among prokaryotic regulators. Proteins belonging to the LTTR family participate in the regulation of diverse physiological processes, such as degradation of aromatic compounds, carbon and nitrogen metabolism, and the oxidative stress responses (Schell 1993; Tropel and van der Meer 2004; Maddocks and Oyston 2008). In solution, LTTRs are homodimers or homotetramers, and their DNA-binding forms are suggested to be tetramers (Maddocks and Oyston 2008). LTTRs associate with two distinct binding sites at the target promoter. The recognition binding site (RBS) contains the LTTR consensus binding sequence (T-N_{11}-A) and is commonly essential for the binding of LTTR. The activation binding site (ABS) located between RBS and a target promoter is important for transcriptional activation (Porrúa et al. 2007). A large number of LTTRs have been shown to induce DNA bending upon binding of the protein. Binding of inducers provokes a conformational change and typically alters the binding region and DNA bending angle (Tropel and van der Meer 2004).

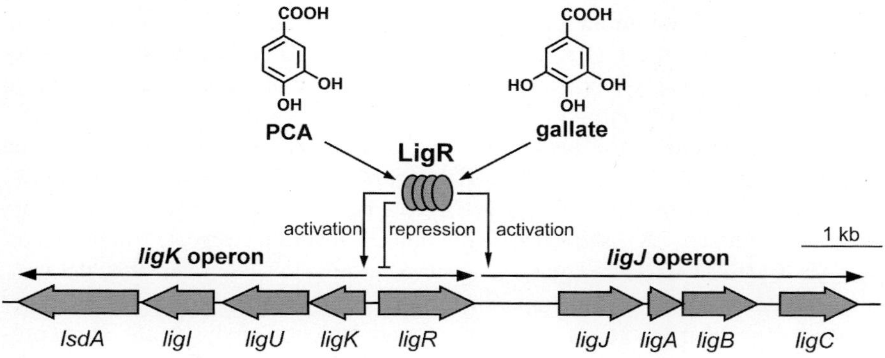

Fig. 10.5 The transcriptional regulatory system of the PCA45 pathway genes controlled by LigR in *Sphingobium* sp. strain SYK-6. LigR activates transcription of the *ligK* and *ligJ* operons in the presence of PCA or gallate, while repressing the transcription of its own gene. LigR binds to the *ligK* and *ligJ* promoter regions as a dimer or dimer of dimers. A PCA- or gallate-bound dimer of dimers appeared to be the transcriptionally active form

Although LigR is a dimer in solution, this protein binds to DNA as a dimer or dimer of dimers. The T-N_{11}-A motifs in the *ligK* and *ligJ* promoter regions, Box-K and Box-J, are respectively, centered at positions −65 and −69 relative to each transcriptional start site. Box-K and Box-J were thought to function as RBS, since they were essential for the binding of LigR and transcriptional activation of both operons. DNase I footprinting and circular permutation analyses demonstrated that LigR bound to the positions −77 to −51 and −80 to −48 of the *ligK* and *ligJ* promoters, respectively, and induced DNA bending. In the presence of PCA or gallate, the DNA bending on both promoters was enhanced. In addition, the regions between the LigR-binding boxes and the −35 regions were required for the enhancement of DNA bending. On the other hand, the 3′ end of LigR-binding region on the *ligK* promoter was significantly extended from a position −51 to proximal to the −35 region in the presence of an effector molecule, while the region on the *ligJ* promoter was shortened from −48 to −55. These results may suggest that binding of effectors results in different conformational changes in the LigR-DNA complex on the *ligK* and *ligJ* promoters. The behavior of LigR is distinct from the mode of action of well-characterized LTTRs, such as AtzR (Porrúa et al. 2007), CbbR (Dubbs et al. 2004), CbnR (Ogawa et al. 1999), OccR (Wang et al. 1992;Wang and Winans 1995a, b), and OxyR (Toledano et al. 1994), known as the sliding dimer model. In this sliding dimer model, LTTRs cause the effector-dependent shortening of the 3′ end of binding region to proximal to the −35 region and a relaxation of the DNA bending. These events are thought to be important in the release of the recognition site of RNA polymerase and the conformational change of regulator-DNA complex suitable for transcriptional activation. However, the behavior of TrpI in response to effector is similar to that of LigR (binding to the *ligK* promoter). In the presence of effector, TrpI enhanced DNA bending with an extension of the 3′ end of TrpI-binding region from −52 to −32 (Chang and Crawford 1990; Piñeiro et al. 1997). This extension of the binding region proximal to −35 was thought to allow the interaction of TrpI with RNA polymerase (Gao and Gussin 1991). In the case of the *ligK* promoter, the PCA- and gallate-dependent structural alteration of LigR-DNA complex seems to cause the same effect as TrpI on transcriptional activation.

The transcription of the *pmd* genes in *Comamonas* sp. strain E6 is activated during growth in the presence of PCA (Kamimura et al. 2010a). Both PCA and PDC act as the inducer molecules of the *pmd* operon, and PDC was found to be the more effective inducer. A sequence containing inverted repeat (underlined), GCTATGCCTTTGCGGCATAGC, is centered at position −67 relative to the transcription start site of the *pmd* operon; however, no obvious canonical promoter sequences for the σ^{70} factor of *E. coli* was found. Deletion analysis of the *pmd* promoter region suggested that this inverted repeat sequence is indispensable for the induction of the promoter. On the other hand, there is no transcriptional regulator gene within and proximal to the *pmd* operon. In *Comamonas* sp. strain E6, an LTTR gene, *orf4*, is flanked by $pmdA_{II}B_{II}$ (Fig. 10.3); however, this gene is not involved in the transcriptional regulation of the *pmd* operon (Kamimura et al. 2010a). In the

Comamonas-type PCA45 pathway gene clusters reported to date, an ORF related to the transcriptional regulator has been found only upstream of the *pcm* gene cluster of a gram-positive bacterium, *A. keyseri* 12B (*pcmR*) (Fig. 10.3). A database search of genome sequences of *C. testosteroni* KF-1 (NCBI accession number, AAUJ02000001), *C. testosteroni* CNB-2 (NC_013446) (Ma et al. 2009), and *C. testosteroni* S44 (NZ_ADVQ00000000) revealed the presence of LTTRs that showed approximately 30 % identity with *pcmR*. The isolation of the regulator gene(s) for the *pmd* operon will be essential to clarify the transcriptional regulatory mechanism of the *Comamonas*-type PCA45 pathway genes.

10.5 Roles of the PCA45 Pathway in the Catabolism of Lignin-Derived Aromatics

Lignin is the most abundant aromatic substance in nature, and its decomposition represents a key step in the terrestrial carbon cycle. It has been described as a random and three-dimensional network polymer comprised of variously linked phenylpropane units. In nature, peroxidases secreted by fungi, such as lignin peroxidase, manganese peroxidase, laccase, and versatile peroxidase, initiate the degradation of native lignin (Martinez et al. 2005; Hofrichter et al. 2010). These enzymes produce free radicals for depolymerization of lignin. Although several studies have reported the direct wood degradation by bacteria, bacteria generally play a main role in the mineralization of large amounts of lignin-derived low-molecular-weight compounds in soils (Vicuña 1988; Zimmermann 1990). *Sphingobium* sp. strain SYK-6 is one of the best-characterized degraders of lignin-derived aromatic compounds, and this strain is able to utilize various lignin-derived biaryls, including β-aryl ether, biphenyl, and phenylcoumaran (Masai et al. 2007). In SYK-6, lignin-derived biaryls with guaiacyl (4-hydroxy-3-methoxyphenyl) and syringyl (4-hydroxy-3,5-dimethoxyphenyl) moieties are converted to vanillate and syringate, respectively (Fig. 10.6). After O demethylation of vanillate and syringate, PCA and 3-*O*-methylgallate are generated, respectively (Abe et al. 2005; Masai et al. 2004). PCA is further degraded via the PCA45 pathway. On the other hand, 3-*O*-methylgallate is degraded through multiple ring-cleavage pathways involving 3-*O*-methylgallate 3,4-dioxygenase (DesZ), PCA 4,5-dioxygenase (LigAB), and vanillate/3-*O*-methylgallate *O*-demethylase (LigM) and gallate dioxygenase (DesB) (Kasai et al. 2004, 2005, 2007). Eventually, these pathways conjoin with the PCA45 pathway at PDC or OMA; therefore, the PCA45 pathway is also essential for the catabolism of syringate. In order to catabolize syringate via PDC or OMA, the generation of inducer molecules for the *ligK* and *ligJ* operons is essential. In the ring-cleavage pathways of 3-*O*-methylgallate, an inducer molecule, gallate, is formed by O demethylation of 3-*O*-methylgallate catalyzed by LigM. This fact indicates that the degradation pathway for 3-*O*-methylgallate via gallate is crucial for the catabolism of syringyl-type lignin-derived aromatics.

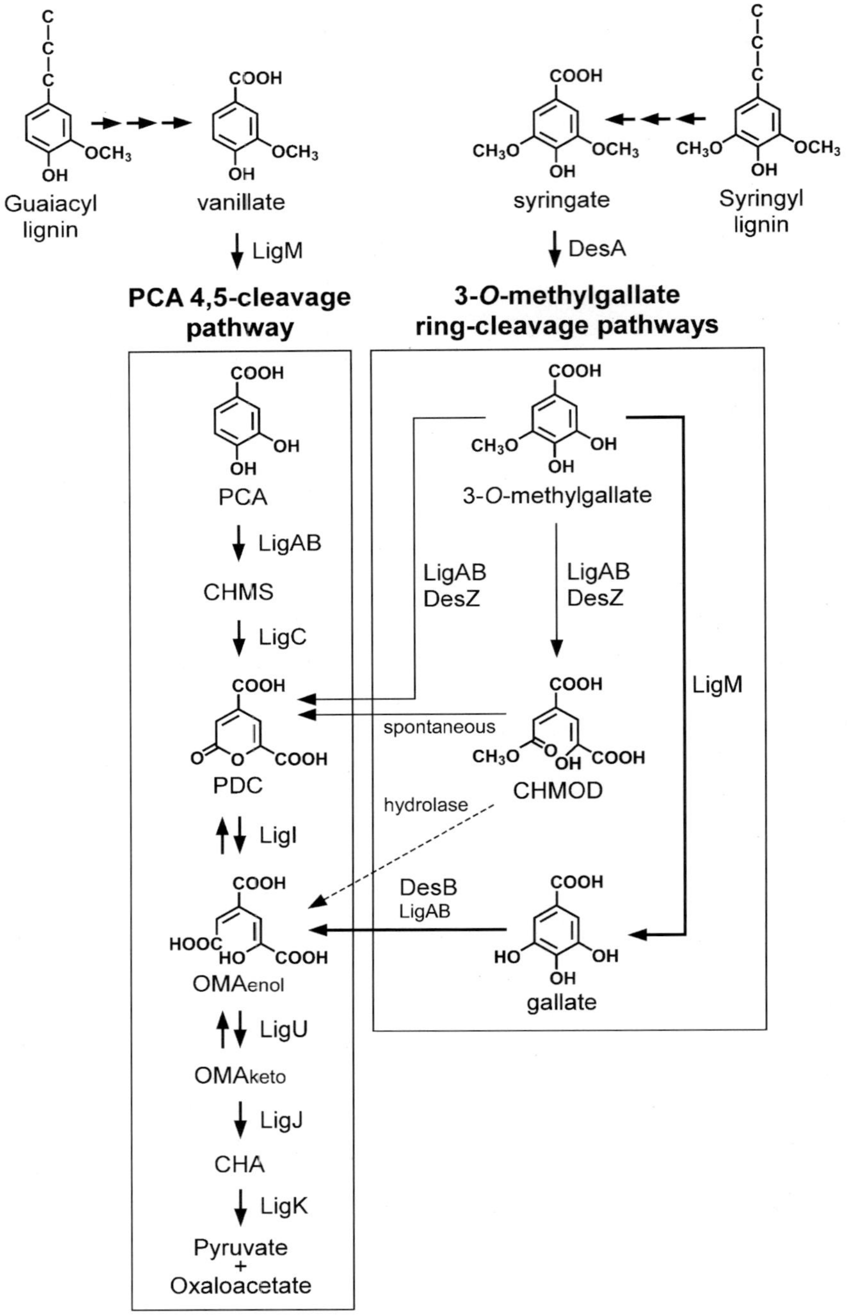

Fig. 10.6 Catabolic pathway for vanillate and syringate in *Sphingobium* sp. strain SYK-6. Enzyme: LigM, vanillate/3-*O*-methylgallate *O*-demethylase; LigA and LigB, α and β subunits of PCA 4,5-dioxygenase; LigC, CHMS dehydrogenase; LigI, PDC hydrolase; LigU, OMA tautomerase; LigJ, OMA hydratase; LigK, CHA aldolase/oxaloacetate decarboxylase; DesA, syringate *O*-demethylase; DesZ, 3-*O*-methylgallate 3,4-dioxygenase; DesB, gallate dioxygenase. Compound: CHMOD, 4-carboxy-2-hydroxy-6-methoxy-6-oxohexa-2,4-dienoate

References

Abe T, Masai E, Miyauchi K, Katayama Y, Fukuda M (2005) A tetrahydrofolate-dependent *O*-demethylase, LigM, is crucial for catabolism of vanillate and syringate in *Sphingomonas paucimobilis* SYK-6. J Bacteriol 187(6):2030–2037. doi:10.1128/JB.187.6.2030-2037.2005

Alhapel A, Darley DJ, Wagener N, Eckel E, Elsner N, Pierik AJ (2006) Molecular and functional analysis of nicotinate catabolism in *Eubacterium barkeri*. Proc Natl Acad Sci USA 103(33):12341–12346. doi:10.1073/pnas.0601635103

Arciero DM, Lipscomb JD (1986) Binding of ^{17}O-labeled substrate and inhibitors to protocatechuate 4,5-dioxygenase-nitrosyl complex. Evidence for direct substrate binding to the active site Fe^{2+} of extradiol dioxygenases. J Biol Chem 261(5):2170–2178

Arciero DM, Lipscomb JD, Huynh BH, Kent TA, Münck E (1983) EPR and Mössbauer studies of protocatechuate 4,5-dioxygenase. Characterization of a new Fe^{2+} environment. J Biol Chem 258(24):14981–14991

Arciero DM, Orville AM, Lipscomb JD (1985) [^{17}O]Water and nitric oxide binding by protocatechuate 4,5-dioxygenase and catechol 2,3-dioxygenase. Evidence for binding of exogenous ligands to the active site Fe^{2+} of extradiol dioxygenases. J Biol Chem 260(26):14035–14044

Arciero DM, Orville AM, Lipscomb JD (1990) Protocatechuate 4,5-dioxygenase from *Pseudomonas testosteroni*. Methods Enzymol 188:89–95

Chang M, Crawford IP (1990) The roles of indoleglycerol phosphate and the TrpI protein in the expression of *trpBA* from *Pseudomonas aeruginosa*. Nucleic Acids Res 18(4):979–988

Chang HK, Zylstra GJ (1998) Novel organization of the genes for phthalate degradation from *Burkholderia cepacia* DBO1. J Bacteriol 180(24):6529–6537

Chen YP, Lovell CR (1994) Purification and properties of a homodimeric protocatechuate 4,5-dioxygenase from *Rhizobium leguminosarum*. Arch Microbiol 161:191–195

Crawford RL (1975) Novel pathway for degradation of protocatechuic acid in *Bacillus* species. J Bacteriol 121(2):531–536

Crawford RL, Bromley JW, Perkins-Olson PE (1979) Catabolism of protocatechuate by *Bacillus macerans*. Appl Environ Microbiol 37(3):614–618

Dagley S, Evans WC, Ribbons DW (1960) New pathways in the oxidative metabolism of aromatic compounds by microorganisms. Nature 188:560–566

Dubbs P, Dubbs JM, Tabita FR (2004) Effector-mediated interaction of $CbbR_I$ and $CbbR_{II}$ regulators with target sequences in *Rhodobacter capsulatus*. J Bacteriol 186(23):8026–8035. doi:10.1128/JB.186.23.8026-8035.2004

Eaton RW (2001) Plasmid-encoded phthalate catabolic pathway in *Arthrobacter keyseri* 12B. J Bacteriol 183(12):3689–3703. doi:10.1128/JB.183.12.3689-3703.2001

Fetzner S (2012) Ring-cleaving dioxygenases with a cupin fold. Appl Environ Microbiol 78(8):2505–2514. doi:10.1128/AEM.07651-11

Gao J, Gussin GN (1991) Mutations in TrpI binding site II that differentially affect activation of the *trpBA* promoter of *Pseudomonas aeruginosa*. EMBO J 10(13):4137–4144

Garvey GS, Rocco CJ, Escalante-Semerena JC, Rayment I (2007) The three-dimensional crystal structure of the PrpF protein of *Shewanella oneidensis* complexed with *trans*-aconitate: insights into its biological function. Protein Sci 16(7):1274–1284. doi:10.1110/ps.072801907

Goyal AK, Zylstra GJ (1997) Genetics of naphthalene and phenanthrene degradation by *Comamonas testosteroni*. J Ind Microbiol Biotechnol 19(5–6):401–407

Hara H, Masai E, Katayama Y, Fukuda M (2000) The 4-oxalomesaconate hydratase gene, involved in the protocatechuate 4,5-cleavage pathway, is essential to vanillate and syringate degradation in *Sphingomonas paucimobilis* SYK-6. J Bacteriol 182(24):6950–6957

Hara H, Masai E, Miyauchi K, Katayama Y, Fukuda M (2003) Characterization of the 4-carboxy-4-hydroxy-2-oxoadipate aldolase gene and operon structure of the protocatechuate 4,5-cleavage pathway genes in *Sphingomonas paucimobilis* SYK-6. J Bacteriol 185(1):41–50

Harwood CS, Parales RE (1996) The β-ketoadipate pathway and the biology of self-identity. Annu Rev Microbiol 50:553–590. doi:10.1146/annurev.micro.50.1.553

Hobbs ME, Malashkevich V, Williams HJ, Xu C, Sauder JM, Burley SK, Almo SC, Raushel FM (2012) Structure and catalytic mechanism of LigI: insight into the amidohydrolase enzymes of cog3618 and lignin degradation. Biochemistry 51(16):3497–3507. doi:10.1021/bi300307b

Hofrichter M, Ullrich R, Pecyna MJ, Liers C, Lundell T (2010) New and classic families of secreted fungal heme peroxidases. Appl Microbiol Biotechnol 87(3):871–897. doi:10.1007/s00253-010-2633-0

Holm L, Sander C (1997) An evolutionary treasure: unification of a broad set of amidohydrolases related to urease. Proteins 28(1):72–82. doi:10.1002/(SICI)1097-0134(199705)28:1

Kamimura N, Aoyama T, Yoshida R, Takahashi K, Kasai D, Abe T, Mase K, Katayama Y, Fukuda M, Masai E (2010a) Characterization of the protocatechuate 4,5-cleavage pathway operon in *Comamonas* sp. strain E6 and discovery of a novel pathway gene. Appl Environ Microbiol 76(24):8093–8101. doi:10.1128/AEM.01863-10

Kamimura N, Takamura K, Hara H, Kasai D, Natsume R, Senda T, Katayama Y, Fukuda M, Masai E (2010b) Regulatory system of the protocatechuate 4,5-cleavage pathway genes essential for lignin downstream catabolism. J Bacteriol 192(13):3394–3405. doi:10.1128/JB.00215-10

Kasai D, Masai E, Miyauchi K, Katayama Y, Fukuda M (2004) Characterization of the 3-*O*-methylgallate dioxygenase gene and evidence of multiple 3-*O*-methylgallate catabolic pathways in *Sphingomonas paucimobilis* SYK-6. J Bacteriol 186(15):4951–4959. doi:10.1128/JB.186.15.4951-4959.2004

Kasai D, Masai E, Miyauchi K, Katayama Y, Fukuda M (2005) Characterization of the gallate dioxygenase gene: three distinct ring cleavage dioxygenases are involved in syringate degradation by *Sphingomonas paucimobilis* SYK-6. J Bacteriol 187(15):5067–5074. doi:10.1128/JB.187.15.5067-5074.2005

Kasai D, Masai E, Katayama Y, Fukuda M (2007) Degradation of 3-*O*-methylgallate in *Sphingomonas paucimobilis* SYK-6 by pathways involving protocatechuate 4,5-dioxygenase. FEMS Microbiol Lett 274(2):323–328. doi:10.1111/j.1574-6968.2007.00855.x

Kasai D, Fujinami T, Abe T, Mase K, Katayama Y, Fukuda M, Masai E (2009) Uncovering the protocatechuate 2,3-cleavage pathway genes. J Bacteriol 191(21):6758–6768. doi:10.1128/JB.00840-09

Kersten PJ, Dagley S, Whittaker JW, Arciero DM, Lipscomb JD (1982) 2-pyrone-4,6-dicarboxylic acid, a catabolite of gallic acids in *Pseudomonas* species. J Bacteriol 152(3):1154–1162

Li S, Kimura M, Takashima T, Hayashi K, Inoue K, Ishiguro R, Sugisaki H, Maruyama K (2007) Role of cysteine residues in 4-oxalomesaconate hydratase from *Pseudomonas ochraceae* NGJ1. Biosci Biotechnol Biochem 71(2):449–457. doi:JST.JSTAGE/bbb/60503 [pii]

Locher HH, Leisinger T, Cook AM (1989) Degradation of *p*-toluenesulphonic acid via sidechain oxidation, desulphonation and *meta* ring cleavage in *Pseudomonas* (*Comamonas*) *testosteroni* T-2. J Gen Microbiol 135(7):1969–1978

Ma YF, Zhang Y, Zhang JY, Chen DW, Zhu Y, Zheng H, Wang SY, Jiang CY, Zhao GP, Liu SJ (2009) The complete genome of *Comamonas testosteroni* reveals its genetic adaptations to changing environments. Appl Environ Microbiol 75(21):6812–6819. doi:10.1128/AEM.00933-09

MacLean AM, MacPherson G, Aneja P, Finan TM (2006) Characterization of the β-ketoadipate pathway in *Sinorhizobium meliloti*. Appl Environ Microbiol 72(8):5403–5413. doi:10.1128/AEM.00580-06

MacLean AM, Anstey MI, Finan TM (2008) Binding site determinants for the LysR-type transcriptional regulator PcaQ in the legume endosymbiont *Sinorhizobium meliloti*. J Bacteriol 190(4):1237–1246. doi:10.1128/JB.01456-07

Maddocks SE, Oyston PC (2008) Structure and function of the LysR-type transcriptional regulator (LTTR) family proteins. Microbiology 154(Pt 12):3609–3623. doi:10.1099/mic.0.2008/022772-0

Mampel J, Providenti MA, Cook AM (2005) Protocatechuate 4,5-dioxygenase from *Comamonas testosteroni* T-2: biochemical and molecular properties of a new subgroup within class III of extradiol dioxygenases. Arch Microbiol 183(2):130–139. doi:10.1007/s00203-004-0755-4

Martinez AT, Speranza M, Ruiz-Duenas FJ, Ferreira P, Camarero S, Guillen F, Martinez MJ, Gutierrez A, del Rio JC (2005) Biodegradation of lignocellulosics: microbial, chemical, and enzymatic aspects of the fungal attack of lignin. Int Microbiol 8(3):195–204

Maruyama K (1979) Isolation and identification of the reaction product of α-hydroxy-γ-carboxymuconic ε-semialdehyde dehydrogenase. J Biochem 86(6):1671–1677

Maruyama K (1983) Purification and properties of 2-pyrone-4,6-dicarboxylate hydrolase. J Biochem 93(2):557–565

Maruyama K (1985) Purification and properties of γ-oxalomesaconate hydratase from *Pseudomonas ochraceae* grown with phthalate. Biochem Biophys Res Commun 128(1):271–277. doi:0006-291X(85)91674-2 [pii]

Maruyama K (1990) Purification and properties of 4-hydroxy-4-methyl-2-oxoglutarate aldolase from *Pseudomonas ochraceae* grown on phthalate. J Biochem 108(2):327–333

Maruyama K, Ariga N, Tsuda M, Deguchi K (1978) Purification and properties of α-hydroxy-γ-carboxymuconic ε-semialdehyde dehydrogenase. J Biochem 83(4):1125–1134

Maruyama K, Shibayama T, Ichikawa A, Sakou Y, Yamada S, Sugisaki H (2004) Cloning and characterization of the genes encoding enzymes for the protocatechuate *meta*-degradation pathway of *Pseudomonas ochraceae* NGJ1. Biosci Biotechnol Biochem 68(7):1434–1441. doi:JST.JSTAGE/bbb/68.1434 [pii]

Masai E, Shinohara S, Hara H, Nishikawa S, Katayama Y, Fukuda M (1999) Genetic and biochemical characterization of a 2-pyrone-4,6-dicarboxylic acid hydrolase involved in the protocatechuate 4,5-cleavage pathway of *Sphingomonas paucimobilis* SYK-6. J Bacteriol 181(1):55–62

Masai E, Momose K, Hara H, Nishikawa S, Katayama Y, Fukuda M (2000) Genetic and biochemical characterization of 4-carboxy-2-hydroxymuconate-6-semialdehyde dehydrogenase and its role in the protocatechuate 4,5-cleavage pathway in *Sphingomonas paucimobilis* SYK-6. J Bacteriol 182(23):6651–6658. doi:10.1128/JB.182.23.6651-6658.2000

Masai E, Sasaki M, Minakawa Y, Abe T, Sonoki T, Miyauchi K, Katayama Y, Fukuda M (2004) A novel tetrahydrofolate-dependent *O*-demethylase gene is essential for growth of *Sphingomonas paucimobilis* SYK-6 with syringate. J Bacteriol 186(9):2757–2765. doi:10.1128/JB.186.9.2757-2765.2004

Masai E, Katayama Y, Fukuda M (2007) Genetic and biochemical investigations on bacterial catabolic pathways for lignin-derived aromatic compounds. Biosci Biotechnol Biochem 71(1):1–15. doi:JST.JSTAGE/bbb/60437 [pii]

Nakatsu CH, Wyndham RC (1993) Cloning and expression of the transposable chlorobenzoate-3,4-dioxygenase genes of *Alcaligenes* sp. strain BR60. Appl Environ Microbiol 59(11):3625–3633

Nakatsu CH, Providenti M, Wyndham RC (1997) The *cis*-diol dehydrogenase *cbaC* gene of Tn*5271* is required for growth on 3-chlorobenzoate but not 3,4-dichlorobenzoate. Gene 196(1–2):209–218. doi:S0378-1119(97)00229-1 [pii]

Ni B, Zhang Y, Chen DW, Wang BJ, Liu SJ (2013) Assimilation of aromatic compounds by *Comamonas testosteroni*: characterization and spreadability of protocatechuate 4,5-cleavage pathway in bacteria. Appl Microbiol Biotechnol. 97(13):6031–6041. doi:10.1007/s00253-012-4402-8

Noda Y, Nishikawa S, Shiozuka K, Kadokura H, Nakajima H, Yoda K, Katayama Y, Morohoshi N, Haraguchi T, Yamasaki M (1990) Molecular cloning of the protocatechuate 4,5-dioxygenase genes of *Pseudomonas paucimobilis*. J Bacteriol 172(5):2704–2709

Nogales J, Canales A, Jiménez-Barbero J, Serra B, Pingarrón JM, García JL, Díaz E (2011) Unravelling the gallic acid degradation pathway in bacteria: the *gal* cluster from *Pseudomonas putida*. Mol Microbiol 79(2):359–374. doi:10.1111/j.1365-2958.2010.07448.x

Ogawa N, McFall SM, Klem TJ, Miyashita K, Chakrabarty AM (1999) Transcriptional activation of the chlorocatechol degradative genes of *Ralstonia eutropha* NH9. J Bacteriol 181(21):6697–6705

Parke D (1996) Characterization of PcaQ, a LysR-type transcriptional activator required for catabolism of phenolic compounds, from *Agrobacterium tumefaciens*. J Bacteriol 178(1):266–272

Piñeiro S, Olekhnovich I, Gussin GN (1997) DNA bending by the TrpI protein of *Pseudomonas aeruginosa*. J Bacteriol 179(17):5407–5413

Porrúa O, García-Jaramillo M, Santero E, Govantes F (2007) The LysR-type regulator AtzR binding site: DNA sequences involved in activation, repression and cyanuric acid-dependent repositioning. Mol Microbiol 66(2):410–427. doi:10.1111/j.1365-2958.2007.05927.x

Providenti MA, Mampel J, MacSween S, Cook AM, Wyndham RC (2001) *Comamonas testosteroni* BR6020 possesses a single genetic locus for extradiol cleavage of protocatechuate. Microbiology 147(Pt 8):2157–2167

Sasoh M, Masai E, Ishibashi S, Hara H, Kamimura N, Miyauchi K, Fukuda M (2006) Characterization of the terephthalate degradation genes of *Comamonas* sp. strain E6. Appl Environ Microbiol 72(3):1825–1832. doi:10.1128/AEM.72.3.1825-1832.2006

Schell MA (1993) Molecular biology of the LysR family of transcriptional regulators. Annu Rev Microbiol 47:597–626. doi:10.1146/annurev.mi.47.100193.003121

Seibert CM, Raushel FM (2005) Structural and catalytic diversity within the amidohydrolase superfamily. Biochemistry 44(17):6383–6391. doi:10.1021/bi047326v

Seo JS, Keum YS, Li QX (2009) Bacterial degradation of aromatic compounds. Int J Environ Res Public Health 6(1):278–309. doi:10.3390/ijerph6010278

Slonczewski JL, Fujisawa M, Dopson M, Krulwich TA (2009) Cytoplasmic pH measurement and homeostasis in bacteria and archaea. Adv Microb Physiol 55:1–79, 317. doi:10.1016/S0065-2911(09)05501-5

Sugimoto K, Senda T, Aoshima H, Masai E, Fukuda M, Mitsui Y (1999) Crystal structure of an aromatic ring opening dioxygenase LigAB, a protocatechuate 4,5-dioxygenase, under aerobic conditions. Structure 7(8):953–965. doi:st7819 [pii]

Tack BF, Chapman PJ, Dagley S (1972) Metabolism of gallic acid and syringic acid by *Pseudomonas putida*. J Biol Chem 247(20):6438–6443

Toledano MB, Kullik I, Trinh F, Baird PT, Schneider TD, Storz G (1994) Redox-dependent shift of OxyR-DNA contacts along an extended DNA-binding site: a mechanism for differential promoter selection. Cell 78(5):897–909. doi:S0092-8674(94)90702-1 [pii]

Tropel D, van der Meer JR (2004) Bacterial transcriptional regulators for degradation pathways of aromatic compounds. Microbiol Mol Biol Rev 68(3):474–500. doi:10.1128/MMBR.68.3.474-500.2004

Vaillancourt FH, Bolin JT, Eltis LD (2006) The ins and outs of ring-cleaving dioxygenases. Crit Rev Biochem Mol Biol 41(4):241–267. doi:10.1080/10409230600817422

Vicuña R (1988) Bacterial degradation of lignin. Enzyme Microbiol Technol 10(11):646–655

Wang L, Winans SC (1995a) High angle and ligand-induced low angle DNA bends incited by OccR lie in the same plane with OccR bound to the interior angle. J Mol Biol 253(1):32–38. doi:10.1006/jmbi.1995.0533

Wang L, Winans SC (1995b) The sixty nucleotide OccR operator contains a subsite essential and sufficient for OccR binding and a second subsite required for ligand-responsive DNA bending. J Mol Biol 253(5):691–702. doi:10.1006/jmbi.1995.0583

Wang L, Helmann JD, Winans SC (1992) The *A. tumefaciens* transcriptional activator OccR causes a bend at a target promoter, which is partially relaxed by a plant tumor metabolite. Cell 69(4):659–667. doi:0092-8674(92)90229-6 [pii]

Wang YZ, Zhou Y, Zylstra GJ (1995) Molecular analysis of isophthalate and terephthalate degradation by *Comamonas testosteroni* YZW-D. Environ Health Perspect 103(Suppl 5):9–12

Wang W, Mazurkewich S, Kimber MS, Seah SY (2010) Structural and kinetic characterization of 4-hydroxy-4-methyl-2-oxoglutarate/4-carboxy-4-hydroxy-2-oxoadipate aldolase, a protocatechuate degradation enzyme evolutionarily convergent with the HpaI and DmpG pyruvate aldolases. J Biol Chem 285(47):36608–36615. doi:10.1074/jbc.M110.159509

Wattiau P, Bastiaens L, van Herwijnen R, Daal L, Parsons JR, Renard ME, Springael D, Cornelis GR (2001) Fluorene degradation by *Sphingomonas* sp. LB126 proceeds through protocatechuic acid: a genetic analysis. Res Microbiol 152(10):861–872. doi:10.1016/S0923-2508(01)01269-4

Zimmermann W (1990) Degradation of lignin by bacteria. J Biotechnol 13(2–3):119–130. doi:10.1016/0168-1656(90)90098-V

Chapter 11
Toluene Tolerance Systems in *Pseudomonas*

Ana Segura and Juan Luis Ramos

Abstract Organic solvents such as monoaromatic hydrocarbons are highly toxic for living cells; however, most *Pseudomonas putida* strains are able to grow in the presence of low concentrations of these compounds. The cellular response toward the presence of organic solvents in the culture media is multifactorial. An early and quick response toward organic solvents is a decreased in the fluidity of the cell membrane via *cis–trans* isomerization of unsaturated fatty acids. This is followed by transcriptional reprogramming, leading to induction of stress-related functions and metabolic pathways to increase energy production. This energy is used by the microbe to extrude toxic compounds to the outer medium via efflux pumps of the RND (resistance, nodulation, and cell division) family—one of the most effective mechanisms for solvent tolerance. All of these processes concur in response to the presence of organic solvents in the majority of *P. putida* strains studied so far. There are a number of *P. putida* strains that have been described as highly solvent tolerant; for these strains, the action of one efflux pump (named TtgGHI in *P. putida* strain DOT-T1E) is required for full tolerance to a second phase of the organic solvent.

Keywords *cis–trans* isomerization • Efflux pumps • Membrane fluidity • *Pseudomonas* • Regulation of efflux pumps • Solvent tolerance • Stress responses

11.1 Introduction

Over the last century high amounts of toxic chemicals have accumulated in the biosphere due to industrial activities. Organic solvents and fuels enter the environment through accidents, solvent evaporation, or loss during storage or production.

A. Segura (✉) • J.L. Ramos
Environmental Protection Department, Estación Experimental del Zaidín (CSIC), Profesor Albareda 1, 18008 Granada, Spain
e-mail: ana.segura@eez.csic.es

H. Nojiri et al. (eds.), *Biodegradative Bacteria: How Bacteria Degrade, Survive, Adapt, and Evolve*, DOI 10.1007/978-4-431-54520-0_11, © Springer Japan 2014

The accumulation of these pollutants represents a serious hazard for plants, animals, and humans, and therefore there is the urgent need to decontaminate affected areas and to prevent further contamination. The use of microorganisms to bioremediate contaminated sites or to prevent the emission of these chemicals to the atmosphere is an environmentally friendly approach that is often limited due to the toxicity of the compounds. Thus, the utilization of robust microorganisms, able to withstand high concentrations of organic solvents, has become a very attractive strategy to fight pollution by toxic chemicals.

Organic solvents are highly toxic for microorganisms because they accumulate in cell membranes, disorganizing them and provoking the loss of ions and metabolites and the malfunction of several membrane-associated proteins that ultimately lead to cell death (Sikkema et al. 1995). Toxicity of an organic solvent depends on the physicochemical properties of the solvent (mainly determined by the $logP_{ow}$ [logarithm of the partition coefficient of the organic solvent in a defined mixture of octanol and water]) (Inoue and Horikoshi 1991; Sikkema et al. 1995) and the intrinsic tolerance of the bacteria (Ramos et al. 2002). Compounds with a $logP_{ow}$ between 1.5 and 4.0 are extremely toxic for microorganisms. Toluene, with a $logP_{ow}$ of 2.69 has been used as model compound in studies carried out to decipher the molecular mechanisms of solvent tolerance in *Pseudomonas,* although the effect of butanol and ethanol especially in *Escherichia coli* strains has become the focus of intense study due to the use of these alcohols in the bioenergy industry (Segura et al. 2011).

In this chapter we will focus on toluene tolerance in *P. putida* because this model has been widely used to unveil the mechanisms of solvent tolerance. Despite the high toxicity of toluene, most *Pseudomonas* are able to grow in its presence when it is added at a concentration below 0.53 g/L (5.75 mM); however, at higher concentrations (above its solubility limit in water), toluene is highly toxic. The presence of low concentrations of toluene in culture media triggers several similar molecular responses independently of the intrinsic solvent tolerance of the strain. Although not all *Pseudomonas* strains have been studied at the same level of detail, most of them respond in a similar way to the presence of toluene in the media. These responses include *cis–trans* isomerization of the unsaturated fatty acids; changes in the saturated to unsaturated fatty acids; changes in phospholipid headgroups; stress-related responses; increased energy demand; and extrusion of toluene via efflux pumps.

In those bacteria able to withstand a double phase of toluene, which have been classified as solvent tolerant (Segura et al. 2003), in addition to the abovementioned mechanisms, the action of one efflux pump has been found to be the main determinant that confers the solvent tolerant phenotype. This efflux pump is known as TtgGHI in *P. putida* DOT-T1E and *P. putida* MTB6 (Rojas et al. 2001; Segura et al. 2003) and SrpABC in *P. putida* S12 (Kieboom et al. 1998) and *P. putida* GM73 (Kim et al. 1998). In these solvent-tolerant strains, tolerance is an inducible process that can be enhanced by pre-incubation of the cells with low concentrations of toluene (Fig. 11.1).

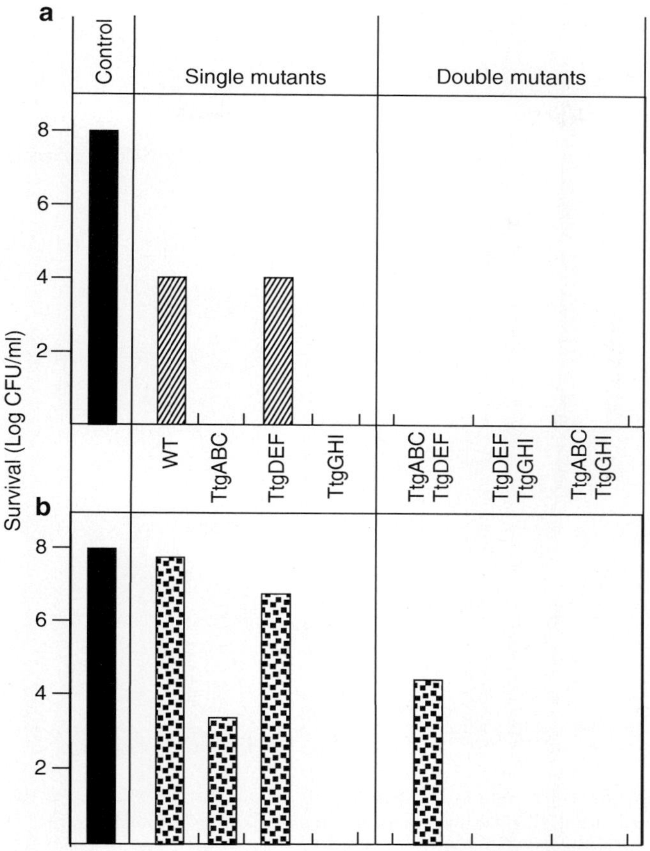

Fig. 11.1 Survival after 0.3 % v/v toluene shock. Cultures of *P. putida* DOT-T1E and isogenic mutants were grown until exponential phase ($O.D._{660nm} \approx 0.8$) in LB (**a**) or LB plus toluene in the gas phase (**b**) and then 0.3 % (v/v) toluene was added to the cultures. No toluene was added to the control culture. Serial dilutions were spread on LB agar plates 10 min after toluene addition and number of viable cells was counted

11.2 General Responses to Low Concentrations of Toluene

11.2.1 Membrane Changes

The main target of the toxicity of toluene is the cellular membrane and there are several mechanisms through which *P. putida* modifies lipid and fatty acid composition in response to the solvent. Most of these mechanisms contribute to a decrease in membrane fluidity in an attempt to counteract the fluidification effect of toluene. Changes in the state of membrane fluidity can be determined measuring fluorescence polarization of a DPH (1,6-diphenyl-1,3,5-hexatriene) probe inserted into the lipid bilayer (Trevors 2003).

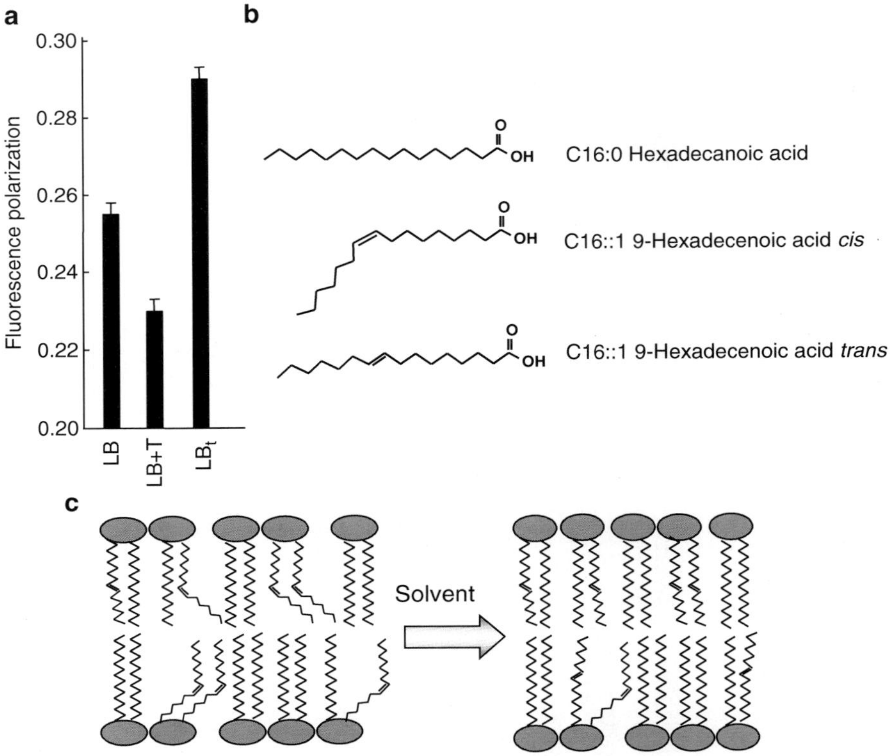

Fig. 11.2 Effect of toluene on the cell membranes. (**a**) Fluorescent polarization values were measured as in Bernal et al. (2007a), with cells grown on either LB (LB); LB with 0.1 % (v/v) toluene added shortly before the DPH probe reading (LB + T); and LB plus 3 mM toluene (LB_t). (**b**) Schematic representation of a saturated fatty acid, a *cis*-unsaturated fatty acid and *trans*-unsaturated fatty acid. (**c**) Schematic representation of the effect of organic solvent on fatty acids within the cell membrane

The sudden addition of a second phase (0.3 % [v/v]) of toluene to a *P. putida* DOT-T1E culture provokes membrane fluidification as detected by a decreased in fluorescence polarization (Bernal et al. 2007a); however incubation of the culture in the presence of 3 mM of toluene provokes an increase in fluorescence polarization when compared with the culture without toluene (Fig. 11.2a). This indicates that in the presence of low concentrations of toluene the bacterial membranes became more rigid than in its absence. Similar effects have been observed in other *Pseudomonas* in response to different organic solvents (Kim et al. 2002).

11.2.1.1 Changes in Fatty Acid Composition

One mechanism that contributes to decreased membrane fluidity is the conversion of the *cis*-unsaturated fatty acids (*cis*-UFAs) to *trans*-UFAs (Heipieper and de Bont 1994; Ramos et al. 1997). The *cis* conformation of the double bond provokes a bending in the acyl chain (Fig. 11.2b) while *trans*-UFAs have long linear structures that are similar to those of saturated fatty acids. This linear conformation allows a tightly packing of the acyl chains in the membranes (Fig. 11.2c) (Seelig and Waespe-Šarcevic 1978; McDonald et al. 1985). Thus, the *cis* to *trans* isomerization provokes an increase in the ordered package of the acyl chains in the membrane and contributes to membrane impermeabilization.

UFAs in *P. putida* are mainly synthesized as *cis*-isomers and under normal laboratory conditions at exponential phase of growth they represent around 30–40 % of the total fatty acids, depending on the strain (with 0.3–5 % of *trans*-UFAs) (Weber et al. 1994; Pini et al. 2009; Rühl et al. 2012). The *cis–trans* isomerization is carried out by a *cis–trans* isomerase that is constitutively expressed (Junker and Ramos 1999; Heipieper et al. 1994; Heipieper et al. 1996; Bernal et al. 2007b) although it is only active in the presence of membrane fluidification agents such as heat or organic solvents (Heipieper and de Bont 1994; Heipieper et al. 1996; Junker and Ramos 1999). As the enzyme is always present in the cell and the modification does not require "de novo" synthesis of the unsaturated fatty acids (Heipieper et al. 1992), the increase in the *trans*-UFAs is a very fast response that occurs within minutes after toluene addition (Ramos et al. 1997).

Although *cis–trans* isomerization is a quick response that contributes to the protective response against low concentrations of toluene, two lines of evidence point to the nonessential role of the *cis–trans* isomerase in solvent tolerance. First, most *P. putida* strains studied so far are able to produce this isomerization independently of their solvent-tolerant or solvent-sensitive character (Heipieper et al. 2003; Rühl et al. 2012), and second a *P. putida* DOT-T1E mutant in the *cis–trans* isomerase is as tolerant as the wild type to the sudden addition of 0.3 % (v/v) toluene (Junker and Ramos 1999). It is quite possible that *cis–trans* isomerization acts as an emergency mechanism that provides the cell time to synthesize other active mechanisms of response.

Denser membranes can also be obtained by increasing the proportion of saturated fatty acids. In some *P. putida* strains an increase in the saturated-to-unsaturated fatty acids ratio has been observed after organic solvent exposure (Pinkart et al. 1996; Loffhagen et al. 2001). This change requires the de novo synthesis and it is therefore considered a long-term adaptation.

Cyclopropane fatty acids (CFAs) are mainly synthesized during the stationary phase of growth (Chang and Cronan 1999; Pini et al. 2009, 2011). Some authors have suggested that CFAs increase membrane stability (McDonald et al. 1985), but their role in membrane fluidity is not fully understood. In fact, membranes of *P. putida* KT2440 showed higher steady-state fluorescence polarization values of the DPH probe at stationary phase than at exponential phase of growth (Muñoz-Rojas et al. 2006), suggesting that at this stage membranes are more rigid. However, a *P. putida* DOT-T1E mutant in the cyclopropane synthase (which is unable to synthesize CFAs) has similar polarization values than the wild-type strain at stationary phase of growth (Pini et al. 2009). This mutant has a slightly reduced tolerance toward sudden exposure to a double phase of toluene than the wild type (Pini et al. 2009), suggesting only a minor role for CFAs in solvent tolerance. Furthermore, the cyclopropane synthase gene expression or the relative amount of CFAs in the membrane does not change in response to toluene (Pini et al. 2011).

11.2.1.2 Changes in Phospholipid Headgroups

The main phospholipids in *P. putida* membranes are phosphatidylethanolamine (PE), phosphatidylglycerol (PG) and diphosphatidyl-glycerol, or cardiolipin (CL). Identification of the changes in phospholipid headgroups after organic solvent addition suggests that different *P. putida* strains developed different strategies for changing the composition of phospholipid headgroups to overcome damage produced by organic solvents (Fang et al. 2000; Rühl et al. 2012). In *P. putida* Idaho the incorporation of ^{32}P after *o*-xylene was higher in PE, suggesting a higher synthesis of this phospholipid headgroup in response to the organic solvent (Pinkart and White 1997). PE has higher melting point than PG and therefore a higher PE content may result in higher membrane stability. In *P. putida* DOT-T1E incorporation of ^{32}P after toluene addition was higher in CL (Ramos et al. 1997). Several studies have suggested that CL is an important structural determinant of the membrane architecture (Mileykovskaya et al. 2003; Kawai et al. 2004). A *P. putida* DOT-T1E *cls* (cardiolipin synthase) knockout mutant (which contained lower levels of CL) is slightly more sensitive to the sudden addition of toluene than the wild-type strain, but its role in solvent tolerance seems to be more related with the correct insertion or stability of the efflux pumps components in the membrane than with an increase in membrane stability (Bernal et al. 2007a).

11.2.1.3 Vesicle Formation

The formation of vesicles in response to organic solvents has been observed in *P. putida* IH-2000 (Kobayashi et al. 2000) and *P. putida* DOT-T1E (Ramos et al. 1995; Baumgarten et al. 2012). In both strains, vesicles consisted of phospholipids, lipopolysaccharides (LPS), and proteins. In *P. putida* IH-2000 the vesicles also contained toluene and vesicle formation has been described as a mechanism to eliminate toluene from the cellular membrane. In *P. putida* DOT-T1E the release of vesicles produces changes in cell hydrophobicity mainly as a consequence of the removal of certain fractions of LPS in the vesicles (Baumgarten et al. 2012).

11.2.2 Stress-Related Responses

Although no differences in growth rates are observed in the solvent-tolerant *P. putida* DOT-T1E (Fig. 11.3) or solvent-sensitive *P. putida* KT2440 in the presence and absence of low concentrations of toluene, the cellular responses identified by proteomic and transcriptomic assays revealed that these cells are undergoing a stress response (Segura et al. 2005; Domínguez-Cuevas et al. 2006). Similar stress-related responses have been described in the solvent-tolerant *P. putida* S12 growing in a chemostat in the presence of toluene (Volkers et al. 2006) and in other *P. putida* and *E. coli* strains using different organic solvents (Roma-Rodrigues et al 2010; Brynildsen and Liao 2009). In these studies several stress-related proteins/genes have been identified that are induced in response to organic solvents suggesting that processes such us oxidative damage or protein

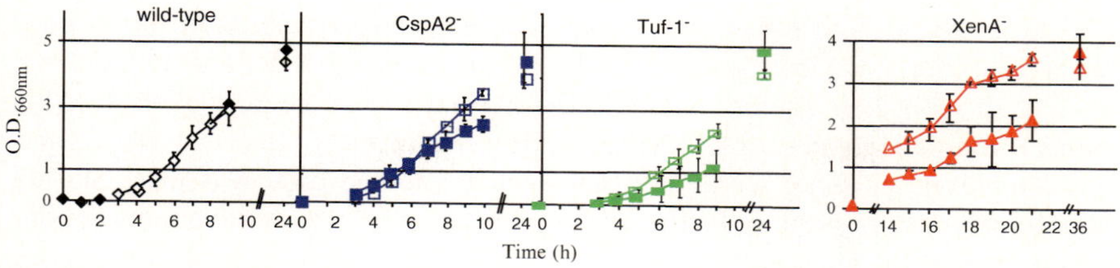

Fig. 11.3 Role of different stress-related proteins in solvent tolerance in *P. putida* DOT-T1E. Growth of the wild-type and mutant strains on M9 minimal medium plus glucose as the carbon source in the absence (*open symbols*) and in the presence of toluene in the gas phase (*solid symbols*)

misfolding may be occurring under these mild conditions. Oxidative damage is probably a consequence of the membrane disorganization and the resulting disruption of the electron transport systems provoked by toluene. Misfolding of proteins due to the presence of toluene in the cytoplasm and the periplasm is also a potential signal for the induction of chaperone genes such us *groES, groL,* or *dnaK.*

Studies have shown that mutants in several stress-related genes in *P. putida* DOT-T1E exhibit retarded growth in the presence of low concentrations of toluene (Fig. 11.3), implicating these genes as important to the solvent response in the presence of low concentrations of toluene (Fig. 11.3).

11.2.3 Energy Requirements

Culture yield in the presence of organic solvents is lower than in its absence, suggesting that energy requirements are higher in the presence of the toxic compound (Ramos et al. 1995; Isken et al. 1999; Neumann et al. 2006). This was confirmed by proteomic analysis, which revealed that even in the presence of low concentrations of toluene, several TCA cycle genes were induced (Segura et al. 2005; Volkers et al. 2006). Furthermore, a number of strains induce the expression of genes that allow metabolism of the toxic compound as way to both remove the compound and gain energy (Mosqueda et al 1999; Segura et al. 2005).

The increased energy demands due to exposure to organic solvents are due both the toxic effects (e.g., increased membrane permeability and loss of ions and proteins) and the adaptive mechanisms by which cells respond to toluene, especially those that depend upon RND efflux pumps (Heipieper et al. 2009).

11.2.4 Efflux Pumps

Extrusion of the toxic compound from the cell is one of the main mechanisms of solvent tolerance (Ramos et al. 2002). This extrusion is carried out mainly by efflux pumps of the RND family (Paulsen et al. 2001; Saier and Paulsen 2001).

The structure of these efflux pumps has been extensively studied in recent years because of their contribution to the multidrug resistance (MDR) phenotype that is associated with their activity, and excellent reviews have been published on the subject (Blair and Piddock 2009; Nikaido 2011). RND transporters consist of tripartite complexes, an inner membrane, an outer membrane protein (OMP), and the periplasmic adaptor protein (PAP). All three elements form a multicomponent complex traversing the inner and the outer membrane. These efflux pumps are energized by proton motive force; it was therefore not surprising that *P. putida* DOT-T1E mutants in *exbBexbDtonB* (system involved in the energy transduction) showed hypersensitivity toward toluene (Godoy et al. 2004).

Most of the efflux pumps described so far have a broad substrate range. Interestingly, although there is no structural similarity among the different substrates and the charge states are also different, they all are relatively lipophilic or contain lipophilic domains (Nikaido 1996).

The best studied efflux pumps of the RND family are the AcrAB efflux pump in *E. coli*—described as able to confer resistance to acriflavine (Ma et al. 1993) and also involved in solvent tolerance (Aono et al. 1998)—and the MexAB-OprM efflux pump of *P. aeruginosa* that confers resistance to fluoroquinolones, chloramphenicol, tetracycline, novobiocin and several β-lactam antibiotics (Li et al. 1995), and organic solvents (Li et al. 1998). The crystal structure of the AcrB (Murakami et al. 2002; Nakashima et al. 2011) and MexB (Sennhauser et al. 2009) efflux pumps, the TolC (Koronakis et al. 2000) and OprM (Akama et al 2004a) outer membrane proteins, and those of AcrA (Mikolosko et al. 2006) and MexA (Akama et al 2004b; Higgins et al 2004) periplasmic adaptor proteins have been solved, allowing the construction of a model of the tripartite efflux pump. AcrB and MexB have twelve transmembrane domains and two large periplasmic loops. Three monomers of the efflux pump are arranged in the final structure, forming a structure with a large periplasmic domain whose upper part forms a funnel whose dimensions would be expected to allow direct contact with the outer membrane protein. The funnel consists of a large cavity (35 Å of diameter) which accommodates the substrates. Substrates can be taken up from the lipid bilayer through an opening between protomers (called vestibules) that lead to the central cavity (Murakami et al. 2002; Yu et al. 2003). The outer membrane protein forms also a trimeric structure that is anchored in the outer membrane and forms a long (100 Å) periplasmic tunnel. The PAP, with an elongated structure, is anchored to the inner membrane and interacts with the efflux pump and the OMP (Husain et al 2004; Touzé et al. 2004; Tamura et al. 2005; Tikhonova et al. 2011). Thus the efflux complex traverses the inner membrane, the periplasmic space, and the outer membrane, such that substrates can be removed from either the cytoplasm or most probably the periplasm (Elkins and Nikaido; 2003), captured in the central cavity and transported directly through the channel to the external medium (Murakami et al. 2002).

Most *P. putida* strains sequenced so far contain several RND efflux pumps in their genomes (Godoy et al. 2010). In the solvent-tolerant strain *P. putida* DOT-T1E three efflux pumps (TtgABC, TtgDEF, and TtgGHI) contribute to solvent tolerance. Of these three efflux pumps, homologs of TtgABC are present in all

P. putida strains sequenced so far regardless of their solvent-tolerant capacities, while TtgDEF homologs are present in all *P. putida* strains that are able to degrade toluene via the toluene dioxygenase (TOD) pathway (Segura et al. 2003). TtgGHI (named SrpABC in *P. putida* S12) is only present in those strains with the highest solvent tolerance (see below).

TtgABC (encoded by the *ttgABC* operon in *P. putida* DOT-T1E) and its homologs ArpABC (encoded by the *arpABC* operon in *P. putida* S12) and MepABC (of a toluene-tolerant mutant derivative of *P. putida* KT2442) are mainly involved in antibiotic resistance. Mutants in these efflux pumps are more sensitive to a range of antibiotics than the corresponding wild-type strains (Ramos et al. 1998; Fukumori et al. 1998; Kieboom and de Bont 2001). Despite high sequence homology between the TtgABC, MepABC, and ArpABC (≈99 %) there are significant functional differences among them; while *ttgB* and *mepB* mutants were more sensitive to toluene, the *arpB* mutant was as tolerant to several organic solvents as the wild-type strain. Furthermore, the *ttgA* promoter was induced by several antibiotics (e.g., chloramphenicol and tetracycline) but not by organic solvents (Terán et al. 2003), while the *arpA* promoter was not induced by chloramphenicol, carbenicillin, or erythromycin (antibiotics that are transported by this efflux pump) or by organic solvents.

The fact that this type of efflux pump is present in solvent-tolerant as well as in solvent-sensitive strains and that it is not induced by solvents suggests that this efflux pump is involved in the innate response of the cells toward the presence of low concentrations of organic solvents.

Unlike the *ttgABC* operon, the *ttgDEF* operon (named *sepABC* in *P. putida* F1) is induced in response to solvents in *P. putida* DOT-T1E and *P. putida* F1 (Mosqueda and Ramos 2000; Phoenix et al. 2003). A *P. putida* DOT-T1E mutant in *ttgD* showed a lower survival rate than the wild type after toluene shock (Fig. 11.1) but only when the cells were first induced by low concentrations of toluene; similar results were obtained by Phoenix et al. (2003) using *P. putida* F1 mutants in *sepB* and *sepC*. These data suggest that this type of efflux pump is involved in the inducible toluene response. Furthermore, the *ttgDEF* operon is considered to have a lateral role in solvent resistance because *P. putida* strains that encode it can be both solvent tolerant and solvent sensitive. Indeed, a *P. putida* DOT-T1E double mutant in which the TtgABC and TtgDEF efflux pumps were inactivated was still able to survive 0.3 % (v/v) toluene (Fig. 11.1) when it was pre-induced with low concentrations of toluene, suggesting that other solvent-tolerance mechanisms in this strain were sufficient for survival (Mosqueda and Ramos 2000).

11.2.4.1 Regulation of *ttgABC* and *ttgDEF*

Located upstream the efflux pump operons *ttgABC* and *ttgDEF* and divergently transcribed, are the genes that encode the corresponding regulatory proteins (named TtgR and TtgT, respectively).

ttgABC Regulation

TtgR (ArpR in *P. putida* S12 and MepR in *P. putida* KT2442TOL) is a repressor of the TtgABC efflux pump that belongs to the TetR family. While TtgR is able to recognize several antibiotics and plant secondary metabolites, it is unable to recognize organic solvents that are substrates of the efflux pump (Terán et al. 2003, 2006; Alguel et al. 2007). A TtgR mutant led to increased expression of the *ttgABC* operon and consequently showed increased antibiotic resistance (Duque et al. 2001). This mutant was as resistant to toluene as the wild type, indicating that overexpression of the TtgABC efflux pump does not provoke an increase in toluene tolerance.

TtgR binds to a pseudo-palindromic site that overlaps with the *ttgR-ttgA* promoters (Terán et al 2003). Two dimers of TtgR bind to the minimal 30 mers DNA TtgR operator. The current model for regulation is that each of the TtgR dimers binds to an 18-bp palindrome (Krell et al. 2007). In the presence of effectors, the repressor protein is released from the operator DNA allowing the binding of the RNA polymerase and the start of transcription. The basal level of expression of the regulator and efflux pump allows for rapid extrusion of antimicrobials, and the high affinity of the TtgR protein for its effectors guarantees responsiveness to toxic chemicals.

ttgDEF Regulation

TtgT and SepR are the regulatory proteins of the *ttgDEF* (in *P. putida* DOT-T1E) and *sepABC* (in *P. putida* F1) operons, respectively. TtgT and SepR belong to the IclR family of regulatory proteins, and they act as repressors of the corresponding efflux pumps. As with other regulatory proteins of efflux pumps SepR and TtgT recognize different substrates (toluene, styrene, *m*-xylene, among others). β-Galactosidase assays suggest that the basal level of expression of the efflux pump operon in both strains is quite low and mutants in the regulatory protein show a twofold increase in the level of efflux pump expression (Phoenix et al. 2003; Terán et al. 2007).

A SepR mutant has a shorter lag phase when growing in toluene than the wild-type strain and cultures of the mutant strain showed higher survival rate when exposed to toluene shock (Phoenix et al. 2003). These effects were not observed in a *P. putida* DOT-T1E TtgT knockout mutant (Terán et al. 2007) because there is another efflux pump, TtgGHI, that masks the physiological role of this efflux pump and also because TtgV, the regulator of the *ttgGHI* operon, also regulates *ttgDEF* expression (discussed below).

11.3 Mechanisms of Solvent Resistance in the Solvent-Tolerant Strains

Only several *P. putida* strains (*P. putida* DOT-T1E, *P. putida* S12, *P. putida* MTB6, and *P. putida* GM73) have been classified as solvent tolerant, which means that these strains are able to grow in the presence of a second phase of organic solvent

(Ramos et al. 1995; Isken and de Bont 1996; Huertas et al 2000; Kim et al 1998). In all these strains a TtgGHI (SrpABC) efflux pump is present, which belongs to the RND family of pumps (along with TtgABC and TtgDEF).

A *P. putida* DOT-T1E mutant deficient in TtgGHI and a *P. putida* S12 mutant lacking the SrpABC efflux pump were unable to thrive in the presence of high concentrations of toluene (Rojas et al. 2001; Kieboom et al. 1998). Furthermore, this strain was sensitive to toluene even when pre-induced with low concentrations of toluene (Fig. 11.1). Considering these results, and because the *ttgGHI* operon is constitutively expressed under normal laboratory conditions and induced in the presence of toluene (and other organic solvents), this efflux pump participates both in the innate and the inducible solvent tolerance.

In *P. putida* DOT-T1E and MTB6 strains, *ttgGHI* genes are located on a megaplasmid named pGRT1 (Rodriguez-Hervá et al 2007). Sequencing of pGRT1 (Molina et al. 2011) revealed that this plasmid is conjugative and that the efflux pump operon is located within a Tn*4653*-like transposon. As such, both plasmid and transposon can be lost at a very low frequency ($<10^{-8}$ per cell per generation) leading to a solvent-sensitive phenotype. Transfer of this plasmid to the solvent-sensitive *P. putida* KT2440 strain results in acquisition of the solvent-tolerant phenotype (Rodriguez-Hervá et al. 2007). These data confirm the hypothesis that the TtgGHI efflux pump is the main determinant for solvent tolerance in DOT-T1E strain.

Contrary to what has been described for other environmental megaplasmids, pGRT1 does not contain any catabolic genes. Instead, many plasmid functions are related to stress resistance; i.e., it contains three genes involved in UV resistance, an ORF that encodes a universal stress protein (UspA), and a transporter involved in iron capture (Molina et al. 2011). Interestingly, this plasmid also encodes two methyl-accepting chemotaxis proteins that are involved in the hyperchemotaxis of *P. putida* DOT-T1E toward toluene (Lacal et al. 2011).

11.3.1 Regulation of *ttgGHI* Efflux Pump

Upstream of the efflux pump operon, two open reading frames encoding putative regulators were found. These putative regulatory proteins were named TtgV and TtgW in *P. putida* DOT-T1E (Rojas et al. 2003), and the homologous proteins were named SrpS and SrpR in *P. putida* S12 (Wery et al. 2001). TtgV and SrpS belong to the IclR family of transcriptional regulators and act as repressors of the corresponding efflux pumps. Under non-induced conditions TtgV binds to the promoter region preventing binding of the RNA polymerase, while in the presence of an inducer the regulatory protein dissociated from its target operator allowing RNA polymerase binding (Guazzaroni et al. 2004). As expected, mutants in TtgV and SrpS are more resistant to toluene than the corresponding wild-type strains (Rojas et al. 2003; Wery et al. 2001).

Despite the high sequence identity between *ttgW* and *srpR* (96 %), SrpR acts as an antirepressor that binds to SrpS releasing it from the DNA, thus allowing transcription (Sun et al. 2011), while the *ttgW* gene has a stop codon that truncates the protein, leading to a nonfunctional protein (Rojas et al. 2003).

There is also another interesting difference between the two strains. In *P. putida* S12 two different insertion sequences, named IS*S12* and IS*S21*, were reported to be located within the *srpS* regulatory protein in *P. putida* S12 variants that grew after sudden addition of high concentrations of toluene (Wery et al. 2001; Sun and Dennis 2009). Hybridization of KpnI cut *P. putida* DOT-T1E genomic DNA with a probe based on IS*S12* failed to identify any band while in *P. putida* S12 seven bands were seen. Furthermore, amplification of *ttgV* gene from cells growing after toluene addition yields a band of the expected size (Segura, unpublished results). These results suggest that the genomic plasticity that allows *P. putida* S12 to maintain high expression of the *srpABC* efflux pump within a significant part of the population is not present in the *P. putida* DOT-T1E strain.

TtgV recognizes a wide variety of effectors such as mono- and biaromatic hydrocarbons, antibiotics, or aliphatic alcohols (Guazzaroni et al. 2005). This property is highly unusual in regulators of the IclR family which generally show high specificity for effector molecules. The *ttgV* gene is transcribed divergently from the *ttgGHI* operon and binds as a dimer of dimers to a 42-bp DNA fragment that covers the −10 and −35 regions of the *ttgG* promoter and the −10 region of the *ttgV* promoter (Lu et al. 2010). Furthermore, binding of TtgV creates a 57° bend that reinforces the steric occlusion of the RNA polymerase binding site (Guazzaroni et al. 2007). Effector binding releases TtgV from its operator sequence and this results in an increased expression of both *ttgV* and *ttgGHI* operons (Rojas et al. 2003; Guazzaroni et al. 2004). TtgV monomer contains an N-terminal domain responsible for DNA binding, a linker helix and a C-terminal domain that contained a hydrophobic pocket proposed to be the effector-binding site (Guazzaroni et al 2005; Lu et al. 2010). The TtgV tetramer exists in two different functional states, a stable symmetric state (with both dimers in the T-state) that is unable to bind to its DNA operator and a less stable asymmetric form (with both dimers in the R-state) that has the correct conformation to bind to DNA (Lu et al. 2010). Effectors bind to each monomer independently (Guazzaroni et al. 2005) stabilizing the T-state. The proposed model suggests that TtgV binds to two DNA sites (within the operator) in a cooperative manner (Guazzaroni et al. 2007; Lu et al. 2010; Fillet et al. 2011).

TtgV is also able to bind to the *ttgT-ttgDEF* intergenic region with higher affinity than the local regulator TtgT (Terán et al. 2007; Fillet et al. 2009). *ttgV* expression is higher than *ttgT*. Consequently, expression of the *ttgDEF* efflux pump operon is mostly controlled by the TtgV regulator. Thus, the TtgV regulator, which is encoded on a plasmid, serves as the key regulator, while chromosomally encoded TtgT takes a secondary role. Therefore, TtgT seems to serve as a backup system for *P. putida* DOT-T1E cells that have lost the pGRT1 plasmid.

Recently, a new mode of regulation of the *ttgGHI* efflux pump was found. Through a still unknown mechanism, ORF32, encoded on the pGRT1 plasmid, is able to regulate the expression of this efflux pump. A mutant in ORF32 displayed higher survival after toluene shock than the wild-type strain. The mutant also showed higher levels of expression from of the *ttgG* promoter, which in turn results from the downregulation of the *ttgV* promoter (Molina et al. 2011). ORF32 has homology with an SdiA-regulated protein and SdiA has been involved in the regulation of the AcrAB efflux pump in *E. coli* (Rahmati et al. 2002).

Regulation of the *ttgGHI* efflux pump is therefore a complex process with at least two different layers of regulation. Furthermore, there is an interplay between the TtgV and TtgT regulatory proteins to regulate plasmid and chromosomal promoters.

11.4 Other Genes Involved in Solvent Tolerance

Several other genes have been involved in solvent tolerance in different *P. putida* strains. These are described below.

Alkyl hydroperoxide reductases (AHPRs) have been identified as general stress proteins and one of them, AhpC, is involved in solvent tolerance in *E. coli* (Ferrante et al. 1995). Fukumori and Kishii (2001) found that AphC is overexpressed in the *P. putida* KT2442TOL solvent-resistant strain. They demonstrated that OxyR, which is the regulatory gene of *ahpC-ahpF* operon, was mutated in the solvent-tolerant strain. This suggests that *oxyR* and *oxyR*-regulated genes affect solvent tolerance in *P. putida*, although these genes are not crucial for survival.

One of the most curious findings is the relationship observed between the flagella biosynthetic pathway and solvent tolerance found in *P. putida* DOT-T1E and *P. putida* S12 (Segura et al. 2001; Kieboom et al. 2001). Several solvent-sensitive mutations in genes involved in flagella biosynthesis have been identified by random mutagenesis in solvent-tolerance assays. In *P. putida* S12 a lower expression of the *srpABC* efflux pump was observed in the solvent-sensitive mutants in *flgK, flaG, fleQ, fliA, fliF, fliC,* and *fliH*. In *P. putida* DOT-T1E however, no differences in the expression of the *ttgGHI* efflux pump was observed in a *fliA* mutant compared with the wild-type strain. Because the flagella apparatus forms a complex structure in the membrane, failure to synthesize this structure may alter the membrane structure, leading to solvent-sensitive strains. It has also been hypothesized that flagella transport system has been parasitized by some solvent-resistance related proteins to allow their insertion into the membrane.

Cytochrome *o* ubiquinol oxidase (*cyoABCDE)* has also been implicated in solvent tolerance (Kobayashi et al. 1999; Duque et al. 2004). Cyo genes are involved in the transport of electrons from the ubiquinone pools, and there are two apparently redundant systems in *P. putida*. The *cyoABCDE* operon is overexpressed in *P. putida* S12 cultures exposed to toluene (Volkers et al. 2009) and a *P. putida* IH-2000 mutant for *cyoC* is unable to grow in the presence of toluene (40 % v/v). The *P. putida* IH-2000 *cyoC* mutant exhibited significant differences in the outer membrane proteins and in the lipid composition and an increase in cell surface hydrophobicity, which may explain loss of the solvent-tolerant phenotype (Kobayashi et al. 1999). A mutant in *cyoB* in *P. putida* DOT-T1E was also hypersensitive to toluene although efflux pump expression and *cis–trans* isomerization were not affected. Proton motive force was also unaffected in this mutant probably as a consequence of the activity of the cytochrome *d* complex (*cyd* system). It was observed, however, that mutant cells showed a number of vesicles on the cell surface together with nonpolar and polar invaginations. So it appears that *cyo* genes are required to maintain cell surface properties necessary for solvent tolerance.

P. putida strains encode several extracytoplasmic sigma factors (ECFs); one of these ECFs, named RpoT, has been implicated in solvent tolerance. RpoT in *P. putida* DOT-T1E controls the expression of around 50 genes including membrane proteins, including components of the respiratory chain and transporters. One of these genes—the *ttgGHI* efflux pump operon—is expressed at levels several fold lower in RpoT mutants versus wild type. Thus it is possible that the solvent sensitivity observed in the RpoT mutant is a consequence of reduced expression of the TtgGHI efflux pump (Duque et al. 2007).

TrgI, a protein of unknown function has also been implicated in solvent tolerance in *P. putida* S12. Using proteomic assays it was found that the TrgI protein is present at lower levels in the presence of toluene (Volkers et al. 2006) and this result was confirmed using transcriptomic assays (Volkers et al. 2009). Furthermore, a TrgI mutant improved survival after 30-min exposure to 1 % (v/v) toluene. Conversely, a strain that overexpressed the gene exhibited a lower survival rate than wild type. The most interesting observation was that the knockout strain was more resistant to toluene-induced lysis than the wild type or the strain overexpressing the gene. This result, together with a high resistance of the mutant strain against β-lactams and other phenotypes consistent with an altered outer membrane, leads to the hypothesis that TrgI is somehow involved in outer membrane impermeabilization (Volkers et al. 2009). Because *trgI* is downregulated almost immediately after toluene addition, it is thought to be one of the first responses in solvent tolerance.

In addition of the efflux pumps described above, other transporters have been implicated in solvent tolerance although their mechanisms of action are still unknown. One of these transporters belongs to the ATP-binding cassette family (ABC), which is present in *P. putida* GM73, KT2440, and DOT-T1E strains (Kim et al 1998; García et al. 2010). In *P. putida* KT2440 proteomic studies have shown that this ABC transporter is induced in the presence of phenol (Roma-Rodrigues et al. 2010). Mutants in this transporter, named Trg2ABC, were hypersensitive to toluene. Another transporter that is a member of the major facilitator superfamily, known as TtgK, is important to solvent tolerance. TtgK was previously shown to be involved in bicyclomycin resistance, and a *P. putida* DOT-T1E TtgK knockout mutant grown in the presence of 0.3 % (v/v) toluene was likewise unable to survive.

The specific mechanisms through which these genes participate in solvent tolerance have not yet been fully elucidated. Interestingly, most of the functions described (except the role of AphC) contribute to the maintenance of cell envelope structure. Thus these findings highlight the importance of cell envelop architecture to coping strategies that provide protection against environmental stressors such as organic solvents.

11.5 Biotechnological Applications of Solvent-Tolerant *Pseudomonas*

Solvent-tolerant bacteria hold much potential as biocatalysts and for the construction of biosensors, and recent advancements have led to the realization of this potential for a number of applications.

Biosensors to detect and follow up contaminant degradation must be highly sensitive and rely on the use of robust strains that are able to resist hazardous environmental conditions. One biosensor that was successfully constructed in *P. putida* F1 takes advantage of the *sep* promoter and the *luxCDABE* gene cassette. A biosensor construct consisting of these genes was chromosomally inserted and the effect of several inducers was studied (Phoenix et al. 2003). One of the main limitations that previously existed for measuring bioluminescence in response to organic solvents is the unspecific activity of the *lux* genes that arises likely due to the induction of fatty acid biosynthesis (which in turn increases the aldehyde levels required for light production). Use of solvent-resistant strains alleviates this problem as no "solvent effect" is observed when the *P. putida* F1 strain is used (Phoenix et al. 2003).

The usefulness of solvent-tolerant bacteria for whole-cell biotechnological processes, especially in double-phase bioreactors in which the reactions involve highly toxic substrates or products, is undeniable. The preferential partitioning of the substrates/products into the organic phase while the cells (biocatalysts) are within the aqueous phase allows higher yields of the toxic compound. Furthermore, higher solubility of the product in the organic phase allows the reduction of the cost of downstream processing and product recovery. Several examples of production of value-added chemicals using double-phase bioreactors have been published recently, including systems for the production of catechols (Husken et al 2001; Rojas et al. 2004); phenol from glucose (Wierckx et al. 2005); cinnamic acid (Nijkamp et al. 2005); and *p*-hydroxybenzoic acid (Ramos-González et al. 2003). With heightened demand for efficient biofuel production, exploitation of solvent-tolerant strains to improve the process of production of butanol and ethanol is a burgeoning field of research.

The study of the regulators and promoters of efflux pumps involved in solvent tolerance has allowed the design of new strategies for the control of therapeutic transgenes (Gitzinger et al. 2009). An example of this was the finding that phloretin, a plant metabolite, is a potent inducer of *ttgABC* efflux pump expression that is recognized by the TtgR regulatory protein. Using this information, a regulatory element was constructed that is able to control reporter protein expression (e.g., secreted alkaline phosphatase). The authors were then able to study the expression of the transgene in different mammalian cell lines as well as the secretion of the reporter protein after application of skin lotions containing phloretin in mice implants.

11.6 Concluding Remarks

Solvent tolerance is a complex process in which many physiological processes conspire to achieve. The first line of defense is constituted by the cell envelop. Not surprisingly, disturbing the cell envelop leads to cells with decreased solvent tolerance. The mechanisms involved in the early response to toluene are typically not regulated by the organic solvent, or if they are, they respond within the first minutes of toluene addition. Efflux pumps are the major determinants in solvent tolerance, and the TtgGHI-type efflux pumps are essential for a highly solvent-tolerant phenotype. Obviously, solvent tolerance is an energy-demanding process—thus, the presence of organic solvents causes much stress to the cell.

Understanding the structural determinants of solvent tolerance will allow researchers to engineer different strains for the creation of better biocatalysts for innovative new green processes. We now know that transfer of the pGRT1 plasmid to solvent-sensitive *Pseudomonas* strains provides them with a solvent-tolerant phenotype. Further manipulation of the plasmid will allow the transfer of the pGRT1 to other relevant strains. With biofuel production currently a key aim of industry, the use of solvent-tolerant strains as gateways for the production of fuels is a timely field of research. The utilization of -omics techniques to optimize microbial strains for the production of value-added chemicals and biofuels has already being initiated (Wierckx et al. 2008; Goodarzi et al. 2010). This work will open new avenues for dramatically improving current biotechnological processes as well as the development of new applications. While much progress has been made, much work is still required in order to realize and successfully translate these breakthroughs from the lab to industry.

References

Akama H, Kanemaki M, Yoshimura M, Tsukihara T, Kashtwagi T, Yoneyama H, Narita S, Nakagawa A, Nakae T (2004a) Crystal structure of the drug discharge outer membrane protein, OprM, of *Pseudomonas aeruginosa*: dual modes of membrane anchoring and occluded cavity end. J Biol Chem 279:52916–52919. doi:10.1074/jbc.C400445200

Akama H, Matsuura T, Kashtwagi T, Yoneyama H, Narita S, Tsukihara T, Nakagawa A, Nakae T (2004b) Crystal structure of the membrane fusion protein, MexA, of the multidrug transporter in *Pseudomonas aeruginosa*. J Biol Chem 279:25939–25942. doi:10.1074/jbc.C400164200

Alguel Y, Meng C, Terán W, Krell T, Ramos JL, Gallegos MT, Zhang X (2007) Crystal structures of multidrug binding protein TtgR in complex with antibiotics and plant antimicrobials. J Mol Biol 369:829–840. doi:10.1016/j.jmb.2007.03.062

Aono R, Tsukagoshi N, Yamamoto M (1998) Involvement of outer membrane protein TolC, a possible member of the *mar-sox* regulon, in maintenance and improvement of organic solvent tolerance of *Escherichia coli* K12. J Bacteriol 180:938–944

Baumgarten T, Vazquez J, Bastisch C, Veron W, Feuilloley MGJ, Nietzsche S, Wick LY, Heipieper HJ (2012) Alkanols and chlorophenols cause different physiological adaptive responses on the level of cell surface properties and membrane vesicle formation in *Pseudomonas putida* DOT-T1E. Appl Microbiol Biotechnol 93:837–845. doi:10.1007/s00253-011-3442-9

Bernal P, Muñoz-Rojas J, Hurtado A, Ramos JL, Segura A (2007a) A *Pseudomonas putida* cardiolipin synthesis mutant exhibits increased sensitivity to drugs related to transport functionality. Environ Microbiol 9:1135–1145. doi:10.1111/j.1462-2920.2006.01236.x

Bernal P, Segura A, Ramos JL (2007b) Compensatory role of the *cis–trans* isomerase and cardiolipin synthase in the membrane fluidity of *Pseudomonas putida* DOT-T1E. Environ Microbiol 9:1658–1664. doi:10.1111/j.1462-2920.2007.01283.x

Blair JMA, Piddock LJV (2009) Structure, function and inhibition of RND efflux pumps in Gram-negative bacteria: an update. Curr Opin Microbiol 12:512–519. doi:10.1016/j.mib.2009.07.003

Brynildsen MP, Liao JC (2009) An integrated network approach identifies the isobutanol response network of *Escherichia coli*. Mol Syst Biol 5:277. doi:10.1038/msb.2009.34

Chang YY, Cronan JE Jr (1999) Membrane cyclopropane fatty acid content is a major factor in acid resistance of *Escherichia coli*. Mol Microbiol 33:249–259. doi:10.1046/j.1365-2958.1999.01456.x
Domínguez-Cuevas P, González-Pastor JE, Marqués S, Ramos JL, deLorenzo V (2006) Transcriptional tradeoff between metabolic and stress-response programs in *Pseudomonas putida* KT2440 cells exposed to toluene. J Biol Chem 281:11981–11991. doi:10.1074/jbc.M509848200
Duque E, Segura A, Mosqueda G, Ramos JL (2001) Global and cognate regulators control expression of the organic solvent efflux pump TtgABC and TtgDEF of *Pseudomonas putida*. Mol Microbiol 39:1100–1106. doi:10.1046/j.1365-2958.2001.02310.x
Duque E, García V, de la Torre J, Godoy P, Bernal P, Ramos JL (2004) Plasmolysis induced by toluene in a *cyoB* mutant of *Pseudomonas putida*. Environ Microbiol 6:1021–1031. doi:10.1111/j.1462-2920.2004.00621.x
Duque E, Rodríguez-Hervá JJ, de la Torre J, Domínguez-Cuevas P, Muñoz-Rojas J, Ramos JL (2007) The RpoT regulon of *Pseudomonas putida* DOT-T1E and its role in stress endurance against solvents. J Bacteriol 189:207–219. doi:10.1128/JB.00950-06
Elkins CA, Nikaido H (2003) 3D structure of AcrB: the archetypal multidrug efflux transporter of *Escherichia coli* likely captures substrates from periplasm. Drug Resist Updat 6:9–13. doi:10.1016/S1368-7646(03)00004-9
Fang J, Barcelona MJ, Alvarez PJJ (2000) Phospholipid compositional changes of five pseudomonad archetypes grown with and without toluene. Appl Microbiol Biotechnol 54:382–389
Ferrante AA, Augliera J, Lewis K, Klibanov AM (1995) Cloning of an organic solvent-resistance gene in *Escherichia coli:* the unexpected role of alkylhydroperoxide reductase. Proc Natl Acad Sci USA 92:7612–7621
Fillet S, Vélez M, Lu D, Zhang X, Gallegos MT, Ramos JL (2009) TtgV represses two different promoters by recognizing different sequences. J Bacteriol 191:1901–1909. doi:10.1128/JB.01504-08
Fillet S, Krell T, Morel B, Lu D, Zhang X, Ramos JL (2011) Intramolecular signal transmission in a tetrameric repressor of the IclR family. Proc Nat Acad Sci USA 108:15372–15377. doi:10.1073/pnas.1018894108
Fukumori F, Kishii M (2001) Molecular cloning and transcriptional analysis of the alkyl hydroperoxide reductase genes from *Pseudomonas putida* KT2442. J Gen Appl Microbiol 47:269–277. doi:10.2323/jgam.47.269
Fukumori F, Hirayama H, Takami H, Inoue A, Horikoshi K (1998) Isolation and transposon mutagenesis of a *Pseudomonas putida* KT2442 toluene-resistant variant: involvement of an efflux system in solvent resistance. Extremophiles 2:395–400
García V, Godoy P, Daniels C, Hurtado A, Ramos JL, Segura A (2010) Functional analysis of new transporters involved in stress tolerance in *Pseudomonas putida* DOT-T1E. Environ Microbiol Rep 2:389–395. doi:10.1111/j.1462-2920.2004.00621.x
Gitzinger M, Kemmer C, El-Baba MD, Weber W, Fussenegger M (2009) Controlling transgene expression in subcutaneous implants using a skin lotion containing the apple metabolite phloretin. Proc Natl Acad Sci USA 106:10638–10643. doi:10.1073/pnas.0901501106
Godoy P, Ramos-González MI, Ramos JL (2004) *Pseudomonas putida* mutants in the *exbBexbDtonB* gene cluster are hypersensitive to environmental and chemical stressors. Environ Microbiol 6:605–610. doi:10.1111/j.1462-2920.2004.00595.x
Godoy P, Molina-Henares AJ, de la Torre J, Duque E, Ramos JL (2010) Characterization of the RND family of multidrug efflux pumps: in silico to in vivo confirmation of four functionally distinct subgroups. Microb Biotechnol 3:691–700. doi:10.1111/j.1751-7915.2010.00189.x
Goodarzi H, Bennett BD, Amini S, Reaves ML, Hottes AK, Rabinowitz JD, Tavazoie S (2010) Regulatory and metabolic rewiring during laboratory evolution of ethanol tolerance in *E. coli*. Mol Syst Biol 6:378
Guazzaroni ME, Terán W, Zhang X, Gallegos MT, Ramos JL (2004) TtgV bound to a complex operator site represses transcription of the promoter for the multidrug and solvent extrusion TtgGHI pump. J Bacteriol 186:2921–2927. doi:10.1128/JB.186.10.2921-2927.2004
Guazzaroni ME, Krell T, Felipe A, Ruiz R, Meng C, Zhang X, Gallegos MT, Ramos JL (2005) The multidrug efflux regulator TtgV recognizes a wide range of structurally different effectors in solution and complexed with target DNA. J Biol Chem 280:20887–20898. doi:10.1074/jbc.M500783200

Guazzaroni ME, Krell T, Gutierrez del Arroyo P, Vélez M, Jiménez M, Rivas G, Ramos JL (2007) The transcriptional repressor TtgV recognizes a complex operator as a tetramer and induces convex DNA bending. J Mol Biol 369:927–939. doi:10.1016/j.jmb.2007.04.022

Heipieper HJ, de Bont JAM (1994) Adaptation of *Pseudomonas putida* S12 to ethanol and toluene at the level of fatty acid composition of membranes. Appl Environ Microbiol 60:4440–4444

Heipieper HJ, Weber FJ, Sikkema J, Keweloh H, de Bont JAM (1994) Mechanisms behind resistance of whole cells to toxic organic solvents. Trends Biotechnol 12:409–415

Heipieper HF, Diefenbach R, Keweloh H (1992) Conversion of *cis*-unsaturated fatty acids to *trans*, a possible mechanism for the protection of phenol-degrading *Pseudomonas putida* P8 from substrate toxicity. Appl Environ Microbiol 58:1847–1852

Heipieper HJ, Meulenbeld G, VanOirschot Q, deBont JAM (1996) Effect of environmental factors on the *trans/cis* ratio of unsaturated fatty acids in *Pseudomonas putida* S12. Appl Environ Microbiol 62:2773–2777

Heipieper HJ, Meinhardt F, Segura A (2003) The *cis-trans* isomerase of unsaturated fatty acids in *Pseudomonas* and *Vibrio*: biochemistry, molecular biology and physiological function of a unique stress adaptive mechanism. FEMS Microbiol Lett 229:1–7. doi:10.1016/S0378-1097(03)00792-4

Heipieper HJ, Cornelissen S, Pepi M (2009) Surface properties and cellular energetics of bacteria in response to the presence of hydrocarbons. In: Timmis KN (ed) Handbook of hydrocarbon and lipid microbiology, vol 2, Part 9. Springer, Berlin, pp 1615–1624

Higgins MK, Bokma E, Koronakis E, Hughes C, Koronakis V (2004) Structure of the periplasmic component of a bacterial drug efflux pump. Proc Natl Acad Sci USA 101:9994–9999. doi:10.1073/pnas.0400375101

Huertas MJ, Duque E, Molina L, Roselló-Mora R, Mosqueda G, Godoy P, Christensen B, Molin S, Ramos JL (2000) Tolerance to sudden organic solvent shocks by soil bacteria and characterization of *Pseudomonas putida* strains isolated from toluene polluted sites. Environ Sci Technol 34:3395–3400. doi:10.1021/es991171a

Husain F, Humbard M, Misra R (2004) Interaction between the TolC and AcrA proteins of a multidrug efflux system of *Escherichia coli*. J Bacteriol 24:8553–8556. doi:10.1128/JB.186.24.8533-8536.2004

Husken LE, Dalm MCF, Tramper J, Wery J, de Bont JAM, Beeftink R (2001) Integrated bioproduction and extraction of 3-methylcatechol. J Biotechnol 88:11–19. doi:10.1016/S0168-1656(01)00252-8

Inoue A, Horikoshi K (1991) Estimation of solvent-tolerance of bacteria by the solvent parameter LogP. J Ferm Bioeng 71:194–196. doi:10.1016/0922-338X(91)90109-T

Isken S, de Bont JAM (1996) Active efflux of toluene in a solvent-resistant bacterium. J Bacteriol 178:6056–6058

Isken S, Derks A, Wolffs PFG, deBont JAM (1999) Effect of organic solvents on the yield of solvent-tolerant *Pseudomonas putida* S12. Appl Environ Microbiol 65:2631–2635

Junker F, Ramos JL (1999) Involvement of the *cis–trans* isomerase CtiT1 in solvent resistance in *Pseudomonas putida* DOT-T1. J Bacteriol 181:5693–5700

Kawai F, Shoda M, Harashima R, Sadaie Y, Hara H, Matsumoto K (2004) Cardiolipin domains in *Bacillus subtilis* marburg membranes. J Bacteriol 186:1475–1483. doi:10.1128/JB.186.5.1475-1483.2004

Kieboom J, de Bont JAM (2001) Identification and molecular characterization of an efflux system involved in *Pseudomonas putida* S12 multidrug resistance. Microbiology 147:43–51

Kieboom J, Dennis JJ, deBont JAM, Zylstra GJ (1998) Identification and molecular characterization of an efflux pump involved in *Pseudomonas putida* S12 solvent tolerance. J Biol Chem 273:85–91. doi:10.1074/jbc.273.1.85

Kieboom J, Bruinenberg R, Keizer-Gunnink I, de Bont JAM (2001) Transposon mutations in the flagella biosynthetic pathway of the solvent-tolerant *Pseudomonas putida* S12 result in a decreased expression of solvent efflux genes. FEMS Microbiol Lett 198:117–122. doi:10.1111/j.1574-6968.2001.tb10628.x

Kim K, Lee S, Lee K, Lim D (1998) Isolation and characterization of toluene-sensitive mutants from the toluene-resistant bacterium *Pseudomonas putida* GM73. J Bacteriol 180:3692–3696

Kim S, Shim JH, Suh YT (2002) Changes in membrane fluidity and fatty acid composition of *Pseudomonas putida* CN-T19 in response to toluene. Biosci Biotechnol Biochem 66:1945–1950. doi:10.1271/bbb.66.1945

Kobayashi H, Takami H, Hirayama H, Kobata K, Usami R, Horikoshi K (1999) Outer membrane changes in a toluene-sensitive mutant of toluene-tolerant *Pseudomonas putida* IH-2000. J Bacteriol 181:4493–4498

Kobayashi H, Uematsu K, Hirayama H, Horikohi K (2000) Novel toluene elimination system in a toluene-tolerant microorganism. J Bacteriol 182:6451–6455. doi:10.1128/JB.182.22.6451-6455.2000

Koronakis V, Sharff A, Koronakis E, Luisi B, Hughes C (2000) Crystal structure of the bacterial membrane protein TolC central to multidrug efflux and protein export. Nature 405:914–919. doi:10.1038/35016007

Krell T, Terán W, Mayorga OL, Rivas G, Jiménez M, Daniels C, Molina-Henares AJ, Martínez-Bueno M, Gallegos MT, Ramos JL (2007) Optimization of the palindromic order of the TtgR operator enhances binding cooperativity. J Mol Biol 369:1188–1199. doi:10.1016/j.jmb.2007.04.025

Lacal J, Muñoz-Martínez F, Reyes-Darías JA, Duque E, Matilla M, Segura A, Ortega Calvo JJ, Jímenez-Sánchez C, Krell T, Ramos JL (2011) Bacterial chemotaxis towards aromatic hydrocarbons in *Pseudomonas*. Environ Microbiol 13:1733–1744. doi:10.1111/j.1462-2920.2011.02493.x

Li XZ, Nikaido H, Poole K (1995) Role of mexA-mexB-oprM in antibiotic efflux in *Pseudomonas aeruginosa*. Antimicrob Agents Chemother 39:1948–1953

Li X, Zhang L, Poole K (1998) Role of the multidrug efflux systems of *Pseudomonas aeruginosa* in organic solvent tolerance. J Bacteriol 180:2987–2991

Loffhagen N, Härtig C, Babel W (2001) *Pseudomonas putida* NCTC 10936 balances membrane fluidity in response to physical and chemical stress by changing the saturation degree and the trans/cis ratio of fatty acids. Biosci Biotechnol Biochem 68:317–323. doi:10.1271/bbb.68.317

Lu D, Fillet S, Meng C, Alguel Y, Kloppsteck P, Bergeron J, Krell T, Gallegos MT, Ramos J, Zhang X (2010) Crystal structure of TtgV in complex with its DNA operator reveals a general model for cooperative DNA binding of tetrameric gene regulators. Genes Dev 24:2556–2565. doi:10.1101/gad.603510

Ma D, Cook DN, Alberti M, Pon NG, Nikaido H, Hearst JE (1993) Molecular cloning and characterization of *acrA* and *acrE* genes of *Escherichia coli*. J Bacteriol 175:6299–6313

McDonald PM, Sykes BD, McElhaney RN (1985) Flourine-19 nuclear magnetic resonance studies of lipid fatty acyl chain order and dynamics in *Acholeplasma laidlawii* membranes, a direct comparison of the effects of *cis* and *trans* cyclopropane ring and double-bond substituents on orientational order. Biochemistry 24:4651–4659

Mikolosko J, Bobyk K, Zgurskaya HI, Ghosh P (2006) Conformational flexibility in the multidrug efflux system protein AcrA. Structure 14:577–587. doi:10.1016/j.str.2005.11.015

Mileykovskaya E, Fishov I, Fu X, Corbin B, Margolin W, Dowhan W (2003) Effects of phospholipids composition on MinD-membrane interactions in vitro and in vivo. J Biol Chem 278:22193–22198. doi:10.1074/jbc.M302603200

Molina L, Duque E, Gómez MJ, Krell T, Lacal J, García-Puente A, García V, Matilla MA, Ramos JL, Segura A (2011) The pGRT1 plasmid of *Pseudomonas putida* DOT-T1E encodes functions relevant for survival under harsh conditions in the environment. Environ Microbiol 13:2315–2327. doi:10.1111/j.1462-2920.2011.02492.x

Mosqueda G, Ramos JL (2000) A set of genes encoding a second toluene efflux system in *Pseudomonas putida* DOT-T1E is linked to the tod genes for toluene metabolism. J Bacteriol 181:937–943. doi:10.1128/JB.182.4.937-943.2000

Mosqueda G, Ramos-Gonzalez MI, Ramos JL (1999) Toluene metabolism by the solvent-tolerant *Pseudomonas putida* DOT-T1 strain, and its role in solvent impermeabilization. Gene 232:69–76. doi:10.1016/S0378-1119(99)00113-4

Muñoz-Rojas J, Bernal P, Duque E, Godoy P, Segura A, Ramos JL (2006) Involvement of cyclopropane fatty acids in the response of *Pseudomonas putida* KT2440 to freeze-drying. Appl Environ Microbiol 72:472–477. doi:10.1128/AEM.72.1.472-477.2006

Murakami S, Nakashima R, Yamashita E, Yamaguchi A (2002) Crystal structure of bacterial multidrug efflux transporter AcrB. Nature 419:587–593. doi:10.1038/nature01050

Nakashima R, Sakurai K, Yamasaki S, Nishino K, Yamaguchi A (2011) Structures of the multidrug exporter AcrB reveal a proximal multisite drug-binding pocket. Nature 480:565–569. doi:10.1038/nature10641

Neumann G, Cornelissen S, van Breukelen F, Hunger S, Lippold H, Loffhagen N, Wick LY, Heipieper HJ (2006) Energetics and surface properties of *Pseudomonas putida* DOT-T1E in a two-phase fermentation system with 1-decanol as second phase. Appl Environ Microbiol 72:4232–4238. doi:10.1128/AEM.02904-05

Nijkamp K, van Luijk N, de Bont JAM, Wery J (2005) The solvent-tolerant *Pseudomonas putida* S12 as host for the production of cinnamic acid from glucose. Appl Microbiol Biotechnol 69:170–177. doi:10.1007/s00253-005-1973-7

Nikaido H (1996) Multidrug efflux pumps of Gram-negative bacteria. J Bacteriol 178:5853–5859

Nikaido H (2011) Structure and mechanism of RND-type multidrug efflux pumps. Adv Enzymol Relat Areas Mol Biol 77:1–60

Paulsen IT, Chen J, Nelson KE, Saier MH Jr (2001) Comparative genomics of microbial drug efflux systems. J Mol Microbiol Biotechnol 3:145–150

Phoenix P, Keane A, Patel A, Bergeron SG, Lau PCK (2003) Characterization of a new solvent-responsive gene locus in *Pseudomonas putida* F1 and its functionalization as a versatile biosensor. Environ Microbiol 5:1309–1327. doi:10.1111/j.1462-2920.2003.00426.x

Pini CV, Bernal P, Godoy P, Ramos JL, Segura A (2009) Cyclopropane fatty acids are involved in organic solvent tolerance but not in acid stress resistance in *Pseudomonas putida* DOT-T1E. Microbiol Biotech 2:253–261. doi:10.1111/j.1751-7915.2009.00084.x

Pini CV, Godoy P, Bernal P, Ramos JL, Segura A (2011) Regulation of the cyclopropane synthase *cfaB* gene in *Pseudomonas putida* KT2440. FEMS Microbiol Lett 321:107–114. doi:10.1111/j.1574-6968.2011.02317.x

Pinkart HC, White DC (1997) Phospholipid biosynthesis and solvent tolerance in *Pseudomonas putida* strains. J Bacteriol 179:4219–4226

Pinkart HC, Wolfram JW, Rodgers R, White DC (1996) Cell envelope changes in solvent-tolerant and solvent-sensitive *Pseudomonas putida* strains following exposure to *o*-xylene. Appl Environ Microbiol 62:1129–1132

Rahmati S, Yang S, Davidson AL, Zechiedrich EL (2002) Control of the AcrAB multidrug efflux pump by quorum-sensing regulator SdiA. Mol Microbiol 43:677–685. doi:10.1046/j.1365-2958.2002.02773.x

Ramos JL, Duque E, Huertas MJ, Haïdour A (1995) Isolation and expansion of the catabolic potential of a *Pseudomonas putida* strain able to grow in the presence of high concentrations of aromatic hydrocarbons. J Bacteriol 177:3911–3916

Ramos JL, Duque E, Rodriguez-Hervá JJ, Haïdour A (1997) Mechanisms for solvent tolerance in bactéria. J Biol Chem 272:3887–3890. doi:10.1074/jbc.272.7.3887

Ramos JL, Duque E, Godoy P, Segura A (1998) Efflux pumps involved in toluene tolerance in *Pseudomonas putida* DOT-T1E. J Bacteriol 180:3323–3329

Ramos JL, Duque E, Gallegos MT, Godoy P, Ramos-Gonzalez MI, Rojas A, Terán W, Segura A (2002) Mechanisms of solvent tolerance in Gram-negative bacteria. Annu Rev Microbiol 56:743–768. doi:10.1146/annurev.micro.56.012302.161038

Ramos-González MI, Ben-Bassat A, Campos MJ, Ramos JL (2003) Genetic engineering of a highly solvent-tolerant *Pseudomonas putida* strain for biotransformation of toluene to *p*-hydroxybenzoate. Appl Environ Microbiol 69:5120–5127. doi:10.1128/AEM.69.9.5120-5127.2003

Rodriguez-Hervá JJ, García V, Hurtado A, Segura A, Ramos JL (2007) The *ttgGHI* solvent efflux pump operon of *Pseudomonas putida* DOT-T1E is located on a large self-transmissible plasmid. Environ Microbiol 9:1550–1561. doi:10.1111/j.1462-2920.2007.01276.x
Rojas A, Duque E, Mosqueda G, Golden G, Hurtado A, Ramos JL (2001) Three efflux pumps are required to provide efficient tolerance to toluene in *Pseudomonas putida* DOT-T1E. J Bacteriol 183:3867–3973. doi:10.1128/JB.183.13.3967-3973.2001
Rojas A, Segura A, Guazzaroni ME, Terán W, Hurtado A, Gallegos MT, Ramos JL (2003) In vivo and in vitro evidence that TtgV is the specific regulator of the TtgGHI multidrug and solvent efflux pump of *Pseudomonas putida*. J Bacteriol 185:4755–4763. doi:10.1128/JB.185.16.4755-4763.2003
Rojas A, Duque E, Schmid A, Hurtado A, Ramos JL, Segura A (2004) Biotransformation in double-phase systems: physiological responses of *Pseudomonas putida* DOT-T1E to double phase made of aliphatic alcohols and biosynthesis of substituted catechols. Appl Environ Microbiol 70:3637–3643. doi:10.1128/AEM.70.6.3637-3643.2004
Roma-Rodrigues C, Santos PM, Benndorf D, Rapp E, Sá-Correia I (2010) Response of *Pseudomonas putida* KT2440 to phenol at the level of membrane proteome. J Proteomics 73:1461–1478. doi:10.1016/j.jprot.2010.02.003
Rühl J, Hein EM, Hayen H, Schmid A, Blank LM (2012) The glycerophospholipid inventory of *Pseudomonas putida* is conserved between strains and enables growth condition-related alterations. Microbiol Biotech 4:45–58. doi:10.1111/j.1751-7915.2011.00286.x
Saier MHJR, Paulsen IT (2001) Phylogeny of multidrug transporters. Semin Cell Dev Biol 12:205–213. doi:10.1006/scdb.2000.0246
Seelig J, Waespe-Šarcevic N (1978) Molecular order in *cis* and *trans* unsaturated phospholipid bilayers. Biochemistry 17:3310–3315
Segura A, Duque E, Hurtado A, Ramos JL (2001) Mutations of genes involved in the flagellar export apparatus of the solvent-tolerant *Pseudomonas putida* DOT-T1E strain impair motility and lead to hypersensitivity to toluene shocks. J Bacteriol 183:4127–4133. doi:10.1128/JB.183.14.4127-4133.2001
Segura A, Rojas A, Hurtado A, Huertas MJ, Ramos JL (2003) Comparative genomic analysis of solvent extrusion pumps in Pseudomonas strains exhibiting different degrees of solvent tolerance. Extremophiles 7:371–376. doi:10.1007/s00792-003-0331-x
Segura A, Godoy P, van Dillewijn P, Hurtado A, Arroyo N, Santacruz S, Ramos JL (2005) Proteomic analysis reveals the participation of energy- and stress-related proteins in the response of *Pseudomonas putida* DOT-T1E to toluene. J Bacteriol 187:5937–5945. doi:10.1128/JB.187.17.5937-5945.2005
Segura A, Molina L, Fillet S, Krell T, Bernal P, Muñoz-Rojas J, Ramos JL (2011) Solvent tolerance in Gram-negative bacteria. Curr Opin Biotech 23:1–7. doi:10.1016/j.copbio.2011.11.015
Sennhauser G, Bukowska MA, Briand C, Grütter MG (2009) Crystal structure of bacterial multidrug efflux transporter MexB from *Pseudomonas aeruginosa*. J Mol Biol 389:134–145. doi:10.1016/j.jmb.2009.04.001
Sikkema J, de Bont JAM, Poolman B (1995) Mechanisms of membrane toxicity of hydrocarbons. Microbiol Rev 59:201–222
Sun X, Dennis JJ (2009) A novel insertion sequence derepresses efflux pump expression and preadapts *Pseudomonas putida* S12 for extreme solvent stress. J Bacteriol 191:6773–6777. doi:10.1128/JB.00832-09
Sun X, Zahir Z, Lynch KH, Dennis JJ (2011) An antirepressor, SrpR, is involved in transcriptional regulation of the SrpABC solvent tolerance efflux pump of *Pseudomonas putida* S12. J Bacteriol 193:2717–2725. doi:10.1128/JB.00149-11
Tamura N, Murakami S, Oyama Y, Ishiguro M, Yamaguchi A (2005) Direct interaction of multidrug efflux transporter AcrB and outer membrane channel TolC detected via site-directed disulfide cross-linking. Biochemistry 44:11115–11121. doi:10.1021/bi050452u
Terán W, Felipe A, Segura A, Rojas A, Ramos JL, Gallegos MT (2003) Antibiotic-dependent induction of *Pseudomonas putida* DOT-T1E TtgABC efflux pump is mediated by the drug

binding repressor TtgR. Antimicrob Agents Chemother 47:3067–3072. doi:10.1128/AAC.47.10.3067-3072.2003

Terán W, Krell T, Ramos JL, Gallegos MT (2006) Effector-repressor interactions, binding of a single effector molecule to the operator-bound TtgR homodimer mediates derepression. J Biol Chem 281:7102–7109. doi:10.1074/jbc.M511095200

Terán W, Felipe A, Fillet S, Guazzaroni ME, Krell T, Ruiz R, Ramos JL, Gallegos MT (2007) Complexity in efflux pump control: cross-regulation by the paralogues TtgV and TtgT. Mol Microbiol 66:1416–1428. doi:10.1111/j.1365-2958.2007.06004.x

Tikhonova EB, Yamada Y, Zgurskaya HI (2011) Sequential mechanism of assembly of multidrug efflux pump AcrAB-TolC. Chem Biol 18:454–463. doi:10.1016/j.chembiol.2011.02.011

Touzé T, Eswaran J, Bokma E, Koronakis E, Hughes C, Koronakis V (2004) Interactions underlying assembly of the *Escherichia coli* AcrAB-TolC multidrug efflux system. Mol Microbiol 53:697–706. doi:10.1111/j.1365-2958.2004.04158.x

Trevors JT (2003) Fluorescent probes for bacterial cytoplasmic membrane research. J Biochem Biophys Methods 57:87–103. doi:10.1016/S0165-022X(03)00076-9

Volkers RJ, de Jong AL, Hulst AG, van Baar BL, deBont JA, Wery J (2006) Chemostat-based proteomic analysis of toluene-affected *Pseudomonas putida* S12. Environ Microbiol 8:1674–1678. doi:10.1111/j.1462-2920.2006.01056.x

Volkers RJ, Ballerstedt H, Ruijssenaars H, de Bont JAM, de Winde JH, Wery J (2009) TrgI, toluene repressed gene I, a novel gene involved in toluene-tolerance in *Pseudomonas putida* S12. Extremophiles 13:283–297. doi:10.1007/s00792-008-0216-0

Weber FJ, Isken S, de Bont JAM (1994) *Cis/trans* isomerization of fatty acids as a defense mechanism of *Pseudomonas putida* strains to toxic concentrations of toluene. Microbiology 140:2013–2017. doi:10.1099/13500872-140-8-2013

Wery J, Hidayat B, Kieboom J, de Bont JAM (2001) An insertion sequence prepares *P. putida* S12 for severe solvent stress. J Biol Chem 276:5700–5706. doi:10.1074/jbc.M007687200

Wierckx NJ, Ballerstedt H, de Bont JAM, Wery J (2005) Engineering of solvent-tolerant *Pseudomonas putida* S12 for bioproduction of phenol from glucose. Appl Environ Microbiol 71:8221–8227. doi:10.1128/AEM.71.12.8221-8227.2005

Wierckx NJ, Ballerstedt H, de Bont JAM, Wery J (2008) Transcriptome analysis of a phenol producing *Pseudomonas putida* S12 construct: Genetic and physiological basis for improved production. J Bacteriol 190:2822–2830. doi:10.1128/JB.01379-07

Yu EW, Aires JR, Nikaido H (2003) AcrB multidrug efflux pump of *Escherichia coli*: composite substrate-binding cavity of exceptional flexibility generates its extremely wide substrate specificity. J Bacteriol 185:5657–5664. doi:10.1128/JB.185.19.5657-5664.2003

Chapter 12
Diversity and Evolution of Aromatic Degradation Pathway Enzymes in an Activated Sludge

Kentaro Miyazaki

Abstract Metagenomics has emerged as an alternative approach to conventional microbial screening that allows exhaustive screening of microbial genomes in their natural environments. We adopted a functional metagenomic approach to screen for catabolic pathway genes for aromatic compounds. A fosmid library was created in *Escherichia coli* using an environmental DNA from an activated sludge used to treat industrial wastewater contaminated with aromatic compounds as a source of metagenome. By screening the library for extradiol dioxygenases (EDOs), we identified 91 positive clones, of which 38 were subjected to sequence analysis. Forty-three genes encoding EDOs were identified, including enzymes belonging to novel subfamilies. In particular, EDOs belonging to the I.2.G subfamily were overrepresented (20 clones). Enzymatic analysis revealed that the I.2.G EDOs were Mn(II)-dependent and had the strongest affinity for catechol reported thus far, which should enable host organisms to utilize the limited carbon sources and may be the basis for their dominance. As for the organization of catabolic pathway genes, unlike to known aromatic-degrading bacteria with complete pathways, the majority of the clones contained only subsets of the genes with novel arrangements. A circular plasmid-like 36.7-kb DNA form, designated pSKYE1, was assembled from the sequences of the clones carrying I.2.G EDOs, which possessed only a small set of genes for phenol degradation. Taking the dominance of this gene set (20 out of 38) into account, pSKYE1 may play a role in detoxification in the environment.

K. Miyazaki (✉)
Bioproduction Research Institute, National Institute of Advanced Industrial Science and Technology (AIST), 2-17-2-1, Tsukisamu-higashi, Toyohira-ku, Sapporo, Hokkaido 062-8517, Japan

Department of Medical Genome Sciences, Graduate School of Frontier Sciences, The University of Tokyo, 2-17-2-1, Tsukisamu-higashi, Toyohira-ku, Sapporo, Hokkaido 062-8517, Japan
e-mail: miyazaki-kentaro@aist.go.jp

H. Nojiri et al. (eds.), *Biodegradative Bacteria: How Bacteria Degrade, Survive, Adapt, and Evolve*, DOI 10.1007/978-4-431-54520-0_12, © Springer Japan 2014

Keywords Aromatic compounds • Catabolic genes • Degradation • Extradiol dioxygenase • Functional screening • Metagenomics

12.1 Introduction

The ability to utilize various aromatic compounds as a source of carbon and energy is widespread in bacteria, and in the natural environment these bacteria contribute greatly to the breakdown of aromatic compounds and the global carbon cycle (Wittich 1998; Furukawa et al. 2004). Most bacterial aromatic degradative processes are aerobic (Gibson and Harwood 2002) and utilize series of enzymes that are usually categorized as "upper"- or "lower"-pathway enzymes (Williams and Sayers 1994). Generally, the upper-pathway enzymes transform aromatic compounds to aromatic vicinal diols. An initial hydroxylation step is performed by a monooxygenase or dioxygenase that incorporates an oxygen atom(s) into the aromatic ring (Gibson and Parales 2000). The second enzyme in the upper pathway is dihydrodiol dehydrogenase, which catalyzes the conversion of dihydrodiol to a dihydroxy compound. In the lower pathway, the resulting dihydroxylated aromatic compounds, or catecholic compounds, are transformed into ring-cleavage products by extradiol dioxygenases (EDOs) or intradiol dioxygenases (IDOs). The subsequent metabolic steps are referred to as *meta* or *ortho* pathways. The ring-cleavage products are further degraded into compounds that can enter the tricarboxylic acid (TCA) cycle. Of these enzymes, EDOs play a key role in determining catabolic pathway specificity (Hirose et al. 1994; Vaillancourt et al. 2004; Fortin et al. 2005). Because the activities of EDOs can be easily identified due to the yellow pigmentation of their products, numerous studies have been carried out to isolate EDOs as well as the entire degradative pathways (e.g., Furukawa et al. 2004; Fortin et al. 2005). On the basis of primary-sequence similarities, EDOs are classified into two evolutionarily independent families, types I and II (Eltis and Bolin 1996). The majority of these enzymes are of type I and may be further divided into subfamilies on the basis of their primary sequences (Eltis and Bolin 1996; Mesarch et al. 2000; Junca et al. 2004). Type II enzymes are rare and lack sequence similarity with each other and with type I enzymes.

The genes encoding upper- and lower-pathway enzymes are often clustered into operons, and the varied combinations of upper and lower pathways provide diverse functions. Point mutations in conjunction with various gene rearrangements (e.g., insertion, deletion, duplication, and inversion) further promote the functional diversity of these genes and gene clusters (van der Meer et al. 1992). The presence of foreign compounds can lead to the selection of mutant bacteria that are capable of metabolizing them. Aromatic catabolic genes are often harbored by mobile genetic elements (e.g., plasmids, transposons, and integrative or conjugative elements) that effectively disseminate the catabolic traits to phylogenetically diverse bacteria (Tsuda et al. 1999; Top and Springael 2003; Nojiri et al. 2004).

However, because the generalizations described above come from studies of bacteria that can be cultivated under standard laboratory conditions, they are probably biased. Since the vast majority (>99 %) of bacteria in the natural environment are difficult to culture in the laboratory and are thought to differ greatly from known cultured ones (Amann et al. 1995), the not-yet cultivated species may possess novel genes, enzymes, or pathways. In addition, the aromatic compound-degrading bacteria thus far analyzed were isolated on the basis of their ability to utilize aromatics as their sole carbon and energy source. This approach can isolate only those bacteria that possess all of the genes necessary to completely degrade aromatic compounds. To complement these drawbacks, we adopted a culture-independent approach, i.e., functional metagenomics, to search for catabolic genes for aromatic compounds.

Metagenomics has emerged as an alternative approach to conventional microbial screening that allows exhaustive screening of microbial genomes in their natural environments. By directly cloning environmental DNA (or metagenome) in a surrogate host, one can exhaustively investigate the metagenome, independent of the culturability of the source organisms. This innovative technology is anticipated to innovate the discovery process of novel genes from microorganisms. In fact, various novel enzymes have been identified that have unique activities and/or sequences (Uchiyama and Miyazaki 2009; Suenaga 2012).

As a target environment, we selected an activated sludge used to treat coke-plant wastewater. The wastewater contains various organic pollutants such as phenols, monocyclic and polycyclic nitrogen-containing aromatics, oxygen- and sulfur-containing heterocyclic compounds, and polycyclic aromatic hydrocarbons (Stamoudis and Luthy 1980); thus, the sludge is considered to contain various microorganisms responsible for decomposing such compounds. We constructed a metagenomic library using the sludge DNA and screened the library for EDO activity. In this chapter, we describe several novel EDOs as well as the genetic organization of aromatic catabolic genes identified through metagenomic analysis.

12.2 Materials and Methods

12.2.1 Construction and Screening of the Metagenomic Fosmid Library

A metagenomic library was constructed using activated sludge treating coke-plant wastewater. The environmental DNA was fragmented into 33–48 kb in length, which was inserted into a fosmid as a vector (pCC1FOS, Epicentre). An advantage to using fosmid is that it can accommodate longer inserts (30–40 kb) than can common plasmid vectors (<5 kb) (Entcheva et al. 2001; Daniel 2005). As known catabolic operons (gene clusters) for aromatic compounds often range from 10 to 30 kb, we used a fosmid so that we could clone whole-gene clusters.

Transformants were selected on LB agar plates containing 12.5 μg/mL chloramphenicol (LB/Cm), and colonies were picked and re-suspended in separated wells of 96-well plates containing 200 μL of LB/Cm; each well contained 100 independent clones. A total of 96,000 colonies were used to construct a library in ten 96-well plates.

Library cells were then grown at 37 °C for 2 h, harvested by centrifugation, resuspended in 50 mM phosphate buffer (pH 7.5), and disrupted by addition of 150 μL of BugBuster plus Benzonase Nuclease (Novagen). For each sample, cell debris was removed by centrifugation and the supernatant transferred to a clean plate. Substrate (0.5 mM catechol) was added to 100 μL of each cell extract, and the reaction plates were incubated with mild agitation (250 rpm) at 25 °C. Positive clones were identified after incubation for 1 h or 16 h.

12.3 Functional Screening of a Metagenomic Library for Genes Involved in the Microbial Degradation of Aromatic Compounds

12.3.1 Construction and Screening of the Metagenomic Library

An *E. coli* library was created using activated sludge used to treat coke-plant wastewater as a source of DNA and a fosmid as a vector with an average insert size of 33 kb. The library comprised approximately 96,000 clones, which covered 3.2 Gb of metagenomic DNA. Approximately 100 clones were mixed in each well to reduce the amount of labor required for screening to yield ten 96-well plates. The entire screening process involved growing recombinant *E. coli* in liquid medium, extracting the proteins from the cells, mixing the proteins with catechol, and detecting EDO activity by monitoring the development of a yellow color. Initial screening identified 85 positive wells; subsequent screening resulted in the retrieval of 91 positive fosmid clones from the 85 wells.

12.3.2 Identification of EDOs Belonging to New Subfamilies

On the basis of the substrate specificities of the 91 fosmid clones, we selected 38 of the clones and subjected them to shotgun DNA sequencing. BLAST search (Altschul et al. 1997) was carried out to identify at least one type I EDO gene was identified in each fosmid as well as other components in the flanking regions, such as phenol hydroxylases. Five fosmids carried different EDOs in their gene clusters, yielding a

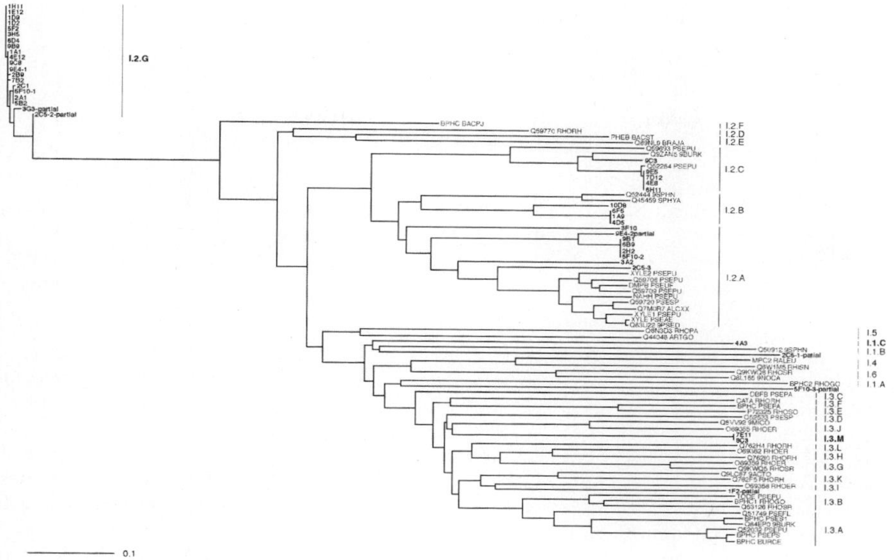

Fig. 12.1 Phylogenetic tree of metagenomic EDOs with known type I EDOs. Sequences isolated in this study from metagenomic library are indicated in *bold*

total of 43 genes (40 full length and three partial). The failure to identify type II EDO genes suggests their scarcity in the environment. Figure 12.1 illustrates the phylogenetic relationship between the metagenomic and known EDOs. Eight EDOs (3 F10, 9E4-2 partial, 9B1, 6B9-2, 2H2, 5 F10-2, 3A2, and 2C5-2) belonging to subfamily I.2.A were identified, where a majority of EDOs obtained through cultivation-based methods belong. However, a greater number (35) of EDOs belonging to other subfamilies were identified, including those belonging to the new subfamilies I.1.C, I.2.G, 1.3.M, and I.3.N. Approximately half (20) of the EDOs fell into subfamily I.2.G, suggesting a specific role for these EDOs in the environment.

12.4 The Molecular Basis for Adaptive Evolution in Novel EDOs Retrieved from the Metagenome

The 20 EDOs belonging to the I.2.G subfamily were not identical; they were separated by single-nucleotide polymorphisms (SNPs) and were classified into six groups. Using combinations of the SNPs, a possible evolutionary lineage of the I.2.G clones was reconstructed as shown in Fig. 12.2. We propose that these genes evolved from a common ancestor (group 1) and diverged via the accumulation of various mutations (groups 2–6). To investigate the reason for the dominance of the

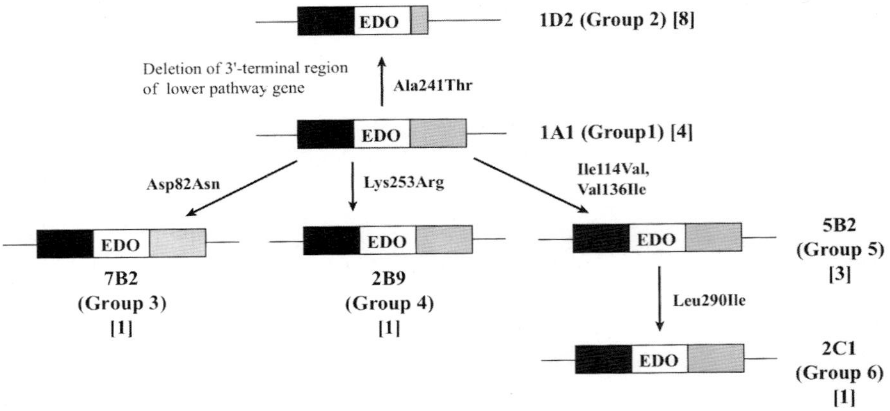

Fig. 12.2 Evolutionary lineage of EDOs in the I.2.G subfamily. The clone names, the groups, and the frequencies of the retrieved EDO clones are indicated in the *boxes*. Amino acid substitutions are indicated by the *arrows*. This figure was taken from Fig. 2 in Suenaga et al. (2007) and Fig. 1 in Suenaga et al. (2009a) with the publisher's permission

I.2.G EDOs and the roles of the amino acid changes observed in the I.2.G enzymes, we selected eight new EDOs: six from I.2.G and one each from the I.3.M (clone 7E11) and I.3.N (clone 1F2) subfamilies. These enzymes were purified to homogeneity, and their physical and catalytic properties were investigated.

12.4.1 Metal Dependence

The majority of known EDOs use Fe(II) as a cofactor, and only five EDOs have been reported that utilize metals other than Fe(II). The enzyme from *Klebsiella pneumoniae* uses Mg(II) (Gibello et al. 1994), while the other four enzymes use Mn(II) (Que et al. 1981; Hatta et al. 2003; Viggiani et al. 2004; Miyazawa et al. 2004). We first tested the metal dependence of the metagenomic EDOs. Recombinant *E. coli* were grown in M9 medium supplemented with Fe(II) or Mn(II). The activities of a crude extract prepared from the cells were then assayed using catechol. Enzymes prepared in the absence of metal ions were inactive with both substrates. The I.2.G EDOs displayed activity only when Mn(II) was added to the medium. By contrast, both the 7E11 and 1F2 EDOs were Fe(II)-dependent.

12.4.2 Kinetic Analysis

The known EDOs exhibiting a preference for catechol have K_M values ranging from 1 to 50 μM (Cerdan et al. 1995; Okuta et al. 2003; Junca et al. 2004; Viggiani et al. 2004). However, all of the I.2.G EDOs showed an extremely high affinity for catechol, with

Table 12.1 Kinetic parameters and specific activities of purified EDOs[a]

Enzyme	Subfamily	Group[b]	K_m (μM)[c]	k_{cat} (s^{-1})[c]	k_{cat}/K_m ($\mu M^{-1}\ s^{-1}$)[c]	Relative activity to 2,3-dihydroxy-biphenyl[c]
1A1	I.2.G	1	0.75 ± 0.15	0.36 ± 0.02	0.48 ± 0.07	1[d]
1D2	I.2.G	2	0.33 ± 0.02	0.38 ± 0.03	1.14 ± 0.60	1.29
7B2	I.2.G	3	0.64 ± 0.13	0.22 ± 0.02	0.34 ± 0.04	1.09
2B9	I.2.G	4	0.55 ± 0.17	0.34 ± 0.02	0.63 ± 0.19	1.58
5B2	I.2.G	5	0.46 ± 0.21	0.40 ± 0.03	0.85 ± 0.26	1.05
2C1	I.2.G	6	0.62 ± 0.56	0.32 ± 0.11	0.51 ± 0.19	1.35
7E11	I.3.M	–	45.9 ± 13.1	0.353 ± 0.06	$8.0 \times 10^{-3} \pm 2.0 \times 10^{-3}$	6.91
1F2	I.3.N	–	245 ± 78	1.00 ± 0.26	$4.08 \times 10^{-3} \pm 0.22 \times 10^{-3}$	3.01

[a] Values are expressed as average ±SD of three to five independent experiments
[b] The groups were defined in Fig. 12.1
[c] Activity to catechol
[d] The specific activity of 1A1 was 0.246 U mg^{-1}

K_M values below 1 μM (Table 12.1), which is the lowest value reported thus far. The overall catalytic efficiency, or k_{cat}/K_M, varied from clone to clone. The 1D2 (group 2) and 5B2 (group 5) clones showed significantly higher levels of activity due to a lower K_M and higher k_{cat}. The 7E11 and 1F2 EDOs exhibited much lower catalytic efficiencies (approximately two orders of magnitude) probably because they preferred 2,3-dihydroxybiphenyl to catechol.

12.4.3 Thermostability

Thermostability was determined by measuring the remaining activity after incubation at a high temperature (60, 70, or 80 °C). As shown in Fig. 12.3, 1F2 was the most thermolabile and lost activity at 60 °C within 10 min. The 7E11 enzyme was also thermolabile; incubation at 60 °C for 1 h decreased the activity to 41 % of the preheated level. In contrast, the I.2.G enzymes were quite thermostable. Of the six I.2.G EDOs, four (1A1, 2C1, 5B2, and 7B2) retained almost full activity after treatment at 60 °C for 1 h. Of these, 1A1 and 7B2 were the most thermostable, retaining 25 and 17 % of their activities even after incubation at 70 °C for 1 h, respectively. Note that the two most active clones, 1D2 and 5B2, were the most thermolabile among the I.2.G EDOs.

12.4.4 Effects of Chemical Inhibitors

Table 12.2 shows the tolerance of the EDOs against the chemical inhibitors H_2O_2 and NaCN. The oxidizing agent H_2O_2 (10 mM) caused concentration-dependent

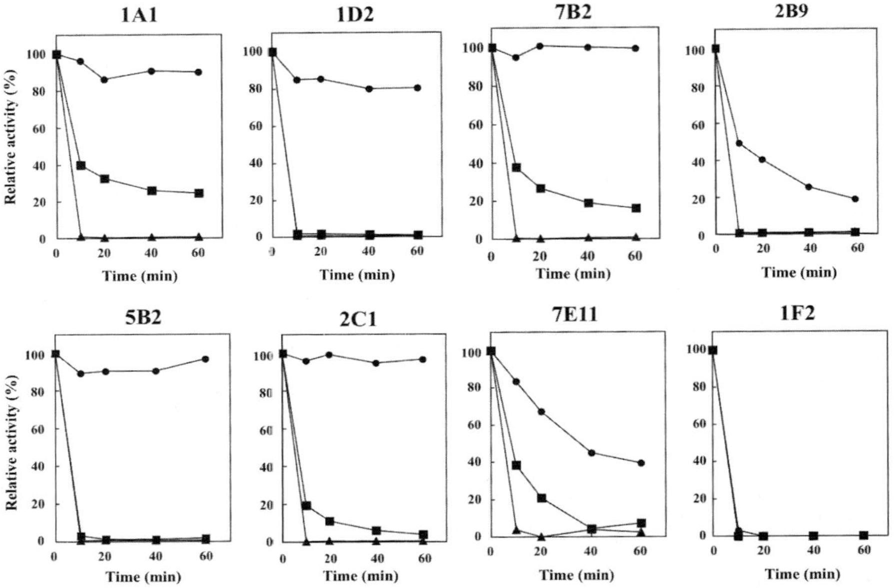

Fig. 12.3 Thermal inactivation profiles of metagenomic EDOs. Purified enzymes were incubated at 60 °C (*closed circle*), 70 °C (*closed square*), or 80 °C (*closed triangle*). At various time intervals, aliquots were recovered, and the remaining activities were determined at 30 °C

Table 12.2 Effects of inhibitors on the activity of purified EDO[a]

			Remaining activity			
			H_2O_2		NaCN	
Enzyme	Subfamily	Group[b]	1 mM	10 mM	1 mM	10 mM
1A1	I.2.G	1	81±6	21±1	98±1	98±2
1D2	I.2.G	2	79±5	26±2	99±1	108±11
7B2	I.2.G	3	70±8	21±4	95±2	113±9
2B9	I.2.G	4	84±6	26±3	88±8	103±12
5B2	I.2.G	5	59±2	24±3	97±9	112±12
2C1	I.2.G	6	63±10	21±4	97±6	113±9
7E11	I.3.M	–	17±4	13±1	87±4	31±4
1F2	I.3.N	–	26±1	16±6	91±10	35±6

[a]Values are the average of three independent experiments
[b]The groups were defined in Fig. 12.1

inactivation and reduced the activities of the I.2.G EDOs to between 21 and 26 % of their initial levels. H_2O_2 decreased the activity of 7E11 to 13 % of its initial value and the activity of 1F2 to 16 % of the level in the preincubated enzyme. The differences in tolerance between the I.2.G EDOs and other enzymes were more significant at 1 mM. The metal-binding reagent NaCN selectively inhibited 7E11 and 1F2 by 65 and 68 % at 10 mM, respectively. No loss of activity was observed for the I.2.G enzymes treated with NaCN under identical conditions. These results and those from our heat inactivation experiments indicate that the I.2.G EDOs were structurally more robust than the 7E11 and 1F2 enzymes.

12.4.5 Basis for the Dominance of the I.2.G EDOs

As described above, unlike many other EDOs, which are Fe(II)-dependent, the I.2.G EDOs preferred Mn(II). We believe that this is the primary reason for the dominance of the I.2.G EDOs in our library. Because coke-plant wastewater contains 0.5–3 mM NaCN (Chao et al. 2006; Chang et al. 2008), the use of Mn(II) instead of Fe(II) should favor enzyme resistance to NaCN stress. The structural robustness of Mn(II)-dependent enzymes is supported by many studies. The Fe(II)-dependent EDO from *Pseudomonas ovalis* is rapidly inactivated by H_2O_2 and NaCN, but this is not the case for Mn(II)-dependent enzymes (Que et al. 1981; Whiting et al. 1996); moreover, the high thermostability of a Mn(II)-dependent EDO was reported by Hatta et al. (2003). Other Mn-dependent enzymes have the capacity to withstand stress (Asada et al. 1975; Lumsden et al. 1976). The I.2.G EDOs were further characterized by their substrate affinities (Table 12.1). The K_M values for the I.2.G EDOs using catechol (<1 μM) were the lowest values reported thus far. Although the actual concentration of catechol in wastewater remains unclear, a lower substrate concentration may enable the host organism to utilize the limited carbon source.

12.4.6 Adaptive Evolution in I.2.G EDOs

Next, we assessed the capacity of the I.2.G EDOs to evolve based on their individual enzymatic properties. The catalytic properties of EDOs are influenced by single amino acid changes (Junca et al. 2004) and are, presumably, positively selected for in the environment in response to pollution. As for the metagenomic I.2.G EDOs, their catalytic activities varied from group to group, apparently due to minor amino acid substitutions (Table 12.1). The ancestral 1A1 was the most thermostable, while two of its descendents, 1D2 and 5B2, had significantly greater catalytic efficiency but reduced stability (Table 12.1 and Fig. 12.3). Hence, an apparent trade-off exists between activity and stability. Because protein stability promotes adaptive protein evolution (Bloom et al. 2006), the more active 1D2 and 5B2 may have arisen through a reduction in 1A1 thermostability. Note that the I.2.G clones with increased activity (groups 2 and 5) were found more frequently among the retrieved I.2.G clones (eight group-2 enzymes and three group-5 enzymes), which probably reflects their allele frequencies in the environment.

12.5 Novel Organization of Aromatic Degradative Pathway Genes in a Microbial Community as Revealed by Metagenomic Analysis

12.5.1 General Features of the Metagenomic Fosmid Inserts

Shotgun DNA sequencing of 38 fosmids resulted in the identification of single contigs in 22 fosmids. The inserted fragments, which collectively amounted to 1.5 Mb, ranged from 29.1 to 41.6 kb in length, from 53.9 to 69.3 % in G+C content, and from 26 to 45 ORFs. In total, there were 1,317 ORFs. When these ORFs were classified on the basis of their predicted functions and compared to those identified in purely cultivated aromatic-utilizing bacteria, the metagenomic DNA fragments exhibited a lower proportion of aromatic-degrading genes and a higher proportion of genes for DNA replication/maintenance/processing, conjugative transfer, DNA transposition, and site-specific recombination, as compared to those in the genomes of the cultivated strains.

12.5.2 Diverse Genetic Organization of Aromatic Degradation Genes in Metagenomic DNA

The genetic organization of representative fosmid inserts is depicted in Fig. 12.4. Most EDO genes were flanked by genes or gene clusters predicted to be involved in the degradation of aromatic compounds. These gene products exhibited moderate to significant levels of identity with well-studied catabolic enzymes from isolated bacteria (Shingler et al. 1992; Duffner et al. 2000; Fukumori and Saint 2001; Zhou et al. 2001; Sota et al. 2006).

12.5.3 Genes in the Fosmids Carrying I.2.G EDO Genes

12.5.3.1 Aromatic Catabolic Genes

In the 38 fosmids, the genes of the novel I.2.G EDO subfamily were overrepresented, constituting 18 clones. The dominance of the I.2.G subfamily suggests that this type of EDO plays an important role in the coke-plant wastewater habitat. Thus, we focused on this unique group to address the basis for their dominance among the retrieved clones and, possibly, in the environment.

Each of the 18 fosmids carried three genes encoding a single-component-type phenol hydroxylase, catechol 2,3-dioxygenase, and 2-hydroxymuconic semialdehyde hydrolase, in that order. The 2-hydroxymuconic semialdehyde hydrolase

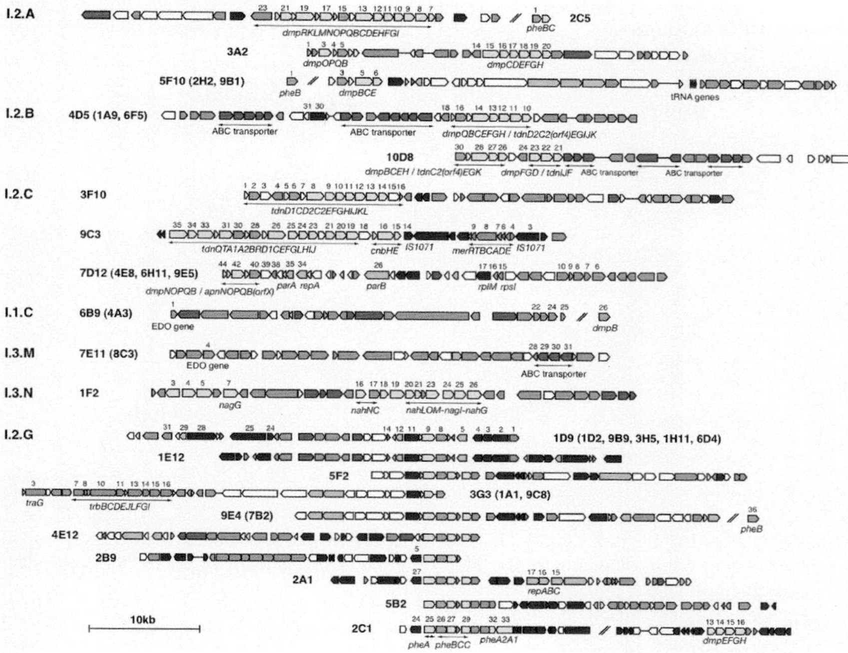

Fig. 12.4 Genetic maps of representative fosmid inserts analyzed in this study. ORFs are indicated by pentagons, and those referred to in the text are shown with their numbers above the symbols. Predicted ORF functions are shown according to color: dark yellow, EDO; *light yellow*, degradation of aromatic compounds; *green*, replication, maintenance, and processing of DNA; *red*, conjugative transfer; *pink*, DNA transposition or site-specific recombination; *dark green*, transport; *light blue*, transcriptional regulation; *grey*, other known functions; and *white*, unknown function. *Horizontal lines* indicate non-scaled spaces between genes. Two contigs in one fosmid insert are separated by double oblique strokes. Fosmids carrying same inserts are described in parentheses. This figure was taken from Fig. 1 in Suenaga et al. (2009b) with the publisher's permission

genes in the fosmids were apparently either intact or truncated at their 3′-ends. All of the phenol hydroxylase genes were very similar to one another, and their deduced products were 46 % identical to PheA from *B. stearothermophilus* BR219, the only known single-component-type phenol hydroxylase (Kim and Oriel 1995). The activities of the putative phenol hydroxylases were confirmed by a coupling assay with EDO; cellular expression of the phenol hydroxylase and EDO resulted in a yellow color in the presence of phenol.

12.5.3.2 Mobile Genetic Elements

Fosmids carrying the I.2.G EDO genes were characterized based on the relative abundance of mobile genetic elements and their remnants. It is noteworthy that the transposase genes for IS*Pme1* (in 1D9 to 7B2; see below) and an IS*Sop9*-related

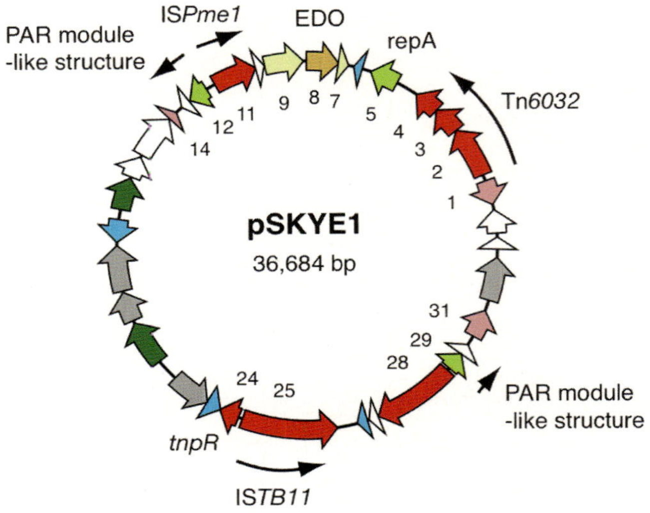

Fig. 12.5 Circular map of plasmid pSKYE1 reconstructed in silico. Predicted ORF functions are shown according to color as in Fig. 12.4. This figure was taken from Fig. 6 in Suenaga et al. (2009b) with the publisher's permission

element (ORF2B9-5, ORF2A1-27, and ORF2C1-24) were located very close to the single-component phenol hydroxylase genes. Fosmid 3G3 contained at least ten genes predicted to be involved in a type IV conjugative transfer system: the gene for coupling protein (TraG) and Trb genes for the mating-pair formation apparatus. The organization of the genes (*trbBCDEJLFGI*) in 3G3 resembles that found in the Tn*4371* genomic island (Toussaint et al. 2003) rather than that found in self-transmissible plasmids (Schröder and Lanka 2005). Taking the high (~90 %) identity to those found on the second chromosome of the α-proteobacterial strain *Ochrobactrum anthropi* ATCC 49188, the transfer-related gene cluster in 3G3 appears to be part of a variant Tn*4371*-type genomic island in which several α-proteobacterial genomes also reside.

The deduced gene products of ORF2A1-17, ORF2A1-16, and ORF2A1-15 were similar to the RepA (ZP_01441940), RepB (ZP_01441941), and RepC (ZP_01038470) of *Roseovarius* sp. This type of *repABC* operon controls replication and the partitioning of large and low copy-number plasmids and has been found only in α-proteobacteria (Cevallos et al. 2008). Therefore, the 2A1 insert appears to originate from an α-proteobacterial plasmid.

12.5.3.3 Reconstitution of a Plasmid-Like Circular DNA from Fosmid Sequences Carrying I.2.G EDO Genes

The fosmid fragments in 1D9, 1D2, 9B9, 3H5, 1H11, and 6D4 had overlapping sequences that were assembled into a circular DNA form 36,684-bp long (Fig. 12.5).

This circular DNA, designated pSKYE1, contained a total of 34 ORFs and was considered to be a plasmid because of the observations described below.

The circular DNA pSKYE1 carried several genes related to plasmid replication (ORF1D9-5), partitioning (ORF1D9-12 and -29), and conjugation (ORF1D9-1, -14, and -31). The deduced gene product of ORF1D9-5 was 82 % identical to RepA of the plasmid pAMI2 in α-proteobacterial *Paracoccus aminophilus*. An IS element and transposon were found in pSKYE1. The former mobile element flanked by 50-bp IR sequences was located very close to the single-component phenol hydroxylase gene and was thought to be IS*Pme1*, an IS element from pMTH1 in *Paracoccus methylutens* DM12 (Bartosik et al. 2008). The latter element, designated Tn*6032*, had 25-bp IR sequences and carried three ORFs encoding polypeptides with moderate sequence identity to TniA, TniB, and TniQ in Tn*5053* from *Xanthomonas* sp. W17 (Kholodii et al. 1995).

With respect to the genes involved in aromatic compound degradation, pSKYE1 contained only two genes for phenol hydroxylase and EDO. This situation is very different from that of known bacterial catabolic plasmids, which are usually large (>50 kb) and carry operonic degradative gene clusters (van der Meer et al. 1992; Dennis 2005). Because of their great lipophilicity and ability to produce reactive oxygen species, phenol and catechol are potentially toxic to bacterial cells (Benndorf et al. 2001). Therefore, pSKYE1, which carries the necessary components for degrading aromatic compounds, is likely to provide a survival advantage to its host cells in this harsh environment. The same environment might host bacterial strains that cannot degrade aromatic compounds by themselves. These bacteria may be fortunate in having this "detoxification apparatus" functioning in their environment.

12.6 Concluding Remarks

Through functional metagenomics, we obtained various new insights into the distribution and organization of aromatic degradation genes. A major finding was that the complete pathways generally recognized as "standard" in previous studies are extremely rare in the sampled natural environment. Instead, various gene subsets were identified that were not similar to previously reported "upper"- or "lower"-pathway modules, suggesting that aromatic compounds are degraded through the concerted actions of various pathways.

Strikingly, a small piece of the aromatic degradation pathway consisting of a single-component-type phenol hydroxylase and EDO was overrepresented. Taking into account the fact that this module was found in pSKYE1, a plasmid-like circular DNA form, we conclude that pSKYE1 plays a crucial role in the detoxification of aromatic compounds in our target environment. These findings were obtained from an analysis of DNA fragments totaling 1.5 Mb. Thus, the sequence analysis of metagenomes exhibiting specific functions is extremely useful in uncovering novel findings concerning targeted biological functions and their evolutionary status across a wide range of bacterial strains.

Similar to our functional or activity-based screening of metagenomic library, Uchiyama et al. (2005) has employed a reporter-based screening method, designated SIGEX. In SIGEX, flow cytometry-compatible gene (i.e., green fluorescence protein) was used as a reporter, which was cloned downstream of the metagenomic inserts. By screening a metagenomic library using a series of aromatic compounds, they have succeeded in the retrieval of several transcriptional regulators that specifically response to benzoate and naphthalene. Because of its high throughput, this approach is complementarily used with activity-based approach.

Acknowledgement This paper is reproduced with permission of John Wiley & Sons, Inc.

References

Altschul SF, Madden TL, Schaffer AA, Zhang J, Zhang Z, Miller W, Lipman DJ (1997) Gapped BLAST and PSI-BLAST: a new generation of protein database search programs. Nucleic Acids Res 25:3389–3402

Amann RI, Ludwig W, Schleifer KH (1995) Phylogenetic identification and in situ detection of individual microbial cells without cultivation. Microbiol Rev 59:143–169

Asada K, Yoshikawa K, Takahashi M, Maeda Y, Enmanji K (1975) Superoxide dismutases from a blue-green alga, *Plectonema boryanum*. J Biol Chem 250:2801–2807

Bartosik D, Putyrski M, Dziewit L, Malewska E, Szymanik M, Jagiello E, Lukasik J, Jadwiga J (2008) Transposable modules generated by a single copy of insertion sequence ISPme1 and their influence on structure and evolution of natural plasmids of *Paracoccus methylutens* DM12. J Bacteriol 190:3306–3313

Benndorf D, Loffhagen N, Babel W (2001) Protein synthesis patterns in *Acinetobacter calcoaceticus* induced by phenol and catechol show specificities of responses to chemostress. FEMS Microbiol Lett 200:247–252

Bloom JD, Labthavikul ST, Otey CR, Arnold FH (2006) Protein stability promotes evolvability. Proc Natl Acad Sci USA 103:5869–5874

Cerdan P, Rekik M, Harayama S (1995) Substrate specificity differences between two catechol 2,3-dioxygenases encoded by the TOL and NAH plasmids from *Pseudomonas putida*. Eur J Biochem 229:113–118

Cevallos MA, Cervantes-Rivera R, Gutiérrez-Ríos RM (2008) The *repABC* plasmid family. Plasmid 60:19–37

Chang E-E, Hsing H-J, Chiang P-C, Chen M-Y, Shyng J-Y (2008) The chemical and biological characteristics of coke-oven wastewater by ozonation. J Hazard Mater 156:560–567

Chao Y-M, Tseng I-C, Chang J-S (2006) Mechanism for sludge acidification in aerobic treatment of coking wastewater. J Hazard Mater 137:1781–1787

Daniel R (2005) The metagenomics of soil. Nat Rev Microbiol 3:470–478

Dennis JJ (2005) The evolution of IncP catabolic plasmids. Curr Opin Biotechnol 16:291–298

Duffner FM, Kirchner U, Bauer MP, Müller R (2000) Phenol/cresol degradation by the thermophilic *Bacillus thermoglucosidasius* A7: cloning and sequence analysis of five genes involved in the pathway. Gene 256:215–221

Eltis LD, Bolin JT (1996) Evolutionary relationships among extradiol dioxygenases. J Bacteriol 178:5930–5937

Entcheva P, Liebl W, Johann A, Hartsch T, Streit WR (2001) Direct cloning from enrichment cultures, a reliable strategy for isolation of complete operons and genes from microbial consortia. Appl Environ Microbiol 67:89–99

Fortin PD, MacPherson I, Neau DB, Bolin JT, Eltis LD (2005) Directed evolution of a ring-cleaving dioxygenase for polychlorinated biphenyl degradation. J Biol Chem 280: 42307–42314

Fukumori F, Saint CP (2001) Complete nucleotide sequence of the catechol metabolic region of plasmid pTDN1. J Gen Appl Microbiol 47:329–333

Furukawa K, Suenaga H, Goto M (2004) Biphenyl dioxygenases: functional versatilities and directed evolution. J Bacteriol 186:5189–5196

Gibello A, Ferrer E, Martín M, Garrido-Pertierra A (1994) 3,4-Dihydroxyphenylacetate 2,3-dioxygenase from *Klebsiella pneumoniae*, a Mg^{2+}-containing dioxygenase involved in aromatic catabolism. Biochem J 301:145–150

Gibson J, Harwood CS (2002) Metabolic diversity in aromatic compound utilization by anaerobic microbes. Annu Rev Microbiol 56:345–369

Gibson DT, Parales RE (2000) Aromatic hydrocarbon dioxygenases in environmental biotechnology. Curr Opin Biotechnol 11:236–243

Hatta T, Mukerjee-Dhar G, Damborsky J, Kiyohara H, Kimbara K (2003) Characterization of a novel thermostable Mn(II)-dependent 2,3-dihydroxybiphenyl 1,2-dioxygenase from a polychlorinated biphenyl- and naphthalene-degrading *Bacillus* sp. JF8. J Biol Chem 278: 21483–21492

Hirose J, Kimura N, Suyama A, Kobayashi A, Hayashida S, Furukawa K (1994) Functional and structural relationship of various extradiol aromatic ring-cleavage dioxygenases of *Pseudomonas* origin. FEMS Microbiol Lett 118:273–278

Junca H, Plumeier I, Hecht HJ, Pieper DH (2004) Difference in kinetic behaviour of catechol 2,3-dioxygenase variants from a polluted environment. Microbiology 150:4181–4187

Kholodii GY, Mindlin SZ, Bass IA, Yurieva OV, Minakhina SV, Nikiforov VG (1995) Four genes, two ends, and a *res* region are involved in transposition of Tn*5053*: a paradigm for a novel family of transposons carrying either a *mer* operon or an integron. Mol Microbiol 17:1189–1200

Kim IC, Oriel PJ (1995) Characterization of the *Bacillus stearothermophilus* BR219 phenol hydroxylase gene. Appl Environ Microbiol 61:1252–1256

Lumsden J, Cammack R, Hall DO (1976) Purification and physicochemical properties of superoxide dismutase from two photosynthetic microorganisms. Biochim Biophys Acta 438:380–392

Mesarch MB, Nakatsu CH, Nies L (2000) Development of catechol 2,3-dioxygenase-specific primers for monitoring bioremediation by competitive quantitative PCR. Appl Environ Microbiol 66:678–683

Miyazawa D, Mukerjee-Dhar G, Shimura M, Hatta T, Kimbara K (2004) Genes for Mn(II)-dependent NahC and Fe(II)-dependent NahH located in close proximity in the thermophilic naphthalene and PCB degrader, *Bacillus* sp. JF8: cloning and characterization. Microbiology 150:993–1004

Nojiri H, Shintani M, Omori T (2004) Divergence of mobile genetic elements involved in the distribution of xenobiotic-catabolic capacity. Appl Microbiol Biotechnol 64:154–174

Okuta A, Ohnishi K, Yagame S, Harayama S (2003) Intersubunit interaction and catalytic activity of catechol 2,3-dioxygenases. Biochem J 371:557–564

Que L Jr, Widom J, Crawford RL (1981) 3,4-Dihydroxyphenylacetate 2,3-dioxygenase. A manganese(II) dioxygenase from *Bacillus brevis*. J Biol Chem 256:10941–10944

Schröder G, Lanka E (2005) The mating pair formation system of conjugative plasmids—a versatile secretion machinery for transfer of proteins and DNA. Plasmid 54:1–25

Shingler V, Powlowski J, Marklund U (1992) Nucleotide sequence and functional analysis of the complete phenol/3,4-dimethyl phenol catabolic pathway of *Pseudomonas* sp. strain CF600. J Bacteriol 174:711–724

Sota M, Yano H, Ono A, Miyazaki R, Ishii H, Genka H, Top EM, Tsuda M (2006) Genomic and functional analysis of the IncP-9 naphthalene-catabolic plasmid NAH7 and its transposon Tn*4655* suggests catabolic gene spread by a tyrosine recombinase. J Bacteriol 188:4057–4067

Stamoudis VC, Luthy RG (1980) Determination of biological removal of organic constituents in quench waters from high-BTU coal-gasification pilot plants. Water Res 14:1143–1156

Suenaga H (2012) Targeted metagenomics: a high-resolution metagenomics approach for specific gene clusters in complex microbial communities. Environ Microbiol 14:13–22

Suenaga H, Ohnuki T, Miyazaki K (2007) Functional screening of a metagenomic library for genes involved in microbial degradation of aromatic compounds. Environ Microbiol 9:2289–2297

Suenaga H, Mizuta S, Miyazaki K (2009a) The molecular basis for adaptive evolution in novel extradiol dioxygenases retrieved from the metagenome. FEMS Microbiol Ecol 69:472–480

Suenaga H, Koyama Y, Miyakoshi M, Miyazaki R, Yano H, Sota M, Ohtsubo Y, Tsuda M, Miyazaki K (2009b) Novel organization of aromatic degradation pathway genes in a microbial community as revealed by metagenomic analysis. ISME J 3:1335–1348

Top EM, Springael D (2003) The role of mobile genetic elements in bacterial adaptation to xenobiotic organic compounds. Curr Opin Biotechnol 14:262–269

Toussaint A, Merlin C, Monchy S, Benotmane MA, Leplae R, Mergeay M, Springael D (2003) The biphenyl- and 4-chlorobiphenyl-catabolic transposon Tn*4371*, a member of a new family of genomic islands related to IncP and Ti plasmids. Appl Environ Microbiol 69:4837–4845

Tsuda M, Tan HM, Nishi A, Furukawa K (1999) Mobile catabolic genes in bacteria. J Biosci Bioeng 87:401–410

Uchiyama T, Miyazaki K (2009) Functional metagenomics for enzyme discovery: challenges to efficient screening. Curr Opin Biotechnol 20:616–622

Uchiyama T, Abe T, Ikemura T, Watanabe K (2005) Substrate-induced gene-expression screening of environmental metagenome libraries for isolation of catabolic genes. Nat Biotechnol 23:88–93

Vaillancourt FH, Bolin JT, Eltis LD (2004) Ring-cleavage dioxygenases. In: Ramos JL (ed) Pseudomonas, vol 3. Kluwer Academic/Plenum Publishers, New York, pp 359–395

van der Meer JR, de Vos WM, Harayama S, Zehnder AJB (1992) Molecular mechanisms of genetic adaptation to xenobiotic compounds. Microbiol Rev 56:677–694

Viggiani A, Siani L, Notomista E, Birolo L, Pucci P, Donato AD (2004) The role of the conserved residues His-246, His-199, and Tyr-255 in the catalysis of catechol 2,3-dioxygenase from *Pseudomonas stutzeri* OX1. J Biol Chem 279:48630–48639

Whiting AK, Boldt YR, Hendrich MP, Wackett LP, Que L Jr (1996) Manganese(II)-dependent extradiol-cleaving catechol dioxygenase from *Arthrobacter globiformis* CM-2. Biochemistry 35:160–170

Williams PA, Sayers JR (1994) The evolution of pathways for aromatic hydrocarbon oxidation in *Pseudomonas*. Biodegradation 5:195–217

Wittich R-M (1998) Degradation of dioxin-like compounds by microorganisms. Appl Microbiol Biotechnol 49:489–499

Zhou N-Y, Fuenmayor SL, Williams PA (2001) *nag* genes of *Ralstonia* (formerly *Pseudomonas*) sp. strain U2 encoding enzymes for gentisate catabolism. J Bacteriol 183:700–708

Part III
Bacterial Behavior in Natural Environmental Systems

Chapter 13
Syntrophic Interactions in Biodegradative Consortia

Kazuya Watanabe and Atsushi Kouzuma

Abstract Bioremediation is carried out in nature, where countless microbes coexist and interact with each other. This chapter highlights syntrophic interactions among biodegradative bacteria that may be important to understand how organics (e.g., pollutants) are biodegraded in nature. We particularly focus on methanogenic and dechlorinating biodegradation with the expectation that scientists will more direct their studies toward syntrophic association among biodegradative bacteria.

Keywords Dechlorination • Dehalogenation • Fermentation • Interspecies electron transfer • Methanogenesis • Symbiosis • Syntrophy

13.1 Introduction

After the prominent work of Robert Koch and Louis Pasteur, microbiology has progressed based on pure-culture techniques and nutrient-rich culture media. Pure-culture techniques have been successfully used for identifying pathogens (Woolhouse and Gaunt 2007) and constructing industrially useful microbes (Steele and Stowers 1991). On the other hand, it has been argued that such microbiology has limitation in understanding the ecology of microbes in the natural environment (Amann et al. 1995; Watanabe and Baker 2000), since natural microbial habitats are largely different from laboratory cultures in terms of nutrient conditions and the presence of different microbial species. In the 1990s, molecular techniques have been introduced in microbial ecology, facilitating the detection and phylogenetic

K. Watanabe (✉) • A. Kouzuma
School of Life Sciences, Tokyo University of Pharmacy and Life Sciences, Hachioji, Japan
e-mail: kazuyaw@toyaku.ac.jp

H. Nojiri et al. (eds.), *Biodegradative Bacteria: How Bacteria Degrade, Survive, Adapt, and Evolve*, DOI 10.1007/978-4-431-54520-0_13, © Springer Japan 2014

identification of microbes in natural and engineered environments without isolation and cultivation (Amann et al. 1995). Microbial ecologists have realized that, in most cases, conventional laboratory isolates constitute minor populations in the environment, and there exist yet-uncultured microbes, some of which play major ecological roles (Watanabe and Baker 2000).

Studies on biodegradative bacteria are typically initiated by the isolation of bacteria from environmental samples that are able to grow on a target organic compound as a carbon and/or energy source. Subsequently, isolated strains are investigated for understanding their physiology and activities, followed by the identification of genes coding for degradative pathways (Watanabe 2001). These studies have uncovered a tremendous diversity of biological metabolic pathways, from which many industrially useful catabolic genes have been identified (Parales and Ditty 2005). It should however be noted that biodegradative bacteria studied in such ways are, in many cases, not major players for the biodegradation of organic compounds in natural environments (Watanabe and Hamamura 2003). For instance, a study on phenol-degrading activated sludge has shown that the major phenol-degrading bacteria in sludge could not be isolated after a conventional batch-enrichment culture (Watanabe et al. 1998). It is also noteworthy that a metagenomic screening of oil-contaminated groundwater for obtaining bacterial catabolic genes showed that those obtained by the metagenomic screening were mostly present in segmentalized operon structures (Uchiyama et al. 2005), suggesting that most bacteria in nature may harbor only parts of complete gene sets necessary for biodegradation. This finding also implies that many bacteria in the environment may partially contribute to the entire biodegradation process (from organic compounds to carbon dioxide and/or methane), as has been observed in methanogenic biodegradation of organic matter (Schink 2002, 2006).

This chapter highlights syntrophic interactions among biodegradative bacteria that may be important to understand how organics (e.g., pollutants) are biodegraded in nature. We particularly focus on methanogenic and dechlorinating biodegradation with the expectation that scientists will more direct their studies toward syntrophic association among biodegradative bacteria.

13.2 Methanogenic Biodegradation

Methanogenic biodegradation occurs in the anaerobic environment, where appropriate electron acceptors are not present (Ferry 1993). This situation is common in subsurface environments, such as agricultural soil (Le Mer and Roger 2001), sediment of eutrophic lakes (Lovley and Klug 1983), and subterranean oil fields (Jones et al. 2008), and a substantial portion (~10 %) of photosynthesized organics are considered to be decomposed via methanogenesis (Vogels 1979). In addition, methanogenic biodegradation has been used in anaerobic digesters to treat organic wastes and generate biological fuel gas (methane) from them (Gujer and Zehnder 1983).

A characteristic of methanogenic biodegradation is that different species of bacteria and archaea share respective steps in the decomposition of complex organic

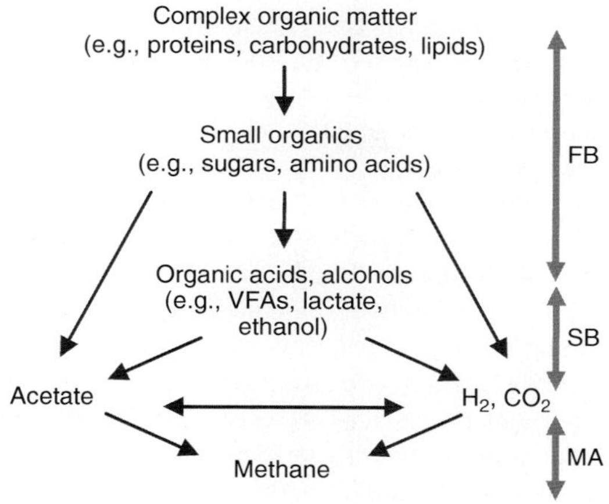

Fig. 13.1 Metabolic interactions in the methanogenic biodegradation process. *FB* fermentative bacteria, *SB* syntrophic bacteria, *MA* methanogenic archaea

matter to produce methane and carbon dioxide (Fig. 13.1); these steps include (1) hydrolysis of complex organic matter to generate monomers (e.g., amino acids and sugars), (2) fermentation of monomers to generate alcohols and volatile fatty acids (VFAs), (3) syntrophic conversion of alcohols and VFAs to acetate, hydrogen, and carbon dioxide, (4) methane production from acetate, and (5) methane production from hydrogen and carbon dioxide. Many microbes that have catalytic activities for respective steps have been isolated, and complete genome sequences for some of them have been determined (Kosaka et al. 2008). Studies have revealed that the first three steps are catalyzed by bacteria, while archaea are responsible for the last two steps. Among these steps, the syntrophic conversion is considered to be the bottleneck step (Schink 1997), since a high load of organic wastes to anaerobic digesters results in the accumulation of VFAs, acidification, and subsequent process breakdown (Ward et al. 2008). Understanding of the syntrophic conversion of VFAs is therefore necessary for the reliable operation of anaerobic digesters.

VFAs are unfavorable substrates for anaerobes, since oxidation of these substrates to hydrogen and carbon dioxide (or formate) is endergonic under standard conditions (i.e., the changes in the Gibbs free energy are positive) and is thermodynamically feasible only when the hydrogen partial pressure (or formate concentration) is kept low. For instance, thermodynamic estimation has predicted that hydrogen partial pressures as low as 10 and100 Pa are necessary for the oxidation of propionate and butyrate, respectively (Schmidt and Ahring 1993; Schink 1997). Since hydrogen and formate are scavenged mainly by the carbonate-respiring methanogenic archaea, syntrophic association of VFA-oxidizing bacteria (called syntrophs) and methanogenic archaea (methanogens) is considered indispensable for efficient VFA oxidation (Thauer et al. 1977; Schink 1997; Ishii et al. 2005).

In order for interspecies hydrogen transfer to occur stably and efficiently, studies have found that syntrophs and methanogens need to be co-aggregated to achieve close physical contacts (Ishii et al. 2005, 2006). Recently, it was uncovered that flagella of syntroph *Pelotomaculum thermopropionicum* specifically adhere to the cell surface of partner methanogen *Methanothermobacter thermautotrophicus*, facilitating their specific contact (Shimoyama et al. 2009). Besides, it was also found that the binding of a flagellum cap protein (FliD) of *P. thermopropionicum* causes transcriptomic shifts in *M. thermautotrophicus*, allowing this methanogen to prepare for the upcoming syntrophic methanogenesis (Shimoyama et al. 2009). This is the first example of specific interspecies communication between prokaryotes, which should be pursued to examine how widely such protein-assisted interspecies communication distributes over prokaryotic communities. It will also be a subject of future studies to develop methods to utilize such molecular mechanisms for facilitating syntrophic conversion of VFA to methane.

Hydrocarbons are also subjected to methanogenic biodegradation in subsurface environments, such as oil reservoirs (Gieg et al. 2008; Jones et al. 2008) and heavily contaminated sites (Kasai et al. 2005; Siddique et al. 2011). Studies analyzing natural oil-contaminated samples have shown linkages between hydrocarbon decay and methane formation, while a direct evidence for methanogenic hydrocarbon degradation was provided using enrichment cultures, in which a consortium comprised of *Deltaproteobacteria* and *Euryarchaeota* transformed ^{14}C-labled hexadecane into ^{14}C-methane at a feasible stoichiometry (Zengler et al. 1999). Aromatic hydrocarbons, such as benzene (Weiner and Lovley 1998), toluene, and xylenes (Edwards and Grbić-Galić 1994), are also reported to be biodegraded under methanogenic conditions. These examples show that methanogenic biodegradation can contribute to the natural attenuation of hydrocarbons and the formation of natural gas in subsurface oil fields.

13.3 Dechlorinating Biodegradation

Chlorinated organic compounds have divergent and useful chemical properties and are widely used as solvents (e.g., chloroform, dichloromethane, dichloroethene, trichloroethene (TCE), and tetrachloroethene [perchloroethene; PCE]), polymers (e.g., polyvinylchloride), insulators (e.g., polychlorinated biphenyls [PCBs]), and pesticides (e.g., pentachlorophenol [PCP]) (Mohn and Tiedje 1992). These compounds are industrially produced in large quantities, in some cases, resulting in accidental and deliberate releases into the environment. In addition, chlorinated organics such as polychlorinated dibenzo-*p*-dioxins (PCDDs) and dibenzo-*p*-furans (PCDFs) are formed as unwanted by-products of manufacturing and incineration processes (Meharg and Osborn 1995). Many of such chlorinated organics are regarded as serious environmental pollutants that affect human health and ecosystems because of their toxicities, environmental persistence, and bioaccumulation properties (Henschler 1994). A wide variety of approaches, including physicochemical and biochemical processes, have therefore been studied and applied for removing these pollutants from the environments. Among them, bioremediation processes utilizing dechlorinating bacteria are regarded as one of the promising approaches to remediate the polluted environments. Studies have reported that many bacterial species and microbial communities are capable of degrading a large variety of halogenated organics with the aid of their metabolic functions and syntrophic interactions.

Several pathways have been reported for bacterial dechlorination of chlorinated organic compounds (e.g., oxidative, fermentative, and reductive dehalogenation pathways) (Fetzner 1998; Smidt and de Vos 2004). Among them, reductive dehalogenation processes have been revealed to play crucial roles as the initial degradation step for chlorinated organic compounds, especially for highly polychlorinated compounds such as PCDDs, PCDFs, PCBs, and PCE, since these compounds are recalcitrant to oxidative dehalogenation and have the tendency to accumulate in aquatic sediments and soils in the anaerobic environment (Bunge and Lechner 2009). Significant attention has therefore been given to the isolation and characterization of anaerobic microbes that are capable of reductive dehalogenation. Studies have indicated that reductive dehalogenation by anaerobic bacteria can be largely divided into two types: co-metabolic (abiotic) dehalogenation and metabolic dehalogenation (halorespiration) (Fetzner 1998). Co-metabolic dehalogenation is proposed to be catalyzed by metal-containing porphyrins, corrinoids, and cytochromes (Egli et al. 1988; Holliger et al. 1992; Picardal et al. 1993). These catalytic substances are produced by a variety of anaerobes, including methanogenic archaea and acetogenic, sulfate-reducing, and iron-reducing bacteria. However, these microbes cannot couple the reductive dehalogenation to energy conservation (i.e., respiration). On the other hand, various bacteria have been isolated and characterized that can conserve energy and grow during reductive dehalogenation. These bacteria, called halorespiring bacteria, utilize halogenated organic compounds as the electron acceptors during anaerobic respiration (i.e., halorespiration). Since halorespiration processes dechlorinate a wide variety of chlorinated compounds, including highly polychlorinated organics, and proceed at rates several orders of magnitude higher than those with co-metabolic processes (Holliger and Schumacher 1994), halorespiring bacteria are assumed to play critical roles in dechlorination and detoxification of these harmful chlorinated compounds in the environment.

In many cases, however, it is reported that isolated halorespiring bacteria are not capable of the complete dehalogenation of polychlorinated organics in pure cultures (Bunge and Lechner 2009). It is also reported that, in natural environments, many non-halorespiring bacteria contribute to the growth of halorespiring bacteria by supplying energy sources and growth factors (Dolfing and Tiedje 1986; Drzyzga and Gottschal 2002; Heimann et al. 2006). These studies have indicated that halorespiring bacteria have developed tight syntrophic relationships with other halorespiring or non-halorespiring bacteria for achieving the effective dehalogenation. Several types of syntrophic interactions have been reported, including interspecies transfer of halogenated intermediates among halorespiring bacteria and interspecies transfer of energy sources (e.g., hydrogen and acetate), growth factors (e.g., vitamins), and other metabolic intermediates between halorespiring and non-halorespiring bacteria (Fig. 13.2).

Interspecies transfer of halogenated intermediates has been observed in dechlorination processes of polychlorinated organics, such as PCE, PCDDs, and PCDFs. Under anaerobic conditions, PCE is consecutively converted to TCE, dichloroethenes (DCEs,

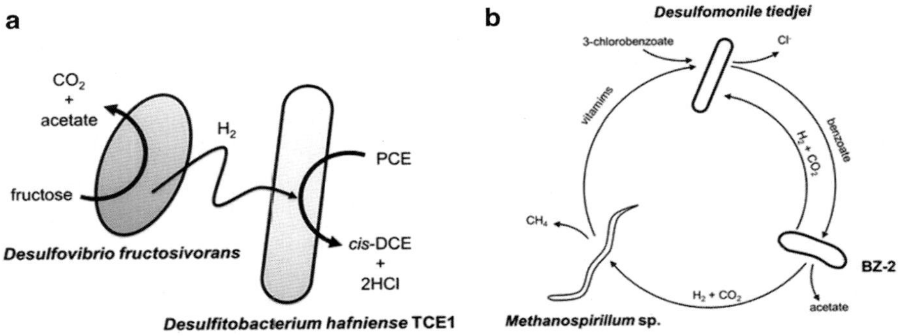

Fig. 13.2 Syntrophic interactions among dechlorinating and non-dechlorinating bacteria. (**a**) Syntrophic dechlorination of PCE by *Desulfovibrio fructosivorans* and *Desulfitobacterium hafniense* TCE1 (Drzyzga and Gottschal 2002). (**b**) Syntrophic interactions among *Desulfomonile tiedjei*, *Methanospirillum* sp., and BZ-2 in a defined consortium growing on 3CB (Mohn and Tiedje 1992)

mainly *cis*-1,2-DCE), vinyl chloride (VC), and ethene during the reductive dehalogenation. Many halorespiring bacteria belonging to the genera *Desulfitobacterium*, *Dehalobacter*, *Sulfurospirillum*, *Desulfuromonas*, and *Geobacter* can dechlorinate PCE to *cis*-1,2-DCE, although these bacteria cannot completely dechlorinate *cis*-1,2-DCE to ethene (Futagami et al. 2008). On the other hand, only bacterial strains phylogenetically related to the genus *Dehalococcoides* have been shown to reductively dechlorinate *cis*-1,2-DCE to ethene (Cupples et al. 2003; He et al. 2003a, b). It is reported that *Dehalococcoides ethenogenes* strain 195 can completely dechlorinate PCE to ethene (Maymó-Gatell et al. 1997). However, this strain cannot utilize VC as the electron acceptor. The dechlorination of VC to ethene by strain 195 is a slow reaction that is assumed to proceed in a co-metabolic manner, therefore resulting in the transient accumulation of this highly toxic intermediate in its pure culture (Maymó-Gatell et al. 1997). However, in the dechlorination processes proceeding in mixed cultures of some halorespiring bacteria, it is reported that various *Dehalococcoides* and related strains contribute to the complete and rapid dechlorination of PCE through tight syntrophic interactions for interspecies transfer of dechlorinated intermediates (Bunge et al. 2007). *Dehalococcoides* and related bacteria are also considered to be the key organisms in reductive dehalogenation of PCDDs and PCDFs. In natural environments, these compounds are also assumed to be degraded through the combination of sequential dechlorinating reactions by interactions of halorespiring bacteria, although details and the extent of such interactions are currently not sufficiently known (Bunge and Lechner 2009).

Most halorespiring bacteria, including *Dehalococcoides* species, mainly utilize hydrogen as the electron donor. Studies have demonstrated that syntrophic relationships for interspecies hydrogen transfer exist between such halorespiring bacteria and hydrogen-producing microbes. For instance, it was reported that PCE-dechlorinating *Desulfitobacterium hafniense* (formerly *Desulfitobacterium frappieri*) strain TCE1 formed a syntrophic relationship with a sulfate-reducing bacterium,

Desulfovibrio fructosivorans, in mixed cultures supplemented with fructose, sulfate, and PCE (Drzyzga and Gottschal 2002). In this mixed culture, hydrogen was produced by *D. fructosivorans* during the fermentation of fructose and was utilized by strain TCE1 as the electron donor for reductive dechlorination of PCE. This type of syntrophic relationship was also found in a mixed culture containing hydrogen-producing archaea (*Methanosarcina*) and VC-dechlorinating bacteria (*Dehalococcoides*) (Heimann et al. 2006). In this mixed culture, *Methanosarcina* produces hydrogen during the oxidization of acetate to CO_2, and the produced hydrogen was utilized for hydrogenotrophic dehalorespiration of VC to ethene by *Dehalococcoides*. Generally, in anaerobic environments where hydrogen is the main source of electrons, hydrogen-utilizing (hydrogenotrophic) microbes such as methanogens, acetogens, and sulfate reducers compete with halorespiring bacteria. However, since dechlorinating bacteria have a high affinity for H_2, it is assumed that they are superior to such hydrogenotrophic competitors, especially in the environment where partial pressures of hydrogen are low (Ballapragada et al. 1997; Fennell and Gossett 1998; Yang and McCarty 1998; Löffler et al. 1999).

Acetate is also one of the main intermediates during the anaerobic degradation of organic substances and has been reported as another electron source for reductively dehalogenating bacteria. Acetate-utilizing (acetotrophic) halorespiring bacteria belonging to the genera *Desulfuromonas*, *Desulfovibrio*, *Trichlorobacter*, and *Anaeromyxobacter* (Krumholz et al. 1996; de Wever et al. 2000; Sun et al. 2000; Sanford et al. 2002) have been isolated from diverse environments, suggesting that interspecies transfer of acetate plays a role in facilitating halorespiration in the natural environment. In addition, He et al. (2002) reported the presence of dechlorinating populations that used acetate but not H_2 as a direct electron donor for dechlorination of PCE to *cis*-DCE in hydrogen-supplemented microcosms. This study indicated that, under that condition, the dechlorinating process depended on syntrophic interspecies transfer of acetate between homoacetogens and acetotrophic dehalorespiring bacteria rather than the activity of hydrogenotrophic dehalogenating bacteria. Similar syntrophic interactions were also reported for dechlorination of 2,3,5,6-tetrachlorobiphenyl in sediment-free enrichment cultures (Cutter et al. 2001).

Essential growth factors for halorespiring bacteria (e.g., vitamins) can also be supplied via syntrophic relationships with non-halorespiring microbes. He et al. (2007) reported that syntrophic interactions with *Acetobacterium woodii*, known for its ability to synthesize vitamin B_{12}, promote the growth of *D. ethenogenes* strain 195. Dolfing and Tiedje (1986) reported syntrophic relationships consisting of three microorganisms in a defined consortium growing on 3-chlorobenzoate (3-CB) as the sole organic substrate. In addition to the key dechlorinating bacterium *Desulfomonile tiedjei*, the consortium contained a benzoate-fermenting bacterium (BZ-2) and a hydrogen-consuming methanogen (*Methanospirillum* sp.). In this consortium, 3-CBwas reductively dechlorinated by *D. tiedjei*, and the released benzoate was fermented by BZ-2 with consequent production of acetate and H_2. Benzoate fermentation was dependent on a low hydrogen partial pressure, which was maintained by the hydrogen consumption by the methanogen and also by the dechlorination of 3-CB by *D. tiedjei*. The rate of benzoate fermentation and consequent hydrogen production were rather simulated by the combination of their hydrogen consumption. *Methanospirillum* sp. also played a role in providing vitamins to *D. tiedjei*.

As mentioned here, syntrophic associations among dehalorespiring bacteria and other members of microbial communities play significant roles in facilitating the dechlorination and detoxification of many chlorinated organic compounds. In addition to the anthropogenic release, abundant and diverse halogenated organics are also produced naturally (Gribble 1996), suggesting that halogenating bacteria and their syntrophic relationships have divergently evolved in the presence of various halogenated organics of natural origin. Although our knowledge of these syntrophic relationships is not yet complete, understanding of such complex microbial interactions would provide useful information for assessing the biodegradability of chlorinating organic compounds in the natural environment.

13.4 Conclusions

This chapter collectively discussed syntrophic interactions in methanogenesis and dechlorination, aiming at highlighting their importance in biodegradation of organic compounds in the environment. We consider that syntrophic interactions among biodegradative microbes ubiquitously contribute to biodegradation of organic matter in the environment, to which we need to pay more attention than hitherto. For instance, it has been shown that degradation of 4-chlorodibenzofuran by *Sphingomonas* sp. RW1 results in the accumulation of a dead-end product 3,5-dichlorosalicylate, while the inoculation with *Burkholderia* sp. JWS enables the cooperative complete degradation (Arfmann et al. 1997). Besides, it has also been reported that a consortium comprised of *Escherichia coli* SD2 and *Pseudomonas putida* KT2440 pSB337 efficiently degrades parathion without accumulation of toxic intermediates (Gilbert et al. 2003). These examples imply possible utilization of syntrophic interactions for bioremediation, which should be investigated in the future.

References

Amann RI, Ludwig W, Schleifer KH (1995) Phylogenetic identification and in situ detection of individual microbial cells without cultivation. Microbiol Rev 59:143–169

Arfmann H, Timmis KN, Wittich R (1997) Mineralization of 4-chlorodibenzofuran by a consortium consisting of *Sphingomonas* sp. strain RW1 and *Burkholderia* sp. strain JWS. Appl Environ Microbiol 63:3458–3462

Ballapragada B, Stensel H, Puhakka J, Ferguson J (1997) Effect of hydrogen on reductive dechlorination of chlorinated ethenes. Environ Sci Technol 31:1728–1734

Bunge M, Lechner U (2009) Anaerobic reductive dehalogenation of polychlorinated dioxins. Appl Microbiol Biotechnol 84:429–444

Bunge M, Kleikemper J, Miniaci C, Duc L, Muusse MG, Hause G, Zeyer J (2007) Benzoate-driven dehalogenation of chlorinated ethenes in microbial cultures from a contaminated aquifer. Appl Microbiol Biotechnol 76:1447–1456

Cupples AM, Spormann AM, McCarty PL (2003) Growth of a *Dehalococcoides*-like microorganism on vinyl chloride and *cis*-dichloroethene as electron acceptors as determined by competitive PCR. Appl Environ Microbiol 69:953–959

Cutter LA, Watts JE, Sowers KR, May HD (2001) Identification of a microorganism that links its growth to the reductive dechlorination of 2,3,5,6-chlorobiphenyl. Environ Microbiol 3:699–709

De Wever H, Cole JR, Fettig MR, Hogan DA, Tiedje JM (2000) Reductive dehalogenation of trichloroacetic acid by *Trichlorobacter thiogenes* gen. nov., sp. nov. Appl Environ Microbiol 66:2297–2301

Dolfing J, Tiedje JM (1986) Hydrogen cycling in a three-tiered food web growing on the methanogenic conversion of 3-chlorobenzoate. FEMS Microbiol Lett 38:293–298

Drzyzga O, Gottschal JC (2002) Tetrachloroethene dehalorespiration and growth of *Desulfitobacterium frappieri* TCE1 in strict dependence on the activity of *Desulfovibrio fructosivorans*. Appl Environ Microbiol 68:642–649

Edwards EA, Grbić-Galić D (1994) Anaerobic degradation of toluene and o-xylene by a methanogenic consortium. Appl Environ Microbiol 60:313–322

Egli C, Tschan T, Scholtz R, Cook AM, Leisinger T (1988) Transformation of tetrachloromethane to dichloromethane and carbon dioxide by *Acetobacterium woodii*. Appl Environ Microbiol 54:2819–2824

Fennell DE, Nijenhuis I, Wilson SF, Zinder SH, Häggblom MM (2004) *Dehalococcoides ethenogenes* strain 195 reductively dechlorinates diverse chlorinated aromatic pollutants. Environ Sci Technol 38:2075–2081

Ferry J (1993) Methanogenesis. Chapman and Hall, New York

Fetzner S (1998) Bacterial dehalogenation. Appl Microbiol Biotechnol 50:633–657

Futagami T, Goto M, Furukawa K (2008) Biochemical and genetic bases of dehalorespiration. Chem Rec 8:1–12

Gieg LM, Duncan KE, Suflita JM (2008) Bioenergy production via microbial conversion of residual oil to natural gas. Appl Environ Microbiol 74:3022–3029

Gilbert ES, Walker AW, Keasling JD (2003) A constructed microbial consortium for biodegradation of the organophosphorus insecticide parathion. Appl Microbiol Biotechnol 61:77–81

Gribble GW (1996) Naturally occurring organohalogen compounds. Acc Chem Res 31:141–152

Gujer W, Zehnder AJB (1983) Conversion processes in anaerobic digestion. Water Sci Technol 15:127–167

He J, Ritalahti KM, Aiello MR, Löffler FE (2003a) Complete detoxification of vinyl chloride by an anaerobic enrichment culture and identification of the reductively dechlorinating population as a *Dehalococcoides* species. Appl Environ Microbiol 69(2):996–1003

He J, Ritalahti KM, Yang KL, Koenigsberg SS, Löffler FE (2003b) Detoxification of vinyl chloride to ethene coupled to growth of an anaerobic bacterium. Nature 424:62–65

He J, Holmes VF, Lee PK, Alvarez-Cohen L (2007) Influence of vitamin B_{12} and cocultures on the growth of *Dehalococcoides* isolates in defined medium. Appl Environ Microbiol 73: 2847–2853

Heimann AC, Batstone DJ, Jakobsen R (2006) *Methanosarcina* spp. drive vinyl chloride dechlorination via interspecies hydrogen transfer. Appl Environ Microbiol 72:2942–2949

Henschler D (1994) Toxicity of chlorinated organic compounds: effects of the introduction of chlorine in organic molecules. Angew Chem Int Ed Engl 33:1920–1935

Holliger C, Schumacher W (1994) Reductive dehalogenation as a respiratory process. Antonie Van Leeuwenhoek 66:239–246

Holliger C, Kengen SW, Schraa G, Stams AJ, Zehnder AJ (1992) Methyl-coenzyme M reductase of *Methanobacterium thermoautotrophicum* delta H catalyzes the reductive dechlorination of 1,2-dichloroethane to ethylene and chloroethane. J Bacteriol 174:4435–4443

Ishii S, Kosaka T, Hori K, Hotta Y, Watanabe K (2005) Coaggregation facilitates interspecies hydrogen transfer between *Pelotomaculum thermopropionicum* and *Methanothermobacter thermautotrophicus*. Appl Environ Microbiol 71:7838–7845

Ishii S, Kosaka T, Hotta Y, Watanabe K (2006) Simulating the contribution of coaggregation to interspecies hydrogen fluxes in syntrophic methanogenic consortia. Appl Environ Microbiol 72:5093–5096

Jones DM, Head IM, Gray ND, Adams JJ, Rowan AK, Aitken CM, Bennett B, Huang H, Brown A, Bowler BF, Oldenburg T, Erdmann M, Larter SR (2008) Crude-oil biodegradation via methanogenesis in subsurface petroleum reservoirs. Nature 451:176–180

Kasai Y, Takahata Y, Hoaki T, Watanabe K (2005) Physiological and molecular characterization of a microbial community established in unsaturated, petroleum-contaminated soil. Environ Microbiol 7:806–818

Kosaka T, Kato S, Shimoyama T, Ishii S, Abe T, Watanabe K (2008) The genome of *Pelotomaculum thermopropionicum* reveals niche-associated evolution in anaerobic microbiota. Genome Res 18:442–448

Krumholz LR, Sharp R, Fishbain SS (1996) A freshwater anaerobe coupling acetate oxidation to tetrachloroethylene dehalogenation. Appl Environ Microbiol 62:4108–4113

Le Mer J, Roger P (2001) Production, oxidation, emission and consumption of methane by soils: a review. Eur J Soil Biol 37:25–50

Löffler FE, Tiedje JM, Sanford RA (1999) Fraction of electrons consumed in electron acceptor reduction and hydrogen thresholds as indicators of halorespiratory physiology. Appl Environ Microbiol 65:4049–4056

Lovley DR, Klug RJ (1983) Methanogenesis from methanol and methylamines and acetogenesis from hydrogen and carbon dioxide in sediments of an eutrophic lake. Appl Environ Microbiol 45:1310–1315

Maymó-Gatell X, Chien Y, Gossett JM, Zinder SH (1997) Isolation of a bacterium that reductively dechlorinates tetrachloroethene to ethene. Science 276:1568–1571

Meharg AA, Osborn D (1995) Dioxins released from chemical accidents. Nature 375:353–354

Mohn WW, Tiedje JM (1992) Microbial reductive dehalogenation. Microbiol Rev 56:482–507

Parales RE, Ditty JL (2005) Laboratory evolution of catabolic enzymes and pathways. Curr Opin Biotechnol 16:315–325

Picardal FW, Arnold RG, Couch H, Little AM, Smith ME (1993) Involvement of cytochromes in the anaerobic biotransformation of tetrachloromethane by *Shewanella putrefaciens* 200. Appl Environ Microbiol 59:3763–3770

Sanford RA, Cole JR, Tiedje JM (2002) Characterization and description of *Anaeromyxobacter dehalogenans* gen. nov., sp. nov., an aryl-halorespiring facultative anaerobic *myxobacterium*. Appl Environ Microbiol 68:893–900

Schink B (1997) Energetics of syntrophic cooperation in methanogenic degradation. Microbiol Mol Biol Rev 61:262–280

Schink B (2002) Synergistic interactions in the microbial world. Antonie Van Leeuwenhoek 81:257–261

Schink B (2006) Syntrophic associations in methanogenic degradation. Prog Mol Subcell Biol 41:1–19

Schmidt JE, Ahring BK (1993) Effects of hydrogen and formate on the degradation of propionate and butyrate in thermophilic granules from an upflow anaerobic sludge blanket reactor. Appl Environ Microbiol 59:2546–2551

Shimoyama T, Kato S, Ishii S, Watanabe K (2009) Flagellum mediates symbiosis. Science 323:1574

Siddique T, Penner T, Semple K, Foght JM (2011) Anaerobic biodegradation of longer-chain n-alkanes coupled to methane production in oil sands tailings. Environ Sci Technol 45: 5892–5899

Smidt H, de Vos WM (2004) Anaerobic microbial dehalogenation. Annu Rev Microbiol 58:43–73

Steele DB, Stowers MD (1991) Techniques for selection of industrially important microorganisms. Annu Rev Microbiol 45:89–106

Sun B, Cole JR, Sanford RA, Tiedje JM (2000) Isolation and characterization of *Desulfovibrio dechloracetivorans* sp. nov., a marine dechlorinating bacterium growing by coupling the oxidation of acetate to the reductive dechlorination of2-chlorophenol. Appl Environ Microbiol 66:2408–2413

Thauer RK, Jungermann K, Decker K (1977) Energy conservation in chemotrophic anaerobic bacteria. Bacteriol Rev 41:100–180

Uchiyama T, Abe T, Ikemura T, Watanabe K (2005) Substrate-induced gene-expression screening of environmental metagenome libraries for isolation of catabolic genes. Nat Biotechnol 23:88–93

Vogels GD (1979) The global cycle of methane. Antonie Van Leeuwenhoek 45:347–352

Ward AJ, Hobbs PJ, Holliman PJ, Jones DL (2008) Optimization of the anaerobic digestion of agricultural resources. Bioresour Technol 99:7928–7940

Watanabe K (2001) Microorganisms relevant to bioremediation. Curr Opin Biotechnol 12:237–241

Watanabe K, Baker PW (2000) Environmentally relevant microorganisms. J Biosci Bioeng 89: 1–11

Watanabe K, Hamamura N (2003) Molecular and physiological approaches to understanding the ecology of pollutant degradation. Curr Opin Biotechnol 143:289–295

Watanabe K, Teramoto M, Futamata H, Harayama S (1998) Molecular detection, isolation, and physiological characterisation of functionally dominant phenol-degrading bacteria in activated sludge. Appl Environ Microbiol 64:4396–4402

Weiner JM, Lovley DR (1998) Rapid Benzene degradation in methanogenic sediments from a petroleum-contaminated aquifer. Appl Environ Microbiol 64:1937–1939

Woolhouse M, Gaunt E (2007) Ecological origins of novel human pathogens. Crit Rev Microbiol 33:231–242

Yang Y, McCarty P (1998) Competition for hydrogen within a chlorinated solvent dehalogenating anaerobic mixed culture. Environ Sci Technol 32:3591–3597

Zengler K, Richnow HH, Rosselló-Mora R, Michaelis W, Widdel F (1999) Methane formation from long-chain alkanes by anaerobic microorganisms. Nature 401:266–269

Chapter 14
Strategies to Reveal Genomic Function in Natural Soil Systems

Yoshiyuki Ohtsubo, Eri Nishiyama, Yoko Ishibashi, Yuji Nagata, and Masataka Tsuda

Abstract IVET (in vivo expression technology) and STM (signature-tagged mutagenesis) are suitable methods for revealing genes that are induced in soil and are essential in soil, respectively. These methods are potentially advantageous over newer methods, such as microarray and RNA-seq, because they allow analyses of bacteria under non-sterile conditions. However, their application to non-sterile soil also presents challenges that must be overcome. In this work, we describe the difficulties we faced when using IVET to identify genes that are upregulated in *Burkholderia multivorans* ATCC 17616 in soil. We also describe the future perspectives of these and other analytical schemes, including the potential use of next-generation sequencing technologies that might allow comprehensive isolation of soil-relevant genes and also might improve the quality of research from a qualitative point of view.

Keywords Illumina • IVET • STM

14.1 Introduction

To utilize bacterial activities in natural environments, including the abilities to degrade man-made recalcitrant compounds, it is crucial to reveal the natural lives of bacteria in the environment. It is especially important to reveal their lives in the soil. It seems to be widely accepted that the transfer of pure-cultured bacteria grown in a laboratory medium into the natural ecosystem eventually results in the

Y. Ohtsubo (✉) • E. Nishiyama • Y. Ishibashi • Y. Nagata • M. Tsuda
Department of Environmental Life Sciences, Graduate School of Life Sciences, Tohoku University, 2-1-1 Katahira, Sendai 980-8577, Japan
e-mail: yohtsubo@ige.tohoku.ac.jp

H. Nojiri et al. (eds.), *Biodegradative Bacteria: How Bacteria Degrade, Survive, Adapt, and Evolve*, DOI 10.1007/978-4-431-54520-0_14,

disappearance of the bacteria and in the failure to observe the expected functions of the augmented bacteria. To overcome this problem, it is important to identify genes that play pivotal roles in the survival of bacteria in the soil environments. As bacterial lives in the soil environments are not well understood, revealing such genes from the genome of each soil strain is also of particular interest from a biological point of view. In this chapter, we describe the two strategies, STM (signature-tagged mutagenesis) and IVET (in vivo expression technology), which have been used to reveal the lives of bacteria in the soil. We also describe the potential use of next-generation sequencing technologies that might aid in the studies in this field. For the general use of IVET and STM, see Rediers et al. (2005) for IVET and Mazurkiewicz et al. (2006) for STM. Also relevant is Saleh-Lakha et al. (2005) that deals with methods for microbial gene expression in soil and environmental factors affecting gene expression in soil.

14.2 Strategies to Reveal the Lives of Bacteria in a Soil Environment

One basic approach to investigate the lives of bacteria in soil environments is to inoculate the bacterial cells of interest, whose genomic sequence might be known, into soil samples. Although the augmentation of cells within a soil sample may lead to an unusual community structure in which the augmented clonal cells occupy the major part of the population, such an experimental scheme is valuable to shed light on the lives of bacteria in soil environments.

The identification of genes that play a pivotal role in natural non-sterile soil samples is compromised because of the nature of soils. First, in the soil samples, there is an immense variety of bacterial cells and eukaryotic creatures, such as fungi and possibly protozoa. This variety itself makes it impossible to isolate RNAs that originated from the augmented bacteria. Even if a sterilized soil sample is used, it is difficult to isolate RNA of sufficient quality and in sufficient amounts to be used for further analyses. This difficulty mainly arises from the presence of humic substances in the soil; these substances have physicochemical properties very similar to those of nucleic acids and are co-purified with RNA, and they interfere with the enzyme reactions and hybridization. The microarray technique was applied to analyze the expression profiles of *Pseudomonas putida* KT2440 in soils by Wang et al. who established a protocol to extract high-quality RNA from sterile soil inoculated with KT2440 (Wang et al. 2011). The other approach is to identify induced genes by adding extracts from soil to a laboratory medium and purify the RNA for further analysis. Yoder-Himes et al. used a next-generation sequencer, illumina, to conduct RNA-seq to identify genes that are expressed in response to soil extracts from in *Burkholderia cenocepacia* HI2424, which was isolated from an agricultural field (Yoder-Himes et al. 2009). Recently, genes upregulated in *Rhodococcus jostii* RHA1, a polychlorinated biphenyl degrader, during its growth in sterile soil, were reported based on microarray analysis of RNAs recovered from the sterilized soil (Iino et al. 2012).

These recent approaches have successfully identified soil-induced or soil-repressed genes. However, these studies are limited because they utilized sterile soil or soil extracts that were added to the laboratory medium. These conditions are different from the in vivo conditions, e.g., there are no competing organisms that might attack the augmented bacteria by predation or continuous production of substances with antimicrobial activities. In this regard, these new technologies allow efficient identification but fail to detect genes that might have been identified by using one of the two more traditional approaches mentioned above. These strategies, IVET and STM, enable the identification of genes that are specifically induced in soils and genes that are essential in soils, respectively. These approaches could be applied to non-sterile environments, although such applications also face difficulties and will require further improvements (see below).

IVET utilizes a positive screening scheme, in which two reporter genes play essential roles. The IVET strategy is rather complicated and there are several variations among the systems. The two major differences are the type of reporters used and the way in which the reporter genes are maintained in the bacterium, i.e., use of a plasmid vector or a specialized integration system (see Rediers et al. (2005) for the details of IVET strategies). For simplicity we here describe the IVET system that we utilized in our previous study to identify genes that are induced in the soil isolate *Burkholderia multivorans* ATCC 17616 within soil but not in a laboratory medium (Nishiyama et al. 2010). In this study, a *dapB* gene and *lacZ* gene were used as the reporter genes. *dapB* encodes an enzyme essential for the biosynthesis of lysine and diaminopimelate (DAP), and a *dapB* mutant of ATCC 17616 required the two substrates. To the *dapB* mutant, we introduced a genomic library of ATCC 17616, which was constructed in *Escherichia coli*. Each plasmid clone in the library carried a DNA region derived from ATCC 17616 fused with a tandem array of *dapB* and *lacZ*. Because of the nature of the R6K *ori*, the plasmid is not replicable in ATCC 17616, and the selection by the tetracycline-resistance gene located on the plasmid resulted in the isolation of strains, in which the plasmid was integrated in the genome by homologous recombination between the DNA region cloned in the plasmid and the corresponding DNA region in the genome (see Fig. 14.1). The IVET library was then inoculated into a soil sample, and incubated for a certain period of time for the elimination of cells not expressing the *dapB* gene (first screen). During the incubation, it was considered that the cells in which the *dapB* gene was transcribed would be able to survive or proliferate in the soil environment. After the incubation, the cell fraction was recovered and spread onto medium containing X-gal as well as DAP and lysine. We chose LacZ$^-$ (i.e., white) colonies to make an output pool (second screen). Each of these clones that formed a white colony had the *dapB-lacZ* cassette in a genomic locus that is expressed in the soil environment but not in the laboratory medium. To identify the genomic regions where the plasmid was integrated, we considered two methods, one based on the retrotransfer of the integrated plasmid to *E. coli* strains, and the other based on determination of the integrated locus by sequencing using the genomic DNAs as templates, and we chose the latter (Shimoda et al. 2008). To obtain the sequencing traces we needed two unique primers that anneal to the region flanking the cloning site of the plasmid. To design such

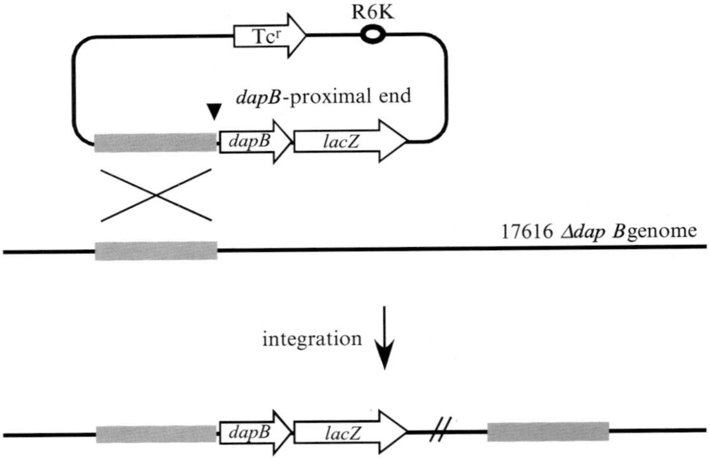

Fig. 14.1 Integration of a reporter plasmid into a genome. The *gray bar* on the plasmid represents an insert DNA derived from the strain to be analyzed. The plasmid is integrated by homologous recombination (*crossing bars*) between the insert DNA and the corresponding DNA in the target genome. In the cells, RNA polymerases are transcribing from left to right through the *dapB-proximal* end of the cloned fragment. Note that the promoter for the transcription might lie in the cloned region or might be located upstream of the cloned fragment

primers, we created a tool named "PrimerFinder," which is now available as an accessory tool of the GenomeMatcher software (Ohtsubo et al. 2008). This tool finds, from the specified region of a replicon, primers whose 3′ N-mer (i.e., 11-mer) are unique among all the DNA sequences of the specified replicons. Once we come up with two good primers to determine the two junctions, we can determine the junctions in a high-throughput manner.

One of the major problems we experienced in the application of the IVET system was the isolation of false-positive clones. The false-positive clones were clones that formed white colonies in the second screen, irrespective of the genomic location where the IVET reporters were integrated. In our preparative experiments, we constructed a negative control strain, in which the IVET plasmid was integrated to a genomic location that should not transcribe the *dapB-lacZ* cassette (the cassette was integrated in the opposite direction of a gene for cytochrome oxidase, *cox*). We also constructed a positive control strain that carries the cassette not in the opposite direction as the *cox* gene. The two strains were mixed and inoculated into the soil, and after certain time intervals the cell fraction was spread onto media containing X-gal. Ideally, the number of white colonies should reach zero after a certain period of incubation. However, white colonies were always observed at a very low, but a steady level, even after 70 days of incubation. We speculated that such white colonies appeared because a fraction of the cells had reached a dormant state after inoculation into the soil sample. There might be many places where a cell can hide itself from the other competing cells, e.g., the hollows of soil particles, leading to the survival of the cell. This speculation was supported by a finding that following the

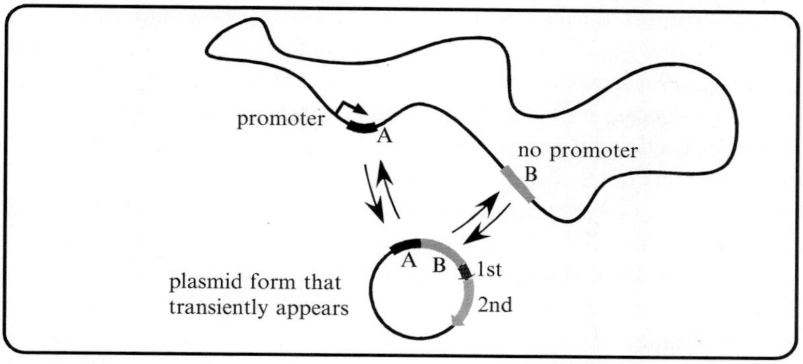

Fig. 14.2 Schematic representation of how chimeric plasmids generate false positives. If two regions (*A* and *B*) from a genome are cloned in a single plasmid, the plasmid can then be integrated into a region *A* or *B* in the genome or transiently exist in the cell in a free plasmid form. Here, it is presumed that transcription of the two reporters (designated as first and second) occurs in the cell when the plasmid is integrated into the region *A* but not into the region *B*. The integrant at region *A* can survive the soil conditions and continuously form the free plasmid. The free plasmid can integrate into region *B*, resulting in an integrant which can pass the second screen of the IVET strategy because it does not express the reporter genes. To avoid false positives, it is essential to determine both ends of the insert DNA

inoculation of the *dapB* mutant into the sterile soil, the CFU of the mutant did not decrease for a long period of time, suggesting that the *dapB* mutant did not grow but could survive in the soil.

The isolation of chimeric clones is more complicated (see Fig. 14.2). The chimeric clones are IVET clones that carry an integrated plasmid whose insert is comprised of more than one DNA region of the genome. The rate of chimeric clones of the IVET library constructed in *E. coli* is low, as exemplified by the plasmid extraction and sequencing of a fraction of clones in the library. However, the rate was significantly higher in the output pool (44 % of isolated clones). This increase in the rate of the chimeric clone was attributed to the chimeric plasmid reaching a complex equilibrium upon introduction to ATCC 17616, a condition under which it could not undergo autonomous replication. Under equilibrium, the plasmid is integrated in the genome at either of the genomic regions that are cloned or in its plasmid form, which might tentatively arise when the plasmid changes the integration location. The presence of such "free" plasmid is evident from the fact that such integrated plasmid could be retrieved to *E. coli* strains by means of conjugative cloning (Rainey et al. 1997). A chimeric IVET plasmid has more than one genomic region for integration, and if one region but not the other region transcribes the two reporters, the clone carrying such a plasmid can pass the first and second screens (see Fig. 14.2). In the soil, cells that carry the integrated plasmid in a genomic locus that transcribes *dapB-lacZ* survive and proliferate. In a very minor fraction of these cells, a free plasmid emerges and is integrated into a non-expressing locus, leading to the emergence of cells not expressing *dapB-lacZ*, which will be chosen in the second screen.

These two features, dormancy and chimeric clones, are the important features of IVET screening that should be taken into consideration to gain insights of biological importance. To overcome the difficulties, we constructed four libraries, and each library was screened in 60 independent tubes (total 240 tubes). Since the dormant cells were supposed to be recovered by chance, genomic loci that were identified from more than one tube were less likely to result from the dormant cells. We also tested whether or not the isolated clones were chimeric by obtaining sequence reads of both of the two junctions. We excluded chimeric clones and listed genomic loci that were identified from at least two independent clones, and loci that were experimentally confirmed to be induced in the soil by LacZ analysis of the cells recovered from the soil sample.

There were several difficulties in applying our IVET system to non-sterile environments. In a control experiment, the *dapB* mutant (LacZ^{-}) and a *dapB-expressing* strain (LacZ^{+}) were mixed and inoculated into the non-sterile soil sample. The colony-forming units (CFUs) of the inoculated cells decreased dramatically, and the ratio of white to blue colonies that were observed when the cell fraction recovered from the soil was plated onto an X-gal-containing agar plate, did not decrease significantly, only reaching about 1/3 in 90 days. This slower progression of competition between the *dapB* mutant and *dapB*-expressing strain indicated that a long incubation period (e.g., 5 years) is required. Although the IVET system has potential to be applied to non-sterile environments, prolonged incubation periods or advances in experimental settings are needed to stimulate the competition and to prevent the initial rapid decrease of CFU. For example, to prevent the initial rapid decrease in the CFU, cells could be inoculated into a small amount of sterile soil for a certain period of time to allow them to adapt to the soil environment before non-sterile soil is added.

It seems that the current IVET systems used thus far are limited in terms of the number of genomic loci they can identify. We spent about 2 years to determine the integration regions of 1,280 clones. Development of further analytical schemes that allow efficient identification of induced genomic loci will accelerate the IVET studies in the future (see also below).

STM is originated from a study that identified genes from *Salmonella typhimurium*, which has been implicated in the virulence to mice (Hensel et al. 1995). Basically, STM utilizes a negative screening scheme, in which transposon mutants that disappeared during the incubation or passage in a specific environment are searched, leading to identification of genes that are essential in the environment. Several modifications have been made to improve the system, resulting in several variations. The two major differences among the resulting systems are (1) the way in which each mutant is tagged with a specific DNA sequence and (2) the way to detect clones that have disappeared during the period of incubation in a given environment (Mazurkiewicz et al. 2006).

To date, except for a study of *Burkholderia vietnamiensis* G4 to identify essential genes in the rhizosphere (O'Sullivan et al. 2007), there has been no study in which STM was applied to reveal the essential genes in soil environments. In our laboratory, a *fur* gene (Yuhara et al. 2008) for the ferric uptake regulator was identified from

B. multivorans ATCC 17616 (our unpublished observation). However, to compile an entire list of genes essential in soil, further screening and confirmatory studies will be needed. The STM techniques will be replaced by new experimental schemes utilizing next-generation sequencers, which should identify essential genes in vivo more efficiently (see below).

14.3 Genes Found to be Induced or Essential in Soil

Table 14.1 lists studies that have identified genes that are induced or essential in soils or the rhizosphere. Although these studies identified tens or hundreds of such genes, in order to understand the function of the genes in soil, further dedicated studies will be needed. Here we describe examples of genes that were identified in the IVET screening and that were analyzed further to gain insight into their function in soil.

An IVET study of *P. fluorescens* Pf0-1 to identify genes important in soil identified 22 genes, including ten genes that were present in an antisense orientation relative to the overlapping protein-coding genes, and two studies followed to reveal the functional significance of the inversely oriented genes. The gene *iiv19* overlaps with the *leuA2* gene, which encodes an enzyme for leucine biosynthesis. A disruption study of *leuA2* resulted in a surprising finding, namely, that the absence of *leuA2* in soil is advantageous when leucine is added exogenously (Kim and Levy 2008).

Table 14.1 Studies pertaining to identify genes that are important in soil or rhizosphere

Strain	Strategy	Growth/incubation conditions used	Reference
Pseudomonas fluorescens SBW25	IVET	Sugar beet seeds soaked with bacteria and recovery of cells from rhizosphere	Rainey (1999)
Pseudomonas fluorescens Pf0-1	IVET	Sterilized sandy loam soil	Silby and Levy (2004)
Pseudomonas putida KT2440	IVET	Maize seeds soaked with bacteria and recovery of cells from rhizosphere	Ramos-Gonzalez et al. (2005)
Burkholderia vietnamiensis G4	STM	Pea rhizosphere	O'Sullivan et al. (2007)
Burkholderia cenocepacia HI2424/AU1054	RNA-seq	Soil medium containing soil extracts	Yoder-Himes et al. (2009)
Burkholderia multivorans ATCC 17616	IVET	Sterilized/non-sterilized brown forest field soil	Nishiyama et al. (2010)
Pseudomonas putida KT2440	Microarray	Sterilized soil in the presence or absence of 3-chlorobenzoic acid	Wang et al. (2011)
Rhodococcus jostii RHA1	Microarray	Sterilized brown forest field soil	Iino et al. (2012)

The gene *iiv8* is upregulated in soil and is antisense to the *ppk* gene that encodes polyphosphate kinase. The induction of antisense RNA encoded by *iiv8* reduced the *ppk* transcript to a level that was 1/5 of the uninduced control, suggesting a post-transcriptional mechanism. It was also suggested that precise control of polyphospate production is important for survival in the soil environment (Silby et al. 2012). As these two studies demonstrated, antisense RNAs might play significant roles in soil, suggesting that the transcriptions that generate transcripts that are complementary to protein-coding mRNA should not be ignored.

A cluster of genes (*andAcAdAbAa*) for anthranilate dioxygenase was identified in the IVET screen of *B. multivorans* ATCC 17616, and the expression level in the soil increased more than 100-fold as determined by measurement of the LacZ activities of cells recovered from the inoculated soil (Nishiyama et al. 2010). The disruption of the *andA* operon resulted in a strain that failed to proliferate in the initial period after inoculation (the initial proliferation started after one week). The expression level of the *andA* locus remained low until, after four days, it started to increase, and then it increased further after two weeks, suggesting that the *andA* expression plays a pivotal role in the initial proliferation of cells after inoculation into the soil (Nishiyama et al. 2012). In laboratory medium, the *andA* operon expression was induced by anthranilate and tryptophan. But no anthranilate or tryptophan was detected in the soil, and thus the origin of the inducer(s) and the physiological significance of the *andA* operon in soil remain a matter of speculation (Nishiyama et al. 2012).

14.4 Factors Controlling Gene Expression in Soil

It is of particular interest to identify the signals that induced the individual genes in soil (see Fig. 14.3 for potential inducers). To date, no global regulator that induces the expression of a set of genes in soil environments has been found. It seems that no soil-specific sigma factor or soil-specific transcriptional regulator exerts its effects to make a global expression profile in the soil. Rather, different independent signals present in the soil induce the respective genes.

Among the possible signal types, one of the most plausible is the low-molecular-weight signal, such as that given by organic compounds or by iron, arsenic, or other ionic forms of metals. In fact, soil extracts prepared by washing the soil with water or by organic solvents such as ethyl acetate have been shown to induce a group of genes in laboratory media (Yoder-Himes et al. 2009; Nishiyama et al. 2010). It is possible that antimicrobial agents, such as antibiotics and bacteriocins produced by co-residing bacterial or eukaryotic cells, are the inducing signals. However, it should be noted that soil is made up of soil particles that do not allow the free diffusion of chemical signals. That is, a chemical compound can be locally accumulated to reach a level sufficient to induce a gene in a bacterial cell present at that very location, but not to a level that upon extraction leads to induction in laboratory media. It is also possible that chemical compounds accumulated in the soil are the inducer.

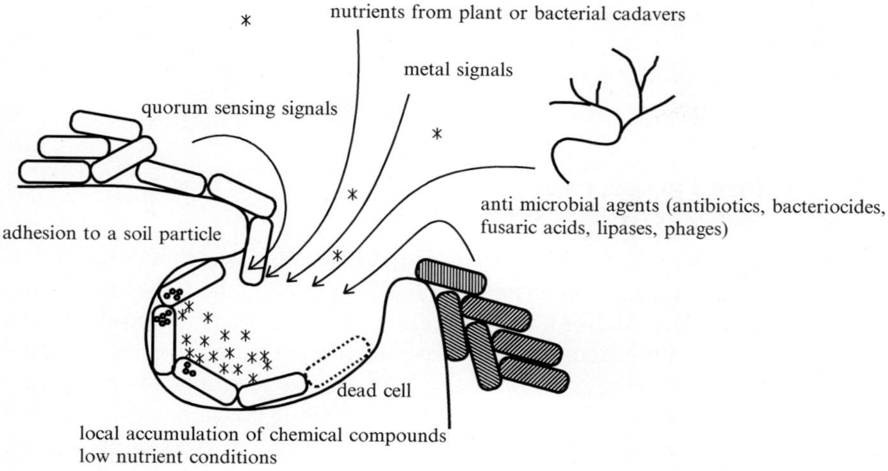

Fig. 14.3 Signals that might generate a soil-specific gene expression profile. White cells drawn in a cavity of a soil particle are the cells for analysis. These cells might be influenced by quorum-sensing signals produced by akin cells or by antimicrobial agents from fungi or other kinds of bacterial cells (hatched cells). Chemical substances could be locally accumulated (*asterisks*). Under low-nutrient conditions, particular substances such as anthranilate or tryptophan may accumulate in the cell (*dots*)

The *andA* genes in ATCC 17616, which encode anthranilate dioxygenase and are involved in anthranilate metabolism, were induced by anthranilate and tryptophan in a laboratory medium. However, neither tryptophan, which is converted to anthranilate by metabolism, nor anthranilate was detected in the soil extract. Moreover, the induction in the soil required several days of incubation, indicating that the anthranilate is not present in the soil sample but accumulated in the cell after the onset of incubation in the soil. These possibilities might account for the failure of induction by the extracts added to the laboratory medium.

Another difficulty with identifying the signal in vivo is that identification of the inducing signal in a laboratory medium might suggest but not prove that the same signal inducer induced the expression. For example, genes for fusaric acid-resistance in ATCC 17616 are upregulated in the soil, and their induction in the laboratory media requires a gene for the LysR-type transcriptional regulator, FusR, and the addition of fusaric acid (our unpublished observation). This finding in laboratory media suggested that fusaric acid is the inducer in the soil; however, the involvement of other inducers and regulatory proteins could not be excluded due to the complex nature of the soil.

Although there should be other types of signals involved in producing a soil-specific expression profile, at present only speculations can be made. It is possible that close proximity to the wall of the soil particles might be recognized by a bacterial cell to change its expression profile. It is also possible that stressful conditions that were brought about by the poor nutrient conditions, contents of dead cells, presence of phages in the soil, and quorum-sensing molecules produced by akin cells might be responsible for generating the soil-specific expression profiles.

14.5 Future Perspectives

The TraSH method, which utilizes microarray technology, was first developed to identify essential genes in *Mycobacterium bovis* on minimal but not rich medium (Sassetti et al. 2001). Like STM, the TraSH procedure utilizes a transposon mutant library, but the identification of the transposon-inserted genes that result in the decreased fitness is done by a microarray co-hybridized with probes generated from input and output pools (Sassetti et al. 2001). In 2009, applications of three new technologies, named HITS (Gawronski et al. 2009), Tn-Seq (van Opijnen et al. 2009), and traDIS (Langridge et al. 2009), were reported. Each utilizes illumina-sequencing technology to identify the location of the transposon cassette in the massive parallel sequencing. The application of the illumina-sequencing greatly increased the efficiency of screening, making it possible to identify the insertion site and simultaneously to qualitatively assess the fitness of the mutants. Since then, several reports that utilize these technologies have been published (Smith et al. 2010; Gallagher et al. 2011; Khatiwara et al. 2012; Eckert et al. 2011), and use of the next-generation sequencing technologies will be the standards in the next decade. These technologies will be replacing the STM strategy in the future because specific tag sequences to discriminate each mutant will no longer be used.

In contrast, although it would be possible to apply the next-generation sequencing technologies to the IVET system, their use will be limited to the identification step of the insertion regions. However, to exclude chimeric clones, both junction sequences must be determined for each clone. Therefore, a new strategy must be developed to identify both junctions of each strain present in a mixture of strains.

To utilize next-generation sequencing technologies to identify soil-induced genes, the next challenge will be to augment a strain of interest into a non-sterile soil sample and prepare the RNA for illumina sequencing. As the illumina produces a huge amount of data, and the data production scale is still increasing, in the near future, it will be possible to collect a great number of sequence reads derived from the input bacteria (i.e., reads that are 100 % matched to the genomic sequence of the input bacteria) while excluding those derived from the indigenous bacteria which should dominate the whole sequence reads, so that the data can be assessed from a quantitative point of view.

The studies of microorganisms in soil are hindered by the fact that different soil samples are so different with respect to the compositions and also to the bacterial flora present in the soil. This makes it difficult to compare the results from different studies that use different soil samples. It would be possible to develop and use a formulated soil made up of known materials that are readily available, such as crushed igneous rocks and leaves of a certain kind of trees, and augment this with a mixture of bacterial cells with known genomic sequences. The other alternative might be to decide to use a representative soil sample and share it among the laboratories. Our laboratory and those of other researchers have shared a brown forest soil sample collected from the Ehime Agricultural Experiment Station (Matsuyama, Japan). The physicochemical properties of the soil have already been reported by Wang et al. (2008).

It will also be challenging to develop an experimental scheme to recover inoculated bacterial cells from a mixture of a huge number of other kinds of cells, possibly by using the surface antigen of the input cells. The endogenous surface antigen could be used or bacteria could be engineered to express an exogenous one. The other idea is to embed the cells of the library in small permeable capsules that allow the diffusion of small molecules across the membrane, but protect the inside cells from being rapidly killed by creatures in soil. The recovery of such capsules from a soil sample would be easy, and the inside cells could be used for further analyses. For example, an RNA sample could be prepared for further analysis including RNA-seq (Mortazavi et al. 2008; Nagalakshmi et al. 2008; Yoder-Himes et al. 2009) or dRNA-seq for transcription start site analyses (Sharma et al. 2010) that utilize next-generation sequencing technologies.

Acknowledgements We are grateful to all of our collaborators. Support for our research has been provided in the form of Grants-in-Aid from the Ministry of Education, Culture, Sports, Science and Technology, Japan.

References

Eckert SE, Dziva F, Chaudhuri RR, Langridge GC, Turner DJ, Pickard DJ, Maskell DJ, Thomson NR, Stevens MP (2011) Retrospective application of transposon-directed insertion site sequencing to a library of signature-tagged mini-Tn5Km2 mutants of *Escherichia coli* O157:H7 screened in cattle. J Bacteriol 193(7):1771–1776. doi:10.1128/JB.01292-10

Gallagher LA, Shendure J, Manoil C (2011) Genome-scale identification of resistance functions in *Pseudomonas aeruginosa* using Tn-seq. mBio 2(1):e00310–e00315. doi:10.1128/mBio.00315-10

Gawronski JD, Wong SM, Giannoukos G, Ward DV, Akerley BJ (2009) Tracking insertion mutants within libraries by deep sequencing and a genome-wide screen for *Haemophilus* genes required in the lung. Proc Natl Acad Sci USA 106(38):16422–16427. doi:10.1073/pnas.0906627106

Hensel M, Shea JE, Gleeson C, Jones MD, Dalton E, Holden DW (1995) Simultaneous identification of bacterial virulence genes by negative selection. Science 269(5222):400–403

Iino T, Wang Y, Miyauchi K, Kasai D, Masai E, Fujii T, Ogawa N, Fukuda M (2012) Specific gene responses of Rhodococcus jostii RHA1 during growth in soil. Appl Environ Microbiol 78(19):6954–6962. doi:10.1128/AEM.00164-12

Khatiwara A, Jiang T, Sung SS, Dawoud T, Kim JN, Bhattacharya D, Kim HB, Ricke SC, Kwon YM (2012) Genome scanning for conditionally essential genes in *Salmonella enterica* serotype typhimurium. Appl Environ Microbiol 78(9):3098–3107. doi:10.1128/AEM.06865-11

Kim W, Levy SB (2008) Increased fitness of *Pseudomonas fluorescens* Pf0-1 leucine auxotrophs in soil. Appl Environ Microbiol 74(12):3644–3651

Langridge GC, Phan MD, Turner DJ, Perkins TT, Parts L, Haase J, Charles I, Maskell DJ, Peters SE, Dougan G, Wain J, Parkhill J, Turner AK (2009) Simultaneous assay of every *Salmonella typhi* gene using one million transposon mutants. Genome Res 19(12):2308–2316. doi:10.1101/gr.097097.109

Mazurkiewicz P, Tang CM, Boone C, Holden DW (2006) Signature-tagged mutagenesis: barcoding mutants for genome-wide screens. Nat Rev Genet 7(12):929–939. doi:10.1038/nrg1984

Mortazavi A, Williams BA, McCue K, Schaeffer L, Wold B (2008) Mapping and quantifying mammalian transcriptomes by RNA-Seq. Nat Methods 5(7):621–628. doi:10.1038/nmeth.1226

Nagalakshmi U, Wang Z, Waern K, Shou C, Raha D, Gerstein M, Snyder M (2008) The transcriptional landscape of the yeast genome defined by RNA sequencing. Science 320(5881): 1344–1349. doi:10.1126/science.1158441

Nishiyama E, Ohtsubo Y, Nagata Y, Tsuda M (2010) Identification of Burkholderia multivorans ATCC 17616 genes induced in soil environment by in vivo expression technology. Environ Microbiol 12(9):2539–2558. doi:10.1111/j.1462-2920.2010.02227.x

Nishiyama E, Ohtsubo Y, Yamamoto Y, Nagata Y, Tsuda M (2012) Pivotal role of anthranilate dioxygenase genes in the adaptation of *Burkholderia multivorans* ATCC 17616 in soil. FEMS Microbiol Lett 330(1):46–55. doi:10.1111/j.1574-6968.2012.02532.x

Ohtsubo Y, Ikeda-Ohtsubo W, Nagata Y, Tsuda M (2008) GenomeMatcher: a graphical user interface for DNA sequence comparison. BMC Bioinformatics 9:376

O'Sullivan LA, Weightman AJ, Jones TH, Marchbank AM, Tiedje JM, Mahenthiralingam E (2007) Identifying the genetic basis of ecologically and biotechnologically useful functions of the bacterium *Burkholderia vietnamiensis*. Environ Microbiol 9(4):1017–1034. doi:10.1111/j.1462-2920.2006.01228.x

Rainey PB (1999) Adaptation of *Pseudomonas fluorescens* to the plant rhizosphere. Environ Microbiol 1(3):243–257

Rainey PB, Heithoff DM, Mahan MJ (1997) Single-step conjugative cloning of bacterial gene fusions involved in microbe-host interactions. Mol Gen Genet 256(1):84–87

Ramos-Gonzalez MI, Campos MJ, Ramos JL (2005) Analysis of *Pseudomonas putida* KT2440 gene expression in the maize rhizosphere: in vivo [corrected] expression technology capture and identification of root-activated promoters. J Bacteriol 187(12):4033–4041

Rediers H, Rainey PB, Vanderleyden J, De Mot R (2005) Unraveling the secret lives of bacteria: use of in vivo expression technology and differential fluorescence induction promoter traps as tools for exploring niche-specific gene expression. Microbiol Mol Biol Rev 69(2):217–261

Saleh-Lakha S, Miller M, Campbell RG, Schneider K, Elahimanesh P, Hart MM, Trevors JT (2005) Microbial gene expression in soil: methods, applications and challenges. J Microbiol Methods 63(1):1–19. doi:10.1016/j.mimet.2005.03.007

Sassetti CM, Boyd DH, Rubin EJ (2001) Comprehensive identification of conditionally essential genes in mycobacteria. Proc Natl Acad Sci USA 98(22):12712–12717. doi:10.1073/pnas.231275498

Sharma CM, Hoffmann S, Darfeuille F, Reignier J, Findeiss S, Sittka A, Chabas S, Reiche K, Hackermuller J, Reinhardt R, Stadler PF, Vogel J (2010) The primary transcriptome of the major human pathogen *Helicobacter pylori*. Nature 464(7286):250–255. doi:10.1038/nature08756

Shimoda Y, Mitsui H, Kamimatsuse H, Minamisawa K, Nishiyama E, Ohtsubo Y, Nagata Y, Tsuda M, Shinpo S, Watanabe A, Kohara M, Yamada M, Nakamura Y, Tabata S, Sato S (2008) Construction of signature-tagged mutant library in *Mesorhizobium loti* as a powerful tool for functional genomics. DNA Res 15(5):297–308

Silby MW, Levy SB (2004) Use of in vivo expression technology to identify genes important in growth and survival of *Pseudomonas fluorescens* Pf0-1 in soil: discovery of expressed sequences with novel genetic organization. J Bacteriol 186(21):7411–7419

Silby MW, Nicoll JS, Levy SB (2012) Regulation of polyphosphate kinase production by antisense RNA in *Pseudomonas fluorescens* Pf0-1. Appl Environ Microbiol 78(12):4533–4537. doi:10.1128/AEM.07836-11

Smith AM, Heisler LE, St Onge RP, Farias-Hesson E, Wallace IM, Bodeau J, Harris AN, Perry KM, Giaever G, Pourmand N, Nislow C (2010) Highly-multiplexed barcode sequencing: an efficient method for parallel analysis of pooled samples. Nucleic Acids Res 38(13):e142. doi:10.1093/nar/gkq368 van Opijnen T, Bodi KL, Camilli A (2009) Tn-seq: high-throughput parallel sequencing for fitness and genetic interaction studies in microorganisms. Nat Methods 6(10):767–772. doi:10.1038/nmeth.1377

Wang Y, Shimodaira J, Miyasaka T, Morimoto S, Oomori T, Ogawa N, Fukuda M, Fujii T (2008) Detection of *bphAa* gene expression of *Rhodococcus* sp. strain RHA1 in soil using a new method of RNA preparation from soil. Biosci Biotechnol Biochem 72(3):694–701

Wang Y, Morimoto S, Ogawa N, Fujii T (2011) A survey of the cellular responses in *Pseudomonas putida* KT2440 growing in sterilized soil by microarray analysis. FEMS Microbiol Ecol 78(2):220–232. doi:10.1111/j.1574-6941.2011.01146.x

Yoder-Himes DR, Chain PS, Zhu Y, Wurtzel O, Rubin EM, Tiedje JM, Sorek R (2009) Mapping the *Burkholderia cenocepacia* niche response via high-throughput sequencing. Proc Natl Acad Sci USA 106(10):3976–3981

Yuhara S, Komatsu H, Goto H, Ohtsubo Y, Nagata Y, Tsuda M (2008) Pleiotropic roles of iron-responsive transcriptional regulator Fur in *Burkholderia multivorans*. Microbiology 154(Pt 6):1763–1774

Chapter 15
Monitoring Microbial Community Dynamics to Evaluate Bioremediation

Brett R. Baldwin, Dora Ogles, and Cindy H. Nakatsu

Abstract Bioremediation can be an effective treatment strategy at contaminated sites when applied appropriately and implemented properly. Thus it is important to have methods that can be used to reliably assess biodegradation potential and ultimately the feasibility and performance of bioremediation as a treatment approach. While an important component of site monitoring, decreasing trends in contaminant concentrations can result from physical processes, do not document that loss is due to biodegradation, and can be difficult to discern in practice. Therefore direct measures of microbes responsible for biodegradation of compounds have been developed. These methods fall into two categories: direct measure of biodegradation functional genes using methods such as quantitative PCR and microarrays and indirect measures in changes in microbial community structure using methods such as PCR-DGGE and T-RFLP when target genes are unknown. This chapter describes the various molecular methods being used and examples where they have been applied for the assessment of environmental samples.

Keywords Dioxygenase • Metagenomic • Microarrays • PCR-DGGE • qPCR • RT-qPCR • T-RFLP

B.R. Baldwin • D. Ogles
Microbial Insights, 10515 Research Dr., Suite 100, Knoxville, TN 37932-2536, USA

C.H. Nakatsu (✉)
Purdue University, 915 West State Street, West Lafayette, IN 47907-2054, USA
e-mail: cnakatsu@purdue.edu

H. Nojiri et al. (eds.), *Biodegradative Bacteria: How Bacteria Degrade, Survive, Adapt, and Evolve*, DOI 10.1007/978-4-431-54520-0_15, © Springer Japan 2014

Abbreviations

BSS	Benzylsuccinate synthase
BTEX	Benzene, toluene, ethylbenzene, and xylenes
DCE	*cis*-1,2-dichloroethene
DGGE	Denaturing gradient gel electrophoresis
FGA	Functional gene array
ISP	Iron–sulfur protein
MNA	Monitored natural attenuation
MPN	Most probable number
NAH	Naphthalene dioxygenase
PAHs	Polycyclic aromatic hydrocarbons
PCE	Tetrachloroethene
PCR	Polymerase chain reaction
PHE	Phenol hydroxylase
qPCR	Quantitative polymerase chain reaction
RMO	Ring-hydroxylating toluene monooxygenase
RT-PCR	Reverse transcriptase-polymerase chain reaction
RT-qPCR	Reverse transcriptase-quantitative polymerase chain reaction
TCE	Trichloroethene
TOD	Toluene/benzene dioxygenase
TOL	Toluene/xylene monooxygenase
T-RFLP	Terminal-restriction fragment length polymorphism

15.1 Introduction

Whether through monitored natural attenuation (MNA) or an engineered strategy, bioremediation has increasingly become the preferred approach for environmental restoration at sites impacted by a wide variety of contaminants including petroleum hydrocarbons and chlorinated solvents (Pandey et al. 2009). Evaluating the feasibility and subsequent performance of a bioremediation strategy for site cleanup should be based on converging lines of chemical, geochemical, and microbiological evidence. Microbial communities can be composed of thousands of species and only a subset is responsible for bioremediation. Therefore strategic approaches must be used to the target-specific populations responsible for bioremediation. In the environmental remediation industry, practitioners have traditionally relied on cultivation-based methods like plate counts (heterotrophic or "specific degrader") or most probable number (MPN) analyses to enumerate target microorganisms and assess biodegradation potential. However, this approach had severe limitations due to the difficulty in identifying conditions suitable for the cultivation of the majority of the microbes present in most communities (Amann et al. 1995; Staley and Konopka 1985). As the biases and limitations of cultivation-based methods have become more apparent and widely known, the use of nucleic acid based techniques has

Table 15.1 Gene-based monitoring techniques used to assess environmental restoration potential by biodegradation

Technique	Application
qPCR (16S rRNA gene)	Quantify specific microorganisms (e.g., *Dehalococcoides* spp.) capable of a particular biodegradation process (e.g., reductive dechlorination)
qPCR (functional genes)	Quantify a gene encoding an enzyme (e.g., toluene dioxygenase) responsible for biodegradation of a specific contaminant
RT-qPCR (functional genes)	Quantify the expression of a functional gene encoding an enzyme in the biodegradation pathway of a specific contaminant as an indicator of activity
Functional gene microarrays	Highly parallel detection of functional genes involved in contaminant biodegradation or geochemical processes (e.g., GeoChip)
PCR-DGGE or T-RFLP (16S rRNA gene)	Genetic fingerprint of microbial community to determine population richness and assess shifts in community structure
Community microarray (16S rRNA gene)	Highly parallel detection of microorganisms present in a community (e.g., PhyloChip) including taxa known to be capable of biodegradation
High-throughput sequencing (metagenomics)	Comprehensive information on microbial community composition and potential functional and other physiological traits encoded in the community

grown substantially not only as research tools but to guide site management decisions (Widada et al. 2002b). With continued advances and increasing availability, molecular biological techniques including quantitative polymerase chain reaction (qPCR), denaturing gradient gel electrophoresis (DGGE), microarrays, and more recently high-throughput DNA sequencing technologies (Table 15.1) will provide more rapid, accurate, and comprehensive characterization of microbial community composition and dynamics at contaminated sites.

15.2 Targeted Functional Gene Analysis

With releases ranging in scale from leaking underground storage tanks at retail gasoline stations to events like the Deepwater Horizon incident, petroleum hydrocarbons are among the most common environmental contaminants. The aromatic compounds benzene, toluene, ethylbenzene, and xylenes (BTEX) as well as naphthalene and larger polycyclic aromatic hydrocarbons (PAHs) are often the contaminants of concern at sites impacted by petroleum products. Biodegradation of aromatic hydrocarbons is mediated by a diverse polyphyletic group of bacteria such that rRNA-based approaches to assess aromatic hydrocarbon catabolism and evaluate contaminant biodegradation in the field are of limited value. However, aerobic and to a lesser extent anaerobic pathways for aromatic hydrocarbon catabolism have been identified allowing development of qPCR assays targeting functional, catabolic genes.

15.2.1 Anaerobic Biodegradation of Petroleum Hydrocarbons

Of the anaerobic pathways, toluene biodegradation is the most understood. The first step in this pathway, mediated by benzylsuccinate synthase (BSS), is the addition of fumarate onto the toluene methyl group to form benzylsuccinate (Biegert et al. 1996). The presence of BSS has been documented in cultures that degrade toluene under denitrifying, iron reducing, sulfate reducing, and methanogenic conditions (Beller and Edwards 2000; Heider et al. 1998; Kane et al. 2002; Spormann and Widdel 2000). Furthermore, while the anaerobic biodegradation of benzene may be initiated by a carboxylase (Abu Laban et al. 2010), the addition of fumarate appears to be a common mechanism for activation of hydrocarbons under anaerobic conditions. BSS and more recently naphthyl-2-methyl-succinate synthase (NMS) have been implicated in the anaerobic biodegradation of a number of methylated aromatic compounds including ethylbenzene (Kniemeyer et al. 2003), *o*- and *m*-xylenes (Achong et al. 2001; Krieger et al. 1999), and 2-methylnaphthalene (Selesi et al. 2010). Finally, activation by addition of fumarate is not limited to aromatic hydrocarbons. Kropp et al. (2000) demonstrated that an analogous enzyme, alkylsuccinate synthase, is responsible for initiating anaerobic biodegradation of alkanes.

15.2.1.1 Quantitative PCR of Anaerobic Degradation Genes

A qPCR assay targeting the α-subunit of the benzylsuccinate synthase gene (*bssA*) to quantify bacteria capable of anaerobic toluene metabolism has been developed (Beller et al. 2002). In microcosms inoculated with aquifer sediments from sites with previous exposure to petroleum hydrocarbons, the number of *bssA* gene copies increased 100–1,000-fold during the period of rapid toluene degradation (Beller et al. 2002). In subsequent laboratory studies, *bssA* gene copies have been found to correspond to (Da Silva and Alvarez 2004) and to correlate with (Kazy et al. 2010) rates of anaerobic toluene degradation. At the field scale, monitoring *bssA* gene copies has been employed to establish that sulfate injection promoted growth of alkylbenzene utilizing populations in a gasoline-impacted aquifer (Sublette et al. 2006). However, the qPCR assay for *bssA* genes was primarily based upon gene sequences closely related to those of *Thauera* and *Azoarcus* species utilizing toluene under denitrifying conditions (Beller et al. 2002; Winderl et al. 2007) and therefore may not capture the in situ diversity of *bssA*-like genes. Winderl et al. (2007) designed alternative PCR primer sets and screened additional pure cultures for *bssA* amplicons to expand the range of reference sequences particularly within the δ-proteobacteria. In subsequent experiments with one of the alternative PCR primer pairs, diverse *bssA*-like genes were amplified from three tar-oil-contaminated sites. At one site, all clones were closely related to *Geobacter* spp. *bssA* genes while detected *bssA*-like genes from a second site appeared to originate from a novel lineage, branching between the reference δ-proteobacterial and β-proteobacterial *bssA* sequences (Winderl et al. 2007). This second group, termed the F1 cluster of unidentified *bssA* genes, was detected in high abundance in a sulfidogenic zone underneath the BTEX plume core (Winderl et al. 2008). Overall, the available evidence demonstrates that monitoring *bssA* genes can provide a direct line of evidence to guide management activities at contaminated sites but cautions that the true environmental diversity of catabolic gene sequences must be continually examined and considered.

15.2.2 Aerobic Biodegradation of Aromatic Hydrocarbons

In contrast to the anaerobic pathways, aerobic biodegradation of aromatic hydrocarbons has been intensively studied and well characterized. In general, aerobic catabolism of aromatic hydrocarbons can be viewed as a collection of upper pathways in which dioxygenases or monooxygenases initiate transformation of aromatic compounds ranging in size from benzene to benzo(a)pyrene into a limited number of common intermediates (e.g., catechols or substituted catechols) that are further metabolized by a few conserved pathways. The ring-hydroxylating aromatic dioxygenases are Rieske nonheme iron oxygenases composed of a terminal oxygenase component (iron-sulfur protein (ISP)) and electron transfer components (Butler and Mason 1996). The subunits of ISPs contain the Rieske-type (2Fe-2S) cluster and a substrate-binding site (for more details see Chap. 2, Sect. 2.2) and are believed to confer substrate specificity (Gibson and Parales 2000; Erickson and Mondello 1993; Parales et al. 2000). Therefore, the overwhelming majority of reported PCR and qPCRs assays target the α subunits of the initial oxygenase genes of the various pathways to quantify bacteria capable of aerobically metabolizing specific aromatic hydrocarbons found in petroleum products. In fact, Iwai et al. (2011) have recently compiled and performed an in silico review of 115 primer sets designed to amplify various aromatic dioxygenase genes.

Aerobic biodegradation of toluene, the model for the BTEX compounds, can proceed via one of five different biochemical pathways based on the mode of the initial oxygenase attack (see Cao et al. (2009) for a review). The first is mediated by toluene/benzene dioxygenase (TOD) that incorporates both atoms of molecular oxygen into the aromatic ring. The second known pathway involves initial monooxygenase attack at the alkyl substituent by a toluene/xylene monooxygenase (TOL). The final three known pathways are initiated by different ring hydroxylating monooxygenases (RMO) which incorporate one atom of molecular oxygen into the aromatic ring at the *ortho* (T2MO), *meta* (T3MO), or *para* (T4MO) position producing arene oxides as unstable intermediates which are converted to phenols. Further oxidation to form substituted catechols can be mediated by phenol hydroxylase (PHE) or by the initial RMO as is the case with the TouA toluene/*o*-xylene monooxygenase (Bertoni et al. 1998). Similarly, some RMOs with greater sequence similarity to phenol hydroxylases such as the TbmD toluene/benzene-2-monooxygenase (Johnson and Olsen 1995) may be responsible for both the initial and second oxidation (Kahng et al. 2001). Thus, the difference between toluene monooxygenases (RMOs) the phenol hydroxylases (PHEs) is not absolute in terms of substrate specificity and catabolic function.

Naphthalene has long served as the model compound for investigating the aerobic biodegradation of polycyclic aromatic hydrocarbons (PAHs). The most well studied naphthalene dioxygenases and catabolic pathways are from the archetypal naphthalene utilizing *Pseudomonas putida* strains PpG7 and NCIMB 9816-4 (Cerniglia 1992). While these *nah* genes have been detected in a wide variety of bacteria and environments (Ahn et al. 1999; Baldwin et al. 2008; DeBruyn et al. 2007; Ferrero et al. 2002; Lloyd-Jones et al. 1999; Widada et al. 2002a; Tuomi et al. 2004), other less closely related genes for naphthalene and PAH biodegradation have been described. For example, the naphthalene dioxygenase from *Ralstonia* sp. strain U2 (*nagA*) and the dinitrotoluene dioxygenase (*dnt*A) from *Burkholderia* sp. DNT show limited sequence identity to *nahA* but clearly belong in the naphthalene dioxygenase family. In addition, the *phn* genes from *Burkholderia* sp. strain RP007 are only distantly related to the *nah* genes (Laurie and Lloyd-Jones 1999) and *nag* genes. Finally, initial oxidation of higher molecular weight PAHs by Gram-positive bacteria including *Mycobacterium* sp. strain PYR-1 is mediated by naphthalene inducible dioxygenases which have little in common with the *nah* genes in terms of sequence identity, regulation, and substrate specificity (Khan et al. 2001).

15.2.2.1 Quantitative PCR of Aerobic Degradation Genes

qPCR quantification of aromatic oxygenase genes can be employed at the site characterization, remedy selection, and performance monitoring stages to guide site management activities. Assays for qPCR quantification of bacteria harboring TOD, TOL, RMO, PHE, and NAH (*nahA*-like, *nagA*-like, and *nidA*-like) genes have been developed by a number of laboratories (see Iwai et al. 2011 for review) and employed to evaluate remedial actions and investigate microbial community dynamics at sites impacted by BTEX, chlorobenzenes, and PAHs. In terms of evaluating the feasibility of MNA, mesocosms deployed in impacted wells have demonstrated colonization by bacteria with *tmoA*-like (RMO) and *xylM/xylE1*-like (TOL) genotypes to document intrinsic biodegradation potential at a BTEX impacted industrial site (Hendrickx et al. 2005). In our laboratories, qPCR quantification of a broad range of aromatic oxygenase genes, most notably PHE and RMO, corresponded with historical trends in contaminant concentrations and allowed identification of a site where MNA would be effective and areas within a second site where engineered corrective actions needed to be considered (Baldwin et al. 2008). While not always the intent of the experiments, the frequent detection of the *nahA*-like naphthalene dioxygenase genes in impacted environmental samples (Ahn et al. 1999; Baldwin et al. 2008; DeBruyn et al. 2007; Ferrero et al. 2002; Lloyd-Jones et al. 1999; Widada et al. 2002a; Tuomi et al. 2004) not only establishes the ecological significance of the genotype but also indicated the potential for aerobic PAH biodegradation at the study sites. Similarly, qPCR quantification of naphthalene dioxygenase genes other than the *nahA*-like subfamily has demonstrated the potential for PAH biodegradation in soil (Laurie and Lloyd-Jones 2000) and freshwater sediments (DeBruyn et al. 2007; DeBruyn et al. 2009; Dionisi et al. 2004). When monitored over time, qPCR enumeration of aromatic oxygenase genes can be used to provide rapid and direct feedback on contaminant degrading populations to predict and evaluate the performance of corrective actions. In column studies, for example, increases in PHE copies corresponded to higher dichlorobenzene removal providing an additional line of evidence suggesting that aerobic biostimulation would promote bioremediation (Dominguez et al. 2008). In our laboratories, quantification of aromatic oxygenase genes was successfully used to assess the impact of the subsurface injection of an oxygen-releasing material. Following injection, PHE and RMO gene copies in groundwater samples increased by 1–3 orders of magnitude and BTEX concentrations decreased substantially indicating that increased oxygen availability promoted growth of aerobic BTEX utilizing bacteria. After depletion of the oxygen-releasing material, PHE and RMO gene copies decreased and BTEX rebounded (Nebe et al. 2009). Thus, the qPCR results corresponded to changes in contaminant concentrations, geochemical conditions, and overall site management activities.

15.2.2.2 Reverse Transcriptase-Quantitative PCR

While enumeration of aromatic oxygenase genes in the aforementioned studies allowed quantitative assessment of biodegradation potential under existing site conditions and in response to corrective actions, qPCR results carry the caveat that the presence of a catabolic gene, even in high copy numbers, does not necessarily indicate expression and activity of the pathway. Organisms harboring aromatic oxygenase genes can and often are present in high abundance at petroleum-contaminated sites but may not be active due to prevailing subsurface conditions such as low oxygen availability. Increased activity of aromatic hydrocarbon-utilizing bacteria in response to site activities (Nebe et al. 2009) or laboratory perturbations (Dominguez et al. 2008; Laurie and Lloyd-Jones 2000; Park and Crowley 2006) at times can be inferred from increases in target gene abundance (i.e., growth). However, with elevated background populations, any enhanced contaminant biodegradation may be due to increased activity rather than increases in degrader populations (Dominguez et al. 2008; Nebe et al. 2009). Detection or quantification of catabolic gene expression would therefore be a more direct avenue to demonstrate in situ biodegradation activity. With early RNA extraction and reverse transcriptase-PCR (RT-PCR) techniques, researchers were able to demonstrate expression of toluene-4-monooxygenase (Ogram et al. 1995) and naphthalene dioxygenase (Wilson et al. 1999) genes directly from environmental samples. With improvements in RNA extraction (Kong and Nakatsu 2010) and field preservation protocols, RT-qPCR is being used more frequently to better evaluate the feasibility and performance of corrective actions at petroleum hydrocarbon-contaminated sites. Recently, RT-qPCR quantification of alkane monooxygenase (*alkB*) and dioxygenase genes involved in PAH metabolism revealed that nutrient addition stimulated microbial activity in Arctic soils impacted by diesel fuel releases (Yergeau et al. 2009). Interestingly, the ratio of PAH dioxygenase gene transcripts from Gram-negative bacteria to those from Gram-positive bacteria decreased over time suggesting a shift as the initial r-strategists (organisms with high growth rate and many offspring) were eventually outcompeted by the K-strategists (organisms with low growth rate and few offspring) capable of utilizing the more persistent high molecular weight PAHs. RT-qPCR has been utilized to monitor gene expression to evaluate the effectiveness of an oxygen infusion system to promote BTEX biodegradation at a gasoline-impacted site (Baldwin et al. 2010). During system operation, transcription of aromatic oxygenase genes (PHE and TOD) at the infusion point increased by as many as five orders of magnitude with concurrent decreases in contaminant concentrations. Moreover, aromatic oxygenase gene transcription was noted at downgradient monitoring points after an initial lag period despite low dissolved oxygen levels. Conversely, oxygenase gene transcripts were not detected at infusion or downgradient monitoring points when the oxygen infusion system was deactivated indicating that enhanced activity of aerobic BTEX degraders was indeed due to system operation. From an overall site management perspective, RT-qPCR has demonstrated in situ expression of target genes which provided site managers with rapid and most importantly direct feedback that the corrective action was indeed stimulating aerobic BTEX degrader activity.

Although qPCR and RT-qPCR quantification of the abundance and expression of aromatic oxygenase genes can provide a direct line of evidence to evaluate the feasibility and performance of corrective actions at sites impacted by aromatic hydrocarbons, what little information is available in the literature regarding the prevalence and selection of known aromatic catabolic genotypes in contaminated environments is often conflicting. In several studies, PHE (Baldwin et al. 2008; Hendrickx et al. 2006b) and RMO (Baldwin et al. 2008; Cavalca et al. 2004; Hendrickx et al. 2005; Hendrickx et al. 2006b; Nebe et al. 2009) were the most frequently detected genotypes or corresponded most closely to site activities (Nebe et al. 2009). In various studies conducted with field samples, TOD was not detected (Hendrickx et al. 2005; Nebe et al. 2009), infrequently detected (Hendrickx et al. 2006b), commonly detected (Baldwin et al. 2008), and routinely detected in high abundances (Dominguez et al. 2008). Similarly, the work of DeBruyn et al. (2007, 2009) suggests that the *nidA* genes from *Mycobacterium* spp. may be particularly important in the biodegradation of high molecular weight PAHs. However, the *nagA*-like and *nahAc*-like naphthalene dioxygenase genes are also commonly detected at PAH-contaminated sites and have been correlated to naphthalene concentrations (Dionisi et al. 2004) and ^{14}C naphthalene mineralization (Tuomi et al. 2004). The selective enrichment of individual genotypes could be related to a number of factors including the spectrum of contaminants present, contaminant concentrations (Duetz et al. 1994; DeBruyn et al. 2007; Yagi and Madsen 2009), and dissolved oxygen concentrations (Leahy and Olsen 1997). However, with the experience gained by performing qPCR enumeration of aromatic oxygenase genes with thousands of environmental samples on a commercial basis, all of the known pathways for aerobic catabolism of aromatic hydrocarbons are ecologically significant and can be prevalent at a petroleum-contaminated site. Additionally, renewed emphasis on evaluating functional gene diversity has shown that the presently characterized aromatic dioxygenases do not adequately reflect in situ diversity (Yeates et al. 2000) and that novel as less well characterized groups are also selected at BTEX-impacted sites

(Ní Chadhain et al. 2006; Taylor and Janssen 2005; Taylor et al. 2002; Witzig et al. 2006; Yeates et al. 2000). With an ever-increasing catalog of applicable gene targets, platforms capable of not only high throughput but also highly parallel quantification will be increasingly important for thorough characterization of contaminant degrader populations, microbial community functional dynamics, and catabolic gene expression.

15.2.3 Functional Gene Arrays

While originally developed for gene expression profiling of pure cultures, functional gene arrays (FGAs) are readily applicable to the environmental restoration industry. With rapid advances in printing technologies, microarrays can now contain hundreds of thousands of probes (Gentry et al. 2006) and therefore offer highly parallel detection/quantification which could provide a more comprehensive characterization of microbial community functional dynamics. For example, the GeoChip has been expanded from 1,662 oligonucleotide probes targeting 2,402 genes (Rhee et al. 2004) to 83,992 probes targeting 152,414 genes in 410 gene categories (Lu et al. 2011). FGAs have been used to investigate biodegradation of petroleum hydrocarbons (Iwai et al. 2008; Lu et al. 2011; Yergeau et al. 2009; Rhee et al. 2004; Rodríguez-Martínez et al. 2006; Hazen et al. 2010) and PCBs (Denef et al. 2003). In recent studies, the GeoChip 4.0 FGA was employed to investigate how deep-sea oil contamination resulting from the Deepwater Horizon incident affected the marine functional composition and to determine which microbial functional genes were enriched in response to the spill (Hazen et al. 2010; Lu et al. 2011). For example, genes involved in the biodegradation of alkanes (e.g., *alkB*), alkenes (e.g., *Xamo*), benzoate (e.g., *bco*), and BTEX (e.g., *catB*) were significantly more abundant in seawater samples obtained from the plume than in samples outside the plume taken from the same depth. Moreover, the abundance of these catabolic genes, most notably those for PAH biodegradation, was correlated with concentrations of various hydrocarbons present in the samples (Hazen et al. 2010). FGA analysis allowed simultaneous evaluation of shifts in microbial processes involved in nitrogen, sulfur, and phosphorus cycles in deep-sea samples (Lu et al. 2011). For example, nitrate reductase (*nasA*) and nitrite reductase (*nir*) for assimilatory nitrate reduction and glutamate dehydrogenase (*gdh*) for ammonia assimilation were more abundant in plume samples. Likewise, the genes encoding exopolyphosphatase (*ppx*) for inorganic polyphosphate degradation were enriched in plume samples suggesting stimulation of organic phosphorus release. Finally, the greater abundance of genes involved in sulfite reduction (*dsrAB*) and iron reduction (cytochrome *c*) as well as anaerobic toluene degradation (*bbs*) in the plume samples suggested enrichment of anaerobic pathways for hydrocarbon degradation. Thus, the highly parallel detection offered by microarray analysis can provide a comprehensive characterization of the impact of a petroleum hydrocarbon release on the functional community composition and selected microbial processes.

15.3 Targeted 16S rRNA Gene Analysis

While trends in contaminant concentrations and subsurface geochemical conditions are important lines of evidence, the most direct and perhaps most rapid means of assessing the feasibility and performance of a bioremediation strategy is to quantify the abundance and activity of target microorganisms capable of biodegradation of the specific contaminants of concern present at a site. With its high degree of specificity and increasing commercial availability, qPCR is regularly employed in the environmental remediation industry to quantify specific microorganisms or functional, catabolic genes involved in contaminant biodegradation. As will be discussed for petroleum hydrocarbons, the ability to catabolize many of the most common environmental pollutants is generally not limited to a few bacterial genera so elucidation of catabolic pathways and development of qPCR assays based on functional genes is required for meaningful results. However, for a few common contaminants, biodegradation appears to be linked to relatively narrow phylogenetic groups. In these cases, qPCR assays targeting 16S rRNA have provided quantification of key microorganisms deemed critical to successful bioremediation in the field. Perhaps the best examples of the latter are members of the genus *Dehalococcoides* and the reductive dechlorination of chloroethenes.

15.3.1 Dehalococcoides *spp. and Reductive Dechlorination*

Chlorinated ethenes including tetrachloroethene (PCE) and trichloroethene (TCE) were widely used as dry cleaning fluids and industrial solvents and are now prominent groundwater contaminants. Under anaerobic conditions, PCE and TCE can undergo sequential reductive dechlorination through the daughter products *cis-1*,2-dichloroethene (DCE) and vinyl chloride to nontoxic ethene (DiStefano et al. 1991; Freedman and Gossett 1989). However, a "stall" where the daughter products accumulate can occur at PCE- and TCE-impacted sites especially those undergoing MNA. While a number of bacterial cultures including *Dehalococcoides*, *Dehalobacter*, *Desulfitobacterium*, and *Desulfuromonas* species capable of reductive dechlorination of PCE and TCE to DCE have been isolated (Field and Sierra-Alvarez 2004), the only bacterial cultures isolated to date that are capable of not only reductive dechlorination of the parent compounds PCE and TCE but also DCE and vinyl chloride belong to the genus *Dehalococcoides* (Maymó-Gatell et al. 1999; also see Chap. 1, Sect. 1.4 for more details). In fact, the presence of *Dehalococcoides* spp. as demonstrated by PCR methods has been associated with complete reductive dechlorination of PCE and TCE to ethene at sites across North America and Europe (Hendrickson et al. 2002). Furthermore, Lu et al. (2006) proposed using a threshold abundance of 1×10^7 *Dehalococcoides* 16S rRNA gene copies/L as a screening criterion to identify sites where reductive dechlorination will yield a generally useful biodegradation rate. The Lu et al. (2006) results are consistent with those of an internal database of *Dehalococcoides* abundance and groundwater monitoring results maintained in our laboratory. In an internal study conducted with nearly 1,000 groundwater samples obtained from sites across the USA, ethene production was observed in approximately 80 % of the samples in which qPCR results were greater than or equal to 1×10^7 *Dehalococcoides* 16S rRNA gene copies/L. Thus, the abundance of *Dehalococcoides* is often the most critical factor in evaluating anaerobic bioremediation at PCE- and TCE-impacted sites.

15.4 Microbial Community Analysis

With the great variety and chemical complexity of contaminants released into the environment, genetic mechanisms underlying biodegradation pathways or the responsible microbes are not always known. As an alternative, or for use as supportive information, changes in microbial community structure can be examined using a variety of methods to obtain an indirect assessment of microbial response to environmental change due to contaminant inputs. Most methods have been chosen because they are relatively high throughput and applicable when small sample sizes or limited microbial biomass is available. The methods that have been used most often are denaturing gradient gel electrophoresis (DGGE) (Muyzer et al. 1993) and terminal-restriction fragment length polymorphism (T-RFLP) (Liu et al. 1997) of PCR products. Less commonly used but a relatively rapid method to determine community composition and relative abundances are microarrays constructed from small subunit rRNA sequences (DeSantis et al. 2007). In the future we will likely see more use of high-throughput sequencing methods that have been decreasing in costs and will soon be a feasible alternative to microarrays and perhaps to genetic fingerprinting. A brief description of these methods and examples of their application to bioremediation assessment is given in the next section.

15.4.1 Genetic Fingerprinting Methods: PCR-DGGE and T-RFLP

Rapid and relatively high-throughput microbial community analysis methods typically are dependent on PCR and produce genetic fingerprints representing community structure. Two of the most commonly used PCR-dependent genetic fingerprinting approaches, DGGE (Muyzer et al. 1993) and T-RFLP (Liu et al. 1997), are performed differently but yield similar results. Fingerprints are produced by DGGE because of differential migration of DNA strands of the same length with different sequence composition when electrophoresed through a linear gradient of denaturant

(Nakatsu and Marsh 2007). For T-RFLP the PCR primer is fluorescently tagged allowing high-resolution detection using automated sequencing system to detect PCR amplicon fragments after restriction enzyme digestion. Details for performing these methods are available elsewhere (Nakatsu and Marsh 2007; Muyzer et al. 1996) but some points should be made to aid in the interpretation of results. Critical to community fingerprinting approaches is choosing the appropriate phylogenetic markers for PCR amplification; typically the 16S rRNA genes are used. By targeting the 16S rRNA gene, the population richness of complex microbial communities may exceed the resolution power of the method and shifts in communities may not be captured. But in highly contaminated sites population richness of communities is less than in non-contaminated locations (Nakatsu et al. 2000) reducing the chances that this will be a problem. Additionally if biodegradation is occurring, the microbes utilizing these compounds as a growth substrate will increase in numbers; these enriched populations will result in changes to the community genetic fingerprint (Huang et al. 2000). Alternatively, to reduce complexity group-specific or functional genes involved in bioremediation have also been targeted using these methods (Hendrickx et al. 2006a).

Genetic fingerprinting methods do not provide detailed information about the members of the communities but this is not the objective for their application. In assessment of bioremediation potential, community analysis using genetic fingerprinting techniques is typically being used for comparisons of many environmental samples to evaluate spatial and temporal differences in the microbial communities in response to applied technologies (Nakatsu et al. 2005; Nebe et al. 2009). They have been used to assess remediation potential both in the laboratory and in the field. Microcosms under controlled laboratory conditions have been used as the first step in assessment of biodegradation of specific compounds or compound mixtures (Bordenave et al. 2007) and to test the efficacy of various remediation technologies (Roling et al. 2004; Stephen et al. 1999). Microcosms constructed from the site of interest are the most appropriate, particularly if remediation potential of particular habitat is being assessed. This maximizes the chance that the indigenous microbes from the field would be present. Also, most field sites are heterogeneous and microcosms provide a means to test specific chemical, physical, and/or biological variables that may promote or hinder bioremediation (Nakatsu et al. 2005). Ultimately these methods are being tested for their value in remediation assessment; therefore final tests must be tested on samples collected from the field. A number of studies have used these methods to evaluate microbial communities involved in bioremediation to identify potential degraders in the field (Duarte et al. 2001) or as a means to assess in situ spatial and temporal changes in communities undergoing remediation treatments (Kao et al. 2010; Ogino et al. 2001; Stephen et al. 1999). Although these fingerprinting techniques do not provide direct proof that enriched populations are directly responsible for biodegradation, other methods such as sequencing and stable isotope probing (SIP) are typically used in conjunction with fingerprinting to identify microbes involved in biodegradation (Busch-Harris et al. 2008; Kasai et al. 2006). Genetic fingerprinting techniques will likely continue to have a role in bioremediation assessment even with the advent of newer high-throughput sequencing methods because they are relatively rapid and less expensive means of screening samples prior to analysis using other more in-depth methods.

15.4.2 Community Microarray Analysis

Microarray analysis can be performed using either directly nucleic acids directly extracted from the sampling site or PCR amplicons to enrich the target 16S rRNA gene to obtain a microbial community profile (Yergeau et al. 2009). Microarrays can only detect the presence of known targets that have been included on the array, a methodological limitation. The most comprehensive phylogenetic microarray is the PhyloChip that includes probes for all known microbial taxa. It can provide substantial and valuable information about microbial communities (DeSantis et al. 2007; DeAngelis et al. 2011). The method was successfully used to assess dynamics of microbial communities and their response to treatments for remediation enhancement in trichloroethene (Lee et al. 2012)- and oil (Hazen et al. 2010) contaminated sites. Similar to the genetic fingerprinting, microarrays cannot directly identify populations responsible for biodegradation. But taxonomic information is available providing a means to identify metabolic groups responding to bioremediation technologies (Mohanty et al. 2011). This method is more costly than the genetic fingerprinting methods, which is possibly one factor contributing to its limited use thus far for bioremediation assessment. With the availability and reduction in costs of high-throughput sequencing supplies and instruments, many genomic facilities are beginning to use sequencing instead of microarrays.

15.4.3 High-Throughput Sequencing

The application of high-throughput sequencing for assessment of biodegradation potential in contaminated sites has just begun. It has been used to directly sequence nucleic acids extracted from environmental samples or targeted genes (e.g., 16S rRNA gene) first enriched using PCR. Metagenomic sequencing has the advantage of providing information on composition, presence of catabolic genes involved in biodegradation, and other physiological traits in the community. However, sufficient funds are needed to obtain sequencing depth to capture the community's diversity; costs increase along with community complexity. Costs can be optimized by prescreening samples using a rapid genetic fingerprinting method and choosing only a subset of representative samples for sequencing (Hamady and Knight 2009). There are a number of instruments and chemical technologies currently in use and in development but this technology is rapidly changing in costs, throughput time, and numbers of reads (Shendure and Ji 2008). Targeted pyrosequencing of the 16S rRNA gene was used to assess potential for degradation oil washed onto beaches after the Deepwater Horizon disaster (Kostka et al. 2011). Most notable was the agreement of intrinsic biodegradation potential found between chemical analysis and cultivation of oil degraders, pyrosequencing, and qPCR. Pyrosequencing has also been used with stable isotope probing to identify populations responding to petroleum remediation efforts (Bell et al. 2011). Much like the other assessment methods there is value in using multiple approaches when attempting to determine biodegradation potential in the field.

15.5 Conclusions

A number of different methods are being used to assess biodegradation potential in contaminated sites. Numerous investigations have shown that approaches targeting functional genes, particularly qPCR, provide a direct measure of biodegradation that is occurring in the field. However, for qPCR to be effective, there is a need to recognize functional genes or microbes that are indicative of biodegradation occurring in the field and not only the laboratory. There is still need to identify the best degraders under field conditions leading to the importance of methods for microbial community analysis in bioremediation assessments. Functional gene analyses and microbial community analyses along with other innovative techniques like stable isotope probing provide valuable and complementary information. Used in conjunction with traditional groundwater monitoring, these techniques for monitoring microbial community dynamics allow better assessment, implementation, and understanding of bioremediation in the field.

References

Abu Laban N, Selesi D, Rattei T, Tischler P, Meckenstock RU (2010) Identification of enzymes involved in anaerobic benzene degradation by a strictly anaerobic iron-reducing enrichment culture. Environ Microbiol 12(10):2783–2796. doi:10.1111/j.1462-2920.2010.02248.x

Achong GR, Rodriguez AM, Spormann AM (2001) Benzylsuccinate synthase of *Azoarcus* sp. strain T: cloning, sequencing, transcriptional organization, and its role in anaerobic toluene and *m*-xylene mineralization. J Bacteriol 183(23):6763–6770. doi:10.1128/jb.183.23.6763-6770.2001

Ahn Y, Sanseverino J, Sayler GS (1999) Analyses of polycyclic aromatic hydrocarbon-degrading bacteria isolated from contaminated soils. Biodegradation 10(2):149–157. doi:10.1023/a:1008369905161

Amann RI, Ludwig W, Schleifer KH (1995) Phylogenetic identification and in situ detection of individual microbial cells without cultivation. Microbiol Rev 59(1):143–169

Baldwin BR, Nakatsu CH, Nies L (2008) Enumeration of aromatic oxygenase genes to evaluate monitored natural attenuation at gasoline-contaminated sites. Water Res 42(3):723–731. doi:10.1016/j.watres.2007.07.052

Baldwin BR, Biernacki A, Blair J, Purchase MP, Baker JM, Sublette K, Davis G, Ogles D (2010) Monitoring gene expression to evaluate oxygen infusion at a gasoline-contaminated site. Environ Sci Technol 44(17):6829–6834. doi:10.1021/es101356t

Bell TH, Yergeau E, Martineau C, Juck D, Whyte LG, Greer CW (2011) Identification of nitrogen-incorporating bacteria in petroleum-contaminated Arctic soils by using (15)N DNA-based stable isotope probing and pyrosequencing. Appl Environ Microbiol 77(12):4163–4171. doi:10.1128/aem.00172-11

Beller HR, Edwards EA (2000) Anaerobic toluene activation by benzylsuccinate synthase in a highly enriched methanogenic culture. Appl Environ Microbiol 66(12):5503–5505. doi:10.1128/aem.66.12.5503-5505.2000

Beller HR, Kane SR, Legler TC, Alvarez PJJ (2002) A real-time polymerase chain reaction method for monitoring anaerobic, hydrocarbon-degrading bacteria based on a catabolic gene. Environ Sci Technol 36(18):3977–3984. doi:10.1021/es025556w

Bertoni G, Martino M, Galli E, Barbieri P (1998) Analysis of the gene cluster encoding toluene/*o-xylene* monooxygenase from *Pseudomonas stutzeri* OX1. Appl Environ Microbiol 64(10): 3626–3632

Biegert T, Fuchs G, Heider J (1996) Evidence that anaerobic oxidation of toluene in the denitrifying bacterium *Thauera aromatica* is initiated by formation of benzylsuccinate from toluene and fumarate. Eur J Biochem 238(3):661–668. doi:10.1111/j.1432-1033.1996.0661w.x

Bordenave S, Goni-Urriza MS, Caumette P, Duran R (2007) Effects of heavy fuel oil on the bacterial community structure of a pristine microbial mat. Appl Environ Microbiol 73(19): 6089–6097. doi:10.1128/aem.01352-07

Busch-Harris J, Sublette K, Roberts KP, Landrum C, Peacock AD, Davis G, Ogles D, Holmes WE, Harris D, Ota C, Yang X, Kolhatkar A (2008) Bio-traps coupled with molecular biological methods and stable isotope probing demonstrate the in situ biodegradation potential of MTBE and TBA in gasoline-contaminated aquifers. Ground Water Monit R 28(4):47–62. doi:10.1111/j.1745-6592.2008.00216.x

Butler CS, Mason JR (1996) Structure-function analysis of the bacterial aromatic ring-hydroxylating dioxygenases. In: Poole RK (ed) Advances in microbial physiology, vol 38. Academic, London, pp 47–84. doi:10.1016/s0065-2911(08)60155-1

Cao B, Nagarajan K, Loh K-C (2009) Biodegradation of aromatic compounds: current status and opportunities for biomolecular approaches. Appl Microbiol Biotechnol 85(2):207–228. doi:10.1007/s00253-009-2192-4

Cavalca L, Dell'Amico E, Andreoni V (2004) Intrinsic bioremediability of an aromatic hydrocarbon-polluted groundwater: diversity of bacterial population and toluene monoxygenase genes. Appl Microbiol Biotechnol 64(4):576–587. doi:10.1007/s00253-003-1449-6

Cerniglia CE (1992) Biodegradation of polycyclic aromatic hydrocarbons. Biodegradation 3(2):351–368. doi:10.1007/bf00129093

Da Silva MLB, Alvarez PJJ (2004) Enhanced anaerobic biodegradation of benzene-toluene-ethylbenzene-xylene-ethanol mixtures in bioaugmented aquifer columns. Appl Environ Microbiol 70(8):4720–4726. doi:10.1128/aem.70.8.4720-4726.2004

DeAngelis KM, Wu CH, Beller HR, Brodie EL, Chakraborty R, DeSantis TZ, Fortney JL, Hazen TC, Osman SR, Singer ME, Tom LM, Andersen GL (2011) PCR amplification-Independent methods for detection of microbial communities by the high-density microarray PhyloChip. Appl Environ Microbiol 77:6313–6322. doi:10.1128/aem.05262-11

DeBruyn JM, Chewning CS, Sayler GS (2007) Comparative quantitative prevalence of *Mycobacteria* and functionally abundant *nidA*, *nahAc*, and *nagAc*dioxygenase genes in coal tar contaminated sediments. Environ Sci Technol 41(15):5426–5432. doi:10.1021/es070406c

DeBruyn JM, Mead TJ, Wilhelm SW, Sayler GS (2009) PAH biodegradative genotypes in Lake Erie sediments: evidence for broad geographical distribution of pyrene-degrading *Mycobacteria*. Environ Sci Technol 43(10):3467–3473. doi:10.1021/es803348g

Denef VJ, Park J, Rodrigues JLM, Tsoi TV, Hashsham SA, Tiedje JM (2003) Validation of a more sensitive method for using spotted oligonucleotide DNA microarrays for functional genomics studies on bacterial communities. Environ Microbiol 5(10):933–943. doi:10.1046/j.1462-2920.2003.00490.x

DeSantis TZ, Brodie EL, Moberg JP, Zubieta IX, Piceno YM, Andersen GL (2007) High-density universal 16S rRNA microarray analysis reveals broader diversity than typical clone library when sampling the environment. Microb Ecol 53(3):371–383. doi:10.1007/s00248-006-9134-9

Dionisi HM, Chewning CS, Morgan KH, Menn F-M, Easter JP, Sayler GS (2004) Abundance of dioxygenase genes similar to *Ralstonia* sp. strain U2 *nagAc*is correlated with naphthalene concentrations in coal tar-contaminated freshwater sediments. Appl Environ Microbiol 70(7):3988–3995. doi:10.1128/aem.70.7.3988-3995.2004

DiStefano TD, Gossett JM, Zinder SH (1991) Reductive dechlorination of high concentrations of tetrachloroethene to ethene by an anaerobic enrichment culture in the absence of methanogenesis. Appl Environ Microbiol 57(8):2287–2292

Dominguez R, da Silva M, McGuire T, Adamson D, Newell C, Alvarez P (2008) Aerobic bioremediation of chlorobenzene source-zone soil in flow-through columns: performance assessment using quantitative PCR. Biodegradation 19(4):545–553. doi:10.1007/s10532-007-9160-4

Duarte GF, Rosado AS, Seldin L, de Araujo W, van Elsas JD (2001) Analysis of bacterial community structure in sulfurous-oil-containing soils and detection of species carrying dibenzothiophene desulfurization (*dsz*) genes. Appl Environ Microbiol 67(3):1052–1062

Duetz WA, de Jong C, Williams PA, van Andel JG (1994) Competition in chemostat culture between *Pseudomonas* strains that use different pathways for the degradation of toluene. Appl Environ Microbiol 60(8):2858–2863

Erickson BD, Mondello FJ (1993) Enhanced biodegradation of polychlorinated biphenyls after site-directed mutagenesis of a biphenyl dioxygenase gene. Appl Environ Microbiol 59(11):3858–3862

Ferrero M, Llobet-Brossa E, Lalucat J, García-Valdés E, Rosselló-Mora R, Bosch R (2002) Coexistence of two distinct copies of naphthalene degradation genes in *Pseudomonas* strains isolated from the western Mediterranean region. Appl Environ Microbiol 68(2):957–962. doi:10.1128/aem.68.2.957-962.2002

Field JA, Sierra-Alvarez R (2004) Biodegradability of chlorinated solvents and related chlorinated aliphatic compounds. Rev Environ Sci Biotechnol 3(3):185–254. doi:10.1007/s11157-004-4733-8

Freedman DL, Gossett JM (1989) Biological reductive dechlorination of tetrachloroethylene and trichloroethylene to ethylene under methanogenic conditions. Appl Environ Microbiol 55(9):2144–2151

Gentry T, Wickham G, Schadt C, He Z, Zhou J (2006) Microarray applications in microbial ecology research. Micro Ecol 52(2):159–175. doi:10.1007/s00248-006-9072-6

Gibson DT, Parales RE (2000) Aromatic hydrocarbon dioxygenases in environmental biotechnology. Curr Opin Biotechnol 11(3):236–243. doi:10.1016/s0958-1669(00)00090-2

Hamady M, Knight R (2009) Microbial community profiling for human microbiome projects: tools, techniques, and challenges. Genome Res 19(7):1141–1152. doi:10.1101/gr.085464.108

Hazen TC, Dubinsky EA, DeSantis TZ, Andersen GL, Piceno YM, Singh N, Jansson JK, Probst A, Borglin SE, Fortney JL, Stringfellow WT, Bill M, Conrad ME, Tom LM, Chavarria KL, Alusi TR, Lamendella R, Joyner DC, Spier C, Baelum J, Auer M, Zemla ML, Chakraborty R, Sonnenthal EL, D'haeseleer P, Holman H-YN, Osman S, Lu Z, Van Nostrand JD, Deng Y, Zhou J, Mason OU (2010) Deep-sea oil plume enriches indigenous oil-degrading bacteria. Science 330(6001):204–208. doi:10.1126/science.1195979

Heider J, Spormann AM, Beller HR, Widdel F (1998) Anaerobic bacterial metabolism of hydrocarbons. FEMS Microbiol Rev 22(5):459–473. doi:10.1111/j.1574-6976.1998.tb00381.x

Hendrickson ER, Payne JA, Young RM, Starr MG, Perry MP, Fahnestock S, Ellis DE, Ebersole RC (2002) Molecular analysis of *Dehalococcoides* 16S ribosomal DNA from chloroethene-contaminated sites throughout North America and Europe. Appl Environ Microbiol 68(2):485–495. doi:10.1128/aem.68.2.485-495.2002

Hendrickx B, Dejonghe W, Boënne W, Brennerova M, Cernik M, Lederer T, Bucheli-Witschel M, Bastiaens L, Verstraete W, Top EM, Diels L, Springael D (2005) Dynamics of an oligotrophic bacterial aquifer community during contact with a groundwater plume contaminated with benzene, toluene, ethylbenzene, and xylenes: an In situ mesocosm study. Appl Environ Microbiol 71(7):3815–3825. doi:10.1128/aem.71.7.3815-3825.2005

Hendrickx B, Dejonghe W, Faber F, Boenne W, Bastiaens L, Verstraete W, Top EM, Springael D (2006a) PCR-DGGE method to assess the diversity of BTEX mono-oxygenase genes at contaminated sites. FEMS Microbiol Ecol 55(2):262–273. doi:10.1111/j.1574-6941.2005.00018.x

Hendrickx B, Junca H, Vosahlova J, Lindner A, Rüegg I, Bucheli-Witschel M, Faber F, Egli T, Mau M, Schlömann M, Brennerova M, Brenner V, Pieper DH, Top EM, Dejonghe W, Bastiaens L, Springael D (2006b) Alternative primer sets for PCR detection of genotypes involved in bacterial aerobic BTEX degradation: distribution of the genes in BTEX degrading isolates and in subsurface soils of a BTEX contaminated industrial site. J Microbiol Meth 64(2):250–265. doi:10.1016/j.mimet.2005.04.018

Huang XJ, Lee LS, Nakatsu C (2000) Impact of animal waste lagoon effluents on chlorpyrifos degradation in soils. Environ Toxicol Chem 19(12):2864–2870. doi:10.1002/etc.5620191202

Iwai S, Kurisu F, Urakawa H, Yagi O, Kasuga I, Furumai H (2008) Development of an oligonucleotide microarray to detect di- and monooxygenase genes for benzene degradation in soil. FEMS Microbiol Lett 285(1):111–121. doi:10.1111/j.1574-6968.2008.01223.x

Iwai S, Johnson TA, Chai B, Hashsham SA, Tiedje JM (2011) Comparison of the specificities and efficacies of primers for aromatic dioxygenase gene analysis of environmental samples. Appl Environ Microbiol 77(11):3551–3557. doi:10.1128/aem.00331-11

Johnson GR, Olsen RH (1995) Nucleotide sequence analysis of genes encoding a toluene/benzene-2-monooxygenase from *Pseudomonas* sp. strain JS150. Appl Environ Microbiol 61(9): 3336–3346

Kahng H-Y, Malinverni JC, Majko MM, Kukor JJ (2001) Genetic and functional analysis of the tbc operons for catabolism of alkyl- and chloroaromatic compounds in *Burkholderia* sp. Strain JS150. Appl Environ Microbiol 67(10):4805–4816. doi:10.1128/aem.67.10.4805-4816.2001

Kane SR, Beller HR, Legler TC, Anderson RT (2002) Biochemical and genetic evidence of benzylsuccinate synthase in toluene-degrading, ferric iron-reducing *Geobacter metallireducens*. Biodegradation 13(2):149–154. doi:10.1023/a:1020454831407

Kao CM, Chen CS, Tsa FY, Yang KH, Chien CC, Liang SH, Yang CA, Chen SC (2010) Application of real-time PCR, DGGE fingerprinting, and culture-based method to evaluate the effectiveness of intrinsic bioremediation on the control of petroleum-hydrocarbon plume. J Hazard Mater 178(1–3):409–416. doi:10.1016/j.jhazmat.2010.01.096

Kasai Y, Takahata Y, Manefield M, Watanabe K (2006) RNA-based stable isotope probing and isolation of anaerobic benzene-degrading bacteria from gasoline contaminated groundwater. Appl Environ Microbiol 72(5):3586–3592

Kazy S, Monier A, Alvarez P (2010) Assessing the correlation between anaerobic toluene degradation activity and *bssA* concentrations in hydrocarbon-contaminated aquifer material. Biodegradation 21(5):793–800. doi:10.1007/s10532-010-9344-1

Khan AA, Wang R-F, Cao W-W, Doerge DR, Wennerstrom D, Cerniglia CE (2001) Molecular cloning, nucleotide sequence, and expression of genes encoding a polycyclic aromatic ring dioxygenase from *Mycobacterium* sp. strain PYR-1. Appl Environ Microbiol 67(8):3577–3585. doi:10.1128/aem.67.8.3577-3585.2001

Kniemeyer O, Fischer T, Wilkes H, Glöckner FO, Widdel F (2003) Anaerobic degradation of ethylbenzene by a new type of marine sulfate-reducing bacterium. Appl Environ Microbiol 69(2):760–768. doi:10.1128/aem.69.2.760-768.2003

Kong W, Nakatsu CH (2010) Optimization of RNA extraction for PCR quantification of aromatic compound degradation genes. Appl Environ Microbiol 76(4):1282–1284. doi:10.1128/aem.01939-09

Kostka JE, Prakash O, Overholt WA, Green SJ, Freyer G, Canion A, Delgardio J, Norton N, Hazen TC, Huettel M (2011) Hydrocarbon-degrading bacteria and the bacterial community response in Gulf of Mexico beach sands impacted by the deepwater horizon oil spill. Appl Environ Microbiol 77(22):7962–7974. doi:10.1128/aem.05402-11

Krieger CJ, Beller HR, Reinhard M, Spormann AM (1999) Initial reactions in anaerobic oxidation of *m*-xylene by the denitrifying bacterium *Azoarcus* sp. strain T. J Bacteriol 181(20):6403–6410

Kropp KG, Davidova IA, Suflita JM (2000) Anaerobic oxidation of n-dodecane by an addition reaction in a sulfate-reducing bacterial enrichment culture. Appl Environ Microbiol 66(12):5393–5398. doi:10.1128/aem.66.12.5393-5398.2000

Laurie AD, Lloyd-Jones G (1999) The *phn* Genes of *Burkholderia* sp. strain RP007 constitute a divergent gene cluster for polycyclic aromatic hydrocarbon catabolism. J Bacteriol 181(2):531–540

Laurie AD, Lloyd-Jones G (2000) Quantification of *phnAc* and *nahAc* in contaminated New Zealand soils by competitive PCR. Appl Environ Microbiol 66(5):1814–1817. doi:10.1128/aem.66.5.1814-1817.2000

Leahy JG, Olsen RH (1997) Kinetics of toluene degradation by toluene-oxidizing bacteria as a function of oxygen concentration, and the effect of nitrate. FEMS Microbiol Ecol 23(1):23–30. doi:10.1111/j.1574-6941.1997.tb00387.x

Lee PKH, Warnecke F, Brodie EL, Macbeth TW, Conrad ME, Andersen GL, Alvarez-Cohen L (2012) Phylogenetic microarray analysis of a microbial community performing reductive dechlorination at a TCE-contaminated site. Environ Sci Technol 46(2):1044–1054. doi:10.1021/es203005k

Liu WT, Marsh TL, Cheng H, Forney LJ (1997) Characterization of microbial diversity by determining terminal restriction fragment length polymorphisms of genes encoding 16S rRNA. Appl Environ Microbiol 63(11):4516–4522

Lloyd-Jones G, Laurie AD, Hunter DWF, Fraser R (1999) Analysis of catabolic genes for naphthalene and phenanthrene degradation in contaminated New Zealand soils. FEMS Microbiol Ecol 29(1):69–79. doi:10.1111/j.1574-6941.1999.tb00599.x

Lu X, Wilson JT, Kampbell DH (2006) Relationship between *Dehalococcoides* DNA in ground water and rates of reductive dechlorination at field scale. Water Res 40(16):3131–3140. doi:10.1016/j.watres.2006.05.030

Lu Z, Deng Y, Van Nostrand JD, He Z, Voordeckers J, Zhou A, Lee Y-J, Mason OU, Dubinsky EA, Chavarria KL, Tom LM, Fortney JL, Lamendella R, Jansson JK, D'Haeseleer P, Hazen TC, Zhou J (2011) Microbial gene functions enriched in the deepwater horizon deep-sea oil plume. ISME J 6(2):451–460. doi:10.1038/ismej.2011.91

Maymó-Gatell X, Anguish T, Zinder SH (1999) Reductive dechlorination of chlorinated ethenes and 1,2-dichloroethane by "*Dehalococcoides ethenogenes*" 195. Appl Environ Microbiol 65(7):3108–3113

Mohanty SR, Kollah B, Brodie EL, Hazen TC, Roden EE (2011) 16S rRNA gene microarray analysis of microbial communities in ethanol-stimulated subsurface sediment. Microbes Environ 26(3):261–265. doi:10.1264/jsme2.ME11111

Muyzer G, de Waal EC, Uitterlinden AG (1993) Profiling of complex microbial populations by denaturing gradient gel electrophoresis analysis of polymerase chain reaction-amplified genes coding for 16S rRNA. Appl Environ Microbiol 59(3):695–700

Muyzer G, Hottentrager S, Teske AP, Wawer C (1996) Denaturing gradient gel electrophoresis of PCR-amplified 16S rDNA-A new molecular approach to analyse the genetic diversity of mixed microbial communities. In: Akkermans A, van Elsas JD, de Bruijn FJ (eds) Molecular microbial ecology manual, vol 3.4.4. Kluwer Academic Publishers, Nowell, pp 1–23

Nakatsu CH, Marsh TL (2007) Analysis of microbial communities with denaturing gradient gel electrophoresis and terminal restriction fragment length polymorphism. In: Reddy CA, Beveridge TL, Breznak JA, Marzluf GA, Schmidt TM, Snyder LR (eds) Methods for general and molecular bacteriology, 3rd edn. ASM, Washington, DC, pp 909–923

Nakatsu CH, Torsvik V, Øvreås L (2000) Soil community analysis using DGGE of 16S rDNA polymerase chain reaction products. Soil Sci Soc Am J 64(4):1382–1388

Nakatsu CH, Carmosini N, Baldwin B, Beasley F, Kourtev P, Konopka A (2005) Soil microbial community responses to additions of organic carbon substrates and heavy metals (Pb and Cr). Appl Environ Microbiol 71(12):7679–7689. doi:10.1128/aem.71.12.7679-7689.2005

Nebe J, Baldwin BR, Kassab RL, Nies L, Nakatsu CH (2009) Quantification of aromatic oxygenase genes to evaluate enhanced bioremediation by oxygen releasing materials at a gasoline-contaminated site. Environ Sci Technol 43(6):2029–2034. doi:10.1021/es900146f

Ní Chadhain SM, Norman RS, Pesce KV, Kukor JJ, Zylstra GJ (2006) Microbial dioxygenase gene population shifts during polycyclic aromatic hydrocarbon biodegradation. Appl Environ Microbiol 72(6):4078–4087. doi:10.1128/aem.02969-05

Ogino A, Koshikawa H, Nakahara T, Uchiyama H (2001) Succession of microbial communities during a biostimulation process as evaluated by DGGE and clone library analyses. J Appl Microbiol 91(4):625–635. doi:10.1046/j.1365-2672.2001.01424.x

Ogram A, Sun W, Brockman FJ, Fredrickson JK (1995) Isolation and characterization of RNA from low-biomass deep-subsurface sediments. Appl Environ Microbiol 61(2):763–768

Pandey J, Chauhan A, Jain RK (2009) Integrative approaches for assessing the ecological sustainability of in situ bioremediation. FEMS Microbiol Rev 33(2):324–375. doi:10.1111/j.1574-6976.2008.00133.x

Parales RE, Resnick SM, Yu C-L, Boyd DR, Sharma ND, Gibson DT (2000) Regioselectivity and enantioselectivity of naphthalene dioxygenase during arene cis-dihydroxylation: control by phenylalanine 352 in the α subunit. J Bacteriol 182(19):5495–5504. doi:10.1128/jb.182.19.5495-5504.2000

Park J-W, Crowley D (2006) Dynamic changes in *nahAc* gene copy numbers during degradation of naphthalene in PAH-contaminated soils. Appl Microbiol Biotech 72(6):1322–1329. doi:10.1007/s00253-006-0423-5

Rhee S-K, Liu X, Wu L, Chong SC, Wan X, Zhou J (2004) Detection of genes involved in biodegradation and biotransformation in microbial communities by using 50-mer oligonucleotide microarrays. Appl Environ Microbiol 70(7):4303–4317. doi:10.1128/aem.70.7.4303-4317.2004

Rodríguez-Martínez E, Pérez E, Schadt C, Zhou J, Massol-Deyá A (2006) Microbial diversity and bioremediation of a hydrocarbon-contaminated aquifer (Vega Baja, Puerto Rico). Int J Environ Res Public Health 3(3):292–300

Roling WFM, Milner MG, Jones DM, Fratepietro F, Swannell RPJ, Daniel F, Head IM (2004) Bacterial community dynamics and hydrocarbon degradation during a field-scale evaluation of bioremediation on a mudflat beach contaminated with buried oil. Appl Environ Microbiol 70(5):2603–2613

Selesi D, Jehmlich N, von Bergen M, Schmidt F, Rattei T, Tischler P, Lueders T, Meckenstock RU (2010) Combined genomic and proteomic approaches identify gene clusters involved in anaerobic 2-methylnaphthalene degradation in the sulfate-reducing enrichment culture N47. J Bacteriol 192(1):295–306. doi:10.1128/jb.00874-09

Shendure J, Ji HL (2008) Next-generation DNA sequencing. Nat Biotechnol 26(10):1135–1145. doi:10.1038/nbt1486

Spormann AM, Widdel F (2000) Metabolism of alkylbenzenes, alkanes, and other hydrocarbons in anaerobic bacteria. Biodegradation 11(2):85–105. doi:10.1023/a:1011122631799

Staley JT, Konopka A (1985) Measurement of in situ activities of nonphotosynthetic microorganisms in aquatic and terrestrial habitats. Ann Rev Microbiol 39:321–346

Stephen JR, Chang YJ, Gan YD, Peacock A, Pfiffner SM, Barcelona MJ, White DC, Macnaughton SJ (1999) Microbial characterization of a JP-4 fuel-contaminated site using a combined lipid biomarker/polymerase chain reaction-denaturing gradient gel electrophoresis (PCR-DGGE)-based approach. Environ Microbiol 1(3):231–241

Sublette K, Peacock A, White D, Davis G, Ogles D, Cook D, Kolhatkar R, Beckmann D, Yang X (2006) Monitoring subsurface microbial ecology in a sulfate-amended, gasoline-contaminated aquifer. Ground Water Monit R 26(2):70–78. doi:10.1111/j.1745-6592.2006.00072.x

Taylor PM, Janssen PH (2005) Variations in the abundance and identity of class II aromatic ring-hydroxylating dioxygenase genes in groundwater at an aromatic hydrocarbon-contaminated site. Environ Microbiol 7(1):140–146. doi:10.1111/j.1462-2920.2004.00679.x

Taylor PM, Medd JM, Schoenborn L, Hodgson B, Janssen PH (2002) Detection of known and novel genes encoding aromatic ring-hydroxylating dioxygenases in soils and in aromatic hydrocarbon-degrading bacteria. FEMS Microbiol Lett 216(1):61–66. doi:10.1111/j.1574-6968.2002.tb11415.x

Tuomi PM, Salminen JM, Jørgensen KS (2004) The abundance of *nahAc* genes correlates with the ^{14}C-naphthalene mineralization potential in petroleum hydrocarbon-contaminated oxic soil layers. FEMS Microbiol Ecol 51(1):99–107. doi:10.1016/j.femsec.2004.07.011

Widada J, Nojiri H, Kasuga K, Yoshida T, Habe H, Omori T (2002a) Molecular detection and diversity of polycyclic aromatic hydrocarbon-degrading bacteria isolated from geographically diverse sites. Appl Microbiol Biotechnol 58(2):202–209. doi:10.1007/s00253-001-0880-9

Widada J, Nojiri H, Omori T (2002b) Recent developments in molecular techniques for identification and monitoring of xenobiotic-degrading bacteria and their catabolic genes in bioremediation. Appl Microbiol Biotechnol 60:45–59. doi:10.1007/s00253-002-1072-y

Wilson MS, Bakermans C, Madsen EL (1999) In situ, real-time catabolic gene expression: extraction and characterization of naphthalene dioxygenase mRNA transcripts from groundwater. Appl Environ Microbiol 65(1):80–87

Winderl C, Schaefer S, Lueders T (2007) Detection of anaerobic toluene and hydrocarbon degraders in contaminated aquifers using benzylsuccinate synthase (*bssA*) genes as a functional marker. Environ Microbiol 9(4):1035–1046. doi:10.1111/j.1462-2920.2006.01230.x

Winderl C, Anneser B, Griebler C, Meckenstock RU, Lueders T (2008) Depth-resolved quantification of anaerobic toluene degraders and aquifer microbial community patterns in distinct redox zones of a tar oil contaminant plume. Appl Environ Microbiol 74(3):792–801. doi:10.1128/aem.01951-07

Witzig R, Junca H, Hecht H-J, Pieper DH (2006) Assessment of toluene/biphenyl dioxygenase gene diversity in benzene-polluted soils: links between benzene biodegradation and genes similar to those encoding isopropylbenzene dioxygenases. Appl Environ Microbiol 72(5):3504–3514. doi:10.1128/aem.72.5.3504-3514.2006

Yagi JM, Madsen EL (2009) Diversity, abundance, and consistency of microbial oxygenase expression and biodegradation in a shallow contaminated aquifer. Appl Environ Microbiol 75(20):6478–6487. doi:10.1128/aem.01091-09

Yeates C, Holmes AJ, Gillings MR (2000) Novel forms of ring-hydroxylating dioxygenases are widespread in pristine and contaminated soils. Environ Microbiol 2(6):644–653. doi:10.1046/j.1462-2920.2000.00147.x

Yergeau E, Arbour M, Brousseau R, Juck D, Lawrence JR, Masson L, Whyte LG, Greer CW (2009) Microarray and real-time PCR analyses of the responses of high-Arctic soil bacteria to hydrocarbon pollution and bioremediation treatments. Appl Environ Microbiol 75(19):6258–6267. doi:10.1128/aem.01029-09

Chapter 16
Selective Stimulation of Aromatic Compound Degradation by the Indigenous Marine Bacterium *Cycloclasticus* for Bioremediation of Oil Spills in the Marine Environment

Noriyuki Iwabuchi

Abstract S-2 EPS, an extracellular polysaccharide produced by *Rhodococcus rhodochrous* S-2, is a functional biopolymer that increases the organism's organic solvent tolerance. We previously demonstrated that addition of S-2 EPS and minerals to oil-contaminated seawater enhanced the degradation of an aromatic fraction (AF) of Arabian light crude oil by *Cycloclasticus*, a polycyclic aromatic hydrocarbon (PAH)-degrading indigenous marine bacterium, and stimulated its growth in AF-contaminated seawater. Moreover, *Cycloclasticus* growth and PAH degradation were selectively activated by the association of S-2 EPS with *Cycloclasticus* cells. In this chapter, the effects of S-2 EPS on the growth and PAH degradation activity of *Cycloclasticus* in hydrocarbon-degrading marine bacterial consortia are summarized. S-2 EPS regulates the interaction between living bacteria and oils, and because it selectively stimulates degradation of aromatic compounds by the indigenous marine bacterium *Cycloclasticus*, S-2 EPS is useful for bioremediation of oil spills in the marine environment.

Keywords Aromatic hydrocarbons • Biodegradation • Bioremediation • *Cycloclasticus* • Extracellular polysaccharides • *Rhodococcus*

16.1 Introduction

Petroleum is composed of more than 1,000 different hydrocarbons and is a major pollutant of marine environments. Polycyclic aromatic hydrocarbons (PAHs) are petroleum constituents that are resistant to biodegradation and therefore may remain

N. Iwabuchi (✉)
Laboratory of Molecular Microbiology, Department of Applied Biological Science, College of Bioresource Sciences, Nihon University, Fujisawa, Kanagawa 252-0880, Japan
e-mail: iwabuchi@brs.nihon-u.ac.jp

H. Nojiri et al. (eds.), *Biodegradative Bacteria: How Bacteria Degrade, Survive, Adapt, and Evolve*, DOI 10.1007/978-4-431-54520-0_16,

at polluted sites for many years (Henson and Hayasaka 1982; Payne and Phillips 1985; Mackay and MacAuliffe 1988). The complete removal of PAHs from oil-contaminated marine sites thus necessitates the development of techniques that will enhance the ability of native marine bacteria to degrade these compounds.

Bacteria of the genus *Cycloclasticus* play a key role in the degradation of various aromatic hydrocarbons in marine environments and as such may be useful in bioremediation efforts (Dyksterhouse et al. 1995; Zaidi and Imam 1999; Maruyama et al. 2003; Dubois et al. 2007; Yakimov et al. 2007; Lozada et al. 2008; Wang et al. 2008; Niepceron et al. 2010). Although *Cycloclasticus* spp. have been detected in hydrocarbon-contaminated seawater and marine sediment through molecular ecological analyses and cloning of genes related to phenanthrene degradation, physiological studies of *Cycloclasticus* are lacking. It is thus necessary to investigate the properties of *Cycloclasticus* species in pure culture and their behavior in marine bacterial consortia if these organisms are to be used for the bioremediation of oil spills in marine environments.

In this chapter, we summarize our efforts to regulate the bacterial community structure using bacterial polysaccharides and discuss selective stimulation of aromatic compound degradation by *Cycloclasticus*. In the first section, we review the role of rhodococcal extracellular polysaccharide (EPS) in the development of oil tolerance. In the next section, we review the effect of S-2 EPS, produced by *Rhodococcus rhodochrous* S-2, on *Cycloclasticus*-mediated degradation of aromatic components in crude oil. Finally, we discuss the properties and growth behavior of *Cycloclasticus* and the adaptive changes in the bacterial community structure in seawater by the addition of various hydrocarbons.

16.2 EPSs Are Functional Biopolymers That Increase Organic Solvent Tolerance

EPSs are found on the surface of many bacteria and are believed to play an important role in regulating interactions between the cell and the surrounding environment (Lindberg 1990; Minnikin 1991). Effective bioremediation efforts require optimization of this interaction in order to maximize pollutant degradation. In this regard, our group has been studying the effects of various EPSs on the biodegradation of oils in the marine environment. Our previous studies indicated that EPSs are important for the degradation of petroleum by rhodococci. In this section, we discuss the importance of EPSs in the biodegradation of oils, based primarily on our research.

Rhodococci are a metabolically diverse group of bacteria. Some rhodococci can degrade a wide variety of organic compounds, including man-made xenobiotics such as PCBs, while others are capable of degrading numerous aliphatic and aromatic hydrocarbons (Finnerty 1992; Warhurst and Fewson 1994; Bell et al. 1998; Whyte et al. 1998). In addition, our group reported that some rhodococci completely

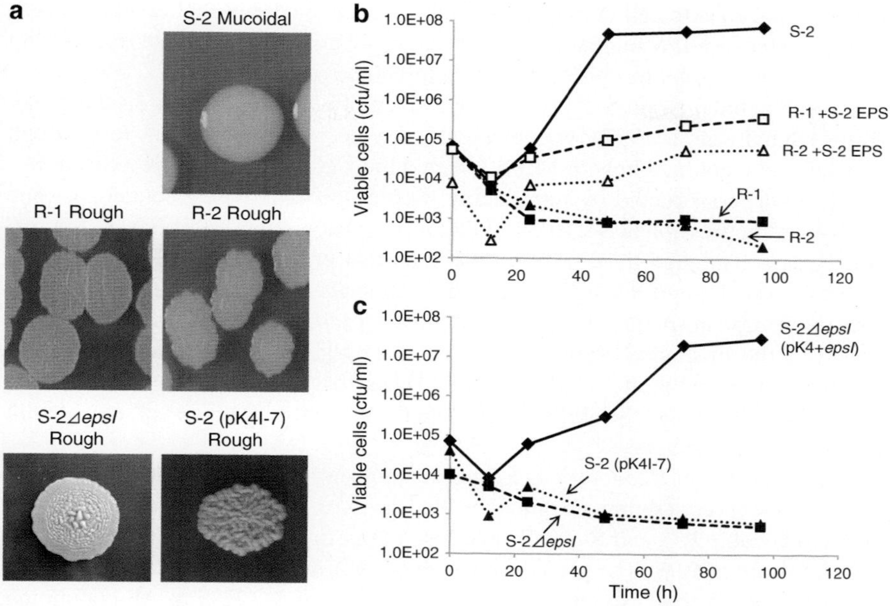

Fig. 16.1 The relationship between colony morphotype and oil tolerance of *Rhodococcus* strains. (**a**) Photographs illustrating the colony morphotypes of rough and mucoidal strains. (**b**) Growth of various colony morphotype strains in seawater-based medium containing AF (SW-AF) with and without S-2 EPS. (**c**) Growth of genetically engineered strains in SW-AF medium with and without S-2 EPS. This figure was modified from that previously reported by Iwabuchi et al. (2000)

translocate to the alkane phase, where they can grow well (Iwabuchi et al. 2009). Rhodococci are thus considered candidates for use in bioremediation efforts.

It is generally believed that PAHs and their derivatives become the major pollutants in oil-contaminated marine environments as a result of photooxidation and/or weathering of spilled oils. Therefore, in a previous study, we screened a number of *Rhodococcus* strains for their ability to degrade AF (Iwabuchi et al. 2000). We found that some *Rhodococcus* strains grow well in seawater-based medium containing a high concentration of AF or various oils. The relationship between colony morphotype and oil tolerance of the *Rhodococcus* strains we examined is summarized in Fig. 16.1. Almost all of the mucoidal strains of *Rhodococcus* exhibited significant growth in medium containing oil, while rough strains did not. To confirm these results, we subsequently examined the growth of colony morphotype mutants derived from *R. rhodochrous* CF222 in AF-containing media. Mucoidal strain S-2 grew better than rough strains R-1 and R-2 in AF-containing media, while strains R-1 and R-2 were more sensitive to organic solvents than was mucoidal strain S-2 (Fig. 16.1b). These data suggest that the mucoidal morphology influences the ability to grow in the presence of AF. A similar observation was reported by Whyte et al. (1995), who demonstrated that *Rhodococcus* sp. Q15 changes its morphology by producing an extracellular polymeric substance when diesel fuel is supplied as a carbon source at a low temperature. These results suggest that a close relationship exists between colony morphology and oil tolerance in *Rhodococcus*.

Since mucoidal strain S-2 produces considerably more EPS than do the rough strains (Sunairi et al. 1997), we genetically engineered 2 strains that form rough colonies and examined how S-2 EPS affected their growth (Fig. 16.1c). Strain S-2 (pK4I-7) was constructed by introducing plasmid pK4I-7, which contains a gene involved in suppression of S-2 EPS production (Iwabuchi et al. 1997). The S-2△*epsI* strain was constructed by disrupting the gene that encodes the TetR family transcriptional regulator of S-2 EPS production. Our analyses showed that the growth of both rough transformants was inhibited in the presence of AF. In contrast, the growth of the mucoidal strain complemented by the *epsI* gene was significantly enhanced compared to the S-2△*epsI* strain. These data indicated that the mucoidal morphology influences the ability to grow in the presence of AF, in agreement with the data described above.

The sensitivity of *Rhodococcus* to hydrocarbons was also examined. Addition of S-2 EPS to AF-containing medium resulted in a 10- to 100-fold increase in the number of viable R-1 and R-2 cells at all sampling times compared to cells grown in the absence of EPS. The growth of rough transformants derived from *R. rhodochrous* S-2 strain was also enhanced by the addition of S-2 EPS to AF-containing media. In addition, survival of strains R-1 and R-2 in the presence of 10 % (w/v) *n*-hexadecane increased approximately 10- to 100-fold after addition of S-2 EPS. The survival of cells of other rough strains was also enhanced by the addition of S-2 EPS. These results suggest that the EPS produced by strain S-2 protects the rough strains from hydrocarbon-induced toxicity (Iwabuchi et al. 2000).

Since S-2 EPS plays an important role in the solvent tolerance and biodegradation capabilities of *Rhodococcus*, we determined the chemical structure of the sugar chain of S-2 EPS (Urai et al. 2006b) and found that S-2 EPS is composed of D-galactose, D-mannose, D-glucose, and D-glucuronic acid, in a 1:1:1:1 molar ratio. The structure also contains 0.8 % (w/w) octadecanoic acid and 2.7 % (w/w) hexadecanoic acid. Chemical and enzymatic analyses, as well as ^{1}H- and ^{13}C-NMR spectroscopy, revealed that the polysaccharide consists of tetrasaccharide repeating units, as shown in Fig. 16.2a.

The sugar chain of S-2 EPS was found to be novel. Several reports by Richards' research group on the chemical structures of five EPSs produced by different serotypes of the equine pathogen *R. equi* indicate that their structures are different from that of S-2 EPS, and they reportedly do not contain fatty acids (Leitch and Richards 1990; Masoud and Richards 1994; Severn and Richards 1990, 1992, 1999).

There have been several reports that indicate bacterial EPS plays a role in hydrocarbon biodegradation. These different EPSs contain long-chain fatty acids (mainly C16 and C18) attached via alkali-labile ester bonds and include emulsan, produced by *Acinetobacter lwoffii* RAG-1 (Belsky et al. 1979), simusan, produced by *Arthrobacter* sp. strain CE-17 (Senchenkova et al. 1995), and two EPSs produced by *R. rhodochrous* ATCC 5396816 and *R. rhodochrous* SM-1 (Urai et al. 2002). These EPSs reportedly contain similar amounts of fatty acids and function as

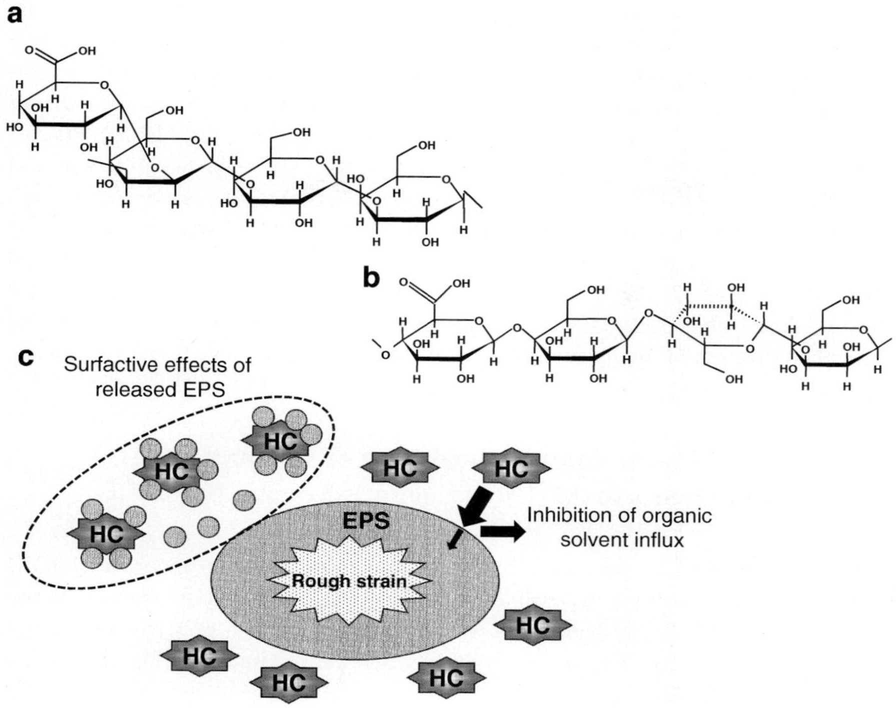

Fig. 16.2 Chemical structures of EPSs. (**a**) S-2 EPS (Urai et al. 2006b). (**b**) 33 EPS (Urai et al. 2006a). (**c**) Schematic illustrating how EPS enhances the oil tolerance of *Rhodococcus*. *HC* indicates hydrocarbon

biosurfactants (Belsky et al. 1979; Senchenkova et al. 1995). The above findings suggest that the biosurfactant activity of EPSs may enhance the degradation capability and organic solvent tolerance of various bacteria.

Our group has also studied the effect of EPS on benzene tolerance in *Rhodococcus* sp. 33 (Aizawa et al. 2005). We found that under both resting and growing conditions, rough mutant strains derived from strain 33 are more sensitive to benzene than the mucoidal parent strain. The rough strains produce little or no EPS, whereas the parental strain 33 produces a large quantity (33 EPS). Supplementation of benzene-containing growth medium with 33 EPS enhanced both the survival and growth of the rough strains. These data indicate that 33 EPS plays an important role in benzene tolerance in *Rhodococcus* strain 33, particularly by helping the cells survive initial challenge with benzene. The effect of 33 EPS on both biodegradation and benzene tolerance is similar to that of S-2 EPS, suggesting that both EPSs share a similar structure. Urai et al. (2006a) reported the chemical structure of 33 EPS (Fig. 16.2b), which consists of tetrasaccharide repeating units containing pyruvic acid: [→4)-β-D-Gal*p*-(1→4)-β-D-Glc*p*-(1→3)-β-D-Man*p*-(1→4)-β-D-Glc*p*-(1→]*n*. The hydrophobic functional groups in the hydrophilic main chain of EPSs are believed to enhance hydrocarbon degradation and/or organic solvent tolerance. The polyphilic structure of EPS is thus considered to play an important role in regulating the response of bacteria to hydrocarbons (Fig. 16.2c). In fact, several reports indicate that addition of S-2 EPS to the culture medium may decrease the cell surface hydrophobicity and increase the cell surface hydrocarbon affinity of many hydrocarbon-sensitive rough strains (Iwabuchi et al. 2000, 2003). In addition, Sunairi et al. (1997) reported that S-2 EPS functions as a hydrophiline, lowering the cell surface hydrophobicity of rough strains (Sunairi et al. 1997). Since the surface of *Rhodococcus* cells is highly hydrophobic due to the presence of mycolic acid (Goodfellow 1989; Roberts 1996), these findings suggest that in order to increase the efficiency of biodegradation, hydrocarbon tolerance must first be enhanced through regulation of the hydrophobicity of the cell surface.

16.3 S-2 EPS Stimulates Degradation of Aromatic Components in AF by Indigenous Marine Bacteria

Since S-2 EPS plays an important role in hydrocarbon tolerance and growth in the presence of hydrocarbons in *Rhodococcus* strains, we subsequently examined the effect of S-2 EPS on the degradation of AF hydrocarbons in seawater by natural microbial populations. The results of that work are summarized in this section (Iwabuchi et al. 2002).

We first examined the effect of S-2 EPS on the ability of native marine bacteria to degrade the AF derived from Arabian light crude oil. Natural seawater medium (NSW medium) containing nitrogen, phosphorus, and iron nutrients was prepared as described previously (Iwabuchi et al. 2002), and the AF was added to a final concentration of 10 mg/mL (NSW-AF medium). Biodegradation of AF components was assessed gravimetrically by determining the amount of AF remaining in culture after 15 days of cultivation. At the start of the determination, the dry weight of the extracted oil was 106.6 ± 5.2 mg. The degradation experiments demonstrated that the relative weight of oil decreased significantly in the presence of S-2 EPS (Fig. 16.3a). In the absence of S-2 EPS, however, the weight of oil remaining after 15 days was 99.2 ± 9.1 mg, which was almost the same as that remaining in the control sample incubated in sterilized seawater.

The degradation of some AF constituents, i.e., antrathene (ant), dibenzothiophene (DBT), phenanthrene (phn), and their alkyl-substituted derivatives was determined using gas chromatography–mass spectrometry (GC–MS), and the results are shown in Fig. 16.3b. In the absence of S-2 EPS, ant, DBT, phn, and their derivatives were not degraded, whereas in the presence of S-2 EPS, all three compounds were degraded. These results were similar to those described above, indicating that AF and its constituents are recalcitrant to degradation by bacteria indigenous to seawater unless S-2 EPS is added.

To determine the effect of S-2 EPS on the seawater bacterial community structure, polymerase chain reaction-denaturing gradient gel electrophoresis (PCR-DGGE)

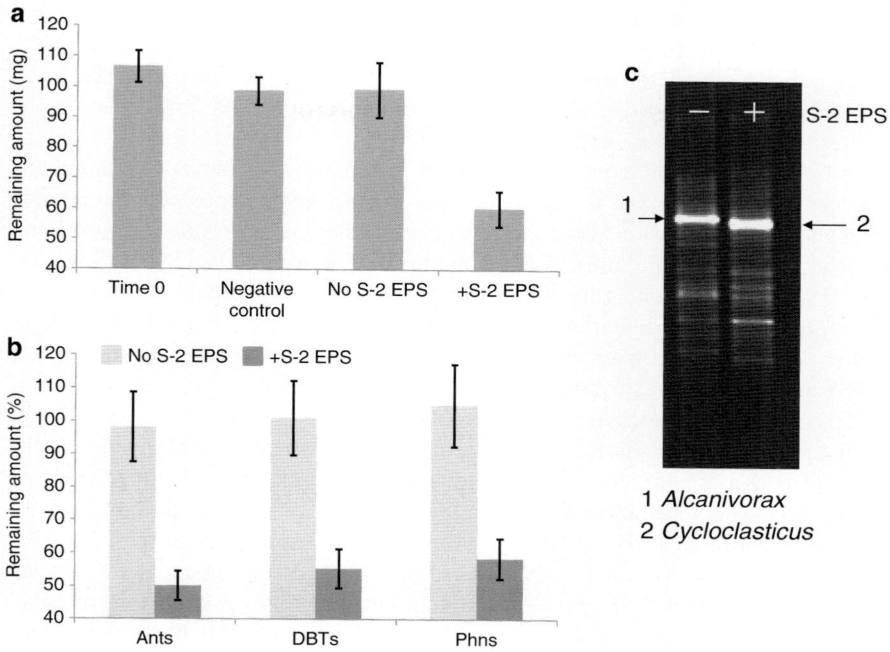

Fig. 16.3 Effect of S-2 EPS on the degradation of AF hydrocarbons in seawater by natural microbial populations. (**a**) Degradation of AF. Time 0, dry weight of the extracted oil at Time 0; negative control, extracted from the control sample incubated in sterilized seawater; no S-2 EPS, extracted from the NSW-AF sample in the absence of S-2 EPS; +S-2 EPS, extracted from the NSW-AF sample in the presence of S-2 EPS. (**b**) Degradation of AF constituents. Relative amounts of remaining oil were determined using GC-MS and are shown here. Ant, degradation average of ant and its derivatives; DBTs, DBT and its derivatives; Phn, phn and its derivatives. (**c**) DGGE profile of the structure of AF-degrading marine bacterial consortia after a 15-day cultivation. This figure was modified from that previously reported by Iwabuchi et al. (2002)

fingerprinting analysis was performed. NSW-AF medium was prepared and cells were cultured for 15 days, after which total DNA was extracted, the V3 region of small-subunit ribosomal DNA was amplified by PCR, and the amplified products were analyzed by DGGE. The DGGE profiles were highly reproducible: the patterns of the major bands were always identical, while those of minor bands were not. A typical electropherogram obtained after a 15-day cultivation is shown in Fig. 16.3c. In the absence of S-2 EPS, major bands for which sequencing analysis indicated a close relationship to *Alcanivorax borkumensis* DMS11573 appeared reproducibly throughout the cultivation period. Many reports have indicated that *Alcanivorax* strains, members of the gamma-proteobacteria, become predominant in oil-contaminated seawater supplemented with nutrients and that these organisms play a key role in the degradation of various aliphatic hydrocarbons in the marine environment (Head et al. 2006; Schneiker et al. 2006; Coulon et al. 2007; Yakimov et al. 2007). In oil-contaminated seawater supplemented with nutrients, saturates of crude oil (i.e., alkanes) rapidly decreased, whereas other major components of crude oil, such as aromatics, resins, and asphaltenes, decreased at much lower rates. *Alcanivorax* strains cannot utilize PAHs for growth (Schneiker et al. 2006) and therefore utilize either alkanes contained in the AF and/or alkyl side chains of PAHs as carbon and energy sources.

In NSW-AF medium containing S-2 EPS, no *Alcanivorax* bands were apparent, but bands closely related to *Cycloclasticus pugetti* PS-1 appeared reproducibly throughout the cultivation. These results suggest that *Cycloclasticus* species dominate the microbial community in AF-contaminated seawater when S-2 EPS is present. *Cycloclasticus* belonging to gamma-proteobacteria is capable of utilizing a variety of aromatic hydrocarbons as a sole carbon source, including toluene, xylene, biphenyl, naphthalene, phenanthrene (Dyksterhouse et al. 1995; Geiselbrecht et al. 1998), and pyrene (pyr) (Wang et al. 2008). Other studies have provided additional evidence that *Cycloclasticus* spp. play a key role in the degradation of various aromatic hydrocarbons in marine environments and thus may be useful in bioremediation efforts (Kasai et al. 2002; Maruyama et al. 2003; Coulon et al. 2007; Yakimov et al. 2007). The wide substrate specificity of *Cycloclasticus* thus accounts for the rapid degradation of PAHs in AF-containing cultures supplemented with S-2 EPS. Although *Cycloclasticus* has been detected in hydrocarbon-contaminated seawater and marine sediment through molecular ecological analyses and cloning of genes related to phenanthrene degradation, evidence indicative of an interaction between living *Cycloclasticus* cells and aromatic compounds contained in oil is lacking.

To clarify the mechanism through which S-2 EPS enhances PAH biodegradation, it is essential that we first understand how S-2 EPS affects the physiology of *Cycloclasticus*. We therefore examined the effect of S-2 EPS on the growth and biodegradation capabilities of *Cycloclasticus* using pure cultures of the type strain and environmental isolates. Although physiological analyses of *Cycloclasticus* are limited by this organism's low plating efficiency, we developed a method to increase the plating efficiency and used it to determine the number of viable biphenyl (bph)-degrading *C. pugetti* PS-1 cells in pure culture. Using our novel spotting inoculation and most probable number (SI-MPN) assay, we demonstrated that S-2 EPS promotes the growth of *Cycloclasticus* and its degradation of light crude oil AF and its constituents. Briefly, the growth of *C. pugetti* PS-1 in seawater-based medium containing AF (SW-AF) was examined in the presence and absence of S-2 EPS, and typical results are shown in Fig. 16.4a. At all sampling times, the number of viable cells was severalfold to 15-fold greater than the number in control cultures incubated in the absence of S-2 EPS. The growth of *C. pugetti* PS-1 in complete medium containing AF was also assessed using the SI-MPN method, and the results indicated that S-2 EPS greatly enhances the growth of *C. pugetti* PS-1. In addition, the growth of various environmental isolates of *Cycloclasticus* in the presence of AF was also enhanced by the addition of S-2 EPS. These results indicated that addition of S-2 EPS to the culture medium enhances the growth of *Cycloclasticus*.

To determine whether enhanced growth of *C. pugetti* PS-1 is associated with increased degradation of AF, we gravimetrically assessed changes in the amount of

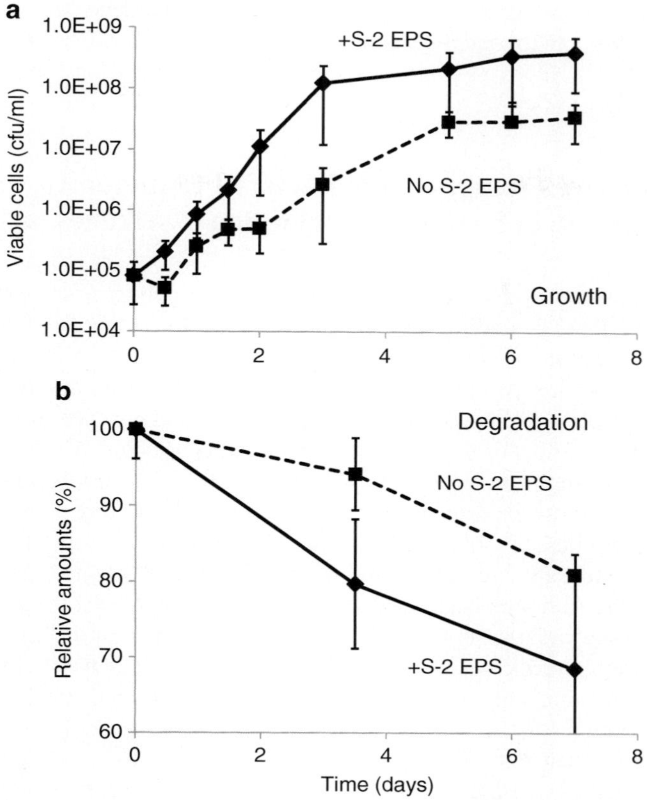

Fig. 16.4 Effect of S-2 EPS on *Cycloclasticus* growth and degradation of AF. (**a**) Growth of *C. pugetti* PS-1 in SW-AF. (**b**) Degradation of AF by *C. pugetti* PS-1 in SW-AF. AF was extracted from the culture and the remaining dry weight amount (%) was determined. Values are the average of three independent experiments. The percent remaining was calculated according to the following formula: percent remaining = dry weight remaining in culture/dry weight remaining in sterilized control × 100. This figure was modified from our unpublished data (Iwabuchi et al. manuscript in preparation)

AF remaining in SW-AF cultures of *C. pugetti* PS-1. The initial dry weight of the extracted AF was 112.9 ± 2.7 mg, and the weight of AF remaining in negative control samples lacking bacteria cultivated for 7 days in the presence and absence of S-2 EPS was almost the same. As shown in Fig. 16.4b, the percentage of AF remaining after 7 days was significantly lower in the presence of S-2 EPS than in its absence. In the absence of S-2 EPS, 80.1 ± 4.1 % of the initial amount of AF remained, which was about 1.25-fold higher than the percentage of AF remaining in cultures containing S-2 EPS. In addition, the degradation of various AF constituents (nph, phn, fluorene, DBT, and alkyl-substituted derivatives of each) was also enhanced in the presence of S-2 EPS. These results indicate that S-2 EPS enhances the biodegradation of AF and its constituents by *C. pugetti* PS-1. The results summarized here suggest that *Cycloclasticus* may become the predominant organism in AF-contaminated seawater supplemented with S-2 EPS and mineral nutrients.

16.4 Adaptive Changes in the Bacterial Community Structure in Seawater by Addition of Alkanes or PAHs

As discussed above, S-2 EPS is a multifunctional biopolymer that stimulates the growth of the PAH-degrading marine bacterium *Cycloclasticus* and enhances its degradation of AF in oil-contaminated seawater containing S-2 EPS and minerals. In this section, we discuss changes in the seawater bacterial community structure by the addition of pure alkanes or PAHs in the presence and absence of S-2 EPS. We first examined the effect of S-2 EPS on the community structure of bacterial alkane-degrading consortia in the presence of *n*-alkanes 10 to 30 carbons in length. The seawater used in these experiments was collected from Heita Bay, Kamaishi, Japan. The DGGE profiles of 16S rDNAs amplified from the cultures are shown in Fig. 16.5a, b. More than ten major and minor bands were detected in each lane of the DGGE gels. Although each experiment was repeated at least three times, there was considerable variation in the DGGE profile, irrespective of the presence of S-2 EPS. In addition, the alkanes were always degraded, independent of changes in the bacterial consortium, suggesting that many marine bacteria are capable of degrading pure alkanes.

In contrast, when pure PAH was added to the seawater-based culture, highly reproducible DGGE profiles were obtained in four independent experiments, and the profiles were not affected by the presence of S-2 EPS (Fig. 16.5c). Nucleotide sequencing of the major band in each lane indicated that it (Fig. 16.5c, arrow) was identical to *C. pugetti* PS-1. These results indicated that *Cycloclasticus* spp. come to dominate the seawater bacterial consortium when pure PAH is present, regardless of the presence or absence of S-2 EPS.

To determine whether the increase in the number of cells is related to the eventual domination of consortia by *Cycloclasticus*, the total number of cells in bph-containing enrichment cultures was assessed over time using DAPI counting, and changes in the bacterial community structure were monitored by PCR-DGGE. As shown in Fig. 16.6a, the total number of cells increased significantly after an initial lag, reaching a maximum of 1.7×10^8 CFU/mL. Those DGGE bands showing the same relative mobility indicated by the arrow in Fig. 16.6b increased in concentration with the increase in total number of cells. Sequencing of the predominant DGGE bands indicated that they were identical to *C. pugetti* PS-1. Addition of S-2 EPS to the medium did not significantly affect the total number of cells or the eventual domination of the community by *Cycloclasticus*. These results suggest that the observed increase in the total number of cells in PAH-contaminated seawater is primarily due to an increase in the *Cycloclasticus* population.

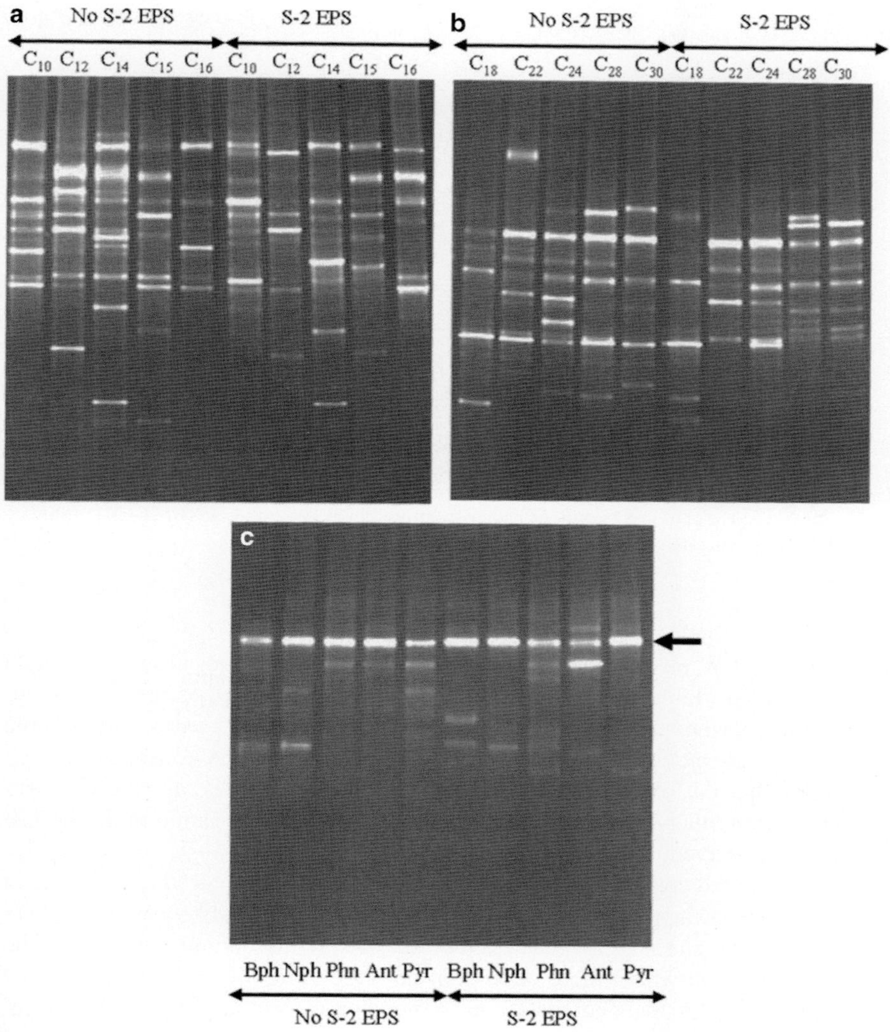

Fig. 16.5 Adaptive changes in the bacterial community structure. (**a**) DGGE profiles of the community structure of bacteria cultured in NSW added alkanes with chain lengths C10 to C16 and with and without S-2 EPS; (**b**) DGGE profiles of the community structure of bacteria cultured in NSW added alkanes with chain lengths C18 to C30 and with and without S-2 EPS; (**c**) DGGE profile of the community structure of bacteria cultured in NSW added PAHs and with and without S-2 EPS

To determine whether the domination of consortia by *Cycloclasticus* sp. in the presence of pure PAH is limited only to Heita Bay, we collected samples of surface and deep seawater from other bays and repeated the above experiments. The degree of *Cycloclasticus* predominance was estimated based on the intensity of DGGE bands that sequencing indicated were closely related to genes of *Cycloclasticus* sp. *Cycloclasticus* sp. came to dominate the consortia when pure bph, nph, phn, or ant

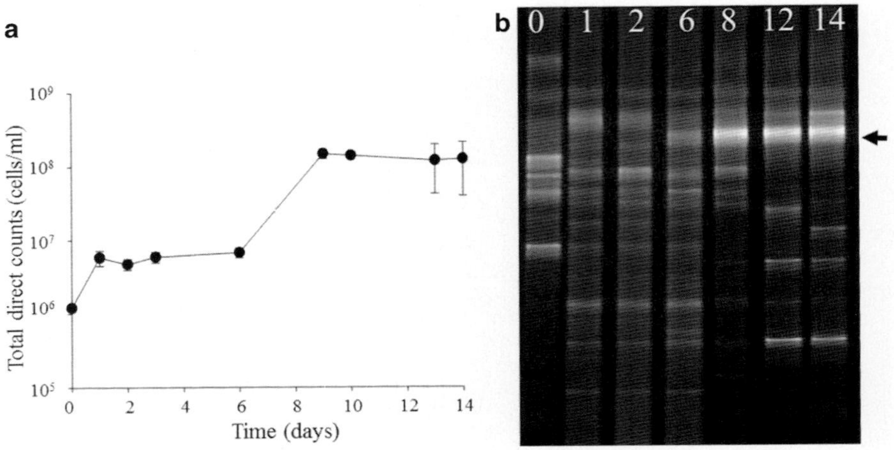

Fig. 16.6 Relationship between the increase in total number of cells and predominance of *Cycloclasticus*. (**a**) Total direct count of bacteria cultured in NSW with bph. (**b**) DGGE profile of the community structure of bacteria cultured in NSW with bph

was added to SW medium as the carbon and energy source, regardless of sample origin or depth (Table 16.1). In contrast, when pyr was added as the carbon and energy source, whether *Cycloclasticus* spp. were detected or not varied with sample origin. *Cycloclasticus* sp. reproducibly came to dominate the consortia of deep seawater collected from Iwanai, Heita, Shiogama, Haruno, and Miyako, but were not detected in other samples. In these cases, considerable variation in the DGGE profile was observed.

With some exceptions, *Cycloclasticus* sp. came to dominate the consortia in many of the seawater samples tested. These results are consistent with the data presented above and suggest that *Cycloclasticus* spp. are common in the marine environment, even where there have been no large-scale oil spills, and that they play a key role in the degradation of a number of aromatic hydrocarbons in seawater.

To determine whether the observed domination of consortia by *Cycloclasticus* sp. is associated with the presence of PAHs, we assessed the degradation of various PAHs by native bacteria in seawater sampled from Heita Bay. The percent of PAH remaining in the sample after 15 days was determined gravimetrically (Table 16.2). The percent remaining decreased to a certain extent for all five of the PAHs tested, suggesting that PAH degradation and dominance of consortia by *Cycloclasticus* are closely related and that degradation of PAHs in seawater is mediated by indigenous *Cycloclasticus* sp.

Table 16.1 Variation by sampling location in *Cycloclasticus* sp. domination of bacterial consortia in PHA-contaminated seawater

Location	Bph	Nph	Phn	Ant	Pyr
Surface seawater					
Iwanai Bay	++	++	++	++	++
Syukutsu Bay	++	++	++	++	–
Heita Bay	++	++	++	++	++
Shiogama Bay	++	++	++	++	++
Tsu	++	++	++	++	–
Toyama Bay	++	++	+	+	–
Haruno Bay	++	++	++	++	++
Deep seawater					
Iwanai	++	++	++	++	–
Miyako	++	++	++	++	++
Toyama	++	++	++	++	–

++ *Cycloclasticus* sp. always predominated consortia; + *Cycloclasticus* sp. usually predominated consortia; – no major or minor DGGE bands identified to *Cycloclasticus* spp. were detected in the seawater sample. Samples of surface seawater were collected from seven locations: Iwanai Bay (Lat. 42°59′12″N, Long. 140°30′35″E), Syukutsu Bay (Lat. 43°13′55″N, Long. 141°0′57″E), Heita Bay (Lat. 39°14′57″N, Long. 141°53′46″E), Shiogama Bay (Lat. 38°19′9″N, Long. 141°2′18″E), Tsu (Lat. 34°45′48″N, Long. 136°32′29″E), Toyama Bay (Lat. 36°46′37″N, Long. 137°20′36″E), and Haruno Bay (Lat. 33°28′7″N, Long. 133°30′′29″E). Samples of deep seawater were collected from three locations: Iwanai (Lat. 42°59′12″N, Long. 140°30′35″E, at a depth of approx. 300 m), Miyako (Lat. 39°54′N, Long. 142°39′35″E, at a depth of 200–300 m), and Toyama (Lat. 39°79′N, Long. 137°32′E, at a depth of approx. 333 m)

Table 16.2 Degradation of PAHs by native marine bacteria

PAH	Amount remaining (%)
Biphenyl	62.9 ± 9.6
Naphthalene	63.3 ± 9.2
Phenanthrene	69.7 ± 9.0
Antrathene	72.3 ± 8.1
Pyrene	78.0 ± 8.5

PAHs were extracted from the culture and the percentage of the dry weight of the compound remaining was determined. Values for the percent remaining were calculated according to the following formula: percent remaining = dry weight remaining in enrichment culture after 15 days of cultivation/dry weight remaining in sterilized control after 15 days of cultivation × 100. Values shown are the average and standard deviation of three independent experiments

16.5 Effect of S-2 EPS on the Change in the Marine Bacterial Community Structure When Alkanes and PAHs Are Supplied Concurrently

As mentioned above, we found that pure PAHs were degraded in NSW medium in the absence of S-2 EPS, in contrast to our previous results. Therefore, we examined the effect of alkanes on the degradation of PAHs by indigenous *Cycloclasticus* since AF contains a small amount of alkanes, branched alkanes, and various alkyl residues (Iwabuchi et al. 2002).

Phenanthrene and *n*-hexadecane (C16) were used as representative compounds to prepare a simple model of AF. Samples of NSW medium containing various concentrations of phn and/or C16 and with or without S-2 EPS were prepared using seawater collected from Heita Bay. The samples were cultured for 15 days, after which the bacterial community structure and degree of phn and C16 degradation were examined using DGGE and GC-MS, respectively. A typical DGGE result is shown in Fig. 16.7. In the absence of C16, a major DGGE band identified to *C. pugetti* PS-1 (band 2) was detected, regardless of the presence of S-2 EPS. In contrast, no bands closely related to *Cycloclasticus* spp. were detected in the absence of phn. These results were similar to those described above (see Fig. 16.5).

In NSW medium containing C16 as a carbon and energy source along with phn, the major band closely related to *Cycloclasticus* was either not detected or was

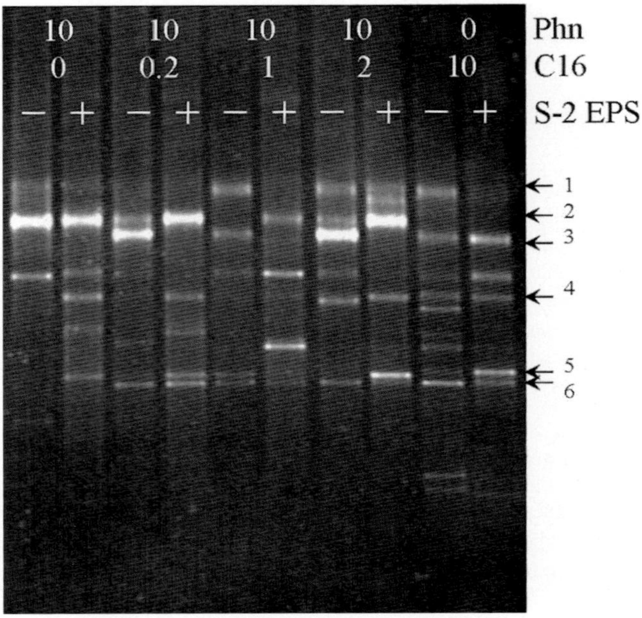

Fig. 16.7 DGGE profile of the community structure of bacteria cultured in NSW with phn and C16 and with and without S-2 EPS

Table 16.3 16S rDNA sequences most similar to the major DGGE bands shown in Fig. 16.7

Band	Closest related match	Accession number	Percentage identity
1	*Pseudoalteromonas* sp. TG4-11	AJ414129	97
2	*Cycloclasticus pugetii* PS-1	U12624	100
3	Uncultured gamma-proteobacterium clone 33-PA143B9	NR_041705	96
4	Uncultured gamma-proteobacterium clone CD2G11	AY038471	96
5	Alpha-proteobacterium GMD29E5	AY162072	96
6	Alpha-proteobacterium Med4C1:1	AF493974	100

Table 16.4 Effect of S-2 EPS on the dominance of bacterial consortia by *Cycloclasticus* in alkane- and PAH-amended seawater from various locations

	S-2 EPS present		S-2 EPS absent	
Location	Predomination frequency (%)	Detection frequency (%)	Predomination frequency (%)	Detection frequency (%)
Iwanai Bay	0	0	100	100
Syukutsu Bay	0	0	80	80
Shiogama Bay	0	0	100	100
Haruno Bay	0	0	75	75

Between three and five independent experiments were repeated using samples collected at each location. Inhibition of *Cycloclasticus* predomination of consortia in the presence of C16 and restoration of *Cycloclasticus* predomination of consortia by the addition of S-2 EPS were assessed using DGGE. Frequency = predomination or detection/total experiments × 100

reduced in intensity in the absence of S-2 EPS, regardless of C16 concentration. In the presence of S-2 EPS, the *Cycloclasticus* band was prominent, regardless of C16 concentration.

The results of database searches for homology with the sequence of the DGGE bands are shown in Table 16.3. Although the major *Cycloclasticus* DGGE band was reproducibly detected in these experiments, the other bands, with the exception of bands 5 and 6, were not consistently detected. In addition, bands 5 and 6 were usually detected when C16 was present in the NSW medium, but not when only phn was present, indicating that the bacteria associated with bands 5 and 6 do not contribute to phn degradation in seawater. These results suggest that a close relationship exists between domination of consortia by *Cycloclasticus* sp. and the presence of S-2 EPS in seawater that also contains phn and C16.

To determine whether the inability of *Cycloclasticus* sp. to dominate consortia in the presence of C16 and the restoration of this ability by the addition of S-2 EPS was limited only to seawater from Heita Bay, we examined changes in the bacterial community structure using NSW-phn+C16 medium prepared with seawater from Syukutsu, Iwanai, Shiogama, and Haruno Bays. The inability of *Cycloclasticus* sp. to dominate consortia in the presence of C16, and the restoration of this ability by the addition of S-2 EPS, was observed with significant reproducibility in NSW-phn+C16 medium prepared using seawater from these other locations (Table 16.4). These results were consistent with those mentioned above; the *Cycloclasticus* band was prominent in the presence of S-2 EPS regardless of C16 concentration.

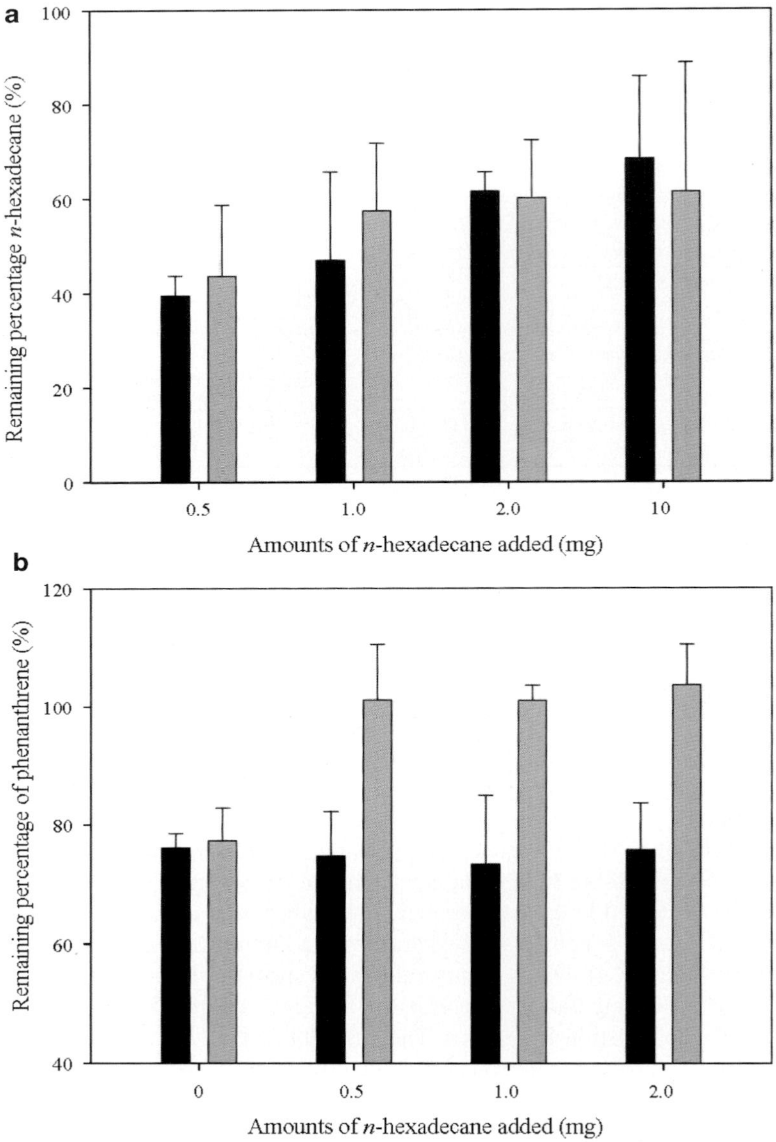

Fig. 16.8 Degradation of C16 and phn by native marine bacteria. (**a**) Degradation of C16. (**b**) Degradation of phn. The *light and dark shaded bars* represent the absence and presence of S-2 EPS, respectively

We then examined the degradation of C16 and phn in order to determine how phn affects the restoration of *Cycloclasticus* sp. as the predominant members of consortia. As shown in Fig. 16.8a, C16 was degraded by indigenous alkane-degrading bacteria regardless of the presence of S-2 EPS. No relationship between degradation enhancement and the presence of S-2 EPS was observed.

In contrast to the results observed in the presence of C16, in the absence of C16, the percentage of phn degraded was unaffected by the presence of S-2 EPS (Fig. 16.8b), indicating that phn degradation was occurred regardless of the presence of S-2 EPS. The percentage of phn remaining in the presence of C16 differed greatly depending upon whether S-2 EPS was present or not. In the absence of S-2 EPS, the percentage of phn remaining under each of the conditions tested was more than 100 %, indicating that essentially no phn degradation occurred, regardless of the amount of C16 present. These results suggest that phn degradation is inhibited in the presence of C16. When S-2 EPS was present, the percentage of phn remaining was about 75 %, a significantly lower amount compared to samples lacking S-2 EPS. Collectively, these results indicate that a close relationship exists between the presence of S-2 EPS and the predominance of *Cycloclasticus* sp. within consortia and degradation of phn. We therefore concluded that degradation of phn in seawater is mediated primarily by indigenous *Cycloclasticus* that are activated by S-2 EPS.

16.6 Adaptive Changes in the Bacterial Community Structure

It is generally held that PAHs in the marine environment are resistant to biodegradation and thus may remain at polluted sites for many years (Henson and Hayasaka 1982; Payne and Phillips 1985; Mackay and MacAuliffe 1988). New techniques that would accelerate PAH degradation by native marine bacteria are required for the success of bioremediation efforts aimed at the removal of PAHs from oil-contaminated marine environments. Since we found that S-2 EPS promotes the degradation of aromatic compounds by indigenous *Cycloclasticus* in marine bacterial consortia, we have attempted to clarify the mechanism through which S-2 EPS enhances PAH biodegradation.

To understand the adaptive changes in native marine bacterial communities that enable degradation of hydrocarbons, we examined how the presence of alkanes or PAHs affects the bacterial community structure. With the exception of long alkanes, the addition of pure alkanes to seawater resulted in a diverse array of adaptive changes in the community, suggesting that a wide variety of bacteria are involved in degradation of these compounds. In contrast, the adaptive changes that occurred when pure PAHs were added to seawater were more uniform and almost always involved *Cycloclasticus* sp. becoming the predominant members of the bacterial community structure. In addition, the predominance of *Cycloclasticus* sp. always coincided with the degradation of PAHs and an increase in the total number of cells, further indicating that *Cycloclasticus* plays an important role in PAH degradation in the marine environment (Maruyama et al. 2003; Coulon et al. 2007; Yakimov et al. 2007; Lozada et al. 2008; Wang et al. 2008; Niepceron et al. 2010). Hence, it is reasonable to conclude that PAH degradation mediated by *Cycloclasticus* sp. is a fundamental characteristic of marine bacterial consortia.

16.7 The Effect of S-2 EPS on *Cycloclasticus*

To clarify the mechanism by which PAH biodegradation is enhanced by S-2 EPS (Iwabuchi et al. 2002), we examined the effect of S-2 EPS on both the predominance of *Cycloclasticus* in the bacterial consortium and the degradation of PAHs using a simplified model medium mimicking AF. The results of these experiments showed that the rise to predominance by *Cycloclasticus* sp. is inhibited by the addition of C16 and that addition of S-2 EPS to NSW medium containing phn and C16 results in restoration of *Cycloclasticus* sp. predominance. These results suggest that the presence of S-2 EPS in seawater contributes significantly to the rise of *Cycloclasticus* sp. to eventual predominance in seawater when C16 and phn are present.

Using model mixed cultures with *A. borkumensis* DMS11573, we recently demonstrated that both *Cycloclasticus* growth and degradation of PAHs are selectively activated by association of S-2 EPS with the cell, suggesting that in seawater S-2 EPS primarily affects *Cycloclasticus*. Our most recent metagenomic analysis involving natural seawater also showed that the presence of S-2 EPS stimulates PAH degradation by *Cycloclasticus* and enhances its growth (Iwabuchi et al. manuscript in preparation). Therefore, it is probable that in seawater S-2 EPS selectively affects the growth of *Cycloclasticus* and enhances its ability to degrade PAHs.

In the present study, C16 inhibited the ability of *Cycloclasticus* to dominate bacterial consortia in seawater also containing phn. Using the SI-MPN assay developed to analyze *Cycloclasticus* viable cells, we examined the effect of C16 on the growth of *Cycloclasticus* sp. in pure culture. The results of this experiment showed that C16 does not inhibit the growth of *C. pugetti* PS-1 in SW-bph medium. In addition, C16 did not inhibit the growth of environmental isolates of *Cycloclasticus*, suggesting that C16 is not toxic to *Cycloclasticus*.

When C16 and minerals were added to seawater, with or without addition of phn or S-2 EPS, many bands were detected upon DGGE analysis, indicating that many bacteria are capable of growing in media containing these components. It is therefore presumed that except for the carbon source, bacteria within the consortium compete for nutrients, such as nitrate, phosphate, and/or trace metals. Schneiker et al. (2006) reported that *Alkanivorax* spp., which play an important role in the degradation of alkanes in marine environments, are highly efficient scavengers of such nutrients, which enables them to dominate bacterial consortia in oil-contaminated marine environments. In fact, Kyoto Encyclopedia of Genes and Genomes (KEGG)-based comparative genomic analyses of *Cycloclasticus* strain S4 and *A. borkumensis* DMS11573 indicated that the predicted nutrient-uptake capability of *Cycloclasticus* strain S4 is lower than that of *A. borkumensis* (Iwabuchi et al. manuscript in preparation). In addition, results similar to those described above were obtained from the comparative genomic analyses of *A. borkumensis* DMS11573 and *Cycloclasticus* P1 (Iwabuchi et al. manuscript in preparation), for which genomic information was recently reported (Lai et al. 2012). The results of our recent proteogenomic analyses of *Cycloclasticus* strain S4 cultured with and without S-2 EPS in pure culture suggest that the expression of some nutrient-uptake transporters is upregulated by S-2 EPS (Iwabuchi et al. manuscript in preparation). It is thus possible that in the presence of C16 and phn, S-2 EPS-induced upregulation of nutrient transporters accelerates the domination of marine consortia by *Cycloclasticus*, thereby enhancing PAH degradation.

In conclusion, S-2 EPS is a functional biopolymer that stimulates the growth of *Cycloclasticus* sp. in seawater, enabling them to dominate the bacterial community and making them a potentially useful tool for the bioremediation of oil-contaminated marine environments. Further detailed investigations are needed, however, to define the relationship between S-2 EPS, nutrient uptake, and domination of the bacterial community structure by *Cycloclasticus*.

References

Aizawa T, Neilan BA, Couperwhite I, Urai M, Anzai H, Iwabuchi N, Nakajima M, Sunairi M (2005) Relationship between extracellular polysaccharide and benzene tolerance of *Rhodococcus* sp. 33. Actinomycetologica 19:1–6

Belsky I, Gutnick DL, Rosenberg E (1979) Emulsifier of arthrobacter RAG-1: determination of emulsifier-bound fatty acids. FEBS Lett 101:175–178

Bell KS, Philip JC, Aw DWJ, Christofi N (1998) The genus *Rhodococcus*. J Appl Microbiol 85:195–210

Coulon F, McKew BA, Osborn AM, McGenity TJ, Timmis KN (2007) Effects of temperature and biostimulation on oil-degrading microbial communities in temperate estuarine waters. Environ Microbiol 9:177–186

Dyksterhouse SE, Gray JP, Herwig RP, Lara JC, Staley JT (1995) *Cycloclasticus pugetii* gen. nov., sp. nov., an aromatic hydrocarbon-degrading bacterium from marine sediments. Int J Syst Bacteriol 45:116–123

Finnerty WR (1992) The biology and genetics of the genus *Rhodococcus*. Ann Rev Microbiol. 46:193–218

Geiselbrecht AD, Hedlund BP, Tichi MA, Staley JT (1998) Isolation of marine polycyclic aromatic hydrocarbon (PAH)-degrading *Cycloclasticus* strains from the Gulf of Mexico and comparison of their PAH degradation ability with that of puget sound *Cycloclasticus* strains. Appl Environ Microbiol 64:4703–4710

Goodfellow M (1989) Nocardioform actinomycetes genus *Rhodococcus*. Williams and Wilkins, Baltimore

Head IM, Jones DM, Roling WF (2006) Marine microorganisms make a meal of oil. Nat Rev Microbiol 4:173–182

Henson JM, Hayasaka SS (1982) Effects of the water-soluble fraction of microbiologically or physically altered crude petroleum on the heterotrophic activity of marine bacteria. Mar Environ Res 6:205–214

Iwabuchi N, Sunairi M, Nakajima M (1997) Cloning of a DNA fragment from *Rhodococcus rhodochrous* which suppresses its mucoidal morphology. Actinomycetologica 11:59–63

Iwabuchi N, Sunairi M, Anzai H, Nakajima M, Harayama S (2000) Relationships between colony morphotypes and oil tolerance in *Rhodococcus rhodochrous*. Appl Environ Microbiol 66: 5073–5077

Iwabuchi N, Sunairi M, Urai M, Itoh C, Anzai H, Nakajima M, Harayama S (2002) Extracellular polysaccharides of *Rhodococcus rhodochrous* S-2 stimulate the degradation of aromatic components in crude oil by indigenous marine bacteria. Appl Environ Microbiol 68:2337–2343

Iwabuchi N, Sunairi M, Anzai H, Morisak H, Nkajima M (2003) Relationships among colony morphotypes, cell-surface properties and bacterial adhesion to substrata in *Rhodococcus*. Colloids Surf. B Biointerfaces 30:51–60

Iwabuchi N, Sharma PK, Sunairi M, Kishi E, Sugita K, van der Mei HC, Nakajima M, Busscher HJ (2009) Role of interfacial tensions in the translocation of *Rhodococcus erythropolis* during growth in a two phase culture. Environ Sci Technol 43:8290–8294

Kasai Y, Kishira H, Harayama S (2002) Bacteria belonging to the genus *Cycloclasticus* play a primary role in the degradation of aromatic hydrocarbons released in a marine environment. Appl Environ Microbiol 68:5625–5633

Lai Q, Li W, Wang B, Yu Z, Shao Z (2012) Complete genome sequence of the pyrene-degrading bacterium *Cycloclasticus* sp. strain P1. J bacterial 194:6677

Leitch RA, Richards JC (1990) Structural analysis of the specific capsular polysaccharide of *Rhodococcus equi* serotype 1. Biochem Cell Biol 68:778–789

Lindberg B (1990) Components of bacterial polysaccharides. Adv Carbohydr Chem Biochem 48:279–318

Lozada M, Riva Mercadal JP, Guerrero LD, Di Marzio WD, Ferrero MA, Dionisi HM (2008) Novel aromatic ring-hydroxylating dioxygenase genes from coastal marine sediments of Patagonia. BMC Microbiol 25:8–50

Mackay D, MacAuliffe CD (1988) Fate of hydrocarbons discharged at sea. Chem Pollut 5:1–20

Maruyama A, Ishiwata H, Kitamura K, Sunamura M, Fujita T, Matsuo M, Higashihara T (2003) Dynamics of microbial populations and strong selection for *Cycloclasticus pugetii* following the Nakhodka oil spill. Microb Ecol 46:442–453

Masoud H, Richards JC (1994) Structural elucidation of the specific capsular polysaccharide of *Rhodococcus equi* serotype 7. Carbohydr Res 252:223–233

Minnikin DE (1991) Chemical principles in the organization of lipid components in the mycobacterial cell envelope. Res Microbiol 142:423–427

Niepceron M, Portet-Koltalo F, Merlin C, Motelay-Massei A, Barray S, Bodilis J (2010) Both *Cycloclasticus* sp. and *Pseudomonas* sp. as PAH-degrading bacteria in the Seine estuary (France). FEMS Microbiol Ecol 71(1):137–147, Erratum in: FEMS Microbiol Ecol 71(3): 479 (Mar 2010)

Payne JR, Phillips CR (1985) Photochemistry of petroleum in water. Environ Sci Technol 19:569–579

Roberts IS (1996) The biochemistry and genetics of capsular polysaccharide production in bacteria. Annu Rev Microbiol 50:285–315

Schneiker S, Martins dos Santos VA, Bartels D, Bekel T, Brecht M, Buhrmester J, Chernikova TN, Denaro R, Ferrer M, Gertler C, Goesmann A, Golyshina OV, Kaminski F, Khachane AN, Lang S, Linke B, McHardy AC, Meyer F, Nechitaylo T, Pühler A, Regenhardt D, Rupp O, Sabirova JS, Selbitschka W, Yakimov MM, Timmis KN, Vorhölter FJ, Weidner S, Kaiser O, Golyshin PN (2006) Genome sequence of the ubiquitous hydrocarbon-degrading marine bacterium *Alcanivorax* borkumensis. Nat Biotechnol 24:997–1004

Senchenkova SN, Knirel YA, Likhosherstov LM, Shashkov AS, Shibaev VN, Starukhina LA, Deryabin VV (1995) Structure of simusan, a new acidic exopolysaccharide from *Arthrobacter* sp. Carbohydr Res 266:103–113

Severn WB, Richards JC (1990) Structural analysis of the specific capsular polysaccharide of *Rhodococcus equi* serotype 2. Carbohydr Res 206:311–332

Severn WB, Richards JC (1992) The acidic specific capsular polysaccharide of *Rhodococcus equi* serotype 3. Structural elucidation and stereochemical analysis of the lactate ether and pyruvate acetal substituents. Can J Chem 70:2664–2676

Severn WB, Richards JC (1999) The structure of the specific capsular polysaccharide of *Rhodococcus equi* serotype 4. Carbohydr Res 320:209–222

Sunairi M, Iwabuchi N, Yoshizawa Y, Murooka H, Morisaki H, Nakajima M (1997) Cell-surface hydrophobicity and scum formation of *Rhodococcus rhodochrous* strains with different colonial morphologies. J Appl Microbiol 82:204–210

Urai M, Anzai H, Iwabuchi N, Sunairi M, Nakajima M (2002) A novel moisture-absorbing extracellular polysaccharide from *Rhodococcus rhodochrous* SM-1. Actinomycetologica 16: 26–31

Urai M, Anzai H, Iwabuchi N, Sunairi M, Nakajima M (2004) A novel viscous extracellular polysaccharide containing fatty acids from *Rhodococcus rhodochrous* ATCC 53968. Actinomycetologica 18:15–17

Urai M, Aizawa T, Anzai H, Ogihara J, Iwabuchi N, Neilan B, Couperwhite I, Nakajima M, Sunairi M (2006a) Structural analysis of an extracellular polysaccharide produced by a benzene tolerant bacterium, *Rhodococcus* sp. 33. Carbohydr Res 341:616–623

Urai M, Anzai H, Ogihara J, Iwabuchi N, Harayama S, Sunairi M, Nakajima M (2006b) Structural analysis of an extracellular polysaccharide produced by *Rhodococcus rhodochrous* strain S-2. Carbohydr Res 341:766–775

Wang B, Lai Q, Cui Z, Tan T, Shao Z (2008) A pyrene-degrading consortium from deep-sea sediment of the West Pacific and its key member *Cycloclasticus* sp. P1. Environ Microbiol 10:1948–1963

Warhurst AM, Fewson CA (1994) Biotransformations catalyzed by the genus *Rhodococcus*. Crit Rev Biotechnol 14:29–73

Whyte LG, Hawari J, Zhou E, Bourbonniere L, Inniss WE, Greer CW (1998) Biodegradation of variable-chain-length alkanes at low temperatures by a psychrotrophic *Rhodococcus* sp. Appl Environ Microbiol 64:2578–2584

Yakimov MM, Timmis KN, Golyshin PN (2007) Obligate oil-degrading marine bacteria. Curr Opin Biotechnol 18:257–266

Zaidi BR, Imam SH (1999) Factors affecting microbial degradation of polycyclic aromatic hydrocarbon phenanthrene in the Caribbean coastal water. Mar Pollut Bull 38:737–742

Chapter 17
Biofilm as a Multicellular Bacterial System

Masanori Toyofuku and Nobuhiko Nomura

Abstract Biofilm is thought to be the primary natural habitat for many bacteria and they are associated with a wide range of human activities, such as wastewater treatment and clinical infections. Current research points out that cells inside biofilms are physiologically distinct from free-floating cells. In biofilms, bacteria are embedded in extracellular polymeric substances (EPS) that have physical and biological functions. One of the main roles of these EPS is to hold the cells together, which leads to the development of multicellular consortia. Once developed, usually the environment inside the biofilm becomes heterogenous which initiates the biofilm to function as a multicellular system. The development of biofilm is well organized following common stages among species. During this process, intercellular and intracellular signaling also takes place. This chapter will give a brief introduction to biofilms to support the understanding of bacterial multicellular systems.

Keywords Biofilm • c-di-GMP • Cell-to-cell communication • Confocal reflection microscopy • Diffusible signal factor (DSF) • EPS • Outer membrane vesicle (OMV)

17.1 Introduction

Although much of the knowledge in microbiology is based on the study of free-floating planktonic cells, it is thought that the primary habitat for many microorganisms is biofilm. Bacterial biofilms are multicellular consortia in which the bacterial cells are embedded in an extracellular matrix or EPS that is often produced by themselves. Microbial mats in the natural environments or plaques that are

M. Toyofuku • N. Nomura (✉)
Graduate School of Life and Environmental Sciences, University of Tsukuba,
1-1-1 Tennodai, Tsukuba, Ibaraki 305-8572, Japan
e-mail: nomura.nobuhiko.ge@u.tsukuba.ac.jp

H. Nojiri et al. (eds.), *Biodegradative Bacteria: How Bacteria Degrade, Survive, Adapt, and Evolve*, DOI 10.1007/978-4-431-54520-0_17, © Springer Japan 2014

developed on the surface of teeth are examples of surface-associated biofilms. Biofilms are not only necessarily attached to the surface but also are present as suspended cell aggregates, also known as flocs and granules (de Beer and Stoodley 2006). Natural biofilms can develop under most conditions were bacteria are present and their relevance in ecosystems has been recognized. Taking advantage of the multicellularity, biofilms are considered as a robust and sustainable system. Currently, biofilms are practically used for biological wastewater treatment. On the other hand, they are also associated with a wide range of problems in industry and also in clinical infections. For instance, biofilms have increased resistance against antimicrobial agents and is hard to eradicate them (Høiby et al. 2010). Given the potential benefits and drawbacks that biofilms can confer, there is high interest in their control.

Accumulating report demonstrates that biofilms are distinct from their free-floating counterparts (López et al. 2010). Cells in biofilms have different gene expression patterns compared to planktonic cells and numerous regulatory mechanisms take place during biofilm formation, suggesting that there is a unique program for biofilm development. Moreover, the gene expressions, phenotypes vary between the cells in a biofilm. This heterogenous nature of biofilm could be a consequence of differential gene expression that is initiated by a different gradient of nutrients, physicochemical factors effecting the cell activates. In addition, emergence of genotypically and phenotypically diverse mutants may lead to the heterogenous nature of biofilm. Remarkably, cells inside biofilms interact physically and chemically with each other to coordinate social behaviors. It is not surprising that the organization and differentiation of biofilm has led some scientists to compare biofilms with higher multicellular organisms.

In this chapter we will give a brief introduction to biofilms and discuss the feature of this multicellular bacterial community.

17.2 Sticking the Cells Together: The EPS Matrix

In order to form a multicellular community, the cells must have a way to connect to each other. EPS is one of the key components in biofilms and it could be simply said that "There is no biofilm without an EPS matrix" (Flemming and Wingender 2010). The EPS consists of different types of biopolymers which are polysaccharides, proteins, lipids, and nucleic acids. Also, polymers from other origin such as humic acid may be embedded into biofilms. It was long believed that polysaccharides are the main constituent of the EPS matrix; however, results from various studies indicate that the main EPS in terms of amounts and their impact on biofilm formation differs among species. The major role of EPS in biofilm is to connect the cells with the substratum or other cells and provide the mechanical stability of the biofilm. Therefore, the maintenance of the structured biofilm depends largely on the production of EPS. Several studies using pure culture have demonstrated that mutants lacking the component of EPS have altered biofilm morphology. For instance, in

Pseudomonas aeruginosa, at least three types of exopolysaccharides that alter biofilm formation are identified. Psl is a mannose- and galactose-rich polysaccharide while Pel is a glucose-rich polysaccharide (Friedman and Kolter 2004a, b). Alginate is the major exopolysaccharide in mucoid strains that are the predominant morphotype of strains isolated from chronic infections (Govan and Deretic 1996); Psl and Pel exopolysaccharides are critical for biofilm formation on abiotic surfaces (Friedman and Kolter 2004a; Jackson et al. 2004; Matsukawa and Greenberg 2004). Further studies have demonstrated extracellular DNA (eDNA) to constitute an important component of the *P. aeruginosa* biofilm matrix (Whitchurch et al. 2002; Allesen-Holm et al. 2006). Although eDNA was initially thought to be a residual material from lysed cells, many groups indicate that it is in fact an integral part of the matrix. In *Rhodovulum* sp., flocculation was disrupted when treated with nucleic acid-degrading enzymes (Watanabe et al. 1998). Similar results are obtained with activated sludge that is composed of multispecies. DNase treatment of the activated sludge resulted in complete dissolution of the flocs in combination with shear force, whereas shear force alone did not (Dominiak et al. 2011). These results demonstrate that eDNA is related to structural integrity. In addition to the role in structural integrity, eDNA in bacteria may retain extracellular proteins as observed in the defensive web of neutrophil extracellular traps (NETs) (Brinkmann et al. 2004).

The presence of proteins in the biofilm matrix has been reported in many samples and they are sometimes even more abundant than polysaccharides in natural cell aggregates (McSwain et al. 2005; Frølund et al. 1996; Urbain et al. 1993). While the functions of these proteins are not fully understood, it is reported that these proteins at least include extracellular enzymes as well as structural proteins. Various extracellular enzymes that are involved in degradation of biopolymers have been detected in the matrix suggesting its role as an external digestion system in addition to the immobilization of the cells (Baty et al. 2001; Flemming and Wingender 2010).

Structural proteins include CdrA which is an extracellular protein in *P. aeruginosa* that is induced upon biofilm formation. This protein cross-links polysaccharides and/or tethers the cells with the polysaccharides providing biofilm stability (Borlee et al. 2010). Carbohydrate-binding proteins, called lectins, are also involved in biofilm formation and stabilization in a wide range of species. Lectins have a very specific recognition of sugars, and therefore, they may be involved in recognizing particular polysaccharides and glycoconjugates on the cell surface of other cells allowing them to coaggregate (Rickard et al. 2003).

Another type of proteins that seems to be ubiquitous among bacterial biofilms are amyloid fibers. Amyloid fibers are defined by their characteristic cross-β-strand structure, where the β-sheets are oriented perpendicular to the fiber axis (Sunde et al. 1997). Amyloids of bacterial origins have been found in various natural biofilms suggesting their importance in the natural habitat (Otzen and Nielsen 2008). In certain species, amyloid proteins such as curli fimbriae in *Escherichia coli* and *Salmonella typhimurium*, TasA in *Bacillus subtilis* have critical roles in biofilm formation together with the polysaccharides (Chapman et al. 2002; Romero et al. 2010; Zogaj et al. 2001).

A group of surface proteins that are named biofilm-associated proteins (Bap) also plays a role in the cell adhesion to biotic and/or abiotic surfaces in Gram-positive and Gram-negative bacteria. These proteins contain core regions of amino acid repeats involved in cellular adhesion (Latasa et al. 2006).

Recent proteomic studies reveal that in addition to these extracellular enzymes, intracellular proteins that are normally localized to the cells are present in the matrix at a certain amount. Examples include membrane proteins, histone-like proteins, ribosomal proteins, and superoxide dismutase (Jiao et al. 2011; Eboigbodin and Biggs 2008). While the functions of the intracellular proteins in the matrix remain to be fully investigated, some reports suggest a role of intracellular proteins outside of the cell. Such proteins that display two distinct functions are known as moonlighting proteins (Jeffery 2003). For example, the elongation factor Tu as well as the chaperonin protein complex GroEL have been identified on the cell surface of lactobacilli and were shown to be involved in adhesion to mucin and human epithelial cells (Bergonzelli et al. 2006; Granato et al. 2004). A recent study suggested that a large amount of these intracellular proteins are delivered to the matrix by membrane vesicles (Toyofuku et al. 2012a).

17.3 The Life Cycle of a Biofilm

Based on the knowledge of a limited number of model organisms, similarity of the biofilm life cycle has been recognized so far. From these observations mainly from mono-species biofilms, it is proposed that the biofilm life cycle consists of initial attachment, production of EPS matrix, maturation to complex colonies, and finally dispersal of cells from the biofilm. Accumulated researches indicate that these steps involve complex regulatory mechanisms that are influenced by environmental conditions and cell interactions. As a consequence, the structure of biofilms becomes highly variable ranging from flat-layered mats to complex mushroomlike structures depending on the nutritional conditions and cell interactions (Davies et al. 1998). Thus, cells undergo flexible and dynamic lifestyles in biofilms.

Biofilm formation begins with attachment to a surface that can be a substratum such as metals, glasses, or a surface of another cell. This step is important for the bacteria so that they settle down in the right environment. Initial attachment to abiotic surfaces is well characterized in the attachment of *Pseudomonas* sp. It was indicated that during the cell attachment, the cells first attach to the surface loosely via adhesins, such as flagellar and pili (O'Toole and Kolter 1998a, b), that may readily detach. Once the cells attach to the surface along their axis, this step is called irreversible attachment, the bacteria commit to their life as a biofilm (Petrova and Sauer 2012). Even this single step in biofilm formation is highly regulated by several factors indicating that biofilm formation is an active biological process, and as discussed later, an intracellular second messenger whose production is controlled by several environmental modulates this step.

While the initial attachment is the beginning of the life as a biofilm, the community comes to an end when cells disperse from the biofilm and seek new environments. The mechanism by which bacteria regulate dispersal is not fully understood and strategies to manipulate biofilm dispersal are of great interest for broad application in controlling biofilm formation. Passive processes for cell dispersal such as the erosion of biofilms by shear stress are observed; however, many bacteria undergo active dispersal processes where the cells convert from sessile to planktonic free-swimming bacteria (Sauer et al. 2002). Several environmental cues such as carbon sources, oxygen depletion, and nitric oxide induce cell dispersal, which allows the cells to respond to changes in environments. As observed in *P. aeruginosa*, not all cells but a subpopulation of cells localized at the center of mature biofilms disperse while the majority of the cells remain in the biofilm. These cells that remain seem to be dead cells which is also an indicative of the heterogenous nature of the cells in the biofilm (Barraud et al. 2006). In the dispersal cells, genes that are related to biofilm formation such as exopolysaccharides are downregulated, whereas genes that are related to motility are induced. In addition to active motility, the cells must disassemble from the EPS matrix. EPS-degrading enzymes such as chitinase and nuclease are also reported that could potentially be involved in the cell dispersion (McDougald et al. 2012).

Several signals that are produced by the cells are also involved in this dispersal event indicating that the cells themselves can initiate cell dispersal in a coordinated fashion. For example, a fatty acid molecule, cis-11-mehyl-2-dodecenoic acid which is known as a diffusible signal factor (DSF), was found to be involved in dispersal in *Xanthomonas campestris* (Dow et al. 2003). DSF in *X. campestris* activate the production of endo-β1, 4-mannanase that disrupts cell aggregates. Further work indicated that DSF reduce the biosynthesis of a putative exopolysaccharide, providing more the detailed mechanism of how DSF regulate dispersal (Tao et al. 2010b). DSF-related molecules are also found in other species implicating its role in interspecies signaling (Zhang and Dong 2004; Ryan and Dow 2011). For instance, DSF produced by *Stenotrophomonas maltophilia* induced stress-related genes in *P. aeruginosa* through a putative sensor kinase whose homologous protein is widespread among pseudomonads (Ryan et al. 2008). *P. aeruginosa* itself also produces a DSF-family signal, *cis*-2-decenoic acid, that induces biofilm dispersion of which the mechanism is not clear (Davies and Marques 2009). Interestingly, DSF-family signals play an interkingdom role where it is recognized by *Candida albicans* (Davies and Marques 2009; Boon et al. 2008).

Signals that induce dispersal are also found from *B. subtilis* biofilms called pellicles that are formed at the air–liquid surface. The cells are held together in the matrix that consists of exopolysaccharide and amyloid fiber which is largely composed of TasA. D-amino acids that are produced during the maturation of biofilm are incorporated into the cell wall peptidoglycan and trigger the release of amyloid fibers from the cell wall. A second biofilm-disassembly factor was found recently from *B. subtilis* that is a polyamine norspermidine. Norspermidine, which is produced in the late life cycle of the pellicle directly, interact directly with the exopolysaccharide component of the matrix (Kolodkin-Gal et al. 2012). Though the mechanism is not fully understood, both D-amino acids and norspermidine can induce cell dispersal in other bacteria, including Gram-negative cells.

17.4 c-di-GMP, An Intracellular Second Messenger, Regulates Biofilm Formation

The observations of the stage-specific gene regulation in biofilms indicate that biofilm development is a biologically programmed process. For instance, once the cells find a suitable surface, they should alter their gene expression so that they can settle down to switch their lifestyle from planktonic cells to biofilm cells. Furthermore, cells respond to environmental cues in dispersal. These phenomena observed universally among bacterial suggest that there is a common mechanism that is involved in these processes. Recent studies have found out that a small molecule, bis-(3'-5')-cyclic dimeric guanosine monophosphate (c-di-GMP), is an (almost) ubiquitous signal that plays a central role in biofilm development of diverse bacteria (Hengge 2009; Römling et al. 2005). C-di-GMP is a second messenger that passes through the information outside the cells into the cytoplasm. In general, high levels of c-di-GMP concentration correlate with the biofilm mode of lifestyle and low concentrations of c-di-GMP correlate with planktonic mode of growth. In other words, a high level of c-di-GMP stimulates the biosynthesis of adhesins and extracellular matrix and inhibits motility, thus allowing the cells to form and maintain biofilms. In dispersal cells, c-di-GMP levels decrease that allow the cells to become motile. The level of c-di-GMP is controlled by the enzymes that produce (diguanylate cyclases) and degrade (phosphodiesterases). Diguanylate cyclases activity is associated with the GGDEF domain, which is named after the highly conserved amino acid sequences they contain, and c-di-GMP-specific phosphodiesterases activity is associated with the EAL or HD-GYP domains. These domains are often found with N-terminal sensory domains, compromising sensor domains as well as phosphoacceptor. The great variety of these sensory domains indicates that c-di-GMP level responds to numerous signals (Hengge 2009). The extracellular signal DSF is also one of the signals that alter c-di-GMP levels by degrading c-di-GMP (Tao et al. 2010a; Deng et al. 2012). Hence, a diverse range of signals are integrated to the c-di-GMP-signaling system enabling the cells to respond to changing environments.

Although the target of c-di-GMP is not fully understood, it has been indicated that c-di-GMP binds to the target component and alter the structure and functions. Several examples demonstrate that the binding components of c-di-GMP are highly diverse that can be proteins as well as riboswitches. For instance, c-di-GMP binding to PilZ proteins is an essential step in c-di-GMP-mediated regulation (Ryjenkov et al. 2006). In riboswitches, c-di-GMP binds to mRNA motifs which alter the expression of the gene (Sudarsan et al. 2008).

17.5 Bacterial Communications that Influence Biofilm Formation

Multicellularity in bacteria can be defined as a group of cells that function as a greater entity. In order to accomplish this, intercellular interactions must play a crucial role to enable cells coordinate their activities (Shapiro 1998). Although it is a still-opening question whether bacteria generally function for the whole in biofilms, scientists have discovered that bacteria can communicate with each other through chemical molecules in order to coordinate the cell behavior. Moreover, cell-to-cell communication has been demonstrated to affect biofilm formation in many species. A typical signaling molecule produced by Gram-negative bacteria is the *N*-acyl homoserine lactones (AHLs). Over 100 species of *Proteobacteria* use AHLs as signal molecules. AHLs usually consist of a fatty chain acid coupled to a homoserine lactone. There are variations in the acyl-group length, substitution at the C3 position (hydrogen, hydroxyl, or oxo group), and the saturation of the fatty acid chain. These variations confer signal specificity that is recognized by their cognate LuxR family transcriptional regulators. AHLs are synthesized by LuxI family proteins. The *luxI* gene is induced in the presence of AHLs, consisting an autoregulatory system.

While the fatty acid side chains were typically found in these homoserine lactone signals, recently, a new class of non-fatty-acid-based AHL signals has been discovered. The phototrophic soil bacterium *Rhodopseudomonas palustris* possesses a pair of LuxIR homologues, RpaIR. RpaI synthesize *p*-coumaroyl-homoserine lactone, from exogenous *p*-coumaric acid, which is thought to be obtained from the plant host. These aryl-homoserine lactone homoserine productions have also been observed in *Bradyrhizobium* sp. BTAi1 and *Silicibacter pomeroyi* DSS-3(Schaefer et al. 2008; Ahlgren et al. 2011).

P. aeruginosa is one of the most studied bacteria in cell-to-cell communication and biofilms. This opportunistic human pathogen produces at least two AHLs that are *N*-butyryl-HSL (C4-HSL) and *N*-(3-oxododecanoyl)-HSL (3-oxo-C12-HSL). Each signal is produced by RhlI and LasI, respectively. The organized structure of biofilms evoked researchers to think that there must be some interactions between cells when forming biofilms. The involvement of AHL on biofilm formation was first reported in 1998, when Davies et al. reported that a *lasI* mutant formed biofilms that were flat compared to the wild type that formed structured microcolonies (Davies et al. 1998). Since then, many groups have confirmed that the effect of cell-to-cell communication on biofilm formation in *P. aeruginosa* depends on strain and experimental conditions (de Kievit 2009).

The mechanism is not clear how cell-to-cell communication influences biofilm formation because AHLs control a wide range of genes. However, several factors that influence biofilm formation are known to be controlled by cell-to-cell communication (de Kievit 2009). These are biofilm matrix, rhamnolipids, and denitrification. In *P. aeruginosa*, at least three types of exopolysaccharide (alginate, Pel, and Psl) that are related to biofilm formation have been confirmed. Thus far, only the *pel*-gene cluster has been identified as being regulated by cell-to-cell communication. Pel is a glucose-rich exopolysaccharide which is essential for biofilm formation (Friedman and Kolter 2004a). It was discovered that the Las system is mainly involved in the transcription of *pel* biosynthetic genes (Sakuragi and Kolter 2007). Another important component of the extracellular matrix, extracellular DNA (eDNA) is also suggested to be influenced by QS. The mechanism by which the DNA is exported outside of the cell is not fully understood. Interestingly, spatial analysis of *P. aeruginosa* biofilms revealed distinct patterns of eDNA localization. In the early stages of biofilm formation, eDNA was shown to accumulate on the surface of microcolonies, while in older biofilms eDNA was shown to accumulate at the junction of the stalk and cap of the typical mushroomlike structure of microcolonies (Allesen-Holm et al. 2006). It was observed that QS-deficient mutants contain less eDNA than the wild type, suggesting that QS induce eDNA (Allesen-Holm et al. 2006).

Rhamnolipids are amphipathic glycolipids that act as biosurfactants. The expression of rhamnolipid synthesis operon *rhlAB* is regulated by the rhl QS system (Pearson et al. 1997). A mutant deficient in rhamnolipid production forms biofilms that are flat and relatively homogenous compared to the structured wild-type biofilms (Davey et al. 2003). From these observations, it is proposed that rhamnolipids are important in forming distinct architecture of the biofilm. In addition to maintaining the biofilm structure, rhamnolipids are necessary for detachment of the cells from the biofilm (Boles et al. 2005). The influence of rhamnolipids on multiple facets of biofilm formation suggests that the production is regulated at particular location and time points during biofilm formation. Supporting this idea, microscopic analysis using *gfp*-fused *rhlAB* genes indicated that rhamnolipid production is maximal in the stalks of a developing microcolony (Lequette and Greenberg 2005).

P. aeruginosa is capable of growing anaerobically by using N-oxides as terminal electron acceptors instead of oxygen (Toyofuku et al. 2012b). Under denitrifying conditions, *P. aeruginosa* forms robust biofilms (Yoon et al. 2002). Inactivation of Rhl QS leads to nitric oxide (NO) accumulation that is toxic to the cells causing cell death. Further study demonstrated that QS repress denitrification activity (Toyofuku et al. 2007, 2008). One of the key factors influencing biofilm formation is NO. While high concentration leads to cell toxicity, sublethal concentrations trigger disperse of the cells from the biofilm (Barraud et al. 2006). Interestingly, the process of biofilm formation under anaerobic condition seems somewhat different from aerobic condition, where cells become filamentous under anaerobic condition (Yawata et al. 2008). The cell elongation is one of the important factors in biofilm formation under anaerobic conditions and it is influenced by the production of NO (Yoon et al. 2011). Hence, fine tune of NO accumulation by QS as well as other factors appears to be important in biofilm formation under anaerobic conditions.

In Gram-positive bacteria amino acid peptides are widely used as signal molecules. Once the precursor signal protein is synthesized within the cell, active signals are generated by cleaving the proteins to short form of peptide chains and these peptides are subject to posttranslational modification in some bacteria resulting in a great diversity of the signal (Dunny and Leonard 1997). The peptide signals are linear or cyclic usually ranging in the size of 5–26 amino acid residues. Interestingly, some signal peptides also function as bacteriocins. These peptide signals that do not diffuse across the cell membrane are secreted by ABC transporters. Once secreted, the cells then sense the peptide signals by a two-component regulatory system. Most typically, the peptide signal is sensed at the surface of the bacteria by a histidine kinase sensor. This results in the autophosphorylation of the sensor kinase and the phosphoryl group gets transferred to a response regulator that regulates the expression of genes.

In the case of *Staphylococcus aureus*, the autoinducing peptide (AIP) is encoded by *agrD*. The AgrB protein processes and exports AIP. The secreted AIP is detected by a histidine kinase AgrC, which activates the response regulator AgrA by phosphorylation. Activated AgrA then modulates gene expression including the positive control of the *agrBDCA* operon. As expected, the *agr* QS system is involved in biofilm formation. In *S. aureus*, the *agr* QS system mutant forms thicker biofilm than the wild type (Vuong et al. 2000). Furthermore, extracellular proteins such as proteases and nucleases that are involved in dispersal of biofilm are upregulated by the QS system. Hence, the *agr* QS system inhibits biofilm formation.

In addition to AHLs and signal peptides, AI-2 is discovered broadly among Gram-negative and Gram-positive bacteria. AI-2 was originally described in *Vibrio harveyi*, where it was involved in the regulation of bioluminescence (Chen et al. 2002). AI-2 is produced by a spontaneous reaction of 4,5-dihydroxy-2,3-pentanedione which is synthesized by LuxS. The AI-2 signal which is recognized by homologues of the *luxS* gene has been identified in over 350 sequenced bacterial genomes (Shrout and Nerenberg 2012). Thus, AI-2 appears to be an interspecies signal. Whether LuxS is involved in biofilm formation or not is not clear. For instance, in *Staphylococcus epidermidis*, LuxS has negative effects on biofilm formation (Li et al. 2008), while this is not the case in *S. aureus* (Doherty et al. 2006). Taking into account that AI-2 is an interspecies signal, AI-2 may be involved in multispecies biofilm formation.

17.6 Multispecies Biofilm

In the natural environment, bacteria form complex multispecies biofilm. Compared to the mono-species biofilm, our understanding of multispecies biofilm is limited. The common feature of these bacterial communities is that normally the cells are dense and physically close to each other compared to the free-living planktonic cells. As a result of active metabolism of the cells in biofilms, nutrient in the liquid surrounding may be consumed by cells in the outer layers and metabolites may accumulate in the biofilms. These gradients create microenvironments that become niches for other bacteria. One of the techniques that are commonly used for observation of complex natural biofilm is fluorescence in situ hybridization (FISH). With this method, localization of bacterial populations can be visualized by hybridizing a fluorescently labeled oligonucleotide probe to the target 16S rRNA (Amann et al. 1990; DeLong et al. 1989). Compartmentalized or mixed structures of species within natural biofilms have been observed based on this method. The stratification of species within the biofilm can mostly be explained by the difference of metabolism among the species. For instance, in a nitrifying activated sludge, nitrifying bacteria are split into layers according to their use of different electron acceptors. During nitrification, ammonia is oxidized by ammonia-oxidizing bacteria and the produced nitrite is further oxidized to nitrate by nitrite-oxidizing bacteria. Consequently, ammonia-oxidizing bacteria locate at the outer layer and the nitrite-oxidizing bacteria colonize immediately below the ammonia oxidizers in nitrifying biofilms (Okabe et al. 1999). Compartmentalized is also explained by computational simulations revealing that these structures result from simple diffusion of growth-limiting nutrient (Nadell et al. 2009). Some model experiments using two bacterial species has also revealed that physicochemical conditions determine the localization of bacterial colonies within the biofilm. For example, *Burkholderia xenovorans* LB400 and *Pseudomonas* sp. B13 are able to degrade a pollutant, 3-chlorobiphenyl, to water and carbon dioxide. In this case, *B. xenovorans* LB400 degrades 3-chlorobiphenyl only to chlorinated benzoate and no further, while *Pseudomonas* sp. B13 further degrades chlorobenzoate to water and carbon dioxide. Since *Pseudomonas* sp. B13 cannot utilize 3-chlorobiphenyl, it is able to grow only if it is together with *B. xenovorans* LB400 when 3-chlorobiphenyl is the sole carbon/energy source. When a mixed-species biofilm was developed by these two strains using 3-chlorobiphenyl as the only available carbon source, *Pseudomonas* sp. B13 was always mixed with *B. xenovorans* LB400, whereas separate colonies were observed when another carbon source that can be used by both strains was given (Nielsen et al. 2000). In another case using *Acinetobacter* sp. C6 and *Pseudomonas putida* KT2400, distinct colonies but not mixed-species colonies were formed even under experimental settings that support host-commensal relationship (Hansen et al. 2007). When benzyl alcohol is given as a sole carbon source, the presence of *P. putida* KT2400 is dependent on *Acinetobacter* sp. C6. *Acinetobacter* sp. C6 metabolize benzyl alcohol to benzoate that is partly excreted and can be utilized by *P. putida* KT2400. Under this condition, *Acinetobacter* sp. C6 and *P. putida* KT2400 form distinct colonies where *Acinetobacter* sp. C6 is loosely surrounded by *P. putida* KT2400. In this case, competition of oxygen might counteract the host-commensal partnership of carbon source. Hence, the examples show us how the final structure could be affected by different interactions that occur among the consortia. Interestingly, extended incubation of the *Acinetobacter* sp. C6 and *P. putida* KT2400 mixed-biofilm under the host-commensal condition led to the emergence of a variant distinct cell from the ancestral type of *P. putida* KT2400 (Hansen et al. 2007). The variant was not detected in chemostat culture with *Acinetobacter* sp. C6 or in single-species biofilms, indicating that the variant can evolve in the microenvironment within biofilms in the presence of *Acinetobacter*. The evolved community was more stable and more productive than the ancestral community. Altogether, these examples demonstrate how spatial organization is important for the function of the community and further leads to the understanding of how a community is optimized in their natural habitat.

In addition to these chemical interactions, direct contact of the cells also plays an important part in mixed-species biofilms. Interspecies electron transfer is known to be important during anaerobic degradation of organic matter to methane. During this process, close physical contact between syntrophs and methanogens is required. Direct observation of granular sludge indicated that these microbes locate close to each other (Harmsen et al. 1996). In *Pelotomaculum thermopropionicum* and *Methanothermobacter thermautotrophicus*, such symbiosis is mediated by a flagellum (Shimoyama et al. 2009). *P. thermopropionicum* can ferment propionate to acetate, bicarbonate, and H_2; however, this conversion is endergonic and the partial pressure of H_2 should be kept low for this process to proceed. In the presence of *M. thermautotrophicus*, H_2 is consumed to CH_4 as an energy source and enable *P. thermopropionicum* to oxidize propionate. When grown together, these two strains form aggregates that are maintained by flagellum of *P. thermopropionicum*. The flagellum not only maintains the symbiosis but also synchronizes their metabolism by inducing gene expression involved in methanogenesis of *M. thermautotrophicus*. More recently, another type of bacterial communication that is based on physical contact was discovered. This communication is mediated by nanotubes that connect neighboring cells in *Bacillus subtilis* (Dubey and Ben-Yehuda 2011). These tubes were not synthesized during planktonic growth but were produced in colonies. These tubes allow bacteria to exchange small molecules including proteins and plasmids even between other species. Although the relevance of this nanotube in the natural environment is yet to be understood, the findings open up a new era of bacterial interaction between species.

Another bacterial-derived structure that potentially has a role in interspecies communication inside biofilms is the outer membrane vesicles (OMVs). OMVs are particles ranging from approximately 20 to 200 nm that are mainly produced in Gram-negative bacteria (Kulp and Kuehn 2010; Tashiro et al. 2012). These vesicles are mainly composed of outer membrane, periplasmic components and also contain DNA. In *P. aeruginosa*, a hydrophobic QS molecule, *Pseudomonas* quinolone signal (PQS) is incorporated in OMVs allowing the PQS to freely diffuse from the cell and shuttle them to the neighboring cells (Mashburn and Whiteley 2005). Consequently, PQS carried by OMVs coordinate social behaviors in *P. aeruginosa*. In addition, PQS are able to induce OMV production in *P. aeruginosa* and other species (Tashiro et al. 2010; Mashburn and Whiteley 2005). Importantly, OMVs have been observed in the EPS fraction of mono-species biofilms such as *P. aeruginosa* and *Myxococcus xanthus* and also mixed-species biofilm of environmental samples (Schooling and Beveridge 2006; Toyofuku et al. 2012a). *M. xanthus* is a bacterium that shows complex social behavior and biofilm (also known as fruiting body) formation. In this bacterium, protein can be exchanged through direct cell contact in which the precise mechanism is not completely understood. As a result of protein exchange, individual cells can endow their neighbors with new characteristics without altering the genome. Direct observation of *M. xanthus* biofilm demonstrated that the intercellular region is packed with OMVs that are tethering on the cell surface (Palsdottir et al. 2009). From these observations it is hypothesized that these OMVs may play a role in the protein exchange of *M. xanthus* (Remis et al. 2010; Konovalova et al. 2010). OMVs are also found extensively in the EPS of *P. aeruginosa* biofilm. It is well studied that OMVs in *P. aeruginosa* are associated with virulence against the host (Tashiro et al. 2012). Moreover, these OMVs derived from *P. aeruginosa* have the potential to fuse and/or attach to other Gram-negative and Gram-positive cells that may confer new properties to the recipient by transporting proteins or DNA. The role of OMVs in biofilms is yet to be investigated while some report indicates that they are involved in the development of mixed-species biofilm. OMVs produced by *Porphyromonas gingivalis* induce co-aggregation of *S. aureus* and *Prevotella intermedia* via an adhesin localized to the OMVs (Kamaguchi et al. 2003), indicating the role of OMVs in interspecies biofilm formation. Given that there are numerous proteins in the EPS, OMVs could also be contributing to place the proteins in the EPS.

17.7 Recently Developed Methods to Study Biofilms

Since biofilm mode of growth differs from planktonic cells, several methods to cultivate and analyze biofilms have been developed. One of the most popular method used in the assay of biofilm development is the 96-well biofilm assay (O'Toole et al. 1999). In this method, biofilm are developed in a static batch culture and the cells attached to the 96-well plate are stained after washing. This method is very powerful for screening for factors that affect biofilm formation and thus has been used in various species by many scientists. Biofilm structures have been revealed by microscopy techniques such as electron microscopy and the bacterial community structure with their localizations is often analyzed by FISH. FISH can be combined with other useful techniques, for instance, with combination of microautoradiography, uptake of different organic substrate can be monitored in order to demonstrate active cells (Lee et al. 1999; Kindaichi et al. 2004). The limitation of these techniques is that they only give a snapshot of the biofilm images while the biofilm development is highly dynamic. The use of time-lapse microscopy with combination of flow-cell system where the intact biofilm development can be monitored for several days has overcome this limitation for some bacteria. By using the flow-cell system, the lifestyle of biofilms became clear that it consists of several stages. Moreover, the use of confocal laser scanning microscopy (CLSM) with the flow-cell system allows the observation of three-dimensional structure of the biofilm during the time course for fluorescent-labeled bacteria. The fluorescent protein can be expressed constitutively and can be used for mono-species and mixed-species biofilms. In addition, fluorescent protein can be expressed under control of a particular promoter which gene is of interest in order to monitor gene expression in biofilms.

Recently, a technique adapting the confocal reflection microscopy (CRM) was used in biofilm. The advantage of this technique is that intact biofilms can be observed directly by confocal laser scanning microscopy without staining, fixation,

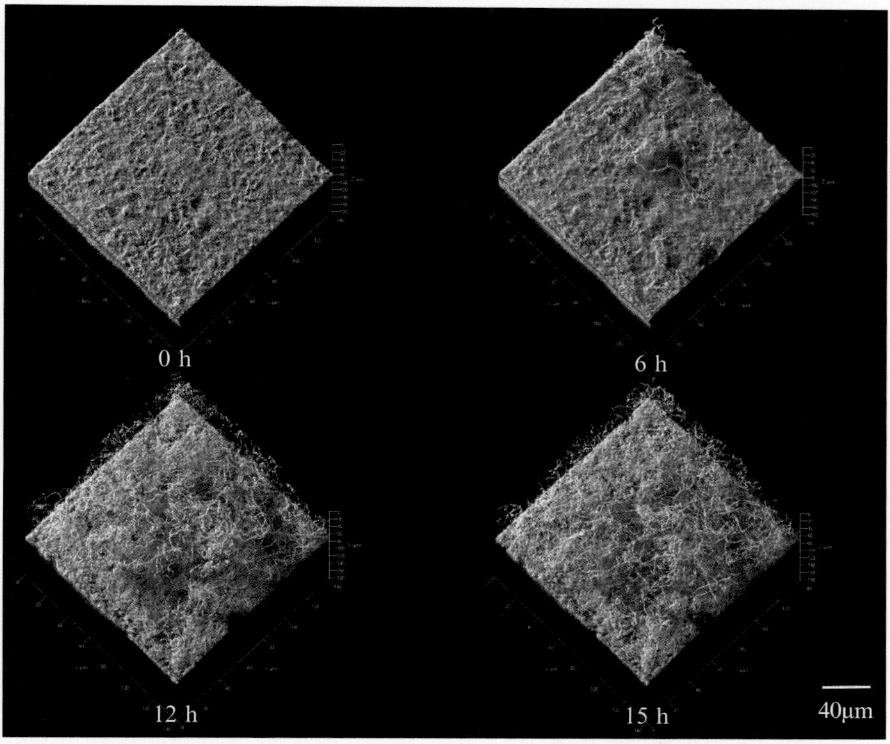

Fig. 17.1 Development of a mixed-species biofilm monitored by COCRM

or labeling with fluorescent proteins. Because CRM relies on the reflected light of the sample, reflection artifact obtained from bright spots such as cover glasses often becomes a problem. To solve this problem, the signal intensity of the cells can be maintained to a constant level along the Z-axis continuously by adjusting the detector gain manually while scanning. This modified mode of CRM called continuous-optimizing-CRM (COCRM) enables direct imaging of mixed-species biofilms and also mono-species biofilms alive during time courses (Yawata et al. 2010) (Fig. 17.1). While microscopy observation enables detailed analysis of the biofilm structure and development, yet there are few nondestructive methods that allow both microscopy analysis and measurement of metabolites. Measurement of metabolites along with the biofilm structure is important in understanding the activity of the biofilm and further understanding how biofilm reacts to the environment as a whole. To this end, flow-cell system was improved that gaseous metabolites can be collected and measured simultaneously with gas chromatograph (Yawata et al. 2008). Another method commonly used to monitor microbial activity is the use of microsensors. Microsensors are used for profiling spatial distributions of specific substrates and products (de Beer and Stoodley 2006). These microsensors can be mounted to chambers that cultivate biofilms and the concentration of certain metabolites can be monitored real-time. Since microsensors are small with a tip size of 1–20 μm it is feasible for small chambers and was adapted to a micro-fluidic device to monitor NH_4^+ consumption of activated sludge with its structure (Toda et al. 2011). Raman microspectrometer is also a powerful tool in chemically characterizing the biofilms noninvasively without any pretreatment of the sample. Biomolecules along with second metabolites can be identified with its space-resolved information, which will give us a more detailed understanding of the activity and heterogeneity of the cells (Venkata et al. 2011; Patzold et al. 2006). These new techniques with others will provide a new insight into the structure functional correlation of biofilms.

Acknowledgement We thank Tomohiro Inaba for kindly providing us the picture.

References

Ahlgren NA, Harwood CS, Schaefer AL, Giraud E, Greenberg EP (2011) Aryl-homoserine lactone quorum sensing in stem-nodulating photosynthetic *bradyrhizobia*. Proc Natl Acad Sci USA 108:7183–7188

Allesen-Holm M, Barken KB, Yang L, Klausen M, Webb JS, Kjelleberg S, Molin S, Givskov M, Tolker-Nielsen T (2006) A characterization of DNA release in *Pseudomonas aeruginosa* cultures and biofilms. Mol Microbiol 59:1114–1128

Amann RI, Krumholz L, Stahl DA (1990) Fluorescent-oligonucleotide probing of whole cells for determinative, phylogenetic, and environmental studies in microbiology. J Bacteriol 172:762–770

Barraud N, Hassett DJ, Hwang SH, Rice SA, Kjelleberg S, Webb JS (2006) Involvement of nitric oxide in biofilm dispersal of *Pseudomonas aeruginosa*. J Bacteriol 188:7344–7353

Baty AM 3rd, Diwu Z, Dunham G, Eastburn CC, Geesey GG, Goodman AE, Suci PA, Techkarnjanaruk S (2001) Characterization of extracellular chitinolytic activity in biofilms. Methods Enzymol 336:279–301

Bergonzelli GE, Granato D, Pridmore RD, Marvin-Guy LF, Donnicola D, Corthésy-Theulaz IE (2006) GroEL of *Lactobacillus johnsonii* La1 (NCC 533) is cell surface associated: potential role in interactions with the host and the gastric pathogen *Helicobacter pylori*. Infect Immun 74:425–434

Boles BR, Thoendel M, Singh PK (2005) Rhamnolipids mediate detachment of *Pseudomonas aeruginosa* from biofilms. Mol Microbiol 57:1210–1223

Boon C, Deng Y, Wang LH, He Y, Xu JL, Fan Y, Pan SQ, Zhang LH (2008) A novel DSF-like signal from Burkholderia cenocepacia interferes with Candida albicans morphological transition. ISME J 2:27–36

Borlee BR, Goldman AD, Murakami K, Samudrala R, Wozniak DJ, Parsek MR (2010) *Pseudomonas aeruginosa* uses a cyclic-di-GMP-regulated adhesin to reinforce the biofilm extracellular matrix. Mol Microbiol 75:827–842

Brinkmann V, Reichard U, Goosmann C, Fauler B, Uhlemann Y, Weiss DS, Weinrauch Y, Zychlinsky A (2004) Neutrophil extracellular traps kill bacteria. Science 303:1532–1535

Chapman MR, Robinson LS, Pinkner JS, Roth R, Heuser J, Hammar M, Normark S, Hultgren SJ (2002) Role of *Escherichia coli* curli operons in directing amyloid fiber formation. Science 295:851–855

Chen X, Schauder S, Potier N, Van Dorsselaer A, Pelczer I, Bassler BL, Hughson FM (2002) Structural identification of a bacterial quorum-sensing signal containing boron. Nature 415:545–549

Davey ME, Caiazza NC, O'Toole GA (2003) Rhamnolipid surfactant production affects biofilm architecture in *Pseudomonas aeruginosa* PAO1. J Bacteriol 185:1027–1236
Davies DG, Marques CN (2009) A fatty acid messenger is responsible for inducing dispersion in microbial biofilms. J Bacteriol 191:1393–1403
Davies DG, Parsek MR, Pearson JP, Iglewski BH, Costerton JW, Greenberg EP (1998) The involvement of cell-to-cell signals in the development of a bacterial biofilm. Science 280:295–298
de Beer D, Stoodley P (2006) Microbial Biofilms. In: Dworkin M, Falkow S, Rosenberg E, Scleifer K, Stackebrandt E (eds) The prokaryotes, vol 1, 3rd edn. Springer, New York
de Kievit TR (2009) Quorum sensing in *Pseudomonas aeruginosa* biofilms. Environ Microbiol 11:279–288
DeLong EF, Wickham GS, Pace NR (1989) Phylogenetic stains: ribosomal RNA-based probes for the identification of single cells. Science 243:1360–1363
Deng Y, Schmid N, Wang C, Wang J, Pessi G, Wu D, Lee J, Aguilar C, Ahrens CH, Chang C, Song H, Eberl L, Zhang LH (2012) Cis-2-dodecenoic acid receptor RpfR links quorum-sensing signal perception with regulation of virulence through cyclic dimeric guanosine monophosphate turnover. Proc Natl Acad Sci USA 109:15479–15484
Doherty N, Holden MT, Qazi SN, Williams P, Winzer K (2006) Functional analysis of *luxS* in *Staphylococcus aureus* reveals a role in metabolism but not quorum sensing. J Bacteriol 188:2885–2897
Dominiak DM, Nielsen JL, Nielsen PH (2011) Extracellular DNA is abundant and important for microcolony strength in mixed microbial biofilms. Environ Microbiol 13:710–721
Dow JM, Crossman L, Findlay K, He YQ, Feng JX, Tang JL (2003) Biofilm dispersal in *Xanthomonas campestris* is controlled by cell-cell signaling and is required for full virulence to plants. Proc Natl Acad Sci USA 100:10995–11000
Dubey GP, Ben-Yehuda S (2011) Intercellular nanotubes mediate bacterial communication. Cell 144:590–600
Dunny GM, Leonard BA (1997) Cell-cell communication in gram-positive bacteria. Annu Rev Microbiol 51:527–564
Eboigbodin KE, Biggs CA (2008) Characterization of the extracellular polymeric substances produced by *Escherichia coli* using infrared spectroscopic, proteomic, and aggregation studies. Biomacromolecules 9:686–695
Flemming HC, Wingender J (2010) The biofilm matrix. Natl Rev Microbiol 8:623–633
Friedman L, Kolter R (2004a) Genes involved in matrix formation in *Pseudomonas aeruginosa* PA14 biofilms. Mol Microbiol 51(3):675–690
Friedman L, Kolter R (2004b) Two genetic loci produce distinct carbohydrate-rich structural components of the *Pseudomonas aeruginosa* biofilm matrix. J Bacteriol 186:4457–4465
Frølund B, Palmgren R, Keiding K, Nielsen PH (1996) Extraction of extracellular polymers from activated sludge using a cation exchange resin. Water Res 30:1749–1758
Govan JR, Deretic V (1996) Microbial pathogenesis in cystic fibrosis: mucoid *Pseudomonas aeruginosa* and *Burkholderia cepacia*. Microbiol Rev 60:539–574
Granato D, Bergonzelli GE, Pridmore RD, Marvin L, Rouvet M, Corthésy-Theulaz IE (2004) Cell surface-associated elongation factor Tu mediates the attachment of *Lactobacillus johnsonii* NCC533 (La1) to human intestinal cells and mucins. Infect Immun 72:2160–2169
Hansen SK, Rainey PB, Haagensen JA, Molin S (2007) Evolution of species interactions in a biofilm community. Nature 445:533–536
Harmsen HJ, Kengen HM, Akkermans AD, Stams AJ, de Vos WM (1996) Detection and localization of syntrophic propionate-oxidizing bacteria in granular sludge by in situ hybridization using 16S rRNA-based oligonucleotide probes. Appl Environ Microbiol 62(5):1656–1663
Hengge R (2009) Principles of c-di-GMP signalling in bacteria. Natl Rev Microbiol 7:263–273
Høiby N, Bjarnsholt T, Givskov M, Molin S, Ciofu O (2010) Antibiotic resistance of bacterial biofilms. Int J Antimicrob Agents 35:322–332

Jackson KD, Starkey M, Kremer S, Parsek MR, Wozniak DJ (2004) Identification of *psl*, a locus encoding a potential exopolysaccharide that is essential for *Pseudomonas aeruginosa* PAO1 biofilm formation. J Bacteriol 186:4466–4475

Jeffery CJ (2003) Moonlighting proteins: old proteins learning new tricks. Trends Genet 19:415–417

Jiao Y, D'Haeseleer P, Dill BD, Shah M, Verberkmoes NC, Hettich RL, Banfield JF, Thelen MP (2011) Identification of biofilm matrix-associated proteins from an acid mine drainage microbial community. Appl Environ Microbiol 77:5230–5237

Kamaguchi A, Nakayama K, Ichiyama S, Nakamura R, Watanabe T, Ohta M, Baba H, Ohyama T (2003) Effect of *Porphyromonas gingivalis* vesicles on coaggregation of *Staphylococcus aureus* to oral microorganisms. Curr Microbiol 47:485–491

Kindaichi T, Ito T, Okabe S (2004) Ecophysiological interaction between nitrifying bacteria and heterotrophic bacteria in autotrophic nitrifying biofilms as determined by microautoradiography-fluorescence in situ hybridization. Appl Environ Microbiol 70:1641–1650

Kolodkin-Gal I, Cao S, Chai L, Bottcher T, Kolter R, Clardy J, Losick R (2012) A self-produced trigger for biofilm disassembly that targets exopolysaccharide. Cell 149:684–692

Konovalova A, Petters T, Søgaard-Andersen L (2010) Extracellular biology of *Myxococcus xanthus*. FEMS Microbiol Rev 34:89–106

Kulp A, Kuehn MJ (2010) Biological functions and biogenesis of secreted bacterial outer membrane vesicles. Annu Rev Microbiol 64:163–184

Latasa C, Solano C, Penades JR, Lasa I (2006) Biofilm-associated proteins. C R Biol 329:849–857

Lee N, Nielsen PH, Andreasen KH, Juretschko S, Nielsen JL, Schleifer KH, Wagner M (1999) Combination of fluorescent in situ hybridization and microautoradiography-a new tool for structure-function analyses in microbial ecology. Appl Environ Microbiol 65:1289–1297

Lequette Y, Greenberg EP (2005) Timing and localization of rhamnolipid synthesis gene expression in *Pseudomonas aeruginosa* biofilms. J Bacteriol 187:37–44

Li M, Villaruz AE, Vadyvaloo V, Sturdevant DE, Otto M (2008) AI-2-dependent gene regulation in *Staphylococcus epidermidis*. BMC Microbiol 8:4

López D, Vlamakis H, Kolter R (2010) Biofilms. Cold Spring Harb Perspect Biol 2:a000398

Mashburn LM, Whiteley M (2005) Membrane vesicles traffic signals and facilitate group activities in a prokaryote. Nature 437(7057):422–425

Matsukawa M, Greenberg EP (2004) Putative exopolysaccharide synthesis genes influence *Pseudomonas aeruginosa* biofilm development. J Bacteriol 186:4449–4456

McDougald D, Rice SA, Barraud N, Steinberg PD, Kjelleberg S (2012) Should we stay or should we go: mechanisms and ecological consequences for biofilm dispersal. Natl Rev Microbiol 10:39–50

McSwain BS, Irvine RL, Hausner M, Wilderer PA (2005) Composition and distribution of extracellular polymeric substances in aerobic flocs and granular sludge. Appl Environ Microbiol 71:1051–1057

Nadell CD, Xavier JB, Foster KR (2009) The sociobiology of biofilms. FEMS Microbiol Rev 33:206–224

Nielsen AT, Tolker-Nielsen T, Barken KB, Molin S (2000) Role of commensal relationships on the spatial structure of a surface-attached microbial consortium. Environ Microbiol 2:59–68

Okabe S, Satoh H, Watanabe Y (1999) In situ analysis of nitrifying biofilms as determined by in situ hybridization and the use of microelectrodes. Appl Environ Microbiol 65:3182–3191

O'Toole GA, Kolter R (1998a) Flagellar and twitching motility are necessary for *Pseudomonas aeruginosa* biofilm development. Mol Microbiol 30:295–304

O'Toole GA, Kolter R (1998b) Initiation of biofilm formation in *Pseudomonas fluorescens* WCS365 proceeds via multiple, convergent signalling pathways: a genetic analysis. Mol Microbiol 28:449–461

O'Toole GA, Pratt LA, Watnick PI, Newman DK, Weaver VB, Kolter R (1999) Genetic approaches to study of biofilms. Methods Enzymol 310:91–109

Otzen D, Nielsen PH (2008) We find them here, we find them there: functional bacterial amyloid. Cell Mol Life Sci 65:910–927

Palsdottir H, Remis JP, Schaudinn C, O'Toole E, Lux R, Shi W, McDonald KL, Costerton JW, Auer M (2009) Three-dimensional macromolecular organization of cryofixed *Myxococcus xanthus* biofilms as revealed by electron microscopic tomography. J Bacteriol 191:2077–2082

Patzold R, Keuntje M, Anders-von Ahlften A (2006) A new approach to non-destructive analysis of biofilms by confocal Raman microscopy. Anal Bioanal Chem 386:286–292

Pearson JP, Pesci EC, Iglewski BH (1997) Roles of Pseudomonas aeruginosa las and rhl quorum-sensing systems in control of elastase and rhamnolipid biosynthesis genes. J Bacteriol 179:5756–5767

Petrova OE, Sauer K (2012) Sticky situations: key components that control bacterial surface attachment. J Bacteriol 194:2413–2425

Remis JP, Costerton JW, Auer M (2010) Biofilms: structures that may facilitate cell-cell interactions. ISME J 4:1085–1087

Rickard AH, Gilbert P, High NJ, Kolenbrander PE, Handley PS (2003) Bacterial coaggregation: an integral process in the development of multi-species biofilms. Trends Microbiol 11:94–100

Romero D, Aguilar C, Losick R, Kolter R (2010) Amyloid fibers provide structural integrity to *Bacillus subtilis* biofilms. Proc Natl Acad Sci USA 107:2230–2234

Römling U, Gomelsky M, Galperin MY (2005) C-di-GMP: the dawning of a novel bacterial signalling system. Mol Microbiol 57:629–639

Ryan RP, Dow JM (2011) Communication with a growing family: diffusible signal factor (DSF) signaling in bacteria. Trends Microbiol 19:145–152

Ryan RP, Fouhy Y, Garcia BF, Watt SA, Niehaus K, Yang L, Tolker-Nielsen T, Dow JM (2008) Interspecies signalling via the Stenotrophomonas maltophilia diffusible signal factor influences biofilm formation and polymyxin tolerance in Pseudomonas aeruginosa. Mol Microbiol 68:75–86

Ryjenkov DA, Simm R, Romling U, Gomelsky M (2006) The PilZ domain is a receptor for the second messenger c-di-GMP: the PilZ domain protein YcgR controls motility in enterobacteria. J Biol Chem 281:30310–30314

Sakuragi Y, Kolter R (2007) Quorum-sensing regulation of the biofilm matrix genes (pel) of *Pseudomonas aeruginosa*. J Bacteriol 189:5383–5386

Sauer K, Camper AK, Ehrlich GD, Costerton JW, Davies DG (2002) *Pseudomonas aeruginosa* displays multiple phenotypes during development as a biofilm. J Bacteriol 184:1140–1154

Schaefer AL, Greenberg EP, Oliver CM, Oda Y, Huang JJ, Bittan-Banin G, Peres CM, Schmidt S, Juhaszova K, Sufrin JR, Harwood CS (2008) A new class of homoserine lactone quorum-sensing signals. Nature 454:595–599

Schooling SR, Beveridge TJ (2006) Membrane vesicles: an overlooked component of the matrices of biofilms. J Bacteriol 188:5945–5957

Shapiro JA (1998) Thinking about bacterial populations as multicellular organisms. Annu Rev Microbiol 52:81–104

Shimoyama T, Kato S, Ishii S, Watanabe K (2009) Flagellum mediates symbiosis. Science 323:1574

Shrout JD, Nerenberg R (2012) Monitoring bacterial twitter: does quorum sensing determine the behavior of water and wastewater treatment biofilms? Environ Sci Technol 46:1995–2005

Sudarsan N, Lee ER, Weinberg Z, Moy RH, Kim JN, Link KH, Breaker RR (2008) Riboswitches in eubacteria sense the second messenger cyclic di-GMP. Science 321:411–413

Sunde M, Serpell LC, Bartlam M, Fraser PE, Pepys MB, Blake CC (1997) Common core structure of amyloid fibrils by synchrotron X-ray diffraction. J Mol Biol 273:729–739

Tao F, He YW, Wu DH, Swarup S, Zhang LH (2010a) The cyclic nucleotide monophosphate domain of *Xanthomonas campestris* global regulator Clp defines a new class of cyclic di-GMP effectors. J Bacteriol 192:1020–1029

Tao F, Swarup S, Zhang LH (2010b) Quorum sensing modulation of a putative glycosyltransferase gene cluster essential for *Xanthomonas campestris* biofilm formation. Environ Microbiol 12:3159–3170

Tashiro Y, Ichikawa S, Nakajima-Kambe T, Uchimura H, Nomura N (2010) *Pseudomonas* quinolone signal affects membrane vesicle production in not only Gram-negative but also Gram-positive bacteria. Microbes Environ 25:120–125

Tashiro Y, Uchiyama H, Nomura N (2012) Multifunctional membrane vesicles in *Pseudomonas aeruginosa*. Environ Microbiol 14:1349–1362

Toda K, Yawata Y, Setoyama E, Fukuda J, Nomura N, Suzuki H (2011) Continuous monitoring of ammonia removal activity and observation of morphology of microbial complexes in a microdevice. Appl Environ Microbiol 77:4253–4255

Toyofuku M, Nomura N, Fujii T, Takaya N, Maseda H, Sawada I, Nakajima T, Uchiyama H (2007) Quorum sensing regulates denitrification in *Pseudomonas aeruginosa* PAO1. J Bacteriol 189:4969–4972

Toyofuku M, Nomura N, Kuno E, Tashiro Y, Nakajima T, Uchiyama H (2008) Influence of the *Pseudomonas* quinolone signal on denitrification in *Pseudomonas aeruginosa*. J Bacteriol 190:7947–7956

Toyofuku M, Roschitzki B, Riedel K, Eberl L (2012a) Identification of proteins associated with the *Pseudomonas aeruginosa* Biofilm extracellular matrix. J Proteome Res 11:4906–4915

Toyofuku M, Uchiyama H, Nomura N (2012b) Social behaviours under anaerobic conditions in *Pseudomonas aeruginosa*. Int J Microbiol 2012. doi:10.1155/2012/405191

Urbain V, Block JC, Manem J (1993) Bioflocculation in activated-sludge: an analytic approach. Water Res 27:829–838

Venkata HNN, Nomura N, Shigeto S (2011) Leucine pools in *Escherichia coli* biofilm discovered by Raman imaging. J Raman Spectrosc 42:1913–1915

Vuong C, Saenz HL, Götz F, Otto M (2000) Impact of the agr quorum-sensing system on adherence to polystyrene in *Staphylococcus aureus*. J Infect Dis 182:1688–1693

Watanabe M, Sasaki K, Nakashimada Y, Kakizono T, Noparatnaraporn N, Nishio N (1998) Growth and flocculation of a marine photosynthetic bacterium Rhodovulum sp. Appl Microbiol Biot 50:682–691

Whitchurch CB, Tolker-Nielsen T, Ragas PC, Mattick JS (2002) Extracellular DNA required for bacterial biofilm formation. Science 295:1487

Yawata Y, Nomura N, Uchiyama H (2008) Development of a novel biofilm continuous culture method for simultaneous assessment of architecture and gaseous metabolite production. Appl Environ Microbiol 74:5429–5435

Yawata Y, Toda K, Setoyama E, Fukuda J, Suzuki H, Uchiyama H, Nomura N (2010) Monitoring biofilm development in a microfluidic device using modified confocal reflection microscopy. J Biosci Bioeng 110:377–380

Yoon SS, Hennigan RF, Hilliard GM, Ochsner UA, Parvatiyar K, Kamani MC, Allen HL, DeKievit TR, Gardner PR, Schwab U, Rowe JJ, Iglewski BH, McDermott TR, Mason RP, Wozniak DJ, Hancock RE, Parsek MR, Noah TL, Boucher RC, Hassett DJ (2002) *Pseudomonas aeruginosa* anaerobic respiration in biofilms: relationships to cystic fibrosis pathogenesis. Dev Cell 3:593–603

Yoon MY, Lee KM, Park Y, Yoon SS (2011) Contribution of cell elongation to the biofilm formation of *Pseudomonas aeruginosa* during anaerobic respiration. PLoS One 6:e16105. doi:10.1371/journal.pone.0016105

Zhang LH, Dong YH (2004) Quorum sensing and signal interference: diverse implications. Mol Microbiol 53:1563–1571

Zogaj X, Nimtz M, Rohde M, Bokranz W, Romling U (2001) The multicellular morphotypes of *Salmonella typhimurium* and *Escherichia coli* produce cellulose as the second component of the extracellular matrix. Mol Microbiol 39:1452–1463

Index

H. Nojiri et al. (eds.), *Biodegradative Bacteria: How Bacteria Degrade, Survive, Adapt, and Evolve*, DOI 10.1007/978-4-431-54520-0, © Springer Japan 2014

E

F

G

H

I

Printed by Printforce, the Netherlands